![Haynes logo] THE BOOK ®

VW Passat & Santana
Service and Repair Manual

I M Coomber

(814-272-11X10)

Models covered

Volkswagen Passat Saloon, Hatchback (inc. GT) & Estate, including limited edition models;
1588 cc, 1595 cc, 1781 cc, 1921 cc and 1994 cc
Volkswagen Santana Saloon; 1595 cc, 1781 cc, 1921 cc, 1994 cc

Does not cover Diesel engine models, or 'Series 3' Passat range introduced May 1988

© Haynes Publishing 1996

A book in the **Haynes Service and Repair Manual Series**

ISBN 1 85960 155 3

British Library Cataloguing in Publication Data
A catalogue record for this book is available from the British Library.

ABCDE
FGHIJ
KLMNO
PQ
2

Printed by **J H Haynes & Co. Ltd, Sparkford, Nr Yeovil, Somerset BA22 7JJ**

Haynes Publishing
Sparkford, Nr Yeovil, Somerset BA22 7JJ, England

Haynes North America, Inc
861 Lawrence Drive, Newbury Park, California 91320, USA

Editions Haynes S.A.
147/149, rue Saint Honoré, 75001 PARIS, France

Contents

LIVING WITH YOUR VOLKSWAGEN

MOT Test Checks

Roadside Repairs

Contents

Introduction to the Volkswagen Passat and Santana

Volkswagen Passat GL5

The 'new' Volkswagen Passat was introduced in September 1981, with revised body and trim features being the main visual differences from the earlier range of models. Mechanically very few alterations were necessary, the engines fitted being of 1.6 and 1.8 (four-cylinder) or 1.9 and 2.0 litre (five-cylinder) capacity. A four- or five-speed manual gearbox or three-speed automatic transmission is fitted according to model, and drive is to the front wheels.

The body, whether Saloon, Hatchback or Estate, has four doors, is of steel monocoque construction and has full underbody protection when new.

Full instrumentation is provided together with electric windows, power steering and a centralised locking system according to model.

Details of later models, and modifications which may affect procedures covered in the main chapters, are given in the Supplement at the end of the manual.

Volkswagen Santana GX5i

Acknowledgements

Thanks are due to the Champion Sparking Plug Company Limited who supplied the illustrations showing the various spark plug conditions, and to Sykes-Pickavant Limited who provided some of the workshop tools. Special thanks are also due to all those people at Sparkford who helped in the production of this manual.

Volkswagen Passat CL Estate

Working on your car can be dangerous. This page shows just some of the potential risks and hazards, with the aim of creating a safety-conscious attitude.

General hazards

Scalding

• Don't remove the radiator or expansion tank cap while the engine is hot.
• Engine oil, automatic transmission fluid or power steering fluid may also be dangerously hot if the engine has recently been running.

Burning

• Beware of burns from the exhaust system and from any part of the engine. Brake discs and drums can also be extremely hot immediately after use.

Crushing

• When working under or near a raised vehicle, always supplement the jack with axle stands, or use drive-on ramps. *Never venture under a car which is only supported by a jack.*
• Take care if loosening or tightening high-torque nuts when the vehicle is on stands. Initial loosening and final tightening should be done with the wheels on the ground.

Fire

• Fuel is highly flammable; fuel vapour is explosive.
• Don't let fuel spill onto a hot engine.
• Do not smoke or allow naked lights (including pilot lights) anywhere near a vehicle being worked on. Also beware of creating sparks (electrically or by use of tools).
• Fuel vapour is heavier than air, so don't work on the fuel system with the vehicle over an inspection pit.
• Another cause of fire is an electrical overload or short-circuit. Take care when repairing or modifying the vehicle wiring.
• Keep a fire extinguisher handy, of a type suitable for use on fuel and electrical fires.

Electric shock

• Ignition HT voltage can be dangerous, especially to people with heart problems or a pacemaker. Don't work on or near the ignition system with the engine running or the ignition switched on.

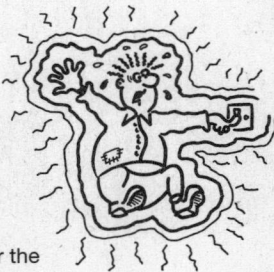

• Mains voltage is also dangerous. Make sure that any mains-operated equipment is correctly earthed. Mains power points should be protected by a residual current device (RCD) circuit breaker.

Fume or gas intoxication

• Exhaust fumes are poisonous; they often contain carbon monoxide, which is rapidly fatal if inhaled. Never run the engine in a confined space such as a garage with the doors shut.
• Fuel vapour is also poisonous, as are the vapours from some cleaning solvents and paint thinners.

Poisonous or irritant substances

• Avoid skin contact with battery acid and with any fuel, fluid or lubricant, especially antifreeze, brake hydraulic fluid and Diesel fuel. Don't syphon them by mouth. If such a substance is swallowed or gets into the eyes, seek medical advice.
• Prolonged contact with used engine oil can cause skin cancer. Wear gloves or use a barrier cream if necessary. Change out of oil-soaked clothes and do not keep oily rags in your pocket.
• Air conditioning refrigerant forms a poisonous gas if exposed to a naked flame (including a cigarette). It can also cause skin burns on contact.

Asbestos

• Asbestos dust can cause cancer if inhaled or swallowed. Asbestos may be found in gaskets and in brake and clutch linings. When dealing with such components it is safest to assume that they contain asbestos.

Special hazards

Hydrofluoric acid

• This extremely corrosive acid is formed when certain types of synthetic rubber, found in some O-rings, oil seals, fuel hoses etc, are exposed to temperatures above 400°C. The rubber changes into a charred or sticky substance containing the acid. *Once formed, the acid remains dangerous for years. If it gets onto the skin, it may be necessary to amputate the limb concerned.*
• When dealing with a vehicle which has suffered a fire, or with components salvaged from such a vehicle, wear protective gloves and discard them after use.

The battery

• Batteries contain sulphuric acid, which attacks clothing, eyes and skin. Take care when topping-up or carrying the battery.
• The hydrogen gas given off by the battery is highly explosive. Never cause a spark or allow a naked light nearby. Be careful when connecting and disconnecting battery chargers or jump leads.

Air bags

• Air bags can cause injury if they go off accidentally. Take care when removing the steering wheel and/or facia. Special storage instructions may apply.

Diesel injection equipment

• Diesel injection pumps supply fuel at very high pressure. Take care when working on the fuel injectors and fuel pipes.

Warning: Never expose the hands, face or any other part of the body to injector spray; the fuel can penetrate the skin with potentially fatal results.

Remember...

DO

• Do use eye protection when using power tools, and when working under the vehicle.
• Do wear gloves or use barrier cream to protect your hands when necessary.
• Do get someone to check periodically that all is well when working alone on the vehicle.
• Do keep loose clothing and long hair well out of the way of moving mechanical parts.
• Do remove rings, wristwatch etc, before working on the vehicle – especially the electrical system.
• Do ensure that any lifting or jacking equipment has a safe working load rating adequate for the job.

DON'T

• Don't attempt to lift a heavy component which may be beyond your capability – get assistance.
• Don't rush to finish a job, or take unverified short cuts.
• Don't use ill-fitting tools which may slip and cause injury.
• Don't leave tools or parts lying around where someone can trip over them. Mop up oil and fuel spills at once.
• Don't allow children or pets to play in or near a vehicle being worked on.

Dimensions, Weights & Capacities

Dimensions

	Passat Saloon	Passat Hatchback	Passat Estate	Santana
Overall length	4530 mm (178.4 in)	4435 mm (174.6 in)	4540 mm (178.7 in)	4545 mm (178.9 in)
Overall width	1710 mm (67.3 in)	1685 mm (66.3 in)	1685 mm (66.3 in)	1695 mm (66.7 in)
Overall height (unladen)	1395 mm (54.9 in)	1385 mm (54.5 in)	1385 mm* (54.5 in)	1400 mm (55.1 in)
Wheelbase	2550 mm (100.3 in)	2550 mm (100.3 in)	2550 mm (100.3 in)	2550 mm (100.3 in)
Track – front	1414 mm (55.6 in)	1414 mm (55.6 in)	1414 mm (55.6 in)	1414 mm (55.6 in)
Track – rear	1422 mm (56.0 in)	1422 mm (56.0 in)	1422 mm (56.0 in)	1422 mm (56.0 in)
*Overall height with roof rails	–	–	1465 mm (57.6 in)	–

Weights

	Passat Saloon	Passat Hatchback	Passat Estate	Santana
Gross vehicle weight:				
1.6 and 1.8 litre	1460 kg (3218 lb)	1460 kg (3218 lb)	1520 kg (3350 lb)	1460 kg (3218 lb)
1.9 and 2.0 litre	1580 kg (3483 lb)	1580 kg (3483 lb)	1620 kg (3571 lb)	1580 kg (3483 lb)
Kerb weight (manual gearbox models*):				
1.6 and 1.8 litre	985 kg (2171 lb)	985/960 kg (2171/2116 lb)	1085 kg (2391 lb)	1005 kg (2215 lb)
1.9 and 2.0 litre	1085 kg (2391 lb)	1085 kg (2391lb)	1105 kg (2436 lb)	1125 kg (2480 lb)

*Add the following for automatic transmission models:
With four-cylinder engine .... 25 kg (55 lb)
With five-cylinder engine .... 15 kg (33 lb)

All models

Maximum roof rack load (gutter rail fixing only) .... 75 kg (165 lb)
Maximum trailer weight (trailer without brakes):
  1.6 litre .... 500 kg (1102 lb)
  1.8 litre .... 500 kg (1102 lb)
  1.9 and 2.0 litre .... 550 kg (1212 lb)
Maximum trailer weight (trailer with brakes) –12% gradient:
  1.6 litre .... 1200 kg (2645 lb)
  1.8 litre .... 1200 kg (2645 lb)
  1.9 and 2.0 litre .... 1500 kg (3306 lb)
Maximum trailer nose weight on ball coupling:
  All models .... 75 kg (165 lb)

Capacities

Engine oil:
  1.6 and 1.8 litre with filter .... 3.0 litre (5.3 Imp pint)
  1.6 and 1.8 litre without filter .... 2.5 litre (4.4 Imp pint)
  1.9 and 2.0 litre with filter .... 3.5 litre (6.2 Imp pint)
  1.9 and 2.0 litre without filter .... 3.0 litre (5.3 Imp pint)
Difference between engine oil dipstick minimum and maximum
  marks .... 1.0 litre (1.8 Imp pint)
Cooling system (with heater):
  1.6 and 1.8 litre .... 6.0 litre (10.6 Imp pint)
  1.9 and 2.0 litre .... 8.0 litre (14.1 Imp pint)
Fuel tank:
  All models .... 60 litre (13.2 Imp gal)
Manual gearbox:
  Type 014 .... 1.7 litre (3 imp pint)
  Type 013 .... 2.0 litre (3.5 imp pint)
  Type 093 .... 2.4 litre (4.1 Imp pint)
Automatic transmission:
  Type 089 and 087 (total) – ATF .... 6.0 litre (10.6 Imp pint)
  Type 089 and 087 (Service) – ATF .... 3.0 litre (5.3 Imp pint)
  Type 089 final drive – SAE 90 .... 0.75 litre (1.3 Imp pint)
  Type 087 final drive – SAE 90 .... 1.0 litre (1.8 Imp pint)

This is a guide to getting your vehicle through the MOT test. Obviously it will not be possible to examine the vehicle to the same standard as the professional MOT tester. However, working through the following checks will enable you to identify any problem areas before submitting the vehicle for the test.

Where a testable component is in borderline condition, the tester has discretion in deciding whether to pass or fail it. The basis of such discretion is whether the tester would be happy for a close relative or friend to use the vehicle with the component in that condition. If the vehicle presented is clean and evidently well cared for, the tester may be more inclined to pass a borderline component than if the vehicle is scruffy and apparently neglected.

It has only been possible to summarise the test requirements here, based on the regulations in force at the time of printing. Test standards are becoming increasingly stringent, although there are some exemptions for older vehicles. For full details obtain a copy of the Haynes publication Pass the MOT! (available from stockists of Haynes manuals).

An assistant will be needed to help carry out some of these checks.

The checks have been sub-divided into four categories, as follows:

1 Checks carried out **FROM THE DRIVER'S SEAT**

2 Checks carried out **WITH THE VEHICLE ON THE GROUND**

3 Checks carried out **WITH THE VEHICLE RAISED AND THE WHEELS FREE TO TURN**

4 Checks carried out on **YOUR VEHICLE'S EXHAUST EMISSION SYSTEM**

1 Checks carried out **FROM THE DRIVER'S SEAT**

Handbrake

☐ Test the operation of the handbrake. Excessive travel (too many clicks) indicates incorrect brake or cable adjustment.
☐ Check that the handbrake cannot be released by tapping the lever sideways. Check the security of the lever mountings.

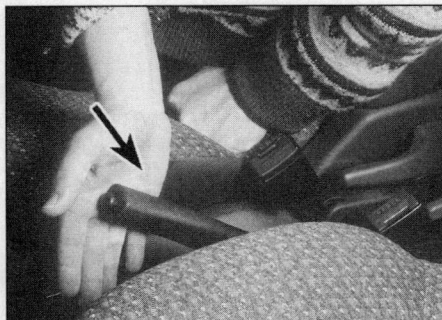

Footbrake

☐ Depress the brake pedal and check that it does not creep down to the floor, indicating a master cylinder fault. Release the pedal, wait a few seconds, then depress it again. If the pedal travels nearly to the floor before firm resistance is felt, brake adjustment or repair is necessary. If the pedal feels spongy, there is air in the hydraulic system which must be removed by bleeding.

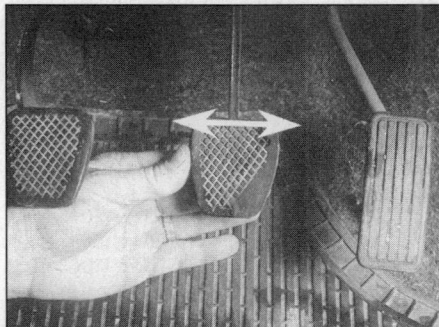

☐ Check that the brake pedal is secure and in good condition. Check also for signs of fluid leaks on the pedal, floor or carpets, which would indicate failed seals in the brake master cylinder.
☐ Check the servo unit (when applicable) by operating the brake pedal several times, then keeping the pedal depressed and starting the engine. As the engine starts, the pedal will move down slightly. If not, the vacuum hose or the servo itself may be faulty.

Steering wheel and column

☐ Examine the steering wheel for fractures or looseness of the hub, spokes or rim.
☐ Move the steering wheel from side to side and then up and down. Check that the steering wheel is not loose on the column, indicating wear or a loose retaining nut. Continue moving the steering wheel as before, but also turn it slightly from left to right.
☐ Check that the steering wheel is not loose on the column, and that there is no abnormal movement of the steering wheel, indicating wear in the column support bearings or couplings.

Windscreen and mirrors

☐ The windscreen must be free of cracks or other significant damage within the driver's field of view. (Small stone chips are acceptable.) Rear view mirrors must be secure, intact, and capable of being adjusted.

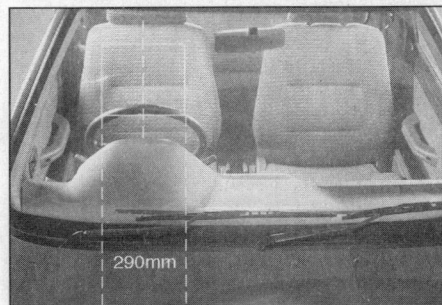

290mm

MOT Test Checks

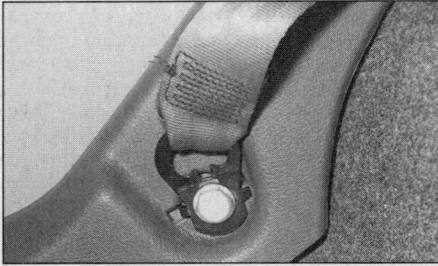

Seat belts and seats

Note: *The following checks are applicable to all seat belts, front and rear.*

☐ Examine the webbing of all the belts (including rear belts if fitted) for cuts, serious fraying or deterioration. Fasten and unfasten each belt to check the buckles. If applicable, check the retracting mechanism. Check the security of all seat belt mountings accessible from inside the vehicle.

☐ The front seats themselves must be securely attached and the backrests must lock in the upright position.

Doors

☐ Both front doors must be able to be opened and closed from outside and inside, and must latch securely when closed.

2 Checks carried out WITH THE VEHICLE ON THE GROUND

Vehicle identification

☐ Number plates must be in good condition, secure and legible, with letters and numbers correctly spaced – spacing at (A) should be twice that at (B).

☐ The VIN plate (A) and homologation plate (B) must be legible.

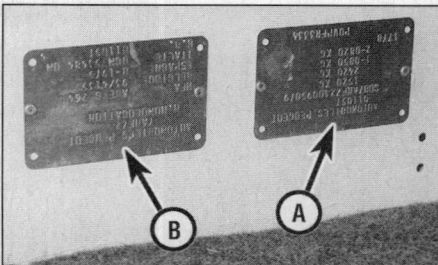

Electrical equipment

☐ Switch on the ignition and check the operation of the horn.

☐ Check the windscreen washers and wipers, examining the wiper blades; renew damaged or perished blades. Also check the operation of the stop-lights.

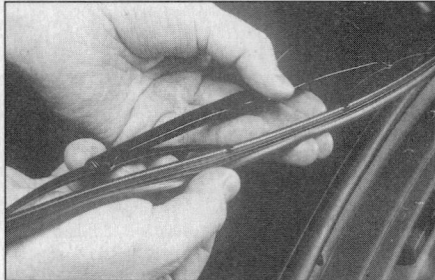

☐ Check the operation of the sidelights and number plate lights. The lenses and reflectors must be secure, clean and undamaged.

☐ Check the operation and alignment of the headlights. The headlight reflectors must not be tarnished and the lenses must be undamaged.

☐ Switch on the ignition and check the operation of the direction indicators (including the instrument panel tell-tale) and the hazard warning lights. Operation of the sidelights and stop-lights must not affect the indicators - if it does, the cause is usually a bad earth at the rear light cluster.

☐ Check the operation of the rear foglight(s), including the warning light on the instrument panel or in the switch.

Footbrake

☐ Examine the master cylinder, brake pipes and servo unit for leaks, loose mountings, corrosion or other damage.

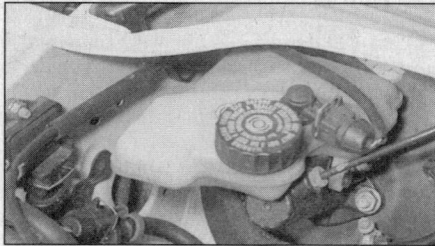

☐ The fluid reservoir must be secure and the fluid level must be between the upper (**A**) and lower (**B**) markings.

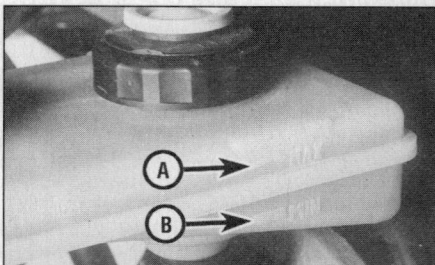

☐ Inspect both front brake flexible hoses for cracks or deterioration of the rubber. Turn the steering from lock to lock, and ensure that the hoses do not contact the wheel, tyre, or any part of the steering or suspension mechanism. With the brake pedal firmly depressed, check the hoses for bulges or leaks under pressure.

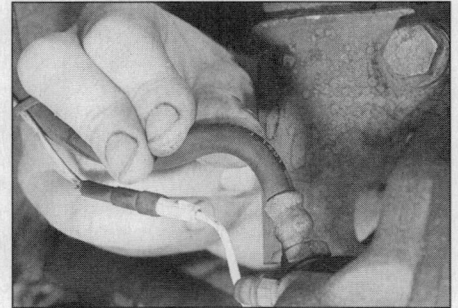

Steering and suspension

☐ Have your assistant turn the steering wheel from side to side slightly, up to the point where the steering gear just begins to transmit this movement to the roadwheels. Check for excessive free play between the steering wheel and the steering gear, indicating wear or insecurity of the steering column joints, the column-to-steering gear coupling, or the steering gear itself.

☐ Have your assistant turn the steering wheel more vigorously in each direction, so that the roadwheels just begin to turn. As this is done, examine all the steering joints, linkages, fittings and attachments. Renew any component that shows signs of wear or damage. On vehicles with power steering, check the security and condition of the steering pump, drivebelt and hoses.

☐ Check that the vehicle is standing level, and at approximately the correct ride height.

Shock absorbers

☐ Depress each corner of the vehicle in turn, then release it. The vehicle should rise and then settle in its normal position. If the vehicle continues to rise and fall, the shock absorber is defective. A shock absorber which has seized will also cause the vehicle to fail.

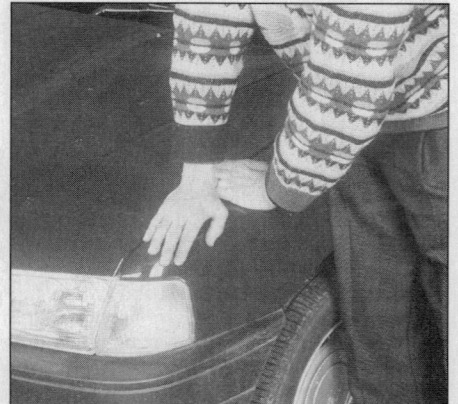

Exhaust system

☐ Start the engine. With your assistant holding a rag over the tailpipe, check the entire system for leaks. Repair or renew leaking sections.

3 Checks carried out WITH THE VEHICLE RAISED AND THE WHEELS FREE TO TURN

Jack up the front and rear of the vehicle, and securely support it on axle stands. Position the stands clear of the suspension assemblies. Ensure that the wheels are clear of the ground and that the steering can be turned from lock to lock.

Steering mechanism

☐ Have your assistant turn the steering from lock to lock. Check that the steering turns smoothly, and that no part of the steering mechanism, including a wheel or tyre, fouls any brake hose or pipe or any part of the body structure.
☐ Examine the steering rack rubber gaiters for damage or insecurity of the retaining clips. If power steering is fitted, check for signs of damage or leakage of the fluid hoses, pipes or connections. Also check for excessive stiffness or binding of the steering, a missing split pin or locking device, or severe corrosion of the body structure within 30 cm of any steering component attachment point.

Front and rear suspension and wheel bearings

☐ Starting at the front right-hand side, grasp the roadwheel at the 3 o'clock and 9 o'clock positions and shake it vigorously. Check for free play or insecurity at the wheel bearings, suspension balljoints, or suspension mountings, pivots and attachments.
☐ Now grasp the wheel at the 12 o'clock and 6 o'clock positions and repeat the previous inspection. Spin the wheel, and check for roughness or tightness of the front wheel bearing.

☐ If excess free play is suspected at a component pivot point, this can be confirmed by using a large screwdriver or similar tool and levering between the mounting and the component attachment. This will confirm whether the wear is in the pivot bush, its retaining bolt, or in the mounting itself (the bolt holes can often become elongated).

☐ Carry out all the above checks at the other front wheel, and then at both rear wheels.

Springs and shock absorbers

☐ Examine the suspension struts (when applicable) for serious fluid leakage, corrosion, or damage to the casing. Also check the security of the mounting points.
☐ If coil springs are fitted, check that the spring ends locate in their seats, and that the spring is not corroded, cracked or broken.
☐ If leaf springs are fitted, check that all leaves are intact, that the axle is securely attached to each spring, and that there is no deterioration of the spring eye mountings, bushes, and shackles.

☐ The same general checks apply to vehicles fitted with other suspension types, such as torsion bars, hydraulic displacer units, etc. Ensure that all mountings and attachments are secure, that there are no signs of excessive wear, corrosion or damage, and (on hydraulic types) that there are no fluid leaks or damaged pipes.
☐ Inspect the shock absorbers for signs of serious fluid leakage. Check for wear of the mounting bushes or attachments, or damage to the body of the unit.

Driveshafts (fwd vehicles only)

☐ Rotate each front wheel in turn and inspect the constant velocity joint gaiters for splits or damage. Also check that each driveshaft is straight and undamaged.

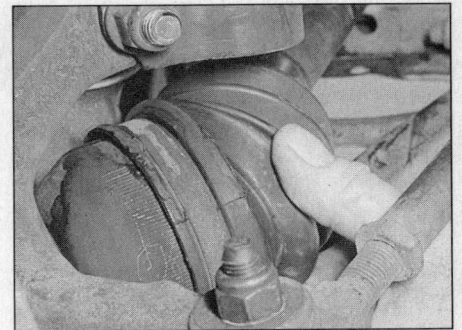

Braking system

☐ If possible without dismantling, check brake pad wear and disc condition. Ensure that the friction lining material has not worn excessively, (A) and that the discs are not fractured, pitted, scored or badly worn (B).

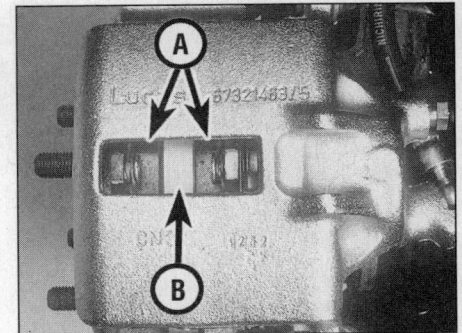

☐ Examine all the rigid brake pipes underneath the vehicle, and the flexible hose(s) at the rear. Look for corrosion, chafing or insecurity of the pipes, and for signs of bulging under pressure, chafing, splits or deterioration of the flexible hoses.
☐ Look for signs of fluid leaks at the brake calipers or on the brake backplates. Repair or renew leaking components.
☐ Slowly spin each wheel, while your assistant depresses and releases the footbrake. Ensure that each brake is operating and does not bind when the pedal is released.

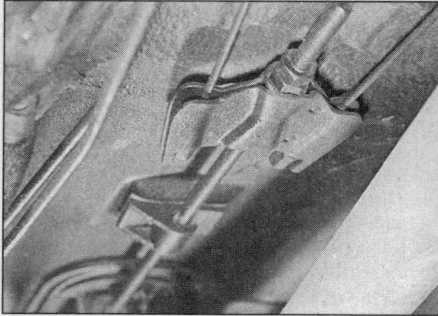

☐ Examine the handbrake mechanism, checking for frayed or broken cables, excessive corrosion, or wear or insecurity of the linkage. Check that the mechanism works on each relevant wheel, and releases fully, without binding.

☐ It is not possible to test brake efficiency without special equipment, but a road test can be carried out later to check that the vehicle pulls up in a straight line.

Fuel and exhaust systems

☐ Inspect the fuel tank (including the filler cap), fuel pipes, hoses and unions. All components must be secure and free from leaks.

☐ Examine the exhaust system over its entire length, checking for any damaged, broken or missing mountings, security of the retaining clamps and rust or corrosion.

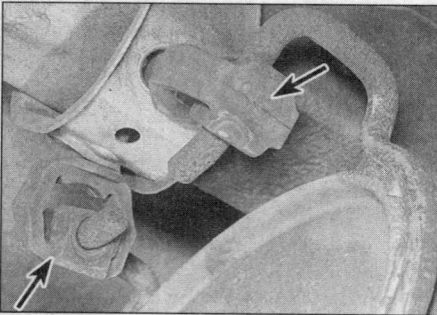

Wheels and tyres

☐ Examine the sidewalls and tread area of each tyre in turn. Check for cuts, tears, lumps, bulges, separation of the tread, and exposure of the ply or cord due to wear or damage. Check that the tyre bead is correctly seated on the wheel rim, that the valve is sound and properly seated, and that the wheel is not distorted or damaged.

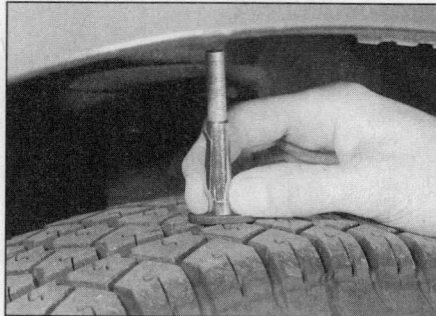

☐ Check that the tyres are of the correct size for the vehicle, that they are of the same size and type on each axle, and that the pressures are correct.

☐ Check the tyre tread depth. The legal minimum at the time of writing is 1.6 mm over at least three-quarters of the tread width. Abnormal tread wear may indicate incorrect front wheel alignment.

Body corrosion

☐ Check the condition of the entire vehicle structure for signs of corrosion in load-bearing areas. (These include chassis box sections, side sills, cross-members, pillars, and all suspension, steering, braking system and seat belt mountings and anchorages.) Any corrosion which has seriously reduced the thickness of a load-bearing area is likely to cause the vehicle to fail. In this case professional repairs are likely to be needed.

☐ Damage or corrosion which causes sharp or otherwise dangerous edges to be exposed will also cause the vehicle to fail.

4 Checks carried out on YOUR VEHICLE'S EXHAUST EMISSION SYSTEM

Petrol models

☐ Have the engine at normal operating temperature, and make sure that it is in good tune (ignition system in good order, air filter element clean, etc).

☐ Before any measurements are carried out, raise the engine speed to around 2500 rpm, and hold it at this speed for 20 seconds. Allow the engine speed to return to idle, and watch for smoke emissions from the exhaust tailpipe. If the idle speed is obviously much too high, or if dense blue or clearly-visible black smoke comes from the tailpipe for more than 5 seconds, the vehicle will fail. As a rule of thumb, blue smoke signifies oil being burnt (engine wear) while black smoke signifies unburnt fuel (dirty air cleaner element, or other carburettor or fuel system fault).

☐ An exhaust gas analyser capable of measuring carbon monoxide (CO) and hydrocarbons (HC) is now needed. If such an instrument cannot be hired or borrowed, a local garage may agree to perform the check for a small fee.

CO emissions (mixture)

☐ At the time or writing, the maximum CO level at idle is 3.5% for vehicles first used after August 1986 and 4.5% for older vehicles. From January 1996 a much tighter limit (around 0.5%) applies to catalyst-equipped vehicles first used from August 1992. If the CO level cannot be reduced far enough to pass the test (and the fuel and ignition systems are otherwise in good condition) then the carburettor is badly worn, or there is some problem in the fuel injection system or catalytic converter (as applicable).

HC emissions

☐ With the CO emissions within limits, HC emissions must be no more than 1200 ppm (parts per million). If the vehicle fails this test at idle, it can be re-tested at around 2000 rpm; if the HC level is then 1200 ppm or less, this counts as a pass.

☐ Excessive HC emissions can be caused by oil being burnt, but they are more likely to be due to unburnt fuel.

Diesel models

☐ The only emission test applicable to Diesel engines is the measuring of exhaust smoke density. The test involves accelerating the engine several times to its maximum unloaded speed.

Note: *It is of the utmost importance that the engine timing belt is in good condition before the test is carried out.*

☐ Excessive smoke can be caused by a dirty air cleaner element. Otherwise, professional advice may be needed to find the cause.

Jacking

The jack supplied in the vehicle tool kit by the manufacturer should only be used for emergency roadside wheel changing unless it is supplemented by safety stands.

The jack supplied is of the half scissors type. Check that the handbrake is fully applied before using the jack, and only jack the vehicle up on firm level ground. If the ground is not firm you will need to position a large flat packing piece under the jack base to provide additional support.

Chock the wheel diagonally opposite the one to be changed. Using the tools provided remove the hub cap where necessary, then loosen the wheel bolts half a turn. Locate the lifting arm of the jack beneath the reinforced seam of the side sill panel (photo) directly beneath the wedge shaped depression nearest to the wheel to be removed. Turn the jack handle until the base of the jack contacts the ground directly beneath the sill (photo) then continue to turn the handle until the wheel is free of the ground. Unscrew the wheel bolts and remove the wheel. On light alloy wheels prise off the centre trim cap and press it into the spare wheel.

Locate the spare wheel on the hub, then insert and tighten the bolts in diagonal sequence. Lower the jack and fully tighten the bolts. Refit the hub cap where necessary, remove the chock and relocate the tool kit, jack and wheel in the luggage compartment.

When jacking up the car with a trolley jack, position the jack beneath the reinforced plate behind the front wheel (see illustration) or beneath the reinforced seam at the rear of the side sill panel. Use the same positions when supporting the car with axle stands. *Never jack up the car beneath the suspension or axle components, the sump, or the gearbox.*

Towing

Towing eyes are fitted to the front and rear of the vehicle (photos) and a tow line should not be attached to any other points. It is preferable to use a slightly elastic tow line, to reduce the strain on both vehicles, either by having a tow line manufactured from synthetic fibre, or one which is fitted with an elastic link.

When towing, the following important precautions must be observed:

(a) *Turn the ignition key of the vehicle being towed, so that the steering wheel is free (unlocked).*

(b) *Remember that when the engine is not running the brake servo will not operate, so that additional pressure will be required on the brake pedal after the first few applications.*

(c) *On vehicles with automatic transmission, ensure that the gear selector lever is at 'N'. Do not tow faster than 30 mph (50 kph), or further than 30 miles (50 km) unless the front wheels are lifted clear of the ground.*

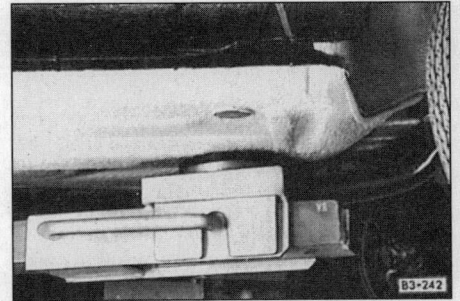

Front jacking point when using a trolley jack

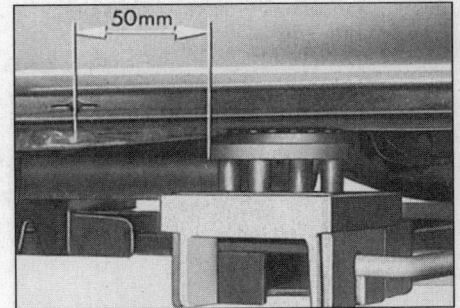

Rear jacking point when using a trolley jack. Place trolley jack head 50 mm (2 in) in front of the side member vertical reinforcement

Front jacking point when using a trolley jack

Front jacking point when using a trolley jack

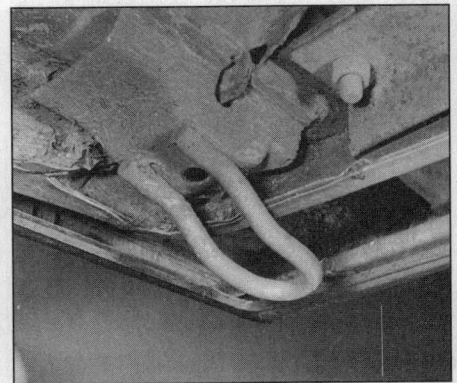

Front jacking point when using a trolley jack

Roadside Repairs

Identifying leaks

Puddles on the garage floor or drive, or obvious wetness under the bonnet or underneath the car, suggest a leak that needs investigating. It can sometimes be difficult to decide where the leak is coming from, especially if the engine bay is very dirty already. Leaking oil or fluid can also be blown rearwards by the passage of air under the car, giving a false impression of where the problem lies.

⚠️ *Warning: Most automotive oils and fluids are poisonous. Wash them off skin, and change out of contaminated clothing, without delay.*

HAYNES HiNT *The smell of a fluid leaking from the car may provide a clue to what's leaking. Some fluids are disctively coloured. It may help to clean the car carefully and to park it over some clean paper overnight as an aid to locating the source of the leak.*
Remember that some leaks may only occur while the engine is running.

Sump oil

Engine oil may leak from the drain plug...

Oil from filter

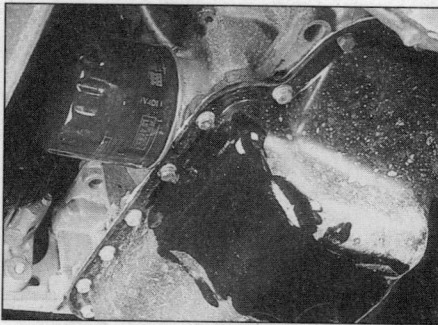

...or from the base of the oil filter.

Gearbox oil

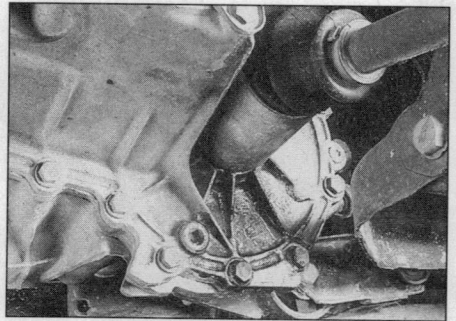

Gearbox oil can leak from the seals at the inboard ends of the driveshafts.

Antifreeze

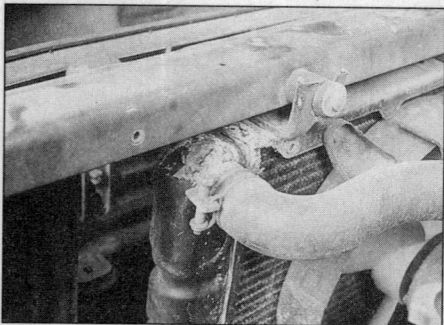

Leaking antifreeze often leaves a crystalline deposit like this.

Brake fluid

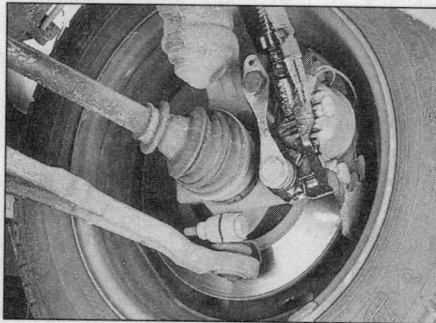

A leak occurring at a wheel is almost certainly brake fluid.

Power steering fluid

Power steering fluid may leak from the pipe connectors on the steering rack.

Booster battery (jump) starting

When jump-starting a car using a booster battery, observe the following precautions:

A) Before connecting the booster battery, make sure that the ignition is switched off.

B) Ensure that all electrical equipment (lights, heater, wipers, etc) is switched off.

C) Make sure that the booster battery is the same voltage as the discharged one in the vehicle.

D) If the battery is being jump-started from the battery in another vehicle, the two vehcles MUST NOT TOUCH each other.

E) Make sure that the transmission is in neutral (or PARK, in the case of automatic transmission).

HAYNES HiNT *Jump starting will get you out of trouble, but you must correct whatever made the battery go flat in the first place. There are three possibilities:*

1 *The battery has been drained by repeated attempts to start, or by leaving the lights on.*

2 *The charging system is not working properly (alternator drivebelt slack or broken, alternator wiring fault or alternator itself faulty).*

3 *The battery itself is at fault (electrolyte low, or battery worn out).*

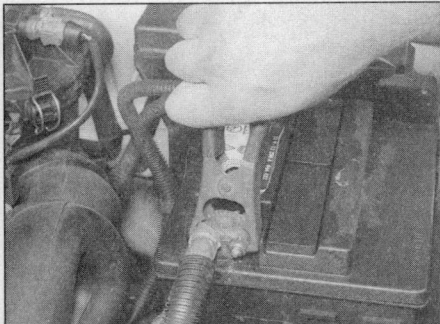

1 Connect one end of the red jump lead to the positive (+) terminal of the flat battery

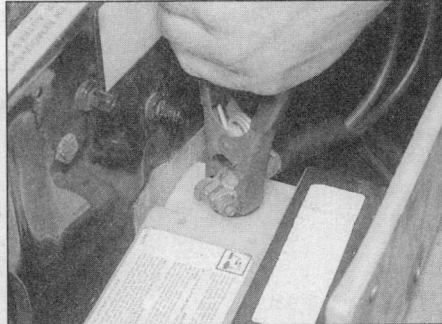

2 Connect the other end of the red lead to the positive (+) terminal of the booster battery.

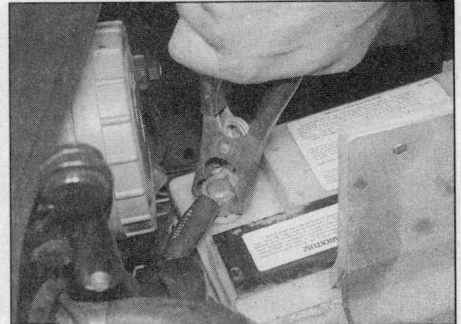

3 Connect one end of the black jump lead to the negative (-) terminal of the booster battery

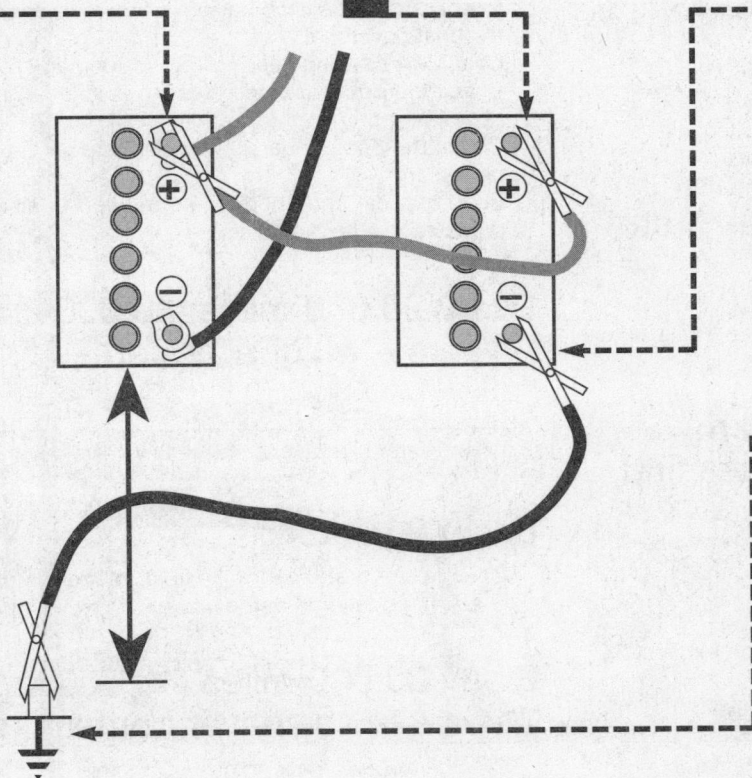

4 Connect the other end of the black jump lead to a bolt or bracket on the engine block, well away from the battery, on the vehicle to be started.

5 Make sure that the jump leads will not come into contact with the fan, drivebelts or other moving parts of the engine.

6 Start the engine using the booster battery, then with the engine running at idle speed, disconnect the jump leads in the reverse order of connection.

Routine Maintenance

Maintenance is essential for ensuring safety and desirable for the purpose of getting the best in terms of performance and economy from your car. Over the years the need for periodic lubrication has been greatly reduced, if not totally eliminated. This has unfortunately tended to lead some owners to think that because no such action is required, the items either no longer exist, or will last for ever. This is certainly not the case; it is essential to carry out regular visual examination as comprehensively as possible in order to spot any possible defects at an early stage, before they develop into major expensive repairs.

Topping up the engine oil level
(four-cylinder engine)

Topping up the coolant level

Every 250 miles (400 km) or weekly – whichever comes first

- [ ] Check the level of engine oil and top up if necessary (photo)
- [ ] Check the coolant level and top up if necessary (photo)
- [ ] Check the battery electrolyte level and top up if necessary
- [ ] Check and adjust the tyre pressures. Don't forget the spare
- [ ] Check that all the lights work
- [ ] Clean the headlamps
- [ ] Check the washer fluid level(s) and top up if necessary (photos)
- [ ] Check the level of fluid in the brake master cylinder reservoir – the level will drop slightly as the disc pads and rear brake linings wear, but if topping up is necessary the hydraulic circuit should be checked for leaks (photo)

MODELS UP TO AND INCLUDING MODEL YEAR 1985

Every 5000 miles (7500 km) or 6 months – whichever comes first

- [ ] Change engine oil
- [ ] Check front disc pads for wear (photo)
- [ ] Lubricate the bonnet lock and hinges
- [ ] Grease the door check straps

Every 10 000 miles (15 000 km) or 12 months – whichever comes first

- [ ] Check antifreeze strength and add as necessary
- [ ] Check drivebelts for condition and tension, and adjust if necessary (photo)
- [ ] Renew the spark plugs
- [ ] Check and adjust the ignition timing
- [ ] Check and adjust the engine idling speed and exhaust CO content
- [ ] Check the engine for oil, coolant and fuel leaks
- [ ] Change engine oil and renew the oil filter
- [ ] Check the exhaust system for security and leakage

- [ ] Adjust clutch play, where applicable
- [ ] Check the driveshaft rubber boots for damage and leaks (photo)
- [ ] Check level of oil/fluid in gearbox/transmission and final drive and top up if necessary
- [ ] Check transmission for leaks
- [ ] Check operation of brake pressure regulator where applicable
- [ ] Visually check the brake hydraulic circuit lines, hoses and unions for damage, deterioration and leaks
- [ ] Check front disc pads and rear brake shoes for wear
- [ ] Check the tyres for wear and tread depth
- [ ] Check the headlight beam alignment and adjust if necessary
- [ ] Check the steering tie-rod end balljoints for wear, and damage to the dust covers
- [ ] Check the suspension balljoints for wear, and damage to the dust covers
- [ ] Check the steering gear bellows for damage and leaks
- [ ] Check the powerassisted steering fluid level and top up if necessary
- [ ] Check the body underseal for damage and apply more if necessary
- [ ] Road test car and check the operation of all systems, in particular brakes and steering

Every 20 000 miles (30 000 km) or 2 years – whichever comes first

- [ ] Renew the fuel filter
- [ ] Renew the air cleaner element and clean the body
- [ ] Adjust valve clearances and renew valve cover gasket

Every 2 years

- [ ] Change the brake hydraulic fluid, bleed the system and check the operation of the brake system warning device

Every 30 000 miles (45 000 km) or 3 years – whichever comes first

- [ ] Change the automatic transmission fluid

MODELS FROM MODEL YEAR 1986 ONWARDS

Every 10 000 miles (15 000 km) or 12 months – whichever comes first

- [ ] Check antifreeze strength and level, and top up if necessary
- [ ] Check battery electrolyte level and top up if necessary
- [ ] Check the engine for oil, coolant and fuel leaks
- [ ] Change engine oil and renew oil filter
- [ ] Check engine idling speed and CO content
- [ ] Check clutch play and adjust if necessary, where applicable
- [ ] Check gearbox and driveshafts for leaks
- [ ] Check automatic transmission fluid level, where applicable
- [ ] Check the brake hydraulic circuit lines, hoses and unions for damage, deterioration and leaks
- [ ] Check disc pads and brake shoes for wear
- [ ] Check brake fluid level and top up if necessary
- [ ] Check operation of footbrake and handbrake
- [ ] Check tyres for wear and damage
- [ ] Check headlight beam alignment and adjust if necessary
- [ ] Check operation of all lights and horn
- [ ] Check operation of windscreen wipers and washers, and top up fluid if necessary
- [ ] Check steering tie-rod ends for wear
- [ ] Check front suspension lower balljoints for wear
- [ ] Check operation of steering
- [ ] Grease door check straps

Every 20 000 miles (30 000 km) or 2 years – whichever comes first

- [ ] Check condition and tension of alternator drivebelt and adjust if necessary
- [ ] Renew the spark plugs
- [ ] Check and if necessary adjust the valve clearances (except on engines with hydraulic tappets)
- [ ] Renew air filter element
- [ ] Renew fuel filter
- [ ] Clean automatic transmission sump and filter, and change fluid
- [ ] Check power steering fluid level and top up where necessary
- [ ] Check underbody sealant and apply where necessary

Every 2 years

- [ ] Renew the brake hydraulic fluid

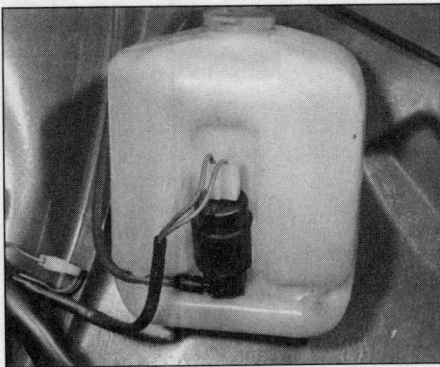

Top up the windscreen/headlamp washer fluid reservoir

Check/top up if necessary the rear window washer fluid reservoir

The brake master cylinder

Check the brake disc pads for wear

Check the drivebelt tension adjustment and the belt for wear

Check the driveshaft rubber boots

Routine Maintenance

Under-bonnet view of the engine compartment – four-cylinder Formel E model shown

1 Engine oil filler cap and dipstick
2 Air filter and carburettor
3 Brake master cylinder and servo unit
4 Ignition coil
5 Ignition distributor
6 Battery
7 Engine oil filter
8 Windscreen washer reservoir
9 Alternator
10 Cooling fan (electric)
11 Alternator drive belt
12 Radiator filler cap

Front suspension and associated components viewed from beneath

1 Brake caliper
2 Gearbox/axle unit
3 Exhaust system
4 Driveshaft
5 Suspension arm
6 Anti-roll bar
7 Engine sump and drain plug

Rear suspension and associated components viewed from beneath

1 Brake pressure regulator unit
2 Rear axle unit
3 Rear drum brakes
4 Shock absorber lower mounting
5 Fuel tank
6 Handbrake cable
7 Handbrake compensator unit

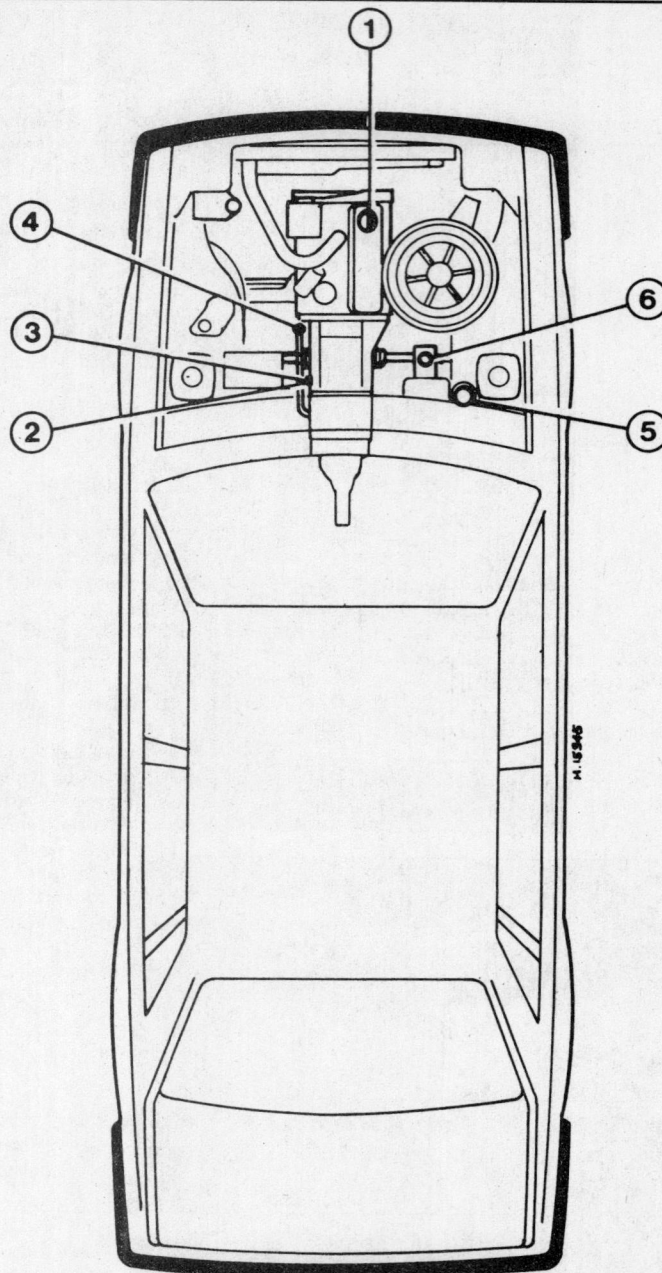

Component or system	Lubricant type/specification
Engine (1)	Multigrade engine oil, viscosity SAE 10W/40 or 15W/40, to VW 500.00 and 505.00. or 501.01and 505.00
Manual gearbox/final drive (2)	Hypoid gear oil, viscosity SAE 80 or 80W/90, to API-GL4
Final drive – automatic transmission (3)	Hypoid gear oil, viscosity SAE 90, to API-GL5, MIL-L-2150B
Automatic transmission (4) and power steering (5)	Dexron type ATF
Brake fluid (6)	Hydraulic fluid to FMVSS 116 DOT 3 or DOT 4
Clutch fluid	Hydraulic fluid to FMVSS 116 DOT 3 or DOT 4

Chapter 1 Engine

For modifications, and information applicable to later models, see Supplement at end of manual

Contents

Degrees of difficulty

Easy, suitable for novice with little experience	**Fairly easy,** suitable for beginner with some experience	**Fairly difficult,** suitable for competent DIY mechanic	**Difficult,** suitable for experienced DIY mechanic	**Very difficult,** suitable for expert DIY or professional

Specifications

Four-cylinder engines

General

	1.6 litre	1.6 litre	1.8 litre
Engine code letters	WV, YN, YP	DT	DS
Capacity	1588 cc (96.9 cu in)	1595 cc (97.29 cu in)	1781 cc (108.6 cu in)
Bore	79.5 mm (3.13 in)	81.0 mm (3.16 in)	81.0 mm (3.18 in)
Stroke	80.0 mm (3.15 in)	77.4 mm (3.02 in)	86.4 mm (3.40 in)
Compression ratio	8.2 : 1	9.0 : 1	10.0 : 1
Firing order	1 – 3 – 4 – 2 (No 1 at timing belt end)		

Cylinder compression – bar (lbf/in²)

	WV,YN, YP	DT	DS
Compression pressure	10 to 12 (145 to 174)	9 to 12 (131 to 174)	10 to 13 (145 to 188)
Minimum allowable pressure	7.5 (109)	7 (102)	7.5 (109)
Maximum pressure difference between cylinders	3 (44)	3 (44)	3 (44)

Lubrication system

Oil type/specification	Muitigrade engine oil, viscosity SAE 10W/40 or 15W/40, to VW 500.00 and 505.00 or 501.01 and 505.00
Oil capacity:	
Including filter	3.0 litre; 5.3 Imp pt
Excluding filter	2.5 litre, 4.4 Imp pt
Difference between dipstick minimum and maximum marks	1.0 litre; 1.8 Imp pt
Oil pressure (minimum) at 2000 rpm	2 bar (29 lbf/in²)
Oil pump:	
Gear backlash	0.05 to 0.20 mm (0.002 to 0.008 in)
Gear endplay	0.15 mm (0.006 in)

Intermediate shaft

Endplay	0.25 mm (0.010 in)

Crankshaft

Needle bearing depth	1.5 mm (0.059 in)
End play:	
New	0.07 to 0.17 mm (0.003 to 0.007 in)
Wear limit	0.25 mm (0.010 in)
Maximum main bearing running clearance	0.17 mm (0.007 in)
Main bearing journal diameter:	
Standard size	53.958 to 53.978 mm (2.1243 to 2.1251in)
1st undersize	53.708 to 53.728 mm (2.1145 to 2.1153 in)
2nd undersize	53.458 to 53.478 mm (2.1047 to 2.1054 in)
3rd undersize	53.208 to 53.228 mm (2.0948 to 2.0956 in)
Big-end bearing journal diameter – 1.6 engine (except 1.6 DT):	
Standard size	45.958 to 45.978 mm (1.8094 to 1.8102 in)
1st undersize	45.708 to 45.728 mm (1.7995 to 1.8003 in)
2nd undersize	45.458 to 45.478 mm (1.7897 to 1.7905 in)
3rd undersize	45.208 to 45.228 mm (1.7798 to 1.7806 in)
Big-end bearing journal diameter – 1.6 DT and 1.8 engine:	
Standard size	47.758 to 47.778 mm (1.8802 to 1.8810 in)
1st undersize	47.508 to 47.528 mm (1.8704 to 1.8712 in)
2nd undersize	47.258 to 47.278 mm (1.8606 to 1.8613 in)
3rd undersize	47.008 to 47.028 mm (1.8507 to 1.8515 in)

Pistons and rings

Piston-to-bore clearance		
New	0.03 mm (0.0012 in)	
Wear limit	0.07 mm (0.0028 in)	
Groove-to-ring clearance:		
New	0.02 to 0.05 mm (0.0008 to 0.0020 in)	
Wear limit	0.15 mm (0.0059 in)	
Piston size- 1.6 engine (except 1.6 DT):	**Piston diameter**	**Bore diameter**
Standard size	79.48 mm (3.1291 in)	79.51 mm (3.1307 in)
1st oversize	79.73 mm (3.1390 in)	79.76 mm (3.1042 in)
2nd oversize	79.98 mm (3.1488 in)	80.01 mm (3.1500 in)
3rd oversize	80.48 mm (3.1685 in)	80.51 mm (3.1697 in)
Piston size – 1.6 DT and 1.8 engine:		
Standard size	80.98 mm (3.1881 in)	81.01 mm (3.1893 in)
1st oversize	81.23 mm (3.1980 in)	81.26 mm (3.1992 in)
2nd oversize	81.48 mm (3.2078 in)	81.51mm (3.2090 in)
3rd oversize	81.98 mm (3.2275 in)	82.01mm (3.2287 in)

Cylinder bore

Maximum allowable deviation	0.08 mm (0.0031 in)
Piston ring endgap clearance (ring 15 mm/0.6 in from bottom of bore):	
New	0.30 to 0.45 mm (0.012 to 0.018 in)
Two section scraper rings	0.25 to 0.40 mm (0.0098 to 0.0157 in)
Three section scraper rings	0.25 to 0.50 mm (0.0098 to 0.0196 in)
Wear limit	1.0 mm (0.040 in)

Connecting rods

Maximum endplay	0.37 mm (0.015 in)
Big-end bearing running clearance:	
New	0.028 to 0.088 mm (0.0011 to 0.0035 in)
Wear limit	0.12 mm (0.0047 in)

Camshaft

Maximum endplay . . . . 0.15 mm (0.006 in)

Valves

Valve clearances:
  Engine warm:
    Inlet . . . . 0.20 to 0.30 mm (0.008 to 0.012 in)
    Exhaust : . . . . 0.40 to 0.50 mm (0.016 to 0.020 in)
  Engine cold:
    Inlet . . . . 0.15 to 0.25 mm (0.006 to 0.010 in)
    Exhaust . . . . 0.35 to 0.45 mm (0.014 to 0.018 in)
Adjusting shim thicknesses . . . . 3.00 to 4.25 mm (0.118 to 0.167 in) in increments of 0.05 mm
Valve guides:
  Maximum valve rock (measured at head):
    Inlet . . . . 1.0 mm (0.039 in)
    Exhaust . . . . 1.3 mm (0.051 in)

Valve timing at 1 mm lift/zero clearance

	1.6 except DT	1.6 DT	1.8 DS
Inlet opens BTDC	4°	5°	1°
Inlet closes ABDC	46°	21 °	37°
Exhaust opens BBDC	44°	41°	42°
Exhaust closes ATDC	6°	3° (BTDC)	2°

Cylinder head

Minimum height (between faces) . . . . 132.60 mm (5.220 in)
Maximum gasket face distortion allowable . . . . 0.1 mm (0.004 in)
Valve seat angle (inlet and exhaust) . . . . 45°

Torque wrench settings

	Nm	lbf ft
Cylinder head bolts:		
Stage 1	40	29
Stage 2	60	43
Stage 3	75	54
Stage 4	Tighten a further quarter turn (90°) – in one movement	

Note: *Do not retighten the socket-head bolts during servicing*

	Nm	lbf ft
Engine to transmission	55	40
Engine mountings	35	25
Engine support to block	25	18
Main bearing caps	65	47
Crankshaft front oil housing	10	7
Intermediate shaft flange	25	18
Crankshaft rear oil seal housing	10	7
Big-end nuts:		
M9 (rigid bolts)	45	33
M8 (stretch bolts):		
Stage 1	30	22
Stage 2	Tighten a further quarter turn (90°) – in one movement	
Camshaft gear	80	58
Timing cover (upper, lower and rear sections)	10	7
Crankshaft pulley to gear	20	14
Crankshaft gear bolt:		
M12 (with thread locking compound)	80	58
M14 (oiled threads)	200	148
Intermediate shaft gear	80	58
Tensioner nut	45	33
Valve (cylinder head) cover	10	7
Camshaft bearing cap	20	14
Oil pressure switch	25	18
Oil filter head	25	18
Oil pump to block	20	14
Oil pump cover	10	7
Sump bolts	20	14
Sump drain plug	30	22
Engine to gearbox cover plate	10	7
Exhaust pipe to manifold	30	22
Starter motor	20	14
Flywheel bolts – early (non-shouldered) type	75	54
Flywheel bolts – late (shouldered) type	100	72

1

1.9 litre five-cylinder engines

General

Engine code	WN
Capacity	1921 cc (117.2 cu in)
Bore	79.5 mm (3.13 in)
Stroke	77.4 mm (3.05 in)
Compression ratio	10.0: 1
Cylinder compression:	
Compression pressure	10.0 to 14.0 bar (145 to 203 lbf/in^2)
Minimum pressure	8 bar (116 lbf/in^2)
Maximum pressure difference between cylinders	3 bar (44 lbf/in^2)
Firing order	1 – 2 – 4 – 5 – 3 (No 1 at timing belt end)

Lubrication system

Oil type/specification	Multigrade engine oil, viscosity SAE 10W/40 or 15W/40, to VW 500.00 and 505.00 or 501.01 and 505.00
Oil capacity:	
Including filter	3.5 litre: 6.1 Imp pt
Excluding filter	3.0 litre; 5.2 Imp pt
Difference between dipstick minimum and maximum marks	1.0 litre; 1.76 Imp pt
Oil pressure (minimum) at idle	1 bar (14.5 lbf/in^2)
Oil pressure (minimum) at 2000 rpm	2 bar (29 lbf/in^2)

Crankshaft

Needle bearing depth	5.5 mm (0.217 in)
Endplay:	
New	0.07 to 0.18 mm (0.003 to 0.007 in)
Wear limit	0.25 mm (0.010 in)
Maximum main bearing running clearance	0.16 mm (0.006 in)
Main bearing journal diameter:	
Standard size	57.958 to 57.978 mm (2.2818 to 2.2825 in)
1st undersize	57.708 to 57.728 mm (2.2719 to 2.2727 in)
2nd undersize	57.458 to 57.478 mm (2.2621 to 2.2629 in)
3rd undersize	57.208 to 57.228 mm (2.2522 to 2.2530 in)
Big-end journal diameter:	
Standard size	45.958 to 45.978 mm (1.8093 to 1.8101 in)
1st undersize	45.708 to 45.728 mm (1.7995 to 1.8003 in)
2nd undersize	45.458 to 45.478 mm (1.7896 to 1.7904 in)
3rd undersize	45.208 to 45.228 mm (1.7798 to 1.7806 in)
Maximum journal out-of-round	0.03 mm (0.0012 in)

Pistons and rings

Piston-to-bore clearance:	
New	0.025 mm (0.001 in)
Wear limit	0.07 mm (0.0028 in)
Groove-to-ring clearance:	
New	0.02 to 0.08 mm (0.0008 to 0.0032 in)
Wear limit	0.1 mm (0.004 in)
Piston rings endgap clearance (ring 15 mm/0.6 in from bottom of bore):	
New	0.25 to 0.50 mm (0.010 to 0.020 in)
Wear limit	1.0 mm (0.040 in)

Piston size:	**Piston diameter**	**Bore diameter**
Standard size	79.48 mm (3.1291 in)	79.51 mm (3.1303 in)
1st oversize	79.73 mm (3.1389 in)	79.76 mm (3.1401 in)
2nd oversize	79.98 mm (3.1488 in)	80.01mm (3.1499 in)
3rd oversize	80.48 mm (3.1684 in)	80.51mm (3.1696 in)

Cylinder bore

Maximum allowable deviation	0.08 mm (0.0031 in)

Connecting rods

Maximum endplay	0.4 mm (0.016 in)
Big-end bearing running clearance:	
New	0.015 to 0.062 mm (0.0006 to 0.0024 in)
Wear limit	0.12 mm (0.0047 in)

Camshaft

Maximum endplay	0.15 mm (0.006 in)

Valves

Valve clearances .... As 1.6 and 1.8 four-cylinder engines
Adjusting shim thicknesses .... 3.00 to 4.25 mm (0.118 to 0.167 in) in increments of 0.05 mm
Valve guides .... As 1.6 and 1.8 four-cylinder engines

Valve timing at 1 mm valve lift/zero clearance

Inlet opens BTDC .... 10°
Inlet closes ABDC .... 36°
Exhaust opens BBDC .... 45°
Exhaust closes ATDC .... 3°

Cylinder head

Minimum height (between faces) .... 132.75 mm (5.2264 in)
Maximum gasket face distortion .... 0.1 mm (0.004 in)
Valve seat angle .... 45°

Torque wrench settings

	Nm	lbf ft
Engine to gearbox:		
M8	20	14
M10	45	32
M12	60	43
Exhaust pipe to manifold	30	21
Subframe to body	70	51
Engine support to mounting	45	32
Cylinder head bolts:		
Stage 1	40	29
Stage 2	60	43
Stage 3	75	54
Stage 4	Tighten a further quarter turn (90°) – in one movement	
Note: *Do not retighten the socket-head bolts during servicing*		
Main bearing caps	65	47
Oil pump to block:		
Bolt (long)	20	14
Stud and bolt (short)	10	7
Oil pick-up	10	7
Crankshaft rear oil seal housing	10	7
Big-end nuts	50	36
Valve (cylinder head) cover	10	7
Timing cover	10	7
Camshaft bearing cap	20	14
Camshaft gear	80	58
Oil pump pressure relief valve	40	29
Crankshaft pulley vibration damper bolt	450	332
Sump bolt	20	14
Sump drain plug	40	29
Flywheel cover bolts	25	18
Flywheel bolts – early (non-shouldered) type	75	54
Flywheel bolts- late (shouldered) type	100	72

2.0 litre five-cylinder engines

At the time of publication, limited information was available for the 2.0 litre engine. With the exception of those items listed below, specifications are generally as for the 1.6 DT and 1.9 engines. If in doubt, contact your dealer for further information.

General

Engine code .... JS (from July 1983 on)
Capacity .... 1994 cc
Bore .... 81 mm (3.19 in)
Stroke .... 77.4 mm (3.05 in)
Compression ratio 10.0: 1

Valve timing at 1 mm valve lift/zero clearance

Inlet opens BTDC .... 1 °
Inlet closes ABDC .... 43°
Exhaust opens BBDC .... 43°
Exhaust closes ATDC .... 1 °

Part A – Four-cylinder engines

1 General description

The engine is of four-cylinder, in-line overhead camshaft type mounted conventionally at the front of the car. The crankshaft is of five-bearing type and the centre main bearing shells incorporate flanged or separate thrust washers to control crankshaft endfloat. The camshaft is driven by a toothed belt from the crankshaft sprocket, and the belt also drives the intermediate shaft which is used to drive the distributor, oil pump and on carburettor engines, the fuel pump. The valves are operated from the camshaft through bucket type tappets, and valve clearances are adjusted by the use of shims located in the top of the tappets.

The engine has a full-flow lubrication system from a gear type oil pump mounted in the sump, and driven by an extension of the distributor which is itself geared to the intermediate shaft. The oil filter is of the cartridge type, mounted on the left-hand side of the cylinder block. The oil pressure switch is located on the rear of the cylinder head (photo).

1.2 Oil pressure switch location on rear face of cylinder head (1.6 litre)

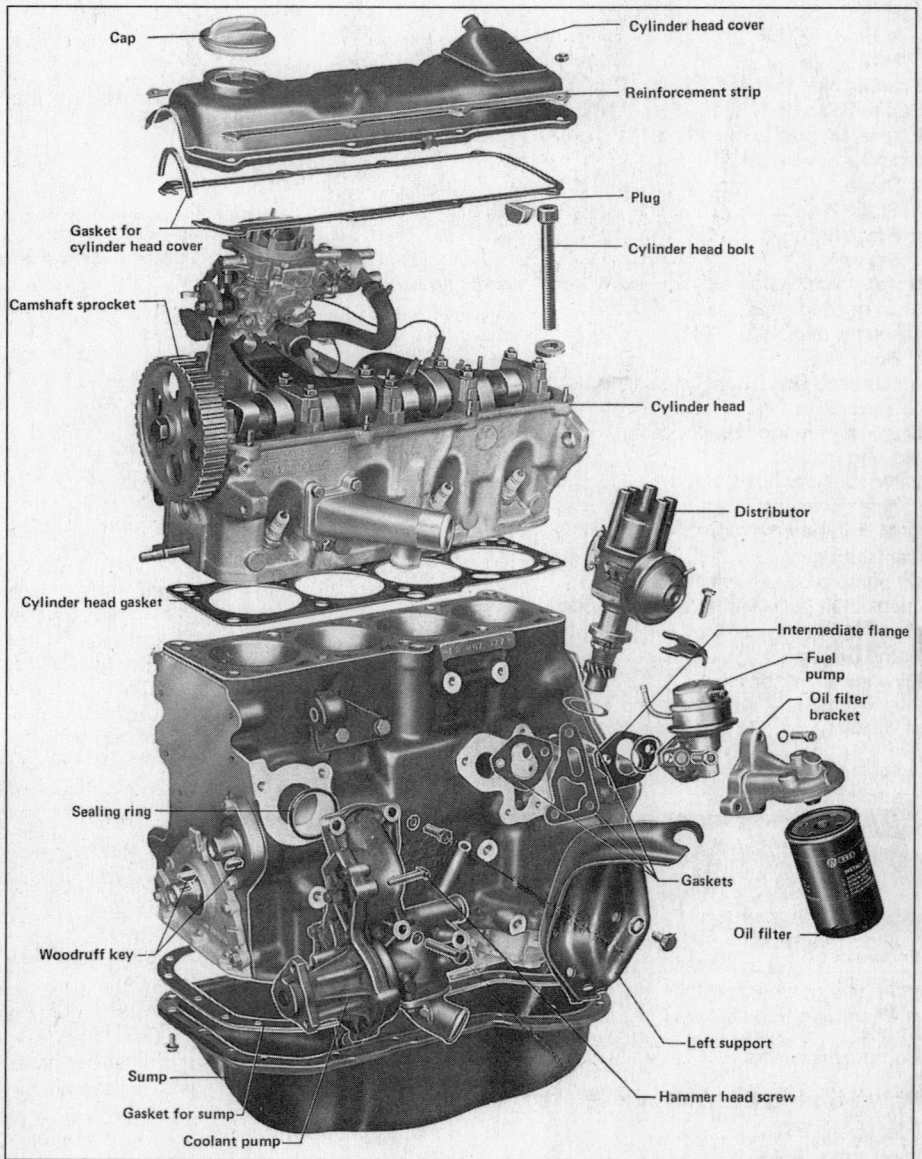

Fig. 1.1 Exploded view of the major components of the four-cylinder engine (Sec 1)

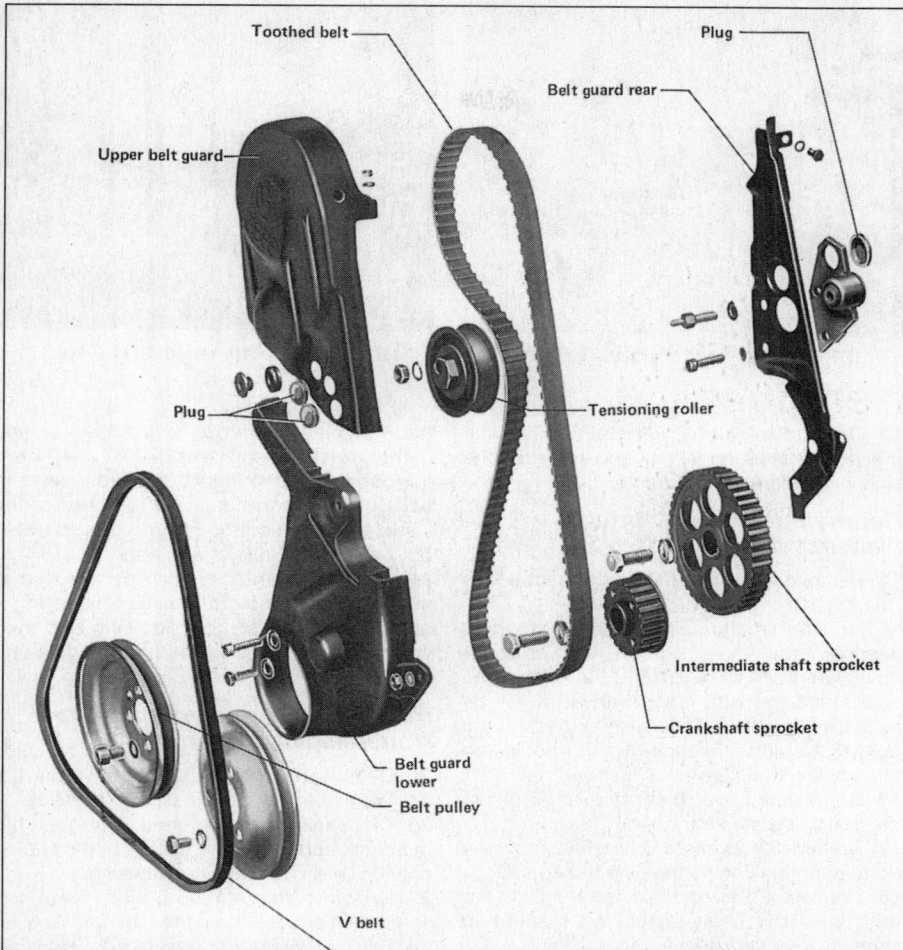

Fig. 1.2 Exploded view of the timing assembly components (four-cylinder engine) (Sec 1)

Labels: Toothed belt, Plug, Belt guard rear, Upper belt guard, Plug, Tensioning roller, Plug, Intermediate shaft sprocket, Crankshaft sprocket, Belt guard lower, Belt pulley, V belt

2.2 Topping up engine oil

3 Major operations possible with the engine in the car

The following operations can be carried out without having to remove the engine from the car:

(a) Removal and servicing of the cylinder head and camshaft
(b) Removal of the timing belt and gears
(c) Removal of the flywheel or driveplate (after first removing the transmission)
(d) Removal of the sump (after first lowering the subframe)
(e) Removal of the oil pump pistons and connecting rods

4 Major operations only possible after removal of the engine from the car

The following operations can only be carried out after removal of the engine from the car:

(a) Removal of the intermediate shaft
(b) Removal of the crankshaft and main bearings

5 Method of engine removal

The engine can be lifted from the car either separately or together with the manual gearbox. On automatic transmission models it is recommended that the engine is removed separately because of the extra weight involved.

6 Engine – removal and refitting

1 Remove the bonnet as described in Chapter 12 then position it out of the way in a safe place, standing it upright on cardboard or rags.
2 Disconnect the battery earth lead.
3 Remove the radiator as described in Chapter 2.
4 Unclip and detach the wiring connector from the rear face of the alternator.

2 Engine – routine maintenance

The following routine maintenance procedures should be undertaken at the specified intervals given at the start of this manual. The intervals given are those for a vehicle used in normal driving conditions.

HAYNES HiNT *Where the vehicle is subject to severe daily use, such as city driving or in a hot dusty climate, then it is advisable to reduce the maintenance intervals accordingly.*

Engine oil check: Check the engine oil with the vehicle parked on level ground, the engine switched off and having been stationary for a short period. This will allow oil in the lubrication circuits to return to the sump to provide a true level reading. The dipstick should be withdrawn, wiped clean, fully inserted then withdrawn again to take the oil level reading. The oil level must be kept between the maximum and minimum markings on the dipstick at all times. If the oil level is down to the minimum mark, it will need a litre (1.7 Imp pint) to raise the oil level to the maximum mark. Do not overfill the engine. Top up through the oil filler neck in the cylinder head cover (photo). Check that the cover is firmly refitted and wipe clean any oil spillage. Recheck the oil level on completion.

Engine oil change: With the vehicle standing on level ground, remove the drain plug from the sump and drain the old engine oil into a container of suitable capacity. Draining is best undertaken directly after the vehicle has been used when the engine oil will be hot and will flow more freely. When draining is complete, refit the drain plug and top up the engine oil to the correct level with the specified grade and quantity of engine oil.

Engine oil filter: The oil filter must be renewed at the specified intervals. The filter is best removed whilst waiting for the engine oil to drain. Renewing the oil filter is described in Section 28 of this Chapter.

Engine general checks: Inspect the engine regularly for any signs of oil, coolant or fuel leaks and if found attend to them without delay.

6.9 Engine mounting (right-side) viewed from below (1.6 litre)

6.11 Engine front support (1.6 litre)

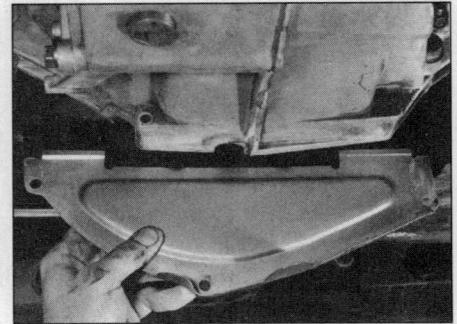

6.16 Cover plate removal (1.6 litre)

5 Remove the air cleaner unit from the carburettor as described in Chapter 3.

6 Identify those coolant, fuel and vacuum hoses affecting engine removal, label or mark them with masking tape then disconnect them. These will include the fuel hose to the fuel pump and the return hose from the filter unit, the vacuum hose at inlet manifold (to servo) and the vacuum line from the reserve chamber to the carburettor.

7 Identify all wiring for location using masking tape, then disconnect those affecting engine removal. These may include the wiring to the oil pressure switch, temperature sender, distributor coil and inlet manifold pre-heater.

8 On manual gearbox models disconnect the clutch cable from the release lever and cable bracket. Disconnect the throttle cable, as described in Chapter 3. Disconnect the hose from the inlet manifold (where applicable) and the heater hoses from the bulkhead. Also disconnect the heat exchanger hose from the outlet connection at the rear of the cylinder head. Be prepared for additional coolant spillage when disconnecting this hose.

9 Unscrew the left-hand side engine mounting nut from the top and the right-hand side engine mounting from the bottom (photo).

10 Unscrew the nuts and separate the exhaust downpipe from the exhaust manifold.

11 Unbolt the support arm from the front of the engine (at the front) (photo).

12 Disconnect the wiring from the starter motor (with reference to Chapter 10).

13 On models with power steering remove the power steering pump and tie it to one side, with reference to Chapter 11.

Removing engine without transmission

14 Remove the starter motor, as described in Chapter 10.

15 On automatic transmission models unscrew the three torque converter-to-driveplate bolts while holding the starter ring gear stationary with a screwdriver. It will be necessary to rotate the engine to position the bolts in the starter aperture, using a socket on the crankshaft pulley bolt (Fig. 1.3).

16 Unbolt the cover plate from the front of the transmission (photo).

17 Detach the exhaust downpipe from the front mounting and tie the pipe to one side.

18 Connect a hoist and take the weight of the engine — the hoist should be positioned centrally over the engine.

19 Support the weight of the transmission with a trolley jack.

20 Unscrew and remove the engine-to-transmission bolts noting the location of the brackets and earth strap.

21 Lift the engine from the mountings and reposition the trolley jack beneath the transmission.

22 Pull the engine from the transmission – make sure on automatic transmission models that the torque converter remains fully engaged with the transmission splines.

23 With the aid of an assistant lift and guide

the engine upwards from the engine compartment taking care not to snag it on surrounding components, the brake lines in particular. With the engine lifted clear of the vehicle, lower it to the ground or preferably onto an engine trolley if available.

24 The rear engine plate can be removed at this stage for protection by simply easing it rearwards off its locating dowels and then withdrawing it from between the engine and flywheel/driveplate.

Removing engine with manual transmission attached

25 Detach the exhaust downpipe from the front mounting and tie the pipe to one side.

26 Disconnect the reverse light/gearshift indicator/econometer wiring multi-plug connector above the clutch housing.

27 Unscrew the nut and disconnect the speedometer cable from the differential case.

28 Disconnect the driveshafts from the drive flanges with reference to Chapter 8. Tie them up and support to the side out of the way.

29 Disconnect the gearshaft rod coupling from the internal selector lever by unscrewing and removing the location bolt then levering the support strut from the balljoint. The shift rod can then be pulled free from the selector lever.

30 Disconnect the earth lead from the upper bellhousing bolt on the left-hand side.

31 Support the transmission with a trolley jack.

32 Loosen the bolt securing the gearbox stay to the underbody. Unscrew and remove the stay inner bolt and pivot the stay from the gearbox.

33 Unbolt the bonded rubber mounting from the gearbox and also unbolt the front gearbox support bracket.

34 Connect a hoist and take the weight of the engine – the hoist should be positioned near the front of the engine so that the engine and gearbox will hang at a steep angle (photo).

35 Lift the engine from the mountings and move it forwards, then lower the trolley jack and lift the engine and gearbox from the engine compartment while turning it, as necessary, to clear the body. Lower the engine and gearbox to the floor (photo).

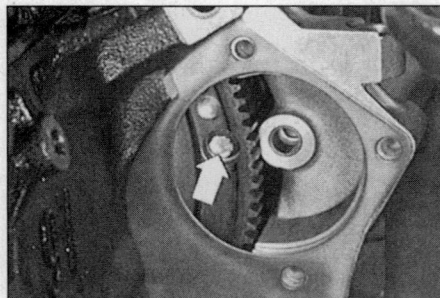

Fig. 1.3 Torque converter to driveplate bolt (automatic transmission model) – remove through starter motor aperture (Sec 6)

6.35 Engine and gearbox removal (1.6 litre)

9.1A Upper timing cover top nuts

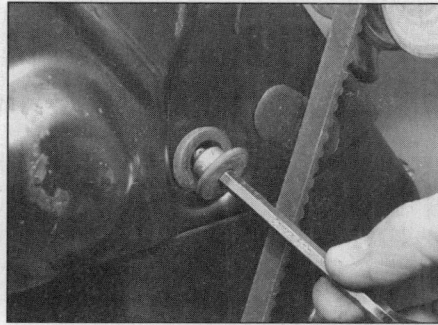
9.1B Upper timing cover bottom nut

9.2A Remove the reinforcement strips . . .

36 If necessary, separate the gearbox from the engine by removing the starter (Chapter 10), gearbox front cover, and engine-to-gearbox bolts. Remove the rear engine plate.

Refitting

37 Refitting is a reversal of the removal procedure, but before starting the engine check that it has been filled with oil and also that the cooling system is full. Make sure that the starter cable is not touching the engine or mounting bracket, and is clear of the exhaust system.

> **HAYNES HiNT** *Delay tightening the engine and gearbox mountings until the engine is idling – this will ensure that the engine is correctly aligned.*

38 Adjust the accelerator cable as described in Chapter 3.
39 Adjust the clutch cable as described in Chapter 5.
40 Tighten the engine to gearbox and engine mounting bolts to the specified torque wrench settings.

7 Engine dismantling – general

1 If possible position the engine on a bench or strong table for the dismantling procedure. Two or three blocks of wood will be necessary to support the engine in an upright position.

2 Cleanliness is most important, and if the engine is dirty, it should be cleaned with paraffin before commencing work.
3 Avoid working with the engine directly on a concrete floor, as grit presents a real source of trouble.
4 As parts are removed, clean them in a paraffin bath. However, do not immerse parts with internal oilways in paraffin as it is difficult to remove, usually requiring a high pressure hose. Clean oilways with nylon pipe cleaners.
5 It is advisable to have suitable containers to hold small items according to their use, as this will help when reassembling the engine and also prevent possible losses.
6 Always obtain complete sets of gaskets when the engine is being dismantled, but retain the old gaskets with a view to using them as a pattern to make a replacement if a new one is not available.
7 When possible, refit nuts, bolts, and washers in their location after being removed, as this helps to protect the threads and will also be helpful when reassembling the engine.
8 Retain unserviceable components in order to compare them with the new parts supplied.

8 Ancillary components – removal and refitting

With the engine removed from the car, the externally mounted ancillary components in the following list can be removed. The removal sequence need not necessarily follow the order given:

Inlet and exhaust manifolds (Chapter 3)
Fuel pump or warm-up regulator, as applicable (Chapter 3)
HT leads and spark plugs (Chapter 4)
Oil filter cartridge (Section 28 of this Chapter)
Distributor (Chapter 4)
Dipstick
Alternator (Chapter 10)
Engine mountings (Section 29 of this Chapter)

9 Cylinder head and camshaft – removal

Note: *If the engine is still in the car first carry out the following operations:*
(a) Disconnect the battery negative lead
(b) Drain the cooling system
(c) Remove the alternator
(d) Remove the inlet and exhaust manifolds
(e) Remove the HT leads and spark plugs
(f) Disconnect all wiring cables and hoses
(g) Disconnect the exhaust downpipe from the exhaust manifold

⚠ **Caution: To avoid possible damage to the valves and piston crowns on 1.8 litre engines do not rotate the crankshaft or camshaft with the timing belt removed.**

1 Unscrew the nuts and lift off the upper timing cover, using an Allen key where necessary (photos).
2 Unscrew the nuts and lift off the valve cover, together with the reinforcement strips and gaskets (photos).

9.2B . . . and valve cover

9.2C Remove the valve cover from the front gasket . . .

9.2D . . . and the rear plug (this must be renewed)

9.3 Cylinder head side outlet

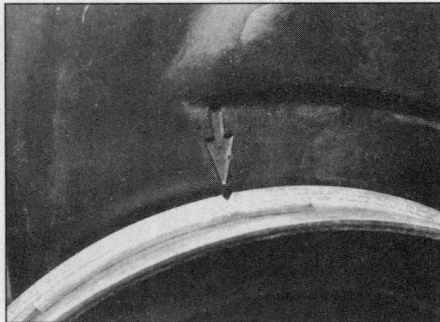

9.4 Crankshaft pulley notch aligned with the TDC arrow on lower timing cover

3 Unbolt the outlet elbows and remove the gaskets if necessary (photo).

4 Using a socket on the crankshaft pulley bolt, turn the engine so that the piston in No 1 cylinder is at TDC (top dead centre) on its compression stroke. The notch in the crankshaft pulley must be in line with the arrow on the lower timing cover (photo), and both No 1 cylinder valves must be closed (ie cam peaks away from the tappets). The notch on the rear of the camshaft gear will also be in line with the top of the timing belt, rear cover.

5 Loosen the nut on the timing belt tensioner, and using an open-ended spanner, rotate the eccentric hub anti-clockwise to release the belt tension.

6 Remove the timing belt from the camshaft gear and tensioner and move it to one side while keeping it in firm contact with the intermediate and crankshaft gears. *The intermediate gear must not be moved otherwise the ignition timing will be lost and the lower timing cover will have to be removed.*

7 Unscrew the nut and remove the timing belt tensioner.

8 Unscrew the bolt securing the timing belt rear cover to the cylinder head.

9 Using a splined socket unscrew the cylinder head bolts a turn at a time in reverse order to that shown in Fig. 1.9.

10 With all the bolts removed lift the cylinder head from the block. If it is stuck, tap it free with a wooden mallet. *Do not insert a lever into the gasket joint.*

11 Remove the cylinder head gasket.

10 Camshaft and tappets – removal

Note: *If the engine is still in the car, first carry out the following operations:*
(a) *Disconnect the battery negative lead*
(b) *Disconnect the wiring, cables and hoses which cross the engine valve cover – remove the air cleaner on carburettor models*
(c) *Remove the alternator/water pump drivebelt*

1 Follow paragraphs 1 to 6 of Section 9, excluding paragraph 3.

2 Unscrew the centre bolt from the camshaft gear while holding the gear stationary with a bar through one of the holes.

10.3A Remove the camshaft gear

3 Withdraw the gear from the camshaft and extract the Woodruff key (photos).

4 Unscrew the nuts from bearing caps 1, 3

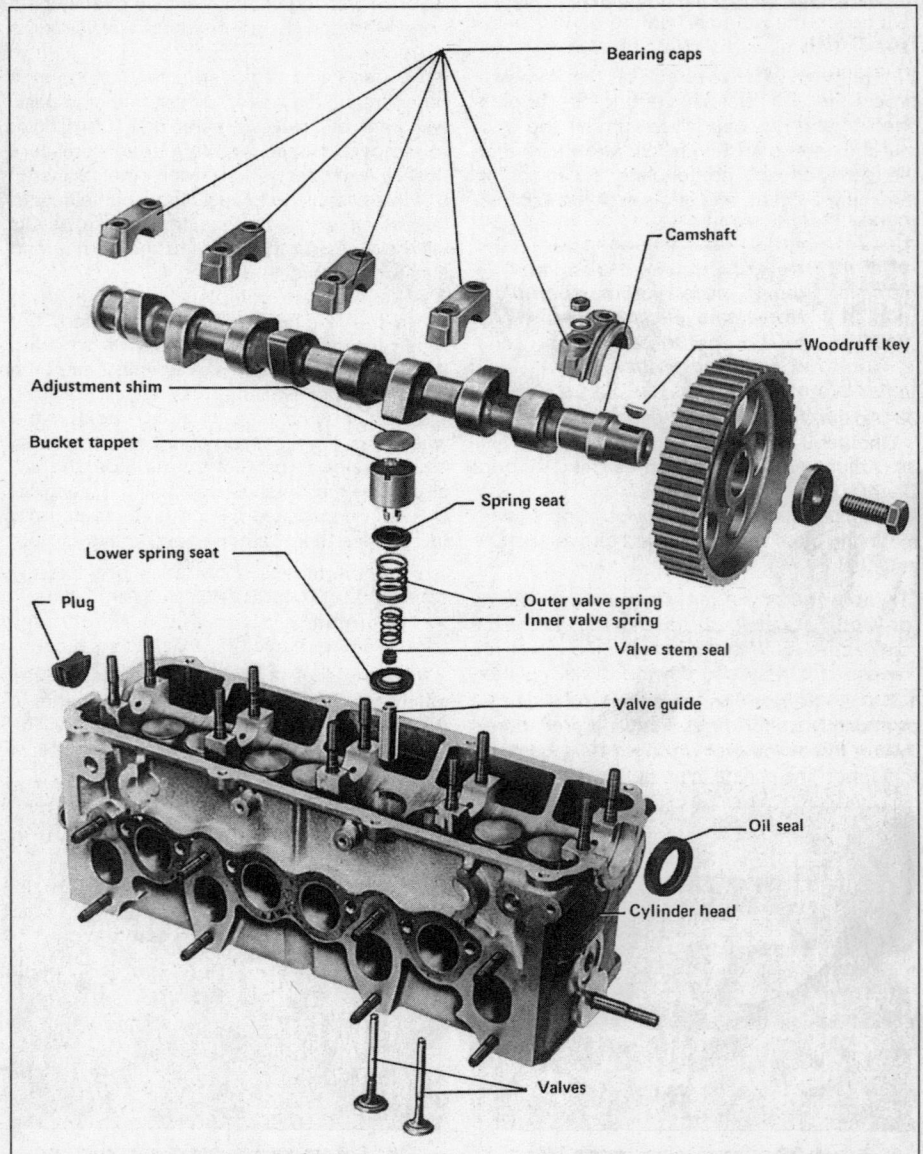

Fig. 1.4 Cylinder head, valves and camshaft components (four-cylinder engine) (Sec 10)

10.3B Camshaft gear Woodruff key

10.4A The number 3 camshaft bearing cap

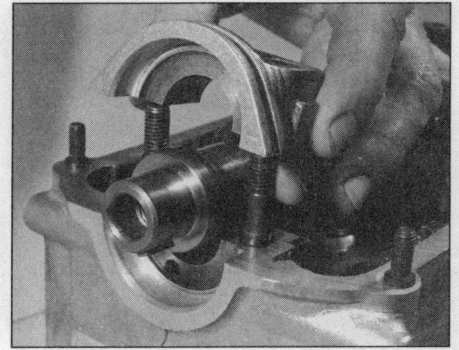
10.4B Removing No 1 camshaft bearing cap

and 5. Identify all the caps for position then remove caps 1, 3 and 5 (photos).

5 Loosen the nuts on bearing caps 2 and 4 evenly until all valve spring tension has been released, then remove the caps.

6 Lift out the camshaft and discard the oil seal (photo).

7 Have ready a board with eight pegs on it, or alternatively use a box with internal compartments, marked to identify cylinder numbers and whether inlet or exhaust or position in the head, numbering from the front of the engine. As each tappet is removed (photo), place it on the appropriate peg, or mark the actual tappet to indicate its position.

Note that each tappet has a shim (disc) fitted into a recess in its top. This shim must be kept with its particular tappet.

10.6 Removing the camshaft

10.7 Removing a tappet together with its shim

11 Valves – removal and renovation

1 With the cylinder head removed as previously described, the valve gear can be dismantled as follows. Because the valves are recessed deeply into the top of the cylinder head, their removal requires a valve spring compressor with long claws or the use of some ingenuity in adapting other types of compressor. One method which can be employed is to use a piece of tubing of roughly the same diameter as the valve spring cover and long enough to reach above the top of the cylinder head. To be able to remove the valve collets, either cut a window in the tube on each side, or cut a complete section away, so that the tube is about three quarters of a circle.

2 Have ready a board with holes in it, into which each valve can be fitted as it is removed, or have a set of labelled containers so that each valve and its associated parts can be identified and kept separate. Inlet valves are Nos. 2-4-5-7, Exhaust valves are Nos 1-3-6-8, numbered from the timing belt end of the engine.

3 Compress each valve spring until the collets can be removed (photo). Take out the collets, release the spring compressor and remove it.

4 Remove the valve spring cover, the outer and inner spring and the valve (photos). Thread the springs over the valve stem to keep them with the valve, and put on the valve spring cover. It is good practice, but not essential, to keep the valve springs the same way up, so that parts are refitted exactly as they were before removal.

5 Prise off the valve stem seals, or pull them

11.3 Compressing a valve spring for collet removal

11.4A Remove the valve spring cover, springs . . .

11.4B . . . and valve

11.5A Remove the valve stem seal . . .

11.5B . . . and spring seat

out the guide and then insert the stem of a new valve into the guide. Because the stem diameters are different, ensure that only an inlet valve is used to check the inlet valve guides, and an exhaust valve for the exhaust valve guides. With the end of the valve stem flush with the top of the valve guide, measure the total amount by which the rim of the valve head can be moved sideways. If the movement exceeds the maximum amount given in the Specifications, new guides should be fitted, but this is a job for a VW dealer, or specialist workshop (Fig. 1.6).

off with pliers and discard them, then lift off the valve spring seat and place it with its valve (photos).

6 Examine the heads of the valves for pitting and burning, paying particular attention to the heads of the exhaust valves. The valve seats should be examined at the same time. If the pitting on the valve and seat is only slight, the marks can be removed by grinding the seats and valves together with coarse and then fine grinding paste. Where bad pitting has occurred, it will be necessary to have the valve seat re-cut and either use a new valve, or have the valve re-faced.

7 The refacing of valves and cutting of valve seats is not expensive and gives a far better result than grinding in valves which are badly pitted. Exhaust valves, if too worn for grinding in, must be discarded. Re-machining of exhaust valves is not permissible.

8 Valve grinding is carried out as follows: Smear a small quantity of coarse carborundum paste around the contact surface of the valve or seat and insert the valve into its guide. Apply a suction grinder tool to the valve head and grind in the valve by semi-rotary motion. This is produced by rolling the valve grinding tool between the palms of the hands. When grinding action is felt to be at an end, extract the valve, turn it and repeat the operation as many times as is necessary to produce a uniform matt grey surface over the whole seating area of the valve head and valve seat. Repeat the process using fine grinding paste.

9 Scrape away all carbon from the valve head and valve stem. Carefully clean away every trace of grinding paste, taking care to leave none in the ports, or in the valve guides. Wipe the valves and valve seats with a paraffin soaked rag and then with a clean dry rag.

12 Cylinder head – examination and renovation

1 Check the cylinder head for distortion, by placing a straight edge across it at a number of points, lengthwise, crosswise and diagonally, and measuring the gap beneath it with feeler gauges. If the gap exceeds the limit given in the Specifications, the head must be refaced by a workshop which is equipped for this work. Re-facing must not reduce the cylinder head height below the minimum dimension given in the Specifications (Fig. 1.5).

2 Examine the cylinder head for cracks. If there are minor cracks of not more than 0.5 mm (0.020 in) width between the valve seats, or at the bottom of the spark plug holes, the head can be re-used, but a cylinder head cannot be repaired or new valve seat inserts fitted.

3 Check the valve guides for wear. First clean

13 Camshaft and bearings – examination and renovation

1 Examine the camshaft for signs of damage or excessive wear. If either the cam lobes or any of the journals have wear grooves, a new camshaft must be fitted.

2 With the camshaft fitted in its bearings, but with the bucket tappets removed so that there is no pressure on the crankshaft, measure the endfloat of the camshaft, which should not exceed the limit given in the Specifications.

3 The camshaft bearings are normally part of the cylinder head and cannot be renewed. The bearing clearance is very small and the clearances can only be checked with a dial gauge. *If there is excessive looseness in the camshaft bearings, do not attempt to decrease it by grinding or filing the bottoms of the bearing caps.*

4 Where an exchange engine or cylinder head has been fitted at some time it may be possible that the camshaft bearings are of shell type. If the shells are to be renewed, be sure to order the correct size as the camshaft may be of standard or undersize type. A cylinder head with an undersize camshaft will have a yellow identification paint mark over the VW – Audi marking.

5 If renewing the camshaft note that the 1.8 engine camshaft can be identified by a blue line between the 1st and 2nd cams, whilst later 1.6 models (from August 1983 on) have a yellow marking in this position (engine code DS), (Fig. 1.7).

Fig. 1.5 Check cylinder head for distortion (Sec 12)

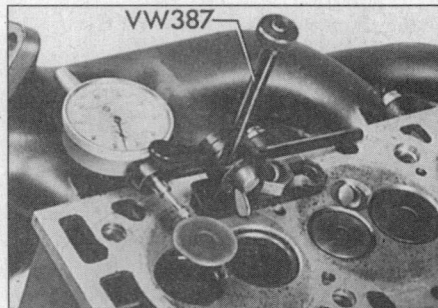

Fig. 1.6 Check valve guide wear by measuring valve movement using a dial gauge (Sec 12)

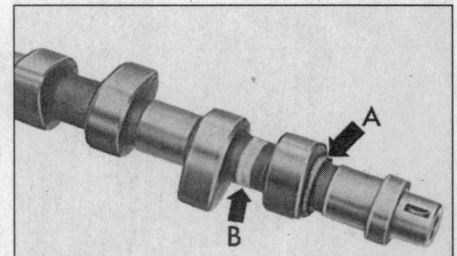

Fig. 1.7 Camshaft identification points on the 1.8 litre engine (Sec 13)

A Shoulder   B Blue line

14 Valves – refitting

1 Locate the valve spring seats over the guides. then press a new seal on to the top of each valve guide. A plastic sleeve should be provided with the valve stem seals, so that the seals are not damaged when the valves are fitted.
2 Apply oil to the valve stem and the stem seal. Fit the plastic sleeve over the top of the valve stem and insert the valve carefully. Remove the sleeve after inserting the valve.

HAYNES HiNT *If there is no plastic sleeve, wrap a piece of thin adhesive tape round the top of the valve stem, so that it covers the recess for the collets and prevents the sharp edges of the recess damaging the seal. Remove the tape after fitting the valve.*

3 Fit the inner and outer valve springs, then the valve spring cover. If renewing springs, they must only be renewed as a pair on any valve.
4 Fit the spring compressor and compress the spring just enough to allow the collets to be fitted. If the spring is pressed right down there is a danger of damaging the stem seal.
5 Fit the collets, release the spring compressor slightly and check that the collets seat properly, then remove the compressor.
6 Tap the top of the valve stem with a soft-headed hammer to ensure that the collets are seated.
7 Repeat the procedure for all the valves.

15 Camshaft and tappets – refitting

⚠️ *Caution: To avoid possible damage to the valves and piston crowns on 1.8 litre engines do not rotate the crankshaft or camshaft with the timing belt removed.*

1 Fit the bucket tappets to their original positions; the adjustment shims on the top of

15.4 Tightening the camshaft bearing cap nuts

15.5 Installing the camshaft oil seal

the tappets must be fitted so that the lettering on them is downward. Lubricate the tappets and the camshaft journals.
2 Lay the camshaft into the lower half of its bearings so that the lowest point of the cams of No 1 cylinder are towards the tappets, and fit the bearing caps in their original positions, making sure that they are the right way round before fitting them over the studs.
3 Fit the nuts to bearing caps Nos 2 and 4 and tighten them in diagonal sequence until the camshaft fully enters its bearings.
4 Fit the nuts to bearing caps Nos 1, 3 and 5, then tighten all the nuts to the specified torque in diagonal sequence (photo).
5 Smear a little oil onto the sealing lip and outer edge of the camshaft oil seal, then locate it open end first in the cylinder head No 1 camshaft bearing cap (photo).
6 Using a metal tube, drive the seal squarely into the cylinder head until flush with the front of the cylinder head – *if the seal is driven in further it will block the oil return hole.*
7 Fit the Woodruff key in its groove and locate the gear on the end of the camshaft.
8 Fit the centre bolt and spacer and tighten the bolt to the specified torque while holding the gear stationary with a bar through one of the holes (photo).
9 Turn the camshaft gear and align the rear notch with the top of the timing belt rear cover (Fig. 1.8).
10 Without disturbing the intermediate gear setting, and keeping the timing belt in firm contact with the intermediate gear and

crankshaft gear, locate the timing belt on the camshaft gear and tensioner. The crankshaft must be positioned with No 1 cylinder at TDC. If the position of the intermediate gear is in doubt, it must be checked with reference to Section 18.
11 Turn the timing belt tensioner clockwise and tension the timing belt until it can just be twisted 90° with the thumb and index finger midway between the camshaft and intermediate gears. Tighten the nut to secure the tensioner.
12 Check and, if necessary, adjust the valve clearances, as described in Section 17.
13 Refit the valve cover and reinforcement strips, together with new gaskets and seals, and tighten the nuts.
14 Refit the upper timing belt cover and tighten the nuts.
15 If the engine is in the car, reverse the preliminary procedures given in Section 10.

16 Cylinder head and camshaft – refitting

⚠️ *Caution: To avoid possible damage to the valves and piston crowns on 1.8 litre engines do not rotate the crankshaft or camshaft with the timing belt removed.*

1 Check that the top of the block is perfectly clean, then locate a new gasket on it with the words OBEN – TOP facing upward (photos).
2 Check that the cylinder head face is

15.8 Tightening the camshaft gear centre bolt

Fig. 1.8 Camshaft gear TDC mark aligned with top of timing belt rear cover (Sec 15)

16.1A Locate new cylinder head gasket onto the block . . .

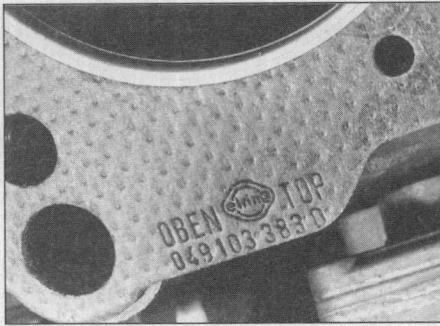

16.1B . . . with words OBEN – TOP uppermost

16.2 Lower cylinder head into position

16.3 Tighten the cylinder head bolts

perfectly clean. Place two long rods or pieces of dowel in two cylinder head bolt holes at opposite ends of the block, to position the gasket and give a location for fitting the cylinder head. Lower the head on to the block (photo), remove the guides and insert the bolts and washers. Do not use jointing compound on the cylinder head joint.

3 Tighten the bolts using the sequence shown in Fig. 1.9 in the four stages given in the Specifications to the specified torque (photo).

4 Insert and tighten the bolt securing the timing belt rear cover to the cylinder head.

5 Refit the timing belt tensioner and fit the nut finger tight.

6 Follow paragraphs 9 to 14 inclusive of Section 15.

7 Refit the outlet elbow together with a new gasket and tighten the bolts.

8 If the engine is in the car, reverse the preliminary procedures given in Section 9.

17 Valve clearances – checking and adjustment

1 Valve clearances are adjusted by inserting the appropriate thickness shim to the top of the tappet. Shims are available in thicknesses from 3.00 to 4.25 mm (0.118 to 0.167 in) in increments of 0.05 mm (0.002 in).

2 Adjust the valve clearances for the initial setting-up after fitting a new camshaft, or grinding in the valves with the engine cold. The valve clearances should be re-checked after 620 miles (1000 km), with the engine warm, and the coolant over 35°C (95°F).

3 On 1.8 litre engines it is important to note that the valve clearance adjustment must not be made when the piston is at TDC position, but at 90° before TDC. This ensures that the valves miss the pistons if the tappets are depressed. Remove the valve cover after removing the upper timing cover.

4 Fit a spanner to the crankshaft pulley bolt and turn the crankshaft until the highest points of the cams for one cylinder are pointing upwards and outwards at similar angles. Use feeler gauges to check the gap between the cam and the tappet and record the dimension (photo).

5 Repeat the operation for all four cylinders and complete the list of clearances. Valves are numbered from the timing belt end of the engine. Inlet valves are Nos 2-4-5-7, exhaust valves are Nos 1-3-6-8.

6 Where any tolerances exceed those given in the Specifications, remove the existing shim by placing a cranked dowel rod with a suitably shaped end between two tappets with the rod resting on the edge of the tappets (photo). With the piston for the relevant cylinder at TDC compression, lever against the camshaft to depress the tappets sufficiently to remove the shim(s) from the top of the tappet(s). Do not overdepress the tappets so that the valves touch the pistons. Note that each tappet incorporates a notch in its upper rim so that a small screwdriver or similar tool can be used to remove the shim (photos).

7 Note the thickness of the shim (engraved on its underside). and calculate the shim thickness required to correct the clearance

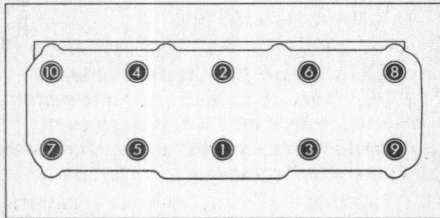

Fig. 1.9 Cylinder head bolt tightening sequence (Sec 16)

17.4 Checking the valve clearances

17.6A Use a cranked dowel rod to depress the tappets

17.6B Removing a shim from a tappet (tappet removed)

17.6C Notches (arrowed) machined in tappet for removing the shim

17.7 Engraved thickness in mm on the tappet shim underside

18.9 Turn the timing belt tensioner anti-clockwise to release the belt tension

18.11 Removing the timing belt tensioner

(photo). For example, if the measured clearance on an inlet valve (warm engine) is 0.35 mm (0.014 in) this is outside the tolerance specified for inlet valves which is 0.20 mm to 0.30 mm (0.008 to 0.012 in). The best adjustment is the mid-point of the range i.e. 0.25 mm (0.010 in) and the measured gap of 0.35 mm (0.014 in) is 0.1 mm (0.004 in) too great. If the shim which is taken out is 3.05 mm (0.120 in), it should be replaced by one of 3.15 mm (0.124 in).

8 Provided they are not worn or damaged, shims which have been removed can be re-used in other positions if they are of the correct thickness.

9 Refit the valve cover and upper timing cover, together with new gaskets, after checking and adjusting the valve clearances.

18 Timing belt and gears – removal and refitting

⚠️ *Caution: To avoid possible damage to the valves and piston crowns on 1.8 litre engines do not rotate the crankshaft or camshaft with the timing belt removed.*

1 If the engine is still in the car, disconnect the battery negative lead.

2 Remove the alternator drivebelt, with reference to Chapter 10.

3 Unscrew the two upper retaining bolts and the single nut (from the front face) and lift off the upper timing cover.

4 Unbolt and remove the pulley from the water pump.

5 Using a socket on the crankshaft pulley bolt, turn the engine so that the piston in No 1 cylinder is at TDC (top dead centre) on its compression stroke. The notch in the crankshaft pulley must be in line with the arrow on the lower timing cover, and both No 1 cylinder valves must be closed (ie cam peaks away from the tappets).

6 If it is required to remove the crankshaft gear, loosen the centre bolt now. Hold the crankshaft stationary with a wide-bladed screwdriver in the starter ring gear (starter motor removed) or engage top gear and apply the handbrake if the engine is still in the car.

7 With the TDC marks aligned, unbolt the crankshaft pulley from the gear.

8 Unbolt and remove the lower timing cover.

9 Loosen the nut on the timing belt tensioner, and using an open-ended spanner, rotate the eccentric hub anti-clockwise to release the belt tension (photo).

10 Remove the timing belt from the crankshaft, camshaft and intermediate gears, and from the tensioner.

11 Unscrew the nut and remove the timing belt tensioner (photo).

12 Unscrew the centre bolt and withdraw the intermediate gear (photo). Remove the

18.12 Removing the gear from the intermediate shaft

Woodruff key. When loosening the centre bolt, hold the gear stationary with a socket and bar on the rear timing cover bolt.

13 On early models, remove the centre bolt and withdraw the crankshaft gear (photos). Remove the Woodruff key. On models produced from October 1982, the crankshaft gear is located by means of a lug which engages with an angled slot on the end of the crankshaft. This modification replaced the Woodruff key engagement method used on earlier models, see Fig. 1.10.

14 Unscrew the centre bolt from the camshaft gear while holding the gear stationary with a bar through one of the holes. Withdraw the gear and remove the Woodruff key.

Fig. 1.10 Crankshaft gear location method on later models (Sec 18)

A Angled slot   B Sprocket lug

18.13A On early models, remove the centre bolt . . .

18.13B . . . and withdraw the crankshaft gear

18.15A Remove the rear timing cover . . .

18.15B . . . and the alternator mounting bracket

18.18 Tightening the crankshaft gear centre bolt

15 Unbolt and remove the rear timing cover together with the alternator mounting bracket from the block and cylinder head (photos).
16 Commence refitting by locating the alternator mounting bracket and rear timing cover on the engine and tightening the bolts.
17 Locate the Woodruff key and camshaft gear on the camshaft, insert the centre bolt and washer, and tighten the bolt.
18 Locate the Woodruff key (earlier models) and crankshaft gear on the crankshaft. On later models engage the crankshaft gear lug with the angled slot in the end of the crankshaft. Where an M12 bolt is used, apply a thread locking compound and insert the bolt and washer; where an M14 bolt is used, apply a little engine oil to the thread then insert the

18.19 Tightening the intermediate shaft gear centre bolt

18.22 Crankshaft pulley TDC notch aligned with the indentation in the intermediate gear (arrowed)

bolt and washer. While holding the crankshaft stationary, *tighten the bolt to the appropriate torque* (photo).
19 Locate the Woodruff key and intermediate gear on the intermediate shaft, then insert the centre bolt and washer, and tighten the bolt (photo).
20 Refit the timing belt tensioner and fit the nut finger tight.
21 Make sure that the notch on the rear of the camshaft gear is aligned with the top of the timing belt rear cover.
22 Temporarily fit the crankshaft pulley to the gear then, with No 1 piston at TDC, turn the intermediate gear so that the indentation is aligned with the notch in the pulley (photo). If the distributor has not been disturbed the rotor arm will point in the direction of No 1 distributor cap segment.
23 Locate the timing belt on the gears and tensioner, turn the tensioner clockwise to pre-tension the timing belt, and check that the TDC marks are still correctly aligned.
24 Turn the tensioner clockwise until the timing belt can just be twisted 90° with the thumb and index finger midway between the camshaft and intermediate gears. Tighten the nut to secure the tensioner.
25 Remove the crankshaft pulley, fit the lower timing cover, and tighten the bolts.
26 Refit the crankshaft pulley and tighten the bolts.
27 Locate the pulley on the water pump and tighten the bolts.

28 Refit the upper timing cover and tighten the nuts.
29 Refit the alternator drivebelt, with reference to Chapter 10.
30 Reconnect the battery negative lead, where necessary.

19 Flywheel/driveplate – removal and refitting

Note: *If the engine is still in the car, first carry out the following operations*:
(a) *On manual gearbox models, remove the gearbox (Chapter 6) and clutch (Chapter 5)*
(b) *On automatic transmission models, remove the automatic transmission (Chapter 7)*

1 The flywheel/driveplate bolts are offset to ensure correct refitting. Unscrew the bolts while holding the flywheel/driveplate stationary (photos).
2 Lift the flywheel/driveplate from the crankshaft (photo). If removing a driveplate note the location of the shim and washer (Fig 1.11).
3 If the flywheel or driveplate are to be renewed you will need to make the appropriate timing mark for your model on the periphery of the new flywheel/driveplate, since the replacement will only be marked with the TDC mark (0). Measure to the left of centre from the TDC index then mark as

19.1A Remove the flywheel bolts

19.1B Use a bar and piece of angle iron to hold the flywheel stationary

Fig. 1.11 Drive plate components (automatic transmission models) (Sec 19)

1 Washer (chamfer to plate)   2 Shim

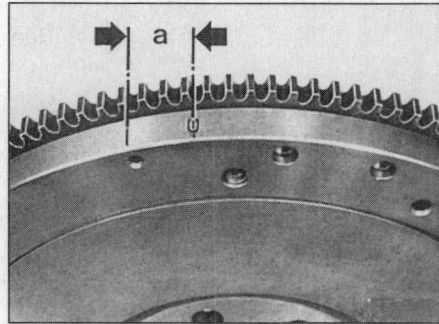

Fig. 1.12 Flywheel/driveplate ignition timing mark position – distance (a) to be 21 mm (0.82 in) left of the TDC mark (Sec 19)

Fig. 1.13 Check driveplate to block rear face distance (a) (Sec 19)

shown by scribing or punching the timing index reference at the set distance specified for your model – see Fig. 1.12.

4 Refitting is a reversal of the removal, but note that the retaining bolts must be renewed. Two bolt types have been fitted, the later type being identified by a shouldered head. On earlier and late models renew with this later type bolt where possible. It should be noted that the torque wrench setting for the two bolt types differs.

5 Coat the bolt threads with a locking agent prior to fitting and tightening to the specified torque.

6 If a replacement driveplate is fitted, its position must be checked and adjusted if necessary. The distance from the rear face of the block to the torque converter mounting face on the driveplate (Fig. 1.13) must be between 30.5 and 32.1 mm (1.20 and 1.26 in). If necessary, remove the driveplate and fit a spacer behind it to achieve the correct dimension.

20 Intermediate shaft – removal and refitting

1 Remove the distributor (Chapter 4), and the fuel pump (Chapter 3).

2 Remove the timing belt and intermediate gear as described in Section 18.

3 Remove the two bolts from the sealing flange, take off the sealing flange and the O-ring (photos).

4 Withdraw the intermediate shaft from the block (photos).

5 With the flange removed, the oil seal can be removed (photo). Fit a new seal with its open face towards the engine and use a block of wood to drive the seal in flush. Oil the lips of the seal before fitting the sealing flange.

6 Refitting is a reversal of removal, but fit a new O-ring and check that the shaft endfloat does not exceed the amount given in the

19.2 Removing the flywheel

20.3A Removing the intermediate shaft sealing flange . . .

20.3B . . . and O-ring

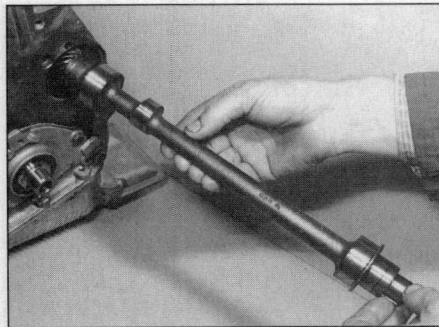

20.4A Withdraw the intermediate shaft

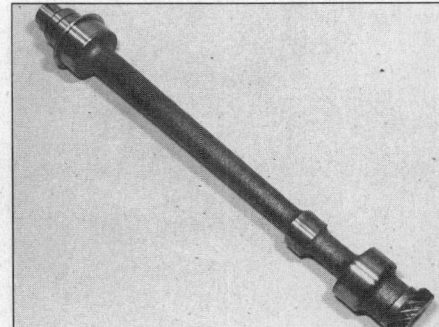

20.4B The intermediate shaft

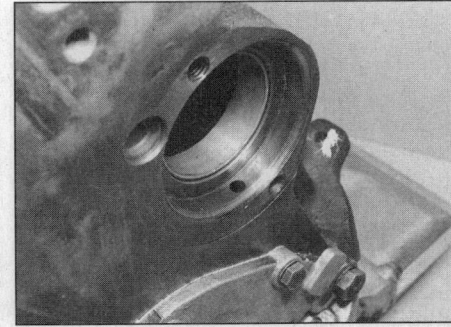

20.4C Showing intermediate shaft front bearing

20.5 Levering the intermediate shaft oil seal out from the flange

21.3 Locating a new sump gasket on the block (do not use any sealant)

22.2 Removing the crankshaft front oil seal housing

Specifications. Refer to Section 18 when refitting the timing belt and intermediate gear, and to Chapters 3 and 4 when refitting the fuel pump and distributor.

21 Sump – removal and refitting

Note: *If the engine is still in the car, first carry out the following operations:*
(a) *Jack up the front of the car and support it on axle stands*
(b) *Support the weight of the engine with a hoist*
(c) *Unbolt and remove the transmission front cover*
(d) *Drain the engine oil*

(e) *Unscrew the subframe front bolts and lower the subframe*

1 Unbolt and remove the sump, using an Allen key where necessary. Remove the dipstick.
2 Remove the gasket.
3 Refitting is a reversal of removal, but use a new gasket without adhesive (photo), and tighten the bolts evenly to the specified torque.

22 Crankshaft oil seals – renewal

Front oil seal

1 Remove the timing belt and crankshaft gear, as described in Section 18.
2 If an extractor tool is available the seal may

be renewed without removing the housing, otherwise unbolt and remove the housing (including the relevant sump bolts) and remove the gasket (photo). If the sump gasket is damaged while removing the housing, it will be necessary to remove the sump and fit a new gasket. However, refit the sump after fitting the housing.
3 Drive the old seal out of the housing, then dip the new seal in engine oil and drive it into the housing with a block of wood or a socket until flush (photos). Make sure that the closed end of the seal is facing outwards.
4 Fit the housing together with a new gasket and tighten the bolts evenly in diagonal sequence.
5 Refit the crankshaft gear and timing belt, as described in Section 18.

Rear oil seal

6 Remove the flywheel or driveplate, as described in Section 19.
7 Follow paragraphs 2 to 4 inclusive (photos).
8 Refit the flywheel or driveplate, as described in Section 19.

23 Oil pump – removal, examination and refitting

1 Remove the sump, as described in Section 21.
2 Using an Allen key unscrew the socket-headed bolts and withdraw the oil pump and strainer from the cylinder block (photo).

22.3A Using a socket to fit a new crankshaft front oil seal

22.3B Crankshaft front oil seal correctly located in its housing

22.7A Removing the crankshaft rear oil seal housing

22.7B Locate the new crankshaft rear oil seal housing gasket on the dowels

23.2 Removing the oil pump and strainer

23.3 Removing the oil pump cover

23.4 Oil pump filter screen

23.6A Checking the oil pump gear backlash . . .

3 Remove the two hexagon-headed bolts from the pump cover and lift off the cover (photo).

4 Bend up the metal rim of the filter plate so that it can be removed and take the filter screen out (photo). Clean the screen thoroughly with paraffin and a brush.

5 Clean the pump casing, cover and gears.

6 Check the backlash of the gears with a feeler gauge (photo) and with a straight edge across the end face of the pump, measure the endplay of the gears (photo). Examine the pump cover grooves worn by the ends of the gears, which will effectively increase the endplay of the gears. If the wear on the pump is beyond the specified limits a new pump should be fitted.

7 Fill the pump housing with engine oil then reassemble and refit it using a reversal of the removal and dismantling procedure. Refer to Section 21 when refitting the sump.

24 Pistons and connecting rods – removal and dismantling

1 Remove the cylinder head (Section 9), timing belt (Section 18) and oil pump (Section 23).

2 Mark each connecting rod and cap in relation to its cylinder and position.

3 Turn the crankshaft so that No 1 piston is at the bottom of its bore, then unscrew the nuts/bolts and remove the big-end bearing cap (photo).

4 Using the handle of a hammer, push the piston and connecting rod out of the top of the cylinder. Put the bearing cap with its connecting rod and make sure that they both have the cylinder number marked on them. If any of the bearing shells become detached while removing the connecting rod and bearing cap, ensure that they are placed with their matching cap or rod (photo).

5 Repeat the procedure given in paragraphs 3 and 4 to remove the remaining pistons and connecting rods.

6 Before removing the pistons from the connecting rods, if necessary mark the connecting rods to show which side of them is towards the front of the engine. The casting marks on the rod and cap face towards the front of the engine (photo).

7 Remove the circlips from the grooves in the gudgeon pin holes and push the pin out enough for the connecting rod to be removed (photo). Do not remove the pins completely unless new ones are to be fitted, to ensure that the pin is not turned end for end when the piston is refitted. If the pin is difficult to push out, heat the piston by immersing it in hot water.

8 New bushes can be fitted to the connecting rods, but as they need to be reamed to size after fitting, the job is best left to a VW agent.

9 Using old feeler gauges, or pieces of rigid plastic inserted behind the piston rings, carefully ease each ring in turn off the piston. Lay the rings out so that they are kept the right way up and so that the top ring can be identified. Carefully scrape the rings free of carbon and clean out the ring grooves on the pistons, using a piece of wood or a piece of broken piston ring.

23.6B . . . and endplay

24.3 Removing a big-end bearing cap

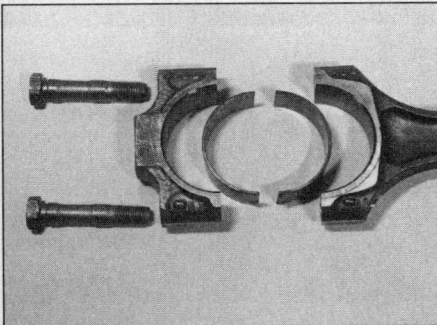

24.4 Big-end bearing components

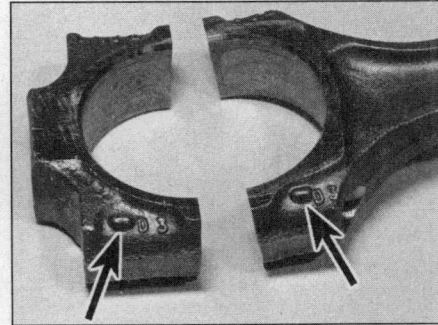

24.6 The casting marks (arrowed) which must face the front of the engine

24.7 Location of gudgeon pin circlip (arrowed) in piston

25.3 Checking a piston ring gap

25.6 Checking the piston ring clearance in its groove

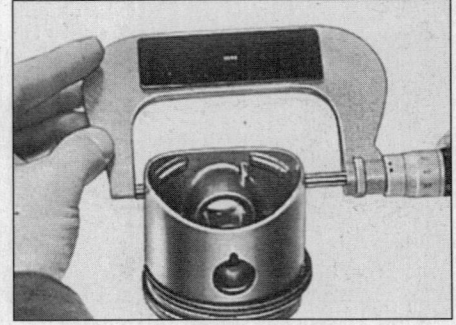

Fig. 1.14 Checking the piston diameter (Sec 25)

25 Pistons and cylinder bores – examination

1 Examine the pistons and the bores for obvious signs of damage and excessive wear. If they appear to be satisfactory, make the following checks.

2 Measure the piston diameter at a position 15 mm (0.60 in) from the lower edge of the skirt and at 90° to the axis of the piston and compare this with the information in the Specifications.

3 Push a piston ring into the cylinder bore and use a piston to push the ring down the bore so that it is square in the bore and about 15 mm (0.6 in) from the bottom of the cylinder. Measure the ring gap using a feeler gauge (photo). If the gap is above the top limit, look for obvious signs of bore wear, or if a new piston ring is available, measure the gap when a new piston ring is fitted to the bore.

4 To measure the bore diameter directly a dial gauge with an internal measuring attachment is required. If one is available, measure each bore in six places and compare the readings with the wear limit given. Bore diameter should be measured 10 mm (0.4 in) from the top of the bore, 10 mm (0.4 in) from the bottom and at the mid-point. At each of the three stations, measure in-line with the crankshaft and at right angles to it. If the bores are worn beyond the limit, they will need to be rebored and new pistons fitted.

5 If one bore is oversize, all four must be rebored and a new set of pistons fitted, otherwise the engine will not be balanced. Connecting rods must only be fitted as complete sets and not be replaced individually.

6 Fit the rings to the pistons and use a feeler gauge to measure the gap between the piston ring and the side of its groove (photo). If the gap is beyond the wear limit, it is more likely that it is the piston groove rather than the ring which has worn, and either a new piston or a proprietary oversize ring will be required. If new piston rings are fitted, the wear ridge at the top of the cylinder bore must be removed, or a stepped top ring used.

26 Piston and connecting rods – reassembly and refitting

1 Heat each piston in hot water, then insert the connecting rod and push in the pin until central. Make sure that the casting marks on the connecting rod and the arrow on the piston crown (photo) are facing the same way, then refit the circlips.

2 Before refitting the piston rings, or fitting new rings, check the gap of each ring in turn in its correct cylinder bore using a piston to push the ring down the bore, as described in the previous Section. Measure the gap between the ends of the piston ring, using feeler gauges. The gap must be within the limits given in the Specifications.

3 If the piston ring gap is too small, carefully file the piston ring end until the gap is sufficient. Piston rings are very brittle, so handle them carefully.

4 When fitting piston rings, look for the word TOP etched on one side of the ring and fit this side so that it is towards the piston crown. The outer recessed edge on the centre ring must face the gudgeon pin (photo).

5 Clean the connecting rods and bearing caps thoroughly and fit the bearing shells so that the tang on the bearing engages in the recess in the connecting rod, or cap, and the ends of the bearing are flush with the joint face.

> **HAYNES HiNT** *Unless the big-end bearing shells are known to be almost new, it is worth fitting a new set when reassembling the engine.*

6 To refit the pistons, first space the joints in the piston rings so that they are at 120° intervals. Oil the rings and grooves generously and fit a piston ring compressor over the piston. To fit the pistons without using a piston ring compressor is difficult and there is a high risk of breaking a piston ring.

7 Oil the cylinder bores and insert the pistons (photo) with the arrow on the piston crown pointing towards the front of the engine. Make sure that the relevant crankpin is at its furthest point from the cylinder.

26.1 The arrow on the piston crown (arrowed) must face the front of the engine

26.4 Piston with all rings fitted correctly

26.7 Inserting a piston into its bore using a ring compressor

26.9 Tightening the big-end bearing cap bolts

27.5A Removing a crankshaft main cap

27.5B Removing the crankshaft

8 When the piston is pushed in flush with the top of the bore, oil the two bearing halves and the crankshaft journal and guide the connecting rod half-bearing on to the crankpin.

9 Fit the big-end bearing cap complete with shell and tighten the nuts/bolts to the specified torque wrench setting (photo). Note that the contact surfaces of the conrod nuts must be lubricated with oil before fitting and tightening. New nuts/bolts are only required if the existing ones are damaged.

10 Rotate the crankshaft to ensure that everything is free, before fitting the next piston and connecting rod.

11 Using feeler gauges between the machined face of each big-end bearing, and the machined face of the crankshaft web, check the endplay, which should not exceed the maximum amount given in the Specifications.

12 Refit the oil pump (Section 23), timing belt (Section 18), and cylinder head (Section 16).

27 Crankshaft – removal, examination and refitting

1 With the engine removed from the car, remove the pistons and connecting rods, as described in Section 24.

2 Reassemble the big-end bearings to their matching connecting rods to ensure correct refitting.

3 Remove the crankshaft oil seals complete with housings, as described in Section 22.

4 Check that each main bearing cap is numbered for position.

5 Remove the bolts from each bearing cap in turn, then remove the caps and lift out the crankshaft (photos).

6 If the bearings are not being renewed, ensure that each half-bearing shell is identified so that it is put back in the same place from which it was removed. This also applies to the thrust washers if fitted – see paragraph 7. If the engine has done a high mileage and it is suspected that the crankshaft requires attention, it is best to seek the opinion of a VW dealer or crankshaft re-finishing specialist for advice on the need for regrinding. Unless the bearing shells (and

thrust washers if applicable) are known to be almost new, it is worth fitting a new set when the crankshaft is refitted. If available, Plastigage may be used to check the running clearance of the existing bearings — a strip of Plastigage is placed across the crankshaft journal and then the bearing is assembled and tightened to the specified torque. After dismantling the bearing, the width of the strip is measured with a gauge supplied with the strip, and the running clearance read off (photo).

7 Clean the crankcase recesses and bearing caps thoroughly and fit the bearing shells so that the tang on the bearing engages in the recess in the crankcase or bearing cap. Make sure that the shells fitted to the crankcase have oil holes, and that these line up with the

27.6 Using Plastigage to check a crankshaft journal running clearance

27.7A Fitting the flanged type centre main bearing to the cap

27.7B Fitting the centre main bearing and separate thrust washers to the cap

27.7C Fitting the flanged type centre main bearing into the crankcase . . .

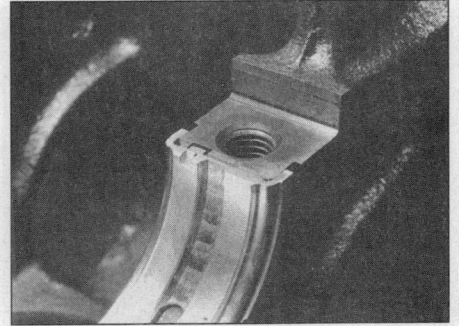

27.7D . . . ensuring that the ends of the bearing are flush with the joint face

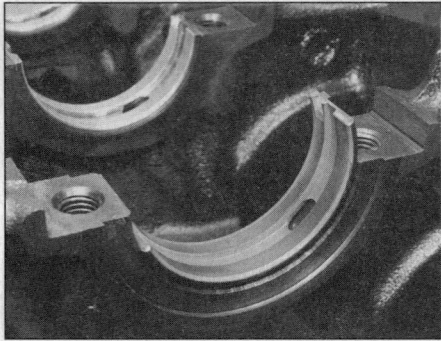

27.7E Fitting the alternative type centre main bearing into the crankcase . . .

27.7 F . . . together with its thrust washers

27.8 Lubricate the main bearing shells

drillings in the bearing housings. The shells fitted to the bearing caps do not have oil holes. Note that the bearing shells of the centre bearing (No 3) may either be flanged to act as thrust washers, or may have separate thrust washers. These should be fitted oil groove outwards as shown. Fit the bearing shells so that the ends of the bearing are flush with the joint face (photos).

8 Oil the bearings and journals (photo) then locate the crankshaft in the crankcase.

9 Fit the main bearing caps (with centre main bearing thrust washers if applicable) in their correct positions.

10 Fit the bolts to the bearing caps and

tighten the bolts of the centre cap to the specified torque (photo), then check that the crankshaft rotates freely. If it is difficult to rotate the crankshaft, check that the bearing shells are seated properly and that the bearing cap is the correct way round. Rotation will only be difficult if something is incorrect, and the fault must be found. Dirt on the back of a bearing shell is sometimes the cause of a tight main bearing.

11 Working out from the centre, tighten the remaining bearing caps in turn, checking that the crankshaft rotates freely after each bearing has been tightened.

12 Check that the endfloat of the crankshaft

is within specification, by inserting feeler gauges between the crankshaft and the centre bearing thrust face/washer while levering the crankshaft first in one direction and then in the other (photo).

13 On manual transmission models only, the rear end of the crankshaft carries a needle roller bearing (photo) which supports the front end of the gearbox input shaft. Inspect the bearing for obvious signs of wear and damage. If the gearbox has been removed and dismantled, fit the input shaft into the bearing to see if there is excessive clearance. If the bearing requires renewing, insert a hook behind the bearing and pull it out of the end of the crankshaft. Install the new bearing with the lettering on the end of the bearing outwards. Press it in until the end of the bearing is 1.5 mm (0.059 in) below the face of the flywheel flange (photo).

14 Fit new crankshaft oil seals (Section 22) then refit the pistons and connecting rods, as described in Section 26.

27.10 Tightening the main bearing cap bolts

27.12 Checking the crankshaft endfloat

28 Oil filter and housing – removal and refitting

1 Place a suitable container beneath the left-hand side of the engine.

2 Unscrew the filter cartridge and discard it – it will be necessary to use a filter strap or special spanner, although if neither of these items is available drive a long screwdriver through the cartridge.

3 If it is required to remove the housing, use an Allen key to unscrew the bolts, then withdraw the housing and gasket (photos).

4 Clean the mating faces of the housing and block then refit the housing, together with a new gasket, and tighten the bolts. Reconnect the oil pressure switch wiring, where applicable.

5 Clean the mating faces of the new oil filter cartridge and housing, and smear a little engine oil on the filter seal.

6 Screw the cartridge onto the housing and tighten it by hand only (photo).

27.13A Location of spigot needle roller bearing in rear end of crankshaft

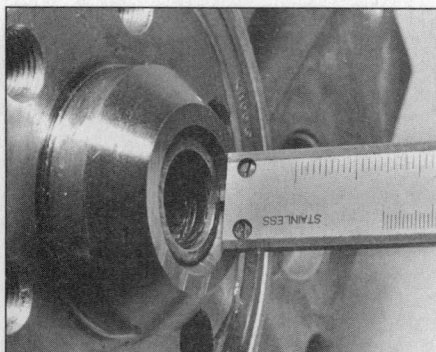

27.1 3B Check position of spigot needle roller bearing

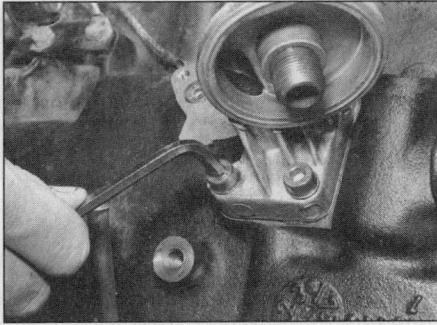
28.3A Unscrew the socket head bolts . . .

28.3B . . . and remove the oil filter housing

28.6 Fitting a new oil filter cartridge

29 Engine mounting – removal and refitting

1 The engine front support bar can be unbolted from the block and removed. When refitting the bar check the engine alignment, as described in Section 6.

2 To remove the left and right engine mountings (photos) first remove the front support bar, then unbolt the mountings using a hoist to support the weight of the engine. When refitting the mountings, check the engine alignment, as described in Section 6.

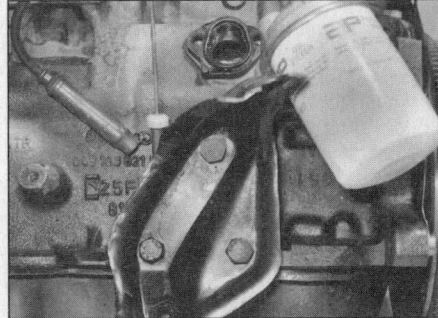
29.2A Left-hand side engine mounting bracket

29.2B Right-hand side engine mounting

Part B – Five-cylinder engines

30 General description

The engine is of five-cylinder, in-line, overhead camshaft type, mounted conventionally at the front of the car. The crankshaft is of six-bearing type and the No 4 (from front) main bearing shells incorporate flanged thrust washers to control crankshaft endfloat. The camshaft is driven by a toothed belt from the crankshaft gear and the belt also drives the water pump mounted on the left-hand side of the block. A gear on the rear of the camshaft drives the distributor, and on carburettor models the camshaft also drives the fuel pump.

The valves are operated from the camshaft through bucket type tappets, and valve clearances are adjusted by the use of shims located in the top of the tappets.

The engine has a full-flow lubrication system. A gear and crescent type oil pump is mounted on the front of the crankshaft. The oil filter is of the cartridge type, mounted on the right-hand side of the cylinder block.

31 Engine routine maintenance

The maintenance procedures for the five-cylinder engine are the same as for the four-cylinder unit described in Section 2 of Part A in this Chapter.

32 Major operations possible with the engine in the car

The following operations can be carried out without having to remove the engine from the car:

(a) Removal and servicing of the cylinder head and camshaft
(b) Removal of the timing belt and gears
(c) Removal of the flywheel or driveplate (after first removing the transmission)
(d) Removal of the sump (after first lowering the subframe)
(e) Removal of the oil pump
(f) Removal of the pistons and connecting rods

33 Major operation only possible after removal of the engine from the car

The following operation can only be carried out after removal of the engine from the car:
Removal of the crankshaft and main bearings

34 Method of engine removal

The engine must be disconnected from the transmission then lifted from the car.

35 Engine – removal and refitting

1 Remove the bonnet, as described in Chapter 12, and stand it on cardboard or rags in a safe place.
2 Disconnect the battery negative lead.
3 Drain the cooling system as described in Chapter 2.
4 Unscrew the bolt securing the coolant pipe to the left-hand side of the engine, then disconnect the pipe from the front hose. On some models the coolant pipe is attached to a transmission bellhousing bolt.
5 Remove the screws and withdraw the upper radiator cover.

Fig. 1.15 Remove pulley cover and power steering pump (where applicable) (Sec 35)

35.10A Oil pressure switch (five-cylinder)

35.10B Warm-up regulator and wiring

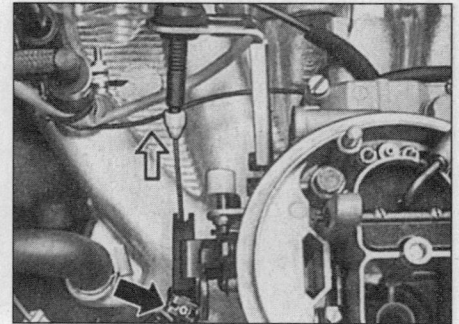

Fig. 1.16 Disconnect the accelerator cable (carburettor engine shown) (Sec 35)

6 Disconnect the top hose from the cylinder head.

7 Prise the one-way valve and hose from the brake vacuum servo unit.

8 Remove the power steering pump (where applicable), with reference to Chapter 11, but leaving the hoses connected. Tie the pump to the bulkhead (Fig 1.15).

9 Disconnect the bottom hose from the thermostat housing on the block.

10 Disconnect the wiring from the oil pressure switch on the left-hand side of the engine (photo) and on fuel injection models from the warm-up regulator. Unclip the wiring and place it to one side (photo).

11 Remove the air cleaner unit as described in Chapter 3 (Part A). On fuel injection models remove the air cleaner and air flow meter as described in Chapter 3 (Part B).

12 Identify all fuel and vacuum hoses using masking tape, then disconnect those affecting engine removal. These will include, where applicable:

Fuel hoses to the fuel pump and non-return valve

Vacuum hose to the warm-up valve (fuel injection)

Vacuum hose to the vacuum reservoir and distributor

Crankcase emission control hoses

13 On fuel injection models detach and remove the warm-up regulator (leaving the fuel lines connected).

14 Identify all wiring for location using masking tape, then disconnect those affecting engine removal. These will include, where applicable:

Fuel injection wiring to cold start valve, thermo-time switch auxiliary air valve and throttle switch

Wiring to temperature sender, distributor, coil, oil temperature switch, inlet manifold preheater and by-pass air cut-off valve

15 On fuel injection models detach the injectors from their seats, remove the cold start valve and the mixture control unit. Plug all fuel lines and connections to prevent the ingress of dirt.

16 Disconnect the accelerator cable as described in Chapter 3 (Fig. 1.16).

17 On manual gearbox models disconnect the clutch cable from the release lever and cable bracket.

18 Remove the alternator. as described in Chapter 10, and where necessary unbolt the alternator bracket from the block.

19 Unscrew and remove the upper transmission-to-engine bolts noting the location of any brackets, but leave the bolt adjacent to the clutch lever in position.

20 Disconnect the heater hoses from the engine block.

21 Unbolt and detach the stop shell from the front crossmember (Fig. 1.17).

22 Loosen off but do not remove at this stage the engine bearer/mounting nuts/bolts.

23 Unbolt and remove the starter motor.

24 Unbolt and detach the exhaust downpipe from the manifold joint and at the gearbox bracket. Tie the exhaust to one side out of the way.

25 On automatic transmission models disconnect the coolant hoses from the oil cooler and remove the flange. Also on automatic transmission models unscrew the three torque converter-to-driveplate bolts while holding the starter ring gear stationary with a screwdriver. It will be necessary to rotate the engine to position the bolts in the starter aperture.

26 For additional room when removing the engine, it is recommended that the vibration damper on the front of the crankshaft is removed. Access to the centre bolt is gained by removing the front grille and the number plate. To remove the damper the crankshaft must be held stationary using VW tool 2084 and the centre bolt loosened (Figs 1.18 and 1.19). The bolt is tightened to a high torque and it is therefore better to use the correct tool. However, it may be possible to loosen the bolt while an assistant engages top gear and applies the brakes (manual gearbox) or by holding the starter ring gear stationary using a wide-bladed screwdriver.

27 With the centre bolt removed, loosen two diagonally opposite bolts using an Allen key and remove the remaining two bolts. Release the vibration damper from the crankshaft gear

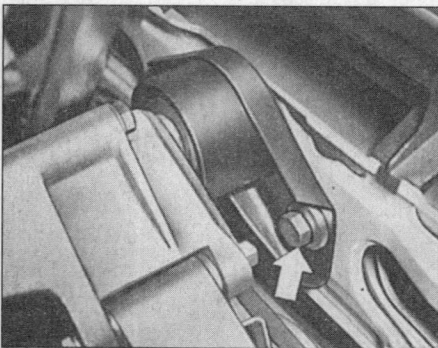

Fig. 1.17 Unbolt the stop shell from the front crossmember (Sec 35)

Fig. 1.18 Crankshaft damper removal. Insert socket and extension through front aperture (Sec 35)

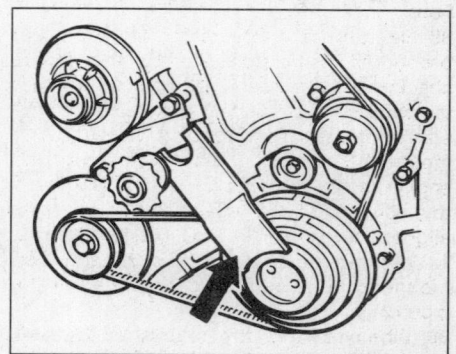

Fig. 1.19 Special tool 2084 attached to crankshaft damper (Sec 35)

Fig. 1.20 Damper to gear retaining bolts (Sec 35)

Fig. 1.21 Engine and gearbox alignment dimensions (Sec 35)

Manual gearbox (a) = 22.2 mm (0.814 in)
Automatic gearbox (a) = 124.7 mm (4.90 in)

Fig. 1.22 Engine and gearbox lateral adjustment (Sec 35)

Manual gearbox (c) = 394.4 ± 1 mm (15.52 ± 0.04 in)
Automatic gearbox (c) = 223.6 ± 1 mm (8.80 ± 0.04 in)

by tapping the bolts and removing them – do not pull off the gear (Fig. 1.20).
28 Connect a hoist and take the weight of the engine – the hoist should be positioned centrally over the engine.
29 Unscrew and remove the subframe front mounting bolts.
30 Support the weight of the transmission with a trolley jack.
31 Unscrew and remove the lower transmission-to-engine bolts.
32 Where necessary unbolt and remove the left-hand side engine mounting bracket – on some models it will first be necessary to remove the radiator panel.
33 Lift the engine from the mounting(s) and reposition the trolley jack beneath the transmission.
34 Further loosen off but do not remove the remaining engine to transmission bolt adjacent to the clutch lever.
35 Pull the engine from the transmission – make sure on automatic transmission models that the torque converter remains fully engaged with the transmission splines.
36 Lift the engine further and then fully remove the remaining engine to transmission bolt, simultaneously raising the jack supporting the transmission.
37 Turn the engine slightly and lift it from the engine compartment then lower it to the floor.
38 Refitting is a reversal of the removal procedure, but before starting the engine check that it has been filled with oil, and that the cooling system is full. Make sure that the starter cable is not touching the engine or mounting bracket. Delay tightening the engine mountings until the engine is idling – this will ensure that the engine is correctly aligned, as shown in Figs. 1.21, 1.22 and 1.23. If necessary loosen the transmission mounting bolts before tightening all of the mountings.
39 When refitting the crankshaft damper align it with the location hole in the crankshaft gear (Fig. 1.24). Coat the threads and contact surfaces of the crankshaft damper centre bolt

with an anti-corrosion compound before inserting it and tightening to the specified torque.

36 Engine dismantling – general

Refer to Part A, Section 7 of this Chapter.

37 Ancillary components – removal and refitting

Refer to Part A, Section B of this Chapter, with the following exceptions:
Oil filter cartridge (Section 56 of this Chapter)
Engine mountings (Section 57 of this Chapter)

38 Camshaft and tappets – removal

Note: *If the engine is still in the car, first carry out the following operations:*

Fig. 1.23 Front engine support (stop shell) check. Stop shell (a) to contact rubber (b), but (b) must not touch crossmember towards the front or side edges of stop shell (Sec 35)

(a) Disconnect the battery negative lead
(b) Remove the air cleaner and fuel pump on carburettor models (Chapter 3)
(c) Disconnect all relevant wiring cables and hoses
(d) Remove the distributor (Chapter 4)
(e) Remove the upper radiator cowl
(f) Disconnect the drivebelts from the crankshaft pulley
(g) Where fitted, remove the power steering pump leaving the hoses connected (Chapter 11)

1 Unscrew the nuts and lift off the valve cover, together with the reinforcement strips and gaskets. Note the location of the HT lead holder.
2 Unscrew the nuts and bolts and remove the timing belt cover(s). The engine has two covers, and an Allen key is required – the lower cover need not be removed.
3 Using a socket on the crankshaft pulley bolt (temporarily refit the bolt and damper if removed during engine removal), turn the engine so that the piston in No 1 cylinder is at TDC (top dead centre) on its compression stroke. The notch in the crankshaft pulley must be in line with the pointer on the oil pump housing alternatively the '0' mark (TDC) on the flywheel/driveplate must be aligned with the pointer in the bellhousing aperture. Both No 1 cylinder valves must be closed (ie

Fig. 1.24 Vibration damper location hole in crankshaft gear (Sec 35)

1

Bearing caps

Camshaft

Woodruff key

Shim

Oil seal

Tappet

Valve cotters

Inner valve spring

Outer valve spring

Valve stem seal

Spring seat

Valve guide

Cylinder head

Valves

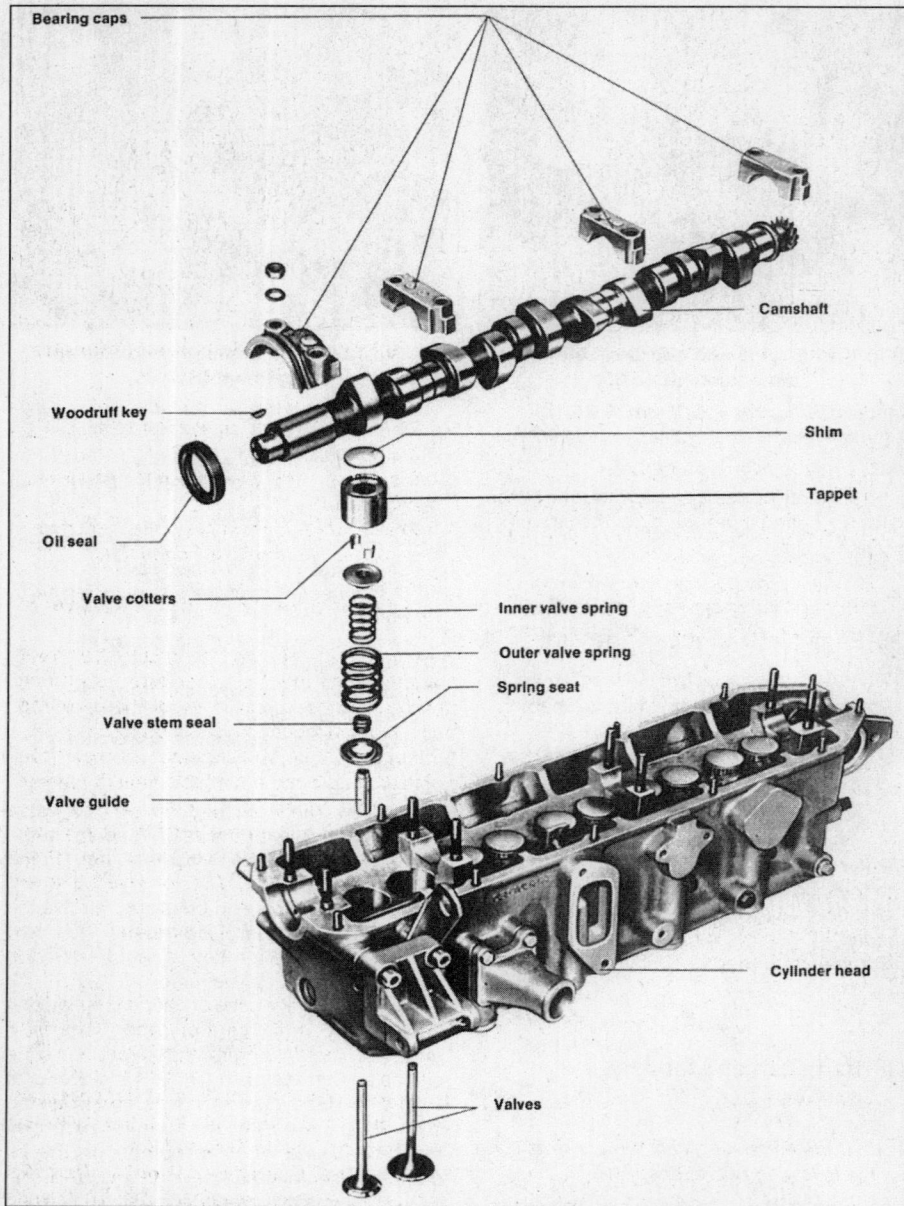

Fig, 1.25 Cylinder head and camshaft components (Sec 38)

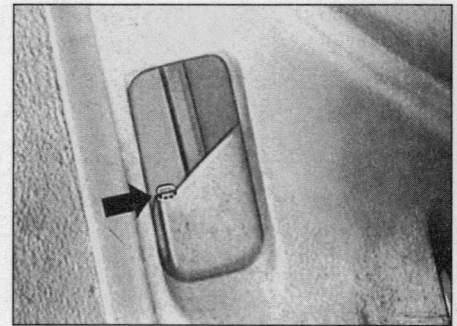

Fig. 1.26 Alignment of TDC marks on flywheel/driveplate (Sec 38)

Fig. 1.27 Alignment of TDC marks on crankshaft pulley and oil pump housing (Sec 38)

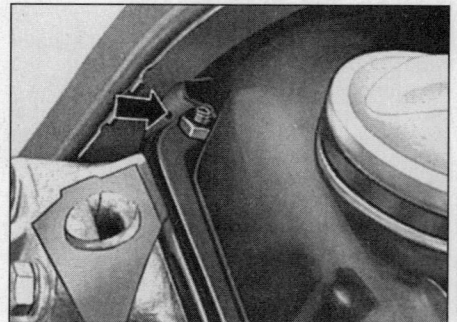

Fig. 1.28 The TDC indentation on rear face of camshaft gear aligned with upper surface of valve cover gasket (Sec 38)

Fig. 1.29 Numerical order of camshaft bearing caps – check they are marked accordingly before removing (Sec 38)

cam peaks away from the tappets) and the indentation on the rear of the camshaft gear in line with the upper surface of the valve cover gasket (temporarily refit the gasket and valve cover, if necessary), see Figs. 1.26, 1.27 and 1.28.

4 Loosen the water pump mounting and adjustment bolts, using an Allen key where necessary, and rotate the pump clockwise to release the tension on the timing belt.

5 Remove the timing belt from the camshaft gear and move it to one side.

6 Unscrew the centre bolt from the camshaft gear while holding the gear stationary with a bar through one of the holes or using a wide-bladed screwdriver.

7 Withdraw the gear from the camshaft and extract the Woodruff key.

8 Check that each bearing cap has its number stamped on it: if not, make an identifying mark to ensure that each cap is put back where it was originally. Note that the caps are offset and can only be fitted one way round (Fig. 1.29).

9 *It is important that the camshaft is removed exactly as described so that there is no danger of it becoming distorted.* Loosen one of the nuts on bearing cap No 2 about two turns and then loosen the diagonally opposite nut on bearing cap No 4 about two turns. Repeat the operations on the other nut of bearing cap No 2 and bearing cap No 4. Continue the

38.10 Removing a camshaft bearing cap

38.11A Removing a tappet (cam follower)

38.11B Keep tappets in correct order after removal

sequence until the nuts are free, then remove them. Loosen and remove the nuts of bearing caps Nos 1 and 3 using a similar diagonal sequence.

10 Lift the bearing caps off (photo) and lift the camshaft out. Discard the oil seal.

11 Withdraw each tappet (cam follower) in turn (photo) and mark its position (1 to 10 numbering from the timing belt end of the engine) using adhesive tape or a box with divisions (photo). Take care to keep the adjustment shims with their respective tappets.

39 Cylinder head – removal

Note: *If the engine is still in the car, first carry out the following operations:*
(a) Disconnect the battery negative lead
(b) Drain the cooling system (Chapter 2)
(c) Remove the inlet and exhaust manifolds (Chapter 3)
(d) Remove the distributor HT leads and spark plugs (Chapter 4)
(e) Disconnect the drivebelts from the crankshaft pulley
(f) Disconnect the relevant wiring cables and hoses
(g) Where fitted, remove the power steering pump leaving the hoses connected (Chapter 11)

1 Follow paragraphs 1 to 7 inclusive of Section 38.

2 Unbolt and remove the timing belt rear cover — note that the water pump pivot bolt must be removed and three of the oil pump bolts. Note the location of the spacers.

3 Using a splined socket unscrew the cylinder head bolts a turn at a time in reverse order to that shown in Fig. 1.32.

4 With all the bolts removed, lift the cylinder head from the block. If it is stuck, tap it free with a wooden mallet. *Do not insert a lever into the gasket joint.*

5 Remove the cylinder head gasket.

40 Valves – removal and renovation

1 Remove the camshaft and tappets, as described in Section 38, and the cylinder head as described in Section 39.

2 Follow the procedure given in Part A, Section 11 — inlet valves are 2-4-5-7-9 and exhaust valves are 1-3-6-8-10, numbered from the timing belt end of the engine.

41 Cylinder head – examination and renovation

Refer to Part A, Section 12 of this Chapter.

42 Camshaft and bearings – examination and renovation

1 Refer to Part A, Section 13, of this Chapter.

2 Check that the oil spray jets located in the top of the cylinder head direct the spray at 90° to the camshaft (Fig. 1.31).

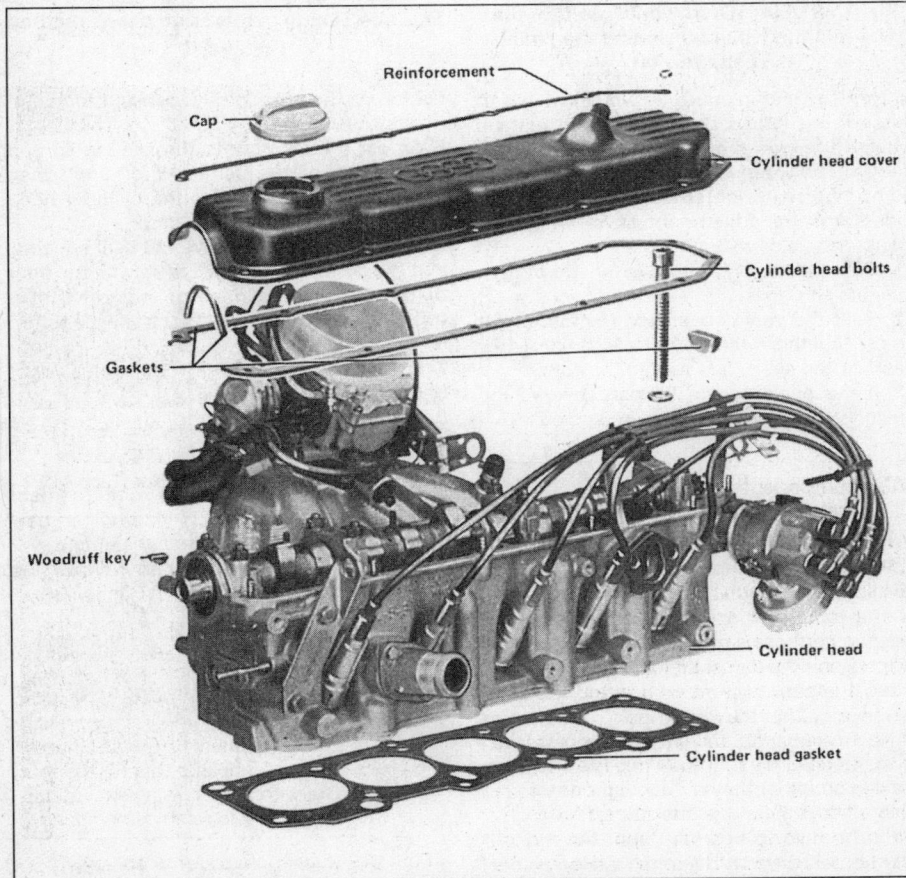

Reinforcement

Cap

Cylinder head cover

Cylinder head bolts

Gaskets

Woodruff key

Cylinder head

Cylinder head gasket

Fig. 1.30 Cylinder head and associated components (carburettor engine) (Sec 39)

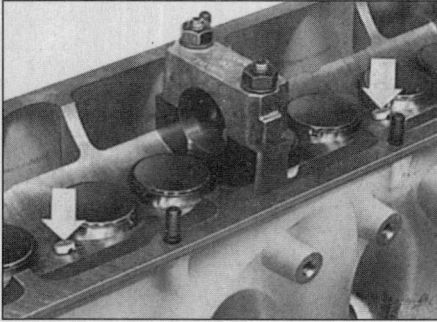

Fig. 1.31 Correct position of oil spray jets (arrowed) in the cylinder head (Sec 42)

44.2 Guide rod positions when fitting the cylinder head

Fig. 1.32 Cylinder head bolt tightening sequence (Sec 44)

43 Valves – refitting

Refer to Part A, Section 14 of this Chapter.

44 Cylinder head – refitting

1 Check that the top of the block is perfectly clean, then locate a new gasket on it with the part number or TOP marking facing upward.
2 Check that the cylinder head face is perfectly clean. Insert two long rods, or pieces of dowel into the cylinder head bolt holes at opposite ends of the block, to position the gasket and to give a location for fitting the cylinder head (photo). Lower the head on to the block, remove the guide dowels and insert the bolts and washers. Do not use jointing compound on the cylinder head joint.
3 Tighten the bolts using the sequence shown in Fig. 1.32 in the four stages given in the Specifications to the specified torque.
4 Locate the timing belt rear cover on the front of the cylinder head and block and tighten the retaining bolts and studs, as applicable. Refit the spacers. Leave the water pump pivot bolt loose.
5 Fit the Woodruff key in its groove and locate the gear on the end of the camshaft.
6 Fit the centre bolt and spacer and tighten the bolt to the specified torque while holding the gear stationary with a bar through one of the holes, or using a wide-bladed screwdriver.
7 Turn the camshaft gear and align the rear indentation with the upper surface of the valve cover gasket (temporarily locate the gasket on the head).
8 Check that No 1 piston is at TDC with the 'O' mark on the flywheel aligned with the pointer in the bellhousing aperture. The notch in the crankshaft pulley will also be in line with the pointer on the oil pump housing.
9 Locate the timing belt on the crankshaft, camshaft and water pump gears, turn the water pump anti-clockwise to pre-tension the belt, then check that the TDC timing marks are still aligned.
10 Use a screwdriver, as shown in Fig. 1.33,

Fig. 1.33 Using a screwdriver to turn the water pump when tensioning the timing belt (Sec 44)

to turn the water pump anti-clockwise and tension the timing belt until it can just be twisted 90° with the thumb and index finger midway between the camshaft and water pump gears. Tighten the water pump mounting and adjustment bolts when the adjustment is correct (Fig. 1.34).
11 Refit the timing belt cover(s) and tighten the nuts and bolts.
12 Refit the valve cover and reinforcement strips, together with the HT lead holder, new gaskets and seals, and tighten the nuts.
13 If the engine is in the car, reverse the preliminary procedures given in Section 39.

45 Camshaft and tappets – refitting

1 Fit the bucket tappets in their original positions – the adjustment shims on the top of the tappets must be fitted so that the lettering on them is downwards. Lubricate the tappets and the camshaft journals.
2 Lay the camshaft into the lower half of its bearings so that the lowest point of the cams of No 1 cylinder are towards the tappets, then fit the bearing caps in their original positions, making sure that they are the right way round before fitting them over the studs.
3 Fit the nuts to bearing caps Nos 2 and 4 and tighten them in diagonal sequence until the camshaft fully enters its bearings.
4 Fit the nuts to bearing caps 1 and 3, then

Fig. 1.34 Timing belt tension check method (Sec 44)

tighten all the nuts to the specified torque in diagonal sequence.
5 Smear a little oil onto the sealing lip and outer edge of the camshaft oil seal, then locate it open end first in the cylinder head and No 1 camshaft bearing cap.
6 Using a metal tube drive the seal squarely into the cylinder head until flush with the front of the cylinder head – do not drive it in further, otherwise it will block the oil return hole.
7 Follow paragraph 5 to 11 inclusive of Section 44.
8 Check and, if necessary, adjust the valve clearances, as described in Section 46, with the engine cold.
9 Refit the valve cover and reinforcement strips, together with the HT lead holder, new gaskets and seals, and tighten the nuts.
10 If the engine is in the car, reverse the preliminary procedures given in Section 38.

46 Valve clearances – checking and adjustment

Refer to Part A, Section 17 of this Chapter. The procedure is identical to that for the four-cylinder engine with the following exceptions:
(a) It is not necessary to remove the timing cover
(b) From the timing belt end of the engine inlet valves are numbered 2-4-5-7-9 and exhaust valves 1-3-6-8-10

47.1 One method of locking the starter ring gear

47.9 Fitting the crankshaft gear and vibration damper to the crankshaft – note location key

48.1 Centre punch alignment marks (arrowed) on the crankshaft and flywheel

47 Timing belt and gears – removal and refitting

Note: *If the engine is still in the car, first carry out the following operations:*
(a) *Disconnect the battery negative lead*
(b) *Where fitted, remove the power steering pump bearing and the hoses connected (Chapter 11)*
(c) *Remove the alternator drivebelt (Chapter 10) and if fitted the air conditioning compressor drivebelt (Chapter 12)*
(d) *Remove the lower radiator grille and number plate*

1 Using VW tool 2084 lock the vibration damper on the front of the crankshaft stationary, then loosen the centre bolt. The bolt is tightened to a high torque and it is recommended that the tool is used if at all possible. However, it may be possible to loosen the bolt while holding the starter ring gear stationary using a wide-bladed screwdriver or the method shown (photo).
2 Follow paragraphs 2 to 7 inclusive of Section 38 (remove both timing covers).
3 Unscrew the centre bolt and withdraw the vibration damper, together with the crankshaft gear and timing belt.
4 Separate the vibration damper from the crankshaft gear by removing the bolts using an Allen key, then reinserting two diagonally opposite bolts on a few threads and tapping them through the damper.
5 **Do not** *turn the crankshaft with the timing belt removed*.
6 If necessary, unbolt and remove the timing belt rear cover – note that the water pump pivot bolt must be removed and three of the oil pump bolts. Note the location of the spacers.
7 Commence refitting by locating the timing belt rear cover on the front of the cylinder head and block and tighten the retaining bolts and studs, as applicable. Refit the spacers. Leave the water pump pivot bolt loose.
8 Reassemble the vibration damper to the crankshaft gear, then insert and tighten the bolts.
9 Align the key on the crankshaft gear with

the slot in the crankshaft, then locate the timing belt on the crankshaft gear and fit the gear to the crankshaft (photo). Take care not to trap the belt between the gear and the oil pump housing.
10 Coat the threads of the centre bolt with an anti-corrosion compound, then insert the bolt and tighten it while holding the crankshaft stationary.
11 Follow paragraphs 4 to 11 inclusive of Section 44.
12 If the engine is in the car, reverse the preliminary procedures at the beginning of this Section.

48 Flywheel/driveplate – removal and refitting

1 The procedure is as given in Part A, Section 19 of this Chapter. However the retaining bolts may not be offset so the flywheel/driveplate and crankshaft should be marked in relation to each other before separation (photo).
2 When refitting a driveplate note that the raised pip must face the torque converter. If a replacement driveplate is to be fitted, its position must be checked and adjusted if necessary. The distance from the rear face of the block to the torque converter *mounting* face on the driveplate (Fig. 1.35) must be between 17.2 and 18.8 mm (0.667 and 0.740 in). If necessary, remove the driveplate and fit

a spacer behind it to achieve the correct dimension.
3 A new flywheel or driveplate will need to have the ignition timing index mark scribed on its periphery in the centre of the TDC '0' mark as shown in Fig. 1.36.
4 New flywheel/driveplate bolts must be used when refitting. Two bolt types have been used, the later type being identified by a shouldered head (see Fig. 1.37). Both early and late models must be fitted with the later type bolts irrespective of which type were fitted previously. Note that the later bolt torque wrench setting differs from that of the earlier type (without the shoulder), see Specifications.

Fig. 1.35 Torque converter-to-cylinder block dimension checking faces – measure in two places for average (Sec 48)

Fig. 1.36 Ignition timing mark position on flywheel/driveplate (Sec 48)

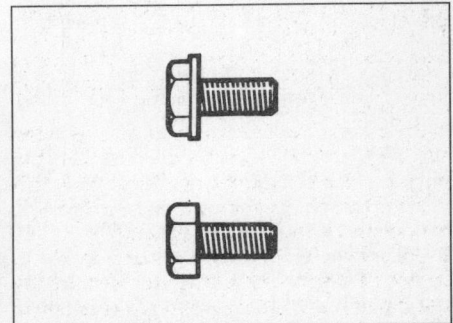

Fig. 1.37 Flywheel/driveplate retaining bolt types (Sec 48)

50.5 Removing the crankshaft rear oil seal and housing

51.4A Oil pump intake pipe

51.4B Oil pump intake pipe flange

49 Sump – removal

Refer to Part A, Section 21 of this Chapter.

50 Crankshaft oil seals – renewal

Front oil seal

1 Remove the timing belt and crankshaft gear, as described in Section 47.
2 If an extractor tool is available the seal may be renewed without removing the oil pump, otherwise refer to Section 51. It is also recommended that VW tool 2080 together with a guide sleeve be used to install the new seal. Dip the seal in engine oil before fitting, and if the old seal has scored the crankshaft, position the new seal on the unworn surface.
3 Refit the crankshaft gear and timing belt, as described in Section 47.

Rear oil seal

4 Remove the flywheel or driveplate, as described in Section 48.
5 If an extractor tool is available the seal may be renewed without removing the housing, otherwise unbolt and remove the housing (including the two sump bolts) and remove the gasket (photo). If the sump gasket is damaged while removing the housing it will be necessary to remove the sump and fit a new gasket. However, refit the sump after fitting the housing.
6 Drive the old seal out of the housing, then dip the new seal in engine oil and drive it into the housing with a block of wood or a socket, until flush. Make sure that the closed end of the seal is facing outwards.
7 Fit the housing together with a new gasket and tighten the bolts evenly in diagonal sequence.
8 Refit the flywheel or driveplate, as described in Section 48.

51 Oil pump – removal, examination and refitting

1 Remove the timing belt and crankshaft gear, as described in Section 47.
2 Remove the sump, with reference to Section 49.
3 Remove the dipstick.
4 Remove the two bolts securing the oil intake pipe stay to the crankcase (photo).

Knock back the tabs of the lockplate on the intake pipe flange (photo), remove the bolts and the intake pipe.
5 Remove the bolts securing the oil pump (photo) and take off the oil pump and gasket (photo).
6 Remove the countersunk screws securing the pump backplate and lift the backplate off, exposing the gears (photo).
7 Check that there is a mark on the exposed face of the gears and if not, make a mark to

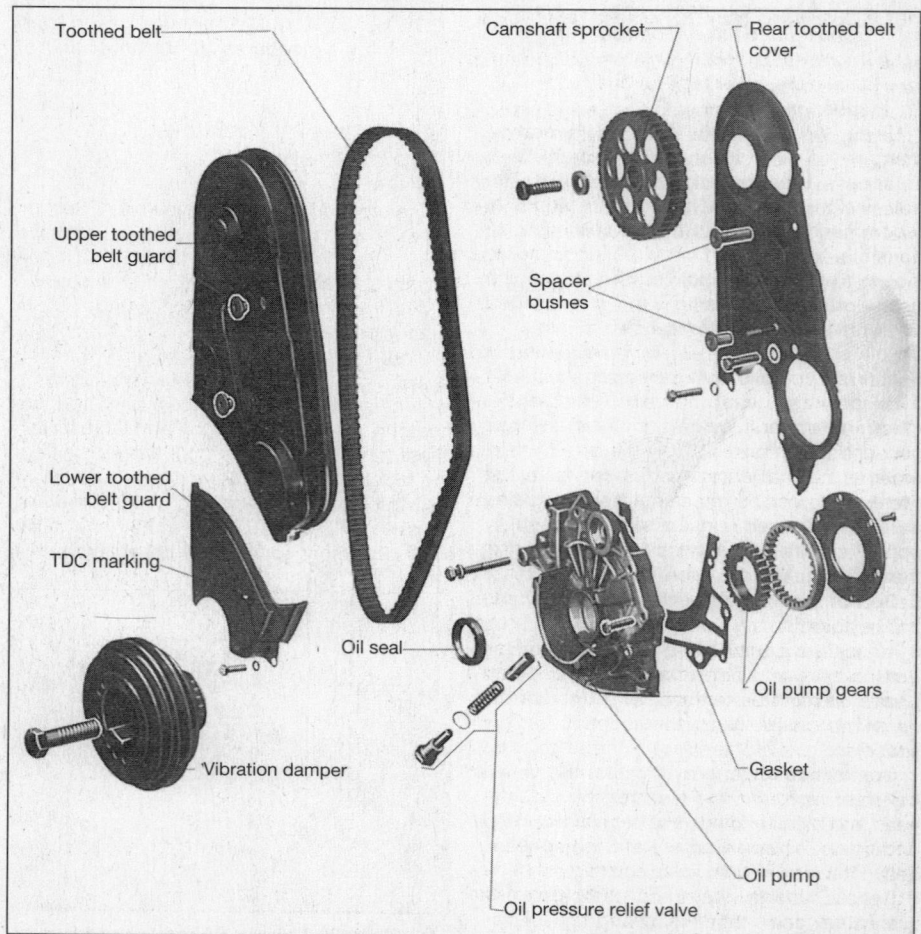
Fig. 1.38 Timing belt and oil pump components (Sec 51)

Toothed belt
Camshaft sprocket
Rear toothed belt cover
Upper toothed belt guard
Spacer bushes
Lower toothed belt guard
TDC marking
Oil seal
Oil pump gears
Vibration damper
Gasket
Oil pressure relief valve
Oil pump

51.5A Oil pump installed

51.5B Removing the oil pump

51.6 Oil pump with backplate removed

51.8 Removing the oil pressure relief valve

show which side of the gears is towards the engine before removing them.

8 Unscrew the pressure relief valve and remove the plug, sealing ring, spring and plunger (photo).

9 Clean all the parts thoroughly and examine the pump casing and backplate for signs of wear or scoring. Examine the pressure relief valve plunger and its seating for damage and wear and check that the spring is not damaged or distorted. Check the gears for damage and wear. New gears may be fitted, but they must be fitted as a pair.

10 Prise out the oil seal from the front of the pump. Oil the lip of the new seal, enter the seal with its closed face outwards and use a block of wood to tap the seal in flush. If there is any scoring on the crankshaft in the area on which the lip of the seal bears, the seal may be pushed to the bottom of its recess so that the lip bears on an undamaged part of the crankshaft.

11 Reassemble the pump by fitting the gears and the backplate. The inner gear has its slotted end towards the crankshaft and although the outer gear can be fitted either way round, it should be fitted the same way round as it was before removal. Some gears have a triangle stamped on them and this mark should be towards the front.

12 Refit the oil pump, together with a new gasket, making sure that the slot on the inner gear engages the dog on the crankshaft.

13 Insert the bolts and tighten them in diagonal sequence to the specified torque.

14 Fit the oil intake pipe, together with a new gasket, tighten the bolts, and bend the lockplate tabs onto the flange.

15 Refit the dipstick, sump (Section 49), and timing belt and crankshaft gear (Section 47).

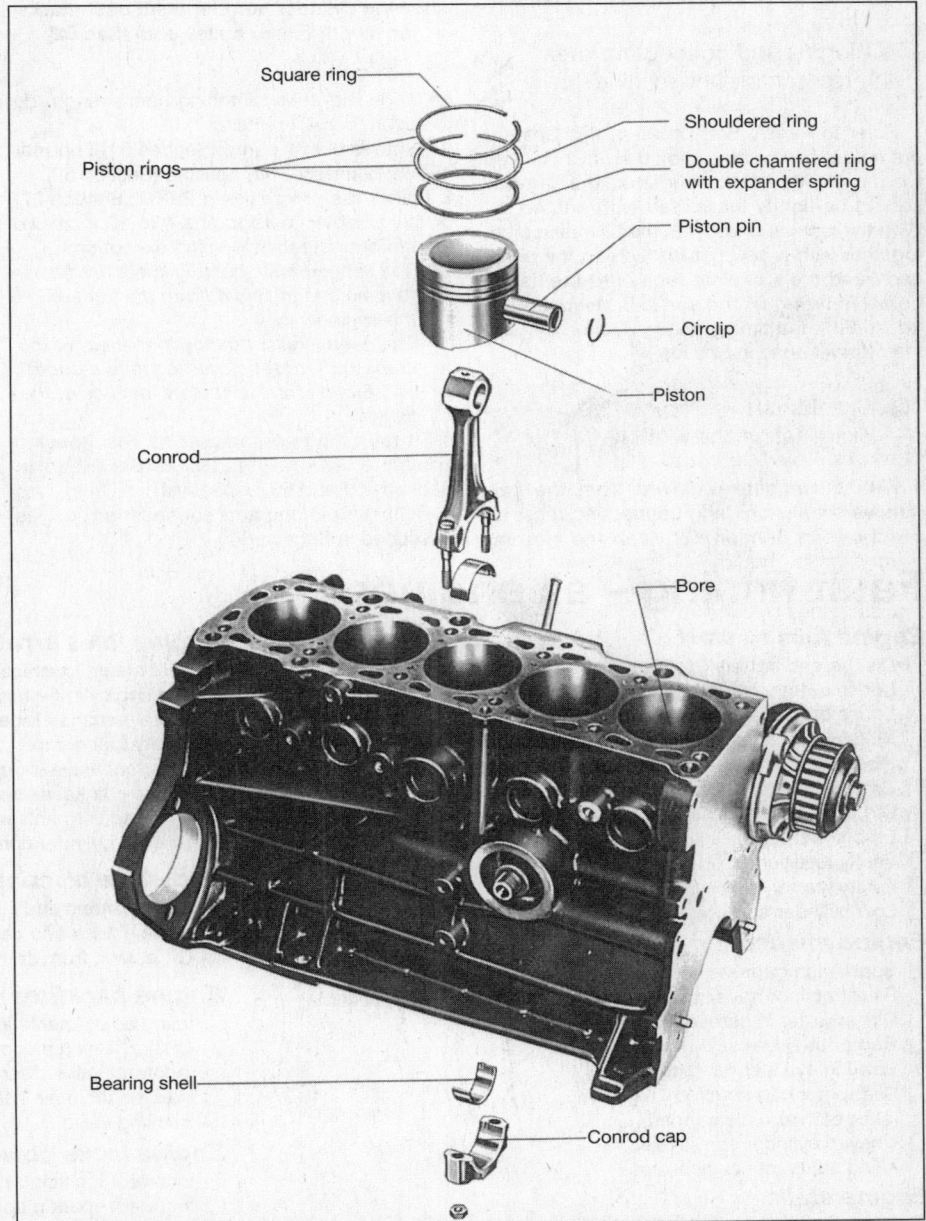

Fig. 1.39 Cylinder block connecting rod and piston assembly components – 1.9 litre engine (Sec 52)

52 Pistons and connecting rods – removal and dismantling

Refer to Part A, Section 24 of this Chapter. Removal of the cylinder head is described in Section 39, and the timing belt in Section 47. Instead of removing the oil pump, remove the sump, as described in Section 49, then unbolt the oil intake pipe, as described in Section 51.

53 Pistons and cylinder bores – examination

Refer to Part A, Section 25 of this Chapter.

54 Pistons and connecting rods – reassembly and refitting

Refer to Part A, Section 26 of this Chapter but note that new nuts should always be used for the big-end bolts and that the threads should be lightly lubricated with engine oil. With the pistons fitted, refit the oil intake pipe, together with a new gasket, tighten the bolts, and bend the lockplate tabs onto the flange bolts. Refitting of the sump is described in Section 49, the timing belt in Section 47, and the cylinder head in Section 44.

55 Crankshaft – removal. examination and refitting

1 With the engine removed from the car, remove the pistons and connecting rods, as described in Section 52. Keep the big-end bearings with their matching connecting rods to ensure correct refitting.
2 Remove the oil pump (Section 51) and rear oil seal complete with housing (Section 50).
3 Follow the procedure in Part A, Section 27, of this Chapter, paragraphs 4 to 13 inclusive (photo) but note the following exceptions:
(a) *The flanged main bearing shells are fitted to main bearing No 4 (from the front of the engine)*
(b) *The needle roller bearing in the rear of the crankshaft must be pressed in to a depth of 5.5 mm (0.217 in) below the face of the flange*
4 Fit the oil pump and rear oil seal housing complete with new seals, as described in Sections 51 and 50 respectively.
5 Refit the pistons and connecting rods, as described in Section 54.

Fig. 1.40 Connecting rod orientation showing cylinder number (with cast marks [arrowed] facing pulley end) (Sec 52)

55.3A Removing No. 4 main bearing cap

56 Oil filter – removal and refitting

1 Place a suitable container beneath the right-hand side of the engine.
2 Unscrew the filter cartridge and discard it – it will be necessary to use a filter strap, although if this is not available drive a long screwdriver through the cartridge.
3 Clean the mating faces of the new oil filter and block, and smear a little engine oil on the filter seal.
4 Screw on the cartridge and seal, and tighten it by hand only.

57 Engine mountings – removal and refitting

Refer to Part A, Section 29 of this Chapter. When refitting the mountings, check the engine alignment as described in Section 35.

Fault finding – all engines

Engine fails to start
☐ Discharged battery
☐ Loose battery connection
☐ Loose or broken ignition leads
☐ Moisture on spark plugs, distributor cap, or HT leads
☐ Incorrect spark plug gap
☐ Cracked distributor cap or rotor
☐ Dirt or water in carburettor (as applicable)
☐ Empty fuel tank
☐ Faulty fuel pump
☐ Faulty starter motor
☐ Low cylinder compression

Engine misfires
☐ Spark plug gaps incorrect
☐ Faulty coil, or transistorised ignition component (as applicable)
☐ Dirt or water in carburettor
☐ Burnt out valve
☐ Leaking cylinder head gasket
☐ Distributor cap cracked
☐ Incorrect valve clearances
☐ Uneven cylinder compressions
☐ Idling adjustments incorrect

Engine stalls
☐ Idling adjustments incorrect
☐ Inlet manifold air leak
☐ Ignition timing incorrect

Engine idles erratically
☐ Inlet manifold air leak
☐ Leaking cylinder head gasket
☐ Worn camshaft lobes
☐ Faulty fuel pump
☐ Incorrect valve clearances
☐ Loose crankcase ventilation hoses
☐ Idling adjustments incorrect
☐ Uneven cylinder compressions

Excessive oil consumption
☐ Worn pistons and cylinder bores
☐ Valve guides and valve stem seals worn
☐ Oil leaking from crankshaft oil seals, valve cover gasket, etc

Engine backfires
☐ Idling adjustments incorrect
☐ Ignition timing incorrect
☐ Incorrect valve clearances
☐ inlet manifold air leak
☐ Sticking valve

Engine lacks power
☐ Incorrect ignition timing
☐ Incorrect spark plug gap
☐ Low cylinder compression
☐ Excessive carbon build up in engine
☐ Air filter choked

Chapter 2 Cooling system

For modifications, and information applicable to later models, see Supplement at end of manual

Contents

Degrees of difficulty

Easy, suitable for novice with little experience	**Fairly easy,** suitable for beginner with some experience	**Fairly difficult,** suitable for competent DIY mechanic	**Difficult,** suitable for experienced DIY mechanic	**Very difficult,** suitable for expert DIY or professional

Specifications

System type .... Pressurized radiator and expansion tank (integral on some models), belt driven water pump, thermostatically controlled electric cooling fan

System coolant capacity (including heater)
1.6 and 1.8 litre engine up to August 1983 .... 5.4 litre (9.5 Imp pint)
1.6 and 1.8 litre engine from August 1983 .... 6.0 litre (10.5 Imp pint)
1.9 and 2.0 litre engine .... 8.0 litre (14.1 Imp pint)

Antifreeze
Type .... Ethylene glycol, with corrosion inhibitor
Concentration for protection down to: **Percent antifreeze by volume**
  –25°C (–14°F) 40
  –30°C (–22°F) 45
  –35°C (–31°F) 50

Filler cap opening pressure
1.6 and 1.8 litre engine .... 1.2 to 1.5 bar (17.4 to 21.7 lbf/in^2)
1.9 and 2.0 litre engine .... 1.2 to 1.35 bar (17.4 to 19.5 lbf/in^2)

Thermostat
Starts to open:
  1.6 and 1.8 litre engine .... 85°C (185°F)
  1.9 and 2.0 litre engine .... 87°C (188°F)
Fully open:
  1.6 and 1.8 litre engine .... 105°C (221°F)
  1.9 and 2.0 litre engine .... 102°C (216°F)
Stroke (minimum):
  1.6 and 1.8 litre engine .... 7.0 mm (0.276 in)
  1.9 and 2.0 litre engine .... 8.0 mm (0.315 in)

Electric cooling fan thermo-switch operating temperatures
Switches on (all engines) .... 93° to 98°C (199° to 208°F)
Switches off (all engines) .... 88° to 93°C (190° to 199°F)

Torque wrench settings

	Nm	lbf ft
Radiator mountings	10	7
Radiator cowl fasteners	10	7
Electric thermo-switch for fan	25	18
Water pump to block	20	14
Water pump to housing (1.6 and 1.8 litre)	10	7
Water pump pulley (1.6 and 1.8 litre)	20	14
Thermostat housing	10	7
Temperature sender unit (1.9 litre)	10	7
Coolant connector housing to cylinder head:		
1.6, 1.8 and 2.0 litre	10	7
1.9 litre	20	14
Coolant pipe mountings:		
1.6 and 1.8 litre (transmission-to-engine bolts)	55	40
1.9 and 2.0 litre	20	14
Electric cooling fan mounting	10	7
Cylinder head rear outlet (heat exchanger) – 1.6 and 1.8 litre	10	7
ATF cooler (automatic transmission)	40	28

1 General description and maintenance

The cooling system is of pressurized type and includes a front (four-cylinder) or side (five-cylinder) mounted radiator, a water pump driven by an external V-belt on four-cylinder engines or by the timing belt on five-cylinder engines, and an electric cooling fan.

All 1.9, 2.0 litre and later 1.6 and 1.8 litre (from August 1983) models have a separate expansion tank. The cooling system thermostat is located in the water pump housing on four-cylinder models, and in the inlet on the left-hand side of the cylinder block on five-cylinder models.

The system functions as follows. With the engine cold, the thermostat is shut and the water pump forces the water through the internal passages then via the bypass hose (and heater circuit if turned on) over the thermostat capsule and to the water pump inlet again. This circulation of water cools the cylinder bores, combustion surfaces and valve seats.: However, when the coolant reaches the predetermined temperature, the thermostat begins to open. The coolant now circulates through the top hose to the top of the radiator. As it passes through the radiator matrix it is cooled by the inrush of air when the car is in forward motion, supplemented by the action of the electric cooling fan when necessary. Finally the coolant is returned to the water pump via the bottom hose and through the open thermostat.

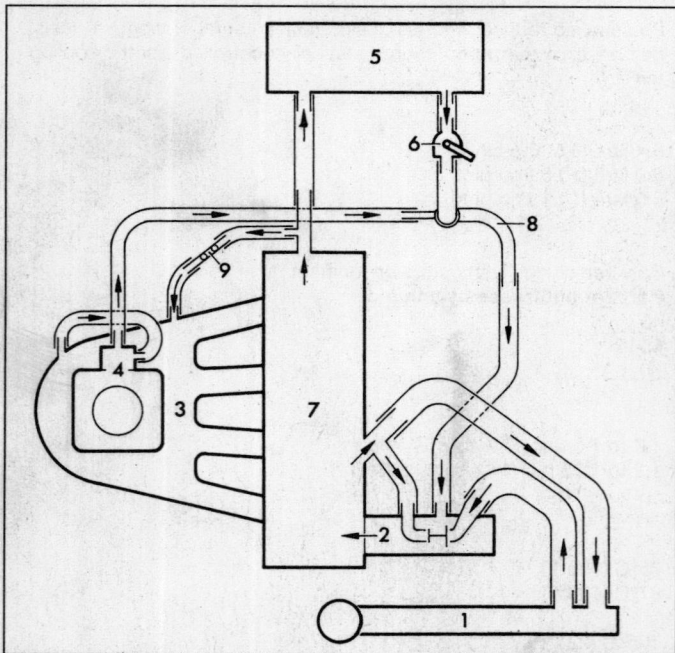

Fig. 2.1 Cooling system hose circuits – 1.6 litre engine (Sec 1)

1 Radiator
2 Water pump and thermostat housing
3 Intake manifold
4 Carburettor automatic choke
5 Heater
6 Heater valve
7 Cylinder block and head
8 Coolant hose
9 Thermo-switch
Note: Later models also have an expansion tank (not shown)

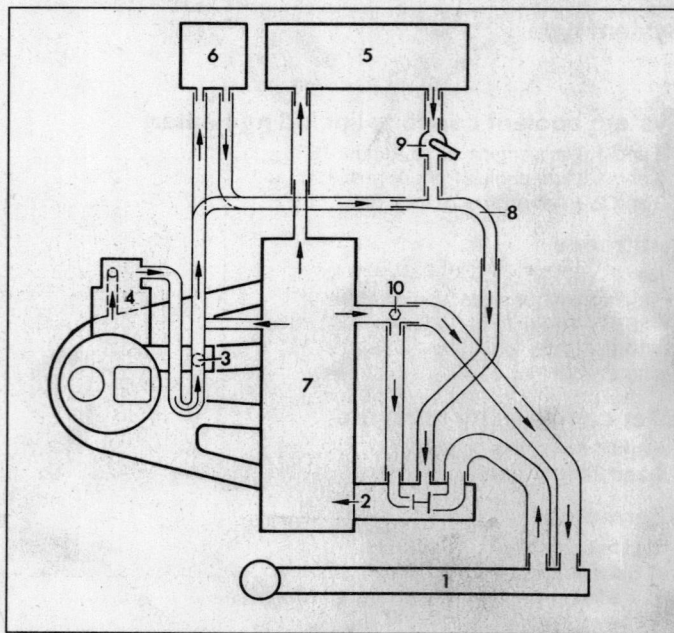

Fig. 2.2 Cooling system hose circuits – 1.8 litre engine (Sec 1)

1 Radiator
2 Water pump and thermostat
3 Expansion element
4 Automatic choke cover
5 Heater
6 ATF cooler (automatic transmission only)
7 Cylinder block and head
8 Coolant hose
9 Heater valve
10 Triple thermo-switch
Note: Later models also have an expansion tank (not shown)

Fig. 2.3 Radiator bottom hose connection (five-cylinder) (Sec 2)

4.2 Topping up the cooling system through radiator filler neck

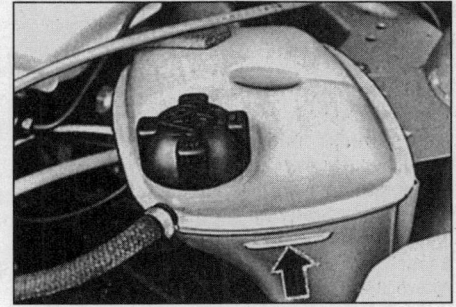

Fig. 2.4 Expansion tank coolant level mark (five-cylinder) (Sec 4)

The electric cooling fan is controlled by a thermo-switch located in the bottom of the radiator. Water temperature is monitored by a sender unit in the cylinder head.

Note: *the electric cooling fan will operate when the temperature of the coolant in the radiator reaches the predetermined level even if the ignition is switched off. Therefore extreme caution should be exercised when working in the vicinity of the fan blades.*

The cooling system must be regularly checked as part of the vehicle's routine maintenance. A weekly check must be made to ensure that the coolant level is correct in the radiator or expansion tank. If a sudden drop in the coolant level occurs, further investigation of the system is necessary to locate the cause and effect any repairs which may be necessary.

Periodically check the system hoses and connections for signs of leakage, deterioration and security.

2 Cooling system – draining

1 It is preferable to drain the cooling system when the engine is cold. If this is not possible, place a cloth over the filler cap and turn it slowly in an anti-clockwise direction until the pressure starts to escape – leave it in this position until all pressure has dissipated.
2 Remove the filler cap.
3 Set the heater controls on the facia to the WARM setting so that the heater control valve is open.
4 Position a suitable container beneath the bottom hose.
5 On four-cylinder engines loosen the clips and disconnect the bottom hose and heater/inlet manifold return hoses from the water pump housing. Drain the coolant into the container.
6 On five-cylinder engines unscrew the bolt retaining the heater return pipe to the cylinder block, then loosen the clip and pull the pipe from the hose. Also loosen the clip and disconnect the hose from the bottom of the radiator. Drain the coolant into the container (Fig. 2.3).

7 Where applicable, on four-cylinder engines loosen the vent screw located on the cylinder head outlet housing.

3 Cooling system – flushing

1 After some time the radiator and engine waterways may become restricted or even blocked with scale or sediment. When this occurs the coolant will appear rusty and dark in colour and the system should then be flushed. In severe cases, reverse flushing may also be required.
2 Drain the cooling system, as described in Section 2.
3 Disconnect the top hose from the radiator, insert a hose in the radiator and allow water to circulate through the matrix and out of the bottom of the radiator until it runs clear.
4 Insert the hose in the expansion tank (where fitted) and allow the water to run through the supply hose.
5 In severe cases of contamination remove the radiator, invert it, and flush it with water until it runs clear.
6 To flush the engine and heater, insert a hose in the top hose and allow the water to circulate through the system until it runs clear from the return hose.
7 The use of chemical cleaners should only be necessary as a last resort; the regular renewal of antifreeze should prevent the contamination of the system (refer to Section 5).

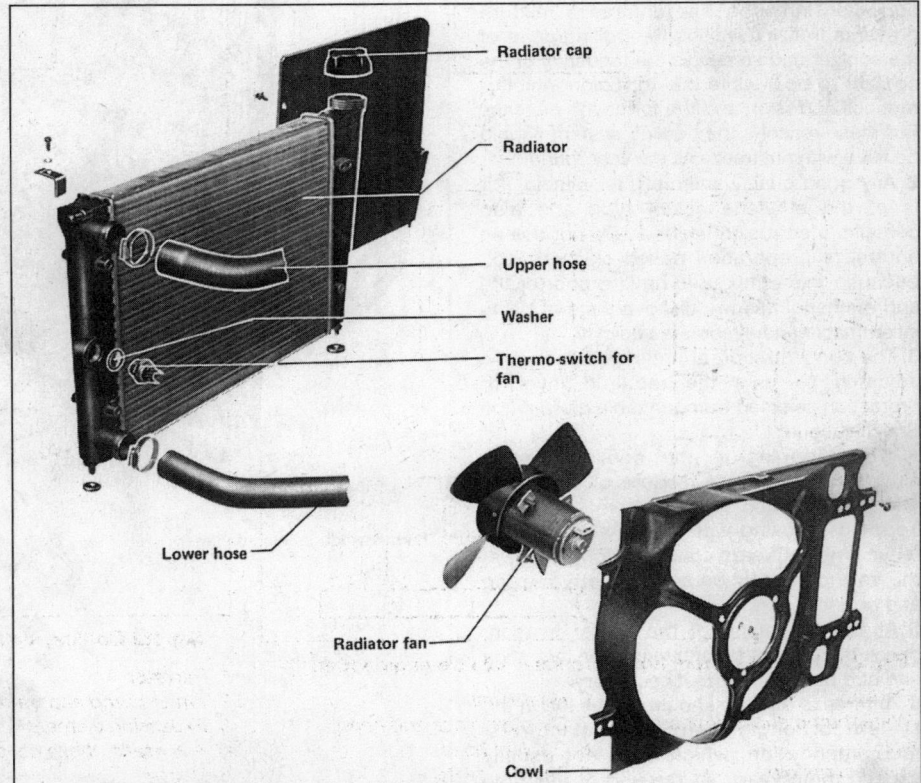

Radiator cap

Radiator

Upper hose

Washer

Thermo-switch for fan

Lower hose

Radiator fan

Cowl

Fig. 2.5 Radiator and associated components (four-cylinder) (Sec 6)

2

4 Cooling system – filling

1 Reconnect all hoses and check that the heater controls are set to WARM. Refit the radiator plug, if applicable.
2 Pour coolant into the radiator or expansion tank until full (photo).
3 Screw on the filler cap and run the engine at a fast idling speed for a few minutes.
4 Stop the engine and check the coolant level. Top up if necessary, and refit the filler cap.
5 Run the engine to normal operating temperature then allow it to cool. On models with an expansion tank the coolant level must reach the tip of the arrow with the engine cold or be a little higher when the engine is warm. On models without an expansion tank, remove the radiator cap with the engine cold and check that the coolant reaches the upper level mark: top up if necessary, and refit the cap (Fig. 2.4).

5 Antifreeze mixture

1 The cooling system is filled at the factory with an antifreeze mixture which contains a corrosion inhibitor. The antifreeze mixture prevents freezing, raises the boiling point of the coolant and so delays the tendency of the coolant to boil, while the corrosion inhibitor reduces corrosion and the formation of scale. For these reasons the cooling system should be filled with antifreeze all the year round.
2 Any good quality antifreeze is suitable, if it is of the ethylene glycol type and also contains a corrosion inhibitor. Do not use an antifreeze preparation based on methanol, because these mixtures have a shorter life and methanol has the disadvantage of being inflammable and evaporates quickly.
3 The concentration of antifreeze should be adjusted to give the required level of protection selected from the table given in the Specifications.
4 When topping-up the cooling system always use the same mixture of water and antifreeze which the system contains. Topping-up using water only will gradually reduce the antifreeze concentration and lower the level of protection against both freezing and boiling.
5 At the beginning of the winter season, check the coolant for antifreeze concentration and add pure antifreeze if necessary.
6 Antifreeze mixture should not be left in the system for longer than its manufacturers' recommendation, which does not usually exceed three years. At the end of this time drain the system and refill with fresh mixture.

6 Radiator – removal, inspection, cleaning and refitting

1 Disconnect the battery negative lead.
2 Drain the cooling system, as described in Section 2.

Four-cylinder engines

3 Remove the front grille referring to Chapter 12 for details.
4 Loosen the hose clips and detach the top and bottom hoses from the radiator (photo).
5 Disconnect the wiring from the electric cooling fan and the fan thermoswitch on the left-hand side of the radiator.
6 Unscrew the two upper radiator retaining bracket bolts and lift the brackets from their slotted locations in the top of the radiator (photo).

7 Unscrew and remove the single screw securing the radiator on the right-hand side (from the front).
8 Lift the radiator from the lower mountings then move it to one side and lift it from the engine compartment, complete with the electric cooling fan and cowling (photo).
9 If necessary, unbolt the cooling fan and cowling from the radiator, then unscrew the nuts and separate the fan and motor from the cowling.

Five-cylinder engines

10 Loosen the clips and disconnect the radiator top and bottom hoses.
11 Disconnect the expansion tank vent hose from the radiator.
12 Disconnect the wiring from the electric cooling fan and the thermoswitch.
13 Remove the nuts and washers from the

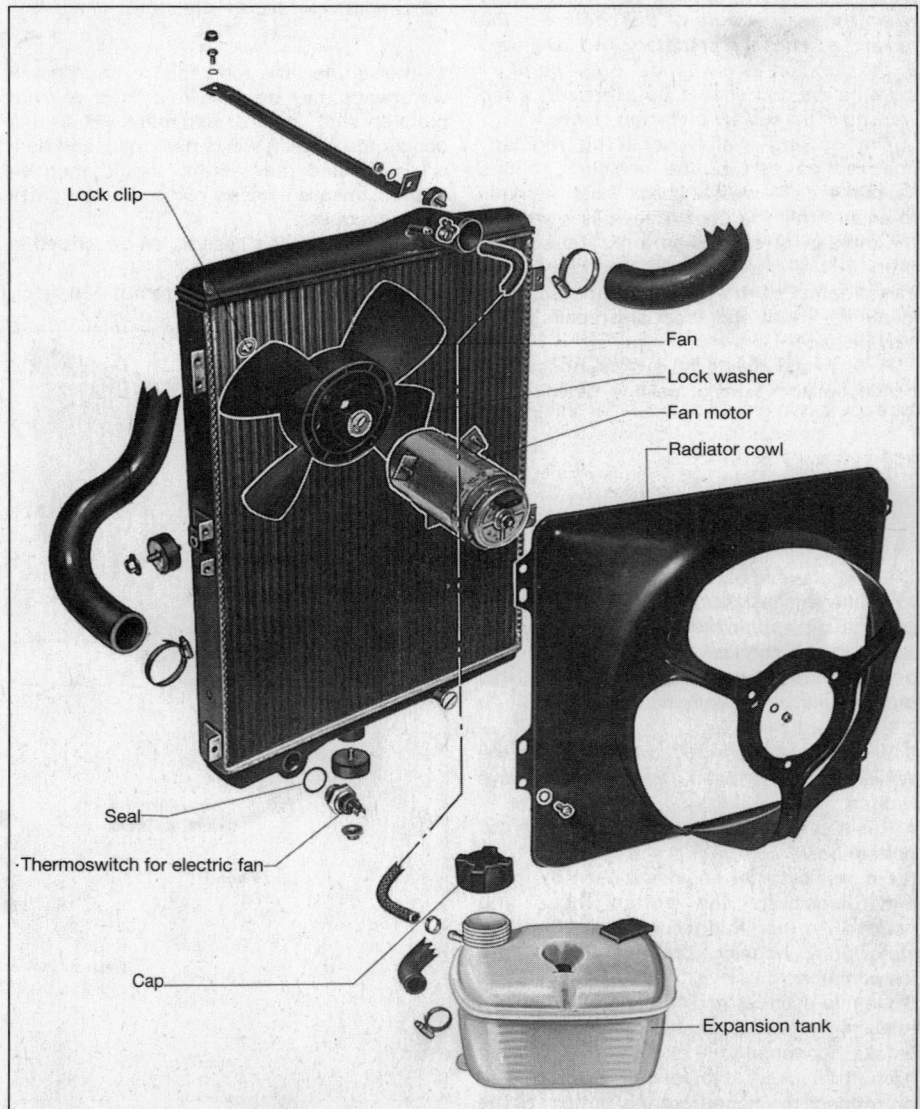

Lock clip

Fan

Lock washer

Fan motor

Radiator cowl

Seal

Thermoswitch for electric fan

Cap

Expansion tank

Fig. 2.6 Radiator and associated components (five-cylinder) (Sec 6)

6.4 Radiator top hose connection (four-cylinder)

6.6 Prise free radiator upper retaining bolt caps, then remove bolts (four-cylinder)

6.8 Lifting out the radiator (four-cylinder)

upper mountings and also unbolt the upper mounting bracket.

14 Unscrew the remaining mounting nut or cowl screws, as applicable, and lift out the radiator complete with the electric cooling fan and cowling.

15 If necessary, unbolt the cooling fan and cowling from the radiator, then unscrew the nuts and separate the fan and motor from the cowling.

All models

16 Radiator repair is best left to a specialist, although minor leaks can be stopped using a proprietary coolant additive. Clear the radiator matrix of flies and small leaves with a soft brush or by hosing.

17 Reverse flush the radiator, as described in Section 3, and renew the hoses and clips if they are damaged or have deteriorated.

18 Refitting is a reversal of removal, but fill the cooling system, as described in Section 4. If the thermo-switch is removed, fit a new sealing washer when refitting it.

7 Thermostat –
removal, testing and refitting

1 On four-cylinder engines the thermostat is located in the bottom of the water pump housing, but on five-cylinder engines it is located behind the water pump on the left-hand side of the cylinder block (photo).

Fig. 2.7 Water pump, thermostat and cooling system hoses (four-cylinder) (Sec 7)

Labels: Thermoswitch; Sender or Thermoswitch for coolant temperature; Gasket; Connection; Upper hose; Water pump; Gasket; Thermostat; Lower hose; Sealing ring; to heat exchanger; from heat exchanger; Gasket; Coolant pipe

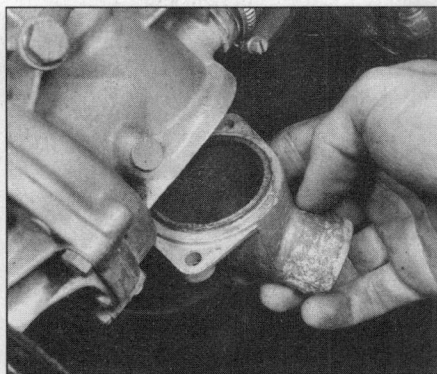

7.1 Thermostat housing at rear of water pump (four-cylinder)

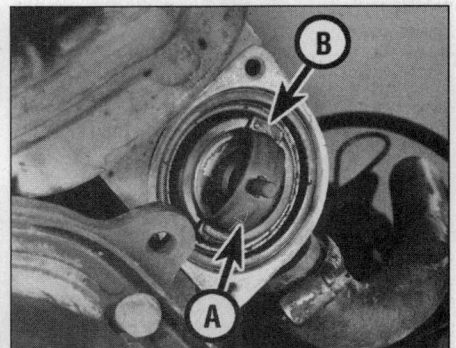

7.3 Removing the thermostat cover (four-cylinder)

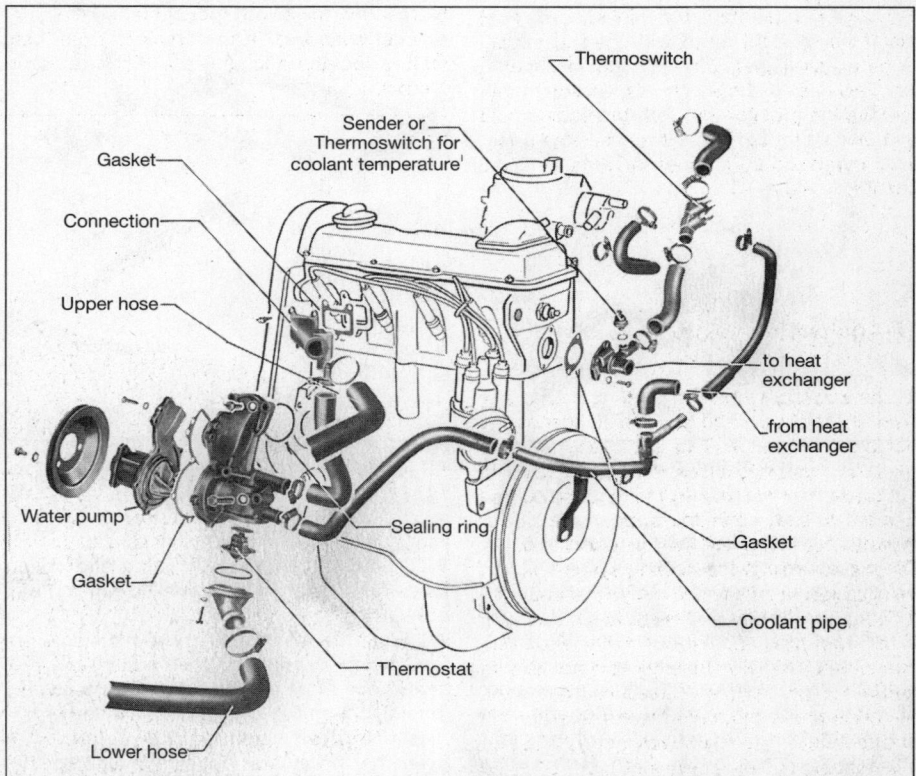

7.7 Thermostat in position showing arrow (A) for fitting direction and sealing ring (B) (four-cylinder)

Fig. 2.9 Thermostat orientation (five-cylinder) (Sec 7)

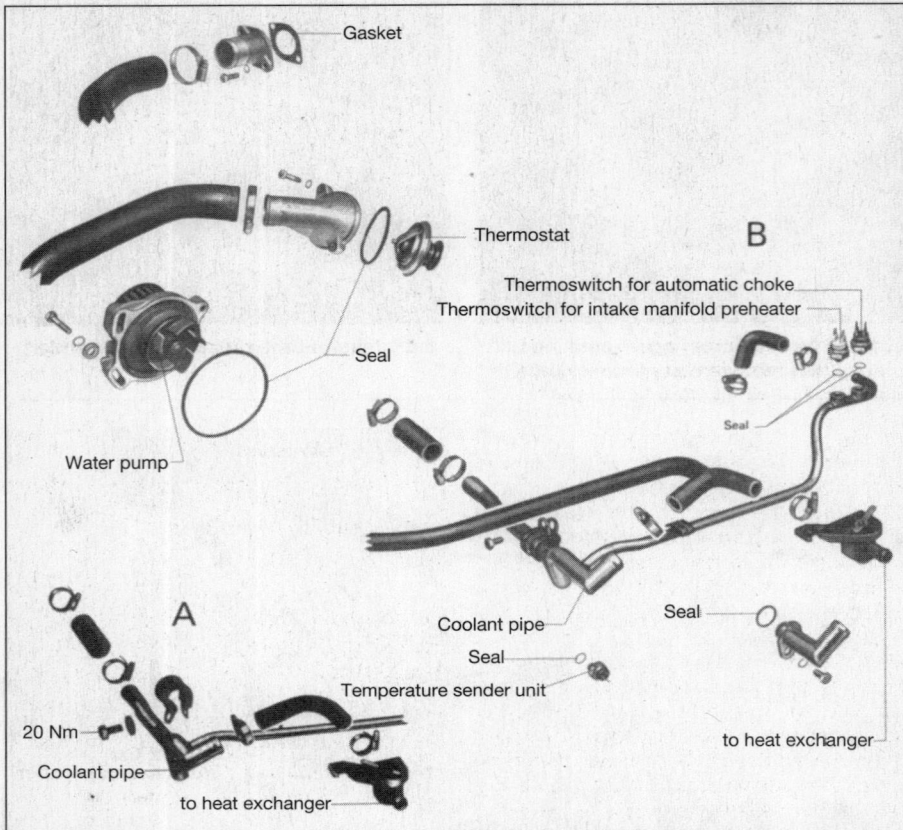

Fig. 2.8 Water pump, thermostat and cooling system hoses (five-cylinder) (Sec 7)

A 2.0 litre   B 1.9 litre

2 To remove the thermostat first drain the cooling system, as described in Section 2.

3 Unbolt and remove the thermostat cover, and remove the sealing ring (photo).

4 Prise the thermostat from its housing.

5 To test whether the unit is serviceable, suspend it with a piece of string in a container of water. Gradually heat the water and note the temperatures at which the thermostat starts to open and is fully open. Remove the thermostat from the water and check that it is fully closed when cold. Renew the thermostat if it fails to operate in accordance with the

information given in the Specifications.

6 Clean the thermostat housing and cover faces, and locate a new sealing ring on the cover.

7 Locate the thermostat in the housing – on four-cylinder engines the arrow on the crosspiece must point away from the engine (photo), but on five-cylinder engines the arrow must point downwards (Fig. 2.9).

8 Fit the thermostat cover and tighten the bolts evenly.

9 Fill the cooling system, as described in Section 4.

8 Water pump – removal and refitting

1 Drain the cooling system, as described in Section 2.

2 On models equipped with power steering, remove the pump leaving the hoses connected and place it to one side, with reference to Chapter 11.

Four-cylinder engines

3 Loosen the clips and disconnect the hoses from the rear of the water pump housing. (See photo 7.1).

4 Unscrew the nut and remove the special bolt retaining the lower timing cover to the water pump assembly.

5 Unbolt the water pump assembly from the cylinder block and remove the sealing ring (photo).

6 Unscrew the bolts and remove the water pump from its housing using a mallet to break the seal. Remove the gasket (photo).

Five-cylinder engines

7 Using an Allen key where necessary, unscrew the nuts and withdraw the timing cover.

8 Set the engine on TDC compression No 1 cylinder, and unbolt the timing belt rear cover from the oil pump, with reference to Chapter 1.

9 Loosen the water pump mounting and adjustment bolts, again using an Allen key

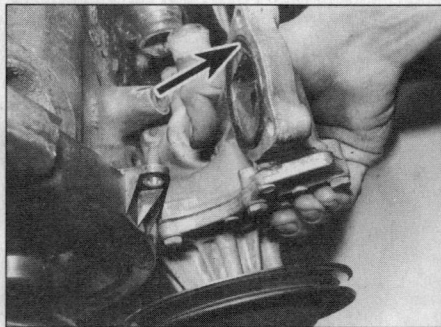

8.5 Water pump removal (four-cylinder). Note sealing ring position (arrowed)

8.6 Unscrew securing bolts to separate water pump from its housing (four-cylinder)

8.10 Water pump and mounting bolts (five-cylinder)

8.11 Removing the water pump (five-cylinder)

10.1 Coolant temperature sender unit location (arrowed) at the rear of the cylinder head (four-cylinder)

11.2 Heater control valve and connections

where necessary, and rotate the pump to release the tension on the timing belt.

10 Unscrew and remove the bolts noting the location of the stay, where applicable (photo).

11 Withdraw the water pump from the cylinder block and remove the sealing ring (photo). *Do not move the crankshaft or camshaft with the timing belt slack.*

All models

12 If the water pump is faulty, renew it, as individual components are not available. Clean the mating faces of the water pump, cylinder block, and pump housing (four-cylinder engines).

13 Refitting is a reversal of removal, but use a new sealing ring and gasket as applicable. On five-cylinder engines tension the timing belt, as described in Chapter 1. At the same time check that the crankshaft and camshaft are still on TDC No 1 cylinder. Fill the cooling system, as described in Section 4. Tension the alternator/power-assisted steering pump drivebelts, as applicable, with reference to Chapters 10 and 11 respectively.

9 Cooling fan thermo-switch – testing, removal and refitting

1 If the thermo-switch located in the bottom of the radiator develops a fault, it is most likely to fail open circuit. This will cause the fan motor to remain stationary even though the coolant reaches the operating temperature.

2 To test the thermo-switch for an open circuit fault, disconnect the wiring and connect a length of wire or suitable metal object between the two wires. The fan should

operate (even without the ignition switched on) in which case the thermo-switch is proved faulty and must be renewed.

3 To remove the thermo-switch first drain the cooling system, as described in Section 2.

4 Disconnect the battery negative lead.

5 Disconnect the wiring, then unscrew the thermo-switch from the radiator and remove the sealing washer.

6 To check the operating temperature of the thermo-switch, suspend it in a pan of water so that only the screwed end of the switch is immersed and the electrical contacts are clear of the water. Either connect an ohmmeter between the switch terminals, or connect up a torch battery and bulb in series with the switch. With a thermometer placed in the pan, heat the water and note the temperature at which the switch contacts close, so that the ohmmeter reads zero, or the bulb lights. Allow the water to cool and note the temperature at which the switch contacts open. Discard the switch and fit a new one if the operating temperatures are not within the specified limits.

7 Refitting is a reversal of removal, but always fit a new sealing washer. Fill the cooling system, as described in Section 4.

10 Coolant temperature sender unit – removal and refitting

1 The temperature sender unit is located at the rear of the cylinder head (photo). To remove it, first drain half of the cooling system, with reference to Section 2.

2 Disconnect the wiring and unscrew the sender unit from the connector or cylinder

12.1 Position of core plugs in the cylinder block (four-cylinder)

head, as applicable. Remove the sealing washer(s).

3 Refitting is a reversal of removal, but always renew the washer(s). Top up the cooling system, with reference to Section 4.

11 Heater valve – removal and refitting

1 Drain the cooling system as described in Section 2.

2 Unhook and detach the control rod from the control unit operating lever and disconnect the cable retainer from the body (photo).

3 Unscrew the inlet and outlet hose clips and detach the hoses. The unit can now be withdrawn.

4 Refitting is a reversal of the removal procedure. If necessary, adjust the heater control as described in Chapter 12, Section 37.

Fault finding overleaf

Fault finding – cooling system

Overheating

- ☐ Low coolant level
- ☐ Faulty pressure cap
- ☐ Thermostat sticking shut
- ☐ Drivebelt broken (four-cylinder)
- ☐ Open circuit thermo-switch
- ☐ Faulty cooling fan motor
- ☐ Clogged radiator matrix
- ☐ Retarded ignition timing

Slow warm-up

- ☐ Thermostat sticking open
- ☐ Short circuit thermo-switch

Coolant loss

- ☐ Deteriorated hose
- ☐ Leaking water pump or cooling system joints
- ☐ Blown cylinder head gasket
- ☐ Leaking radiator
- ☐ Leaking core plugs (photo)

Chapter 3 Fuel and exhaust systems

For modifications, and information applicable to later models, see Supplement at end of manual

Contents

3

Degrees of difficulty

Easy, suitable for novice with little experience	Fairly easy, suitable for beginner with some experience	Fairly difficult, suitable for competent DIY mechanic	Difficult, suitable for experienced DIY mechanic	Very difficult, suitable for expert DIY or professional

Specifications

Part A: Carburettor and associated fuel system components

Air cleaner type .... Renewable paper element; automatic air temperature control on some models

Fuel pump
Type .... Mechanical, diaphragm, operated by eccentric on intermediate shaft (four-cylinder) or camshaft (five-cylinder)

Operating pressure:
1.6 and 1.8 litre engine .... 0.2 to 0.25 bar (2.9 to 3.6 lbf/in²)
1.9 litre engine .... 0.35 to 0.40 bar (5.1 to 5.8 lbf/in²)

Fuel tank capacity .... 60 litres (13.2 Imp gallons)

Fuel octane requirement (minimum)
1.6 litre .... 91 RON
1.8 and 1.9 litre .... 98 RON

Carburettor 1.6 litre – 1B3

	Manual gearbox	Automatic gearbox
Type	Single choke downdraught, automatic choke	
Choke valve gap:		
Initial	2.05 to 2.35 mm (or 1.85 to 2.15 mm*)	2.25 to 2.55 mm
Final	3.35 to 3.65 mm (or 4.15 to 4.45 mm*)	3.15 to 3.45 mm (or 4.15 to 4.45 mm*)
Fast idle speed	3700 to 4100 rpm	3500 to 3900 rpm
Choke cover number	253 (or 234*)	213
Idle speed	900 to 1000 rpm	900 to 1000 rpm
CO content	0.5 to 1.5%	0.5 to 1.5%
Venturi	26 mm	26 mm
Main jet	x125	x122.5
Air correction jet with emulsion tube	100	100
Idle fuel/air jet	50/130	50/130
Auxiliary fuel/air jet	37.5/130	37.5/130
Float needle vaive	1.75 (or 2.0*)	1.75 (or 2.0*)
Enrichment jet	102.5 (or 110*)	102.5 (or 110*)
Pump injection tube	0.40	0.50 (or 0.55*)
Pump injection capacity (slow)	0.75 to 1.05 cc/stroke	0.75 to 1.05 cc/stroke

Models pre October 1981

Carburettor 1.6 litre – 2B5

	Manual gearbox		Automatic gearbox	
Type	Twin progressive choke downdraught automatic choke			
Choke valve gap:				
Initial	1.65 to 1.95 mm		1.65 to 1.95 mm	
Final	3.75 to 4.05 mm		3.55 to 3.85 mm	
Fast idle speed	3350 to 3450 rpm		3550 to 3650 rpm	
Choke cover number	232		232	
Idle speed	900 to 1000 rpm		900 to 1000 rpm	
CO content	0.5 to 1.5%		0.5 to 1.5%	
	Stage 1	**Stage 2**	**Stage 1**	**Stage 2**
Venturi	24 mm	28 mm	24 mm	28 mm
Main jet	x117.5	x125	x117.5	x125
Air correction jet with emulsion tube	135	92.5	135	92.5
Idle fuel/air jet	52.5/135	40/125	52.5/135	40/125
Auxiliary fuel/air jet	42.5/130	–	42.5/130	–
Idle air jet for progression reserve	–	180	–	180
Idle fuel jet for progression reserve	–	130	–	100
Float needle valve	2.0	2.0	2.0	2.0
Float setting	27.0 to 29.0 mm	29.0 to 31.0 mm	27.0 to 29.0 mm	29.0 to 31.0 mm
Enrichment valve	65	–	65	–
Pump injection tube (vertical/horizontal)	0.4/0.4 mm	–	0.4/0.4 mm	–
Injection capacity	0.85 to 1.15 cc/stroke	–	0.75 to 1.05 cc/stroke	–

Carburettor 1.6 litre – 2E2

	Stage 1	Stage 2
Type	Twin progressive choke downdraught, automatic choke	
Choke valve gap:		
Manual gearbox:		
Initial	2.15 to 2.45 mm	
Final	4.55 to 4.85 mm	
Automatic transmission:		
Initial	2.55 to 2.85 mm	
Final	5.55 to 5.85 mm	
Fast idle speed	2800 to 3200 rpm	
Choke cover number	258	
Idle speed	700 to 800 rpm	
CO content	0.5 to 1.5%	
Venturi diameter	22 mm	26 mm
Main jet	x107.5	x127.5
Air correction jet with emulsion tube	80	105
Idle fuel jet	42.5	–
Full load enrichment valve	–	0.7
Pump injection tube:		
Manual gearbox	0.35	–
Automatic transmission	0.5	–
Pump injection capacity (slow)	0.8 to 1.2 cc/stroke	

Carburettor 1.8 litre – 2E2

	Stage 1	Stage 2
Type	Twin progressive choke downdraught, automatic choke	
Choke valve gap:		
Initial	1.65 to 1.95 mm	
Final:		
Manual gearbox	3.85 to 4.15 mm	
Automatic transmission	3.75 to 4.05 mm	
Fast idle speed	2800 to 3200 rpm	
Choke cover number	258	
Idle speed	700 to 800 rpm	
CO content	0.5 to 1.5%	
Venturi diameter	22 mm	26 mm
Main jet	x105	x120
Air correction jet with emulsion tube	100	100
Idle fuel/air jet	40	–
Full load enrichment valve	–	1.25
Pump injection tube	0.35	–
Pump injection capacity (slow)	0.95 to 1.25 cc/stroke	

Carburettor 1.9 litre – 2B5 (pre August 1982)

	Stage 1	Stage 2
Type	Twin progressive choke downdraught, automatic choke	
Choke valve gap:		
Manual gearbox	3.45 to 3.75 mm	
Automatic transmission	3.25 to 3.55 mm	
Fast idle speed:		
Manual gearbox	3500 to 3700 rpm	
Automatic transmission	3600 to 3800 rpm	
Idle speed (all transmissions)	750 to 800 rpm	
CO content	0.8 to 1.2%	
Venturi	24 mm	28 mm
Main jet	x115	x122.5
Air correction jet with emulsion tube	135	115
Idle fuel/air jet	42.5/125	40/125
Auxiliary fuel/air jet	45/125	–
Idle air for progression reserve	–	205
Idle fuel jet for progression reserve	–	95
Float needle valve	2.0	2.0
Float setting	27.0 to 29.0 mm	29.0 to 31.0 mm
Enrichment valve	100	–
Pump injection tube (vertical/horizontal)	0.4/0.55 mm	–
Injection capacity	1.35 to 1.65 cc/stroke	–

Carburettor 1.9 litre – 2B5 (from August 1982)

	Stage 1	Stage 2
Choke valve gap:		
Manual gearbox	3.90 to 4.20 mm	
Automatic transmission	3.70 to 4.00 mm	
Fast idle speed:		
Manual gearbox	3500 to 3700 rpm	
Automatic transmission	3600 to 3800 rpm	
Idle speed (all transmissions)	750 to 800 rpm	
CO content	0.8 to 1.2%	
Venturi	24 mm	28 mm
Main jet	x117.5	x125
Air correction jet with emulsion tube	135	115
Idle fuel/air jet	45/130	40/130
Auxiliary fuel/air jet	42.5/130	–
Idle air jet for progression reverse	–	205
Idle fuel jet for progression reserve , ...	–	95
Float needle diameter	2.0	2.0
Pump injection tube diameter	2 x 04	0.55
Enrichment valve	85	
Float setting	27 to 29	29 to 31
Injection capacity (slow)	1.35 to 1.65 cc/stroke	

Torque wrench settings	Nm	lbf ft
Fuel tank mounting	25	18
Fuel pump	20	15
Non-return valve (1.9 litre)	20	15
Carburettor – 1B3 and 2E2	7	5
Carburettor – 2B5	10	7
Carburettor cover screws	5	4
Automatic choke cover	5	4
Automatic choke centre bolt	10	7
Carburettor throttle housing 12 B 5)	10	7
Carburettor bypass air cut-off valve	5	4
Inlet manifold	25	18
Inlet manifold preheater	10	7
Exhaust manifold:		
1.6 and 1.8 litre	25	18
1.9 litre	20	15
Exhaust downpipe to manifold	30	22
Exhaust system pipe joint clamp(s)	25	18

Part B: Fuel injection system

Type	Continuous fuel injection system (CIS)
System pressure	4.7 to 5.4 bar (68 to 78 lbf/in²)
Warm up regulator heater coil resistance	16 to 22 ohm
Idle speed	920 ± 20 rpm
Fuel tank capacity	As Part A
Fuel octane requirement (minimum)	91 RON
CO content	1 ± 0.2%

Torque wrench settings	Nm	lbf ft
Thermo-time switch	30	22
Auxiliary air valve	10	7
Throttle housing	20	14
Cold start valve	10	7
Warm-up valve	10	7
Metering unit	35	25
Fuel distributor	35	25
Non-return valve	25	18
Fuel line to distributor	10	7
Fuel line to warm-up valve	20	14
Distributor/non-return valve banjo bolts	25	18
Warm-up valve/distributor line banjo bolt	10	7
Inlet manifold	25	18
Exhaust manifold	20	14

Part A: Carburettor and associated fuel system components

1 General description

The fuel system consists of a rear mounted fuel tank, a mechanical diaphragm fuel pump and a downdraught carburettor.

The air cleaner has a renewable paper element and incorporates an automatic temperature control.

A conventional exhaust system is used on all models, being sectionised for ease of replacement.

2 Fuel and exhaust system – routine maintenance

The following routine maintenance procedures are required for the fuel and exhaust system and must be carried out at the specified intervals given at the start of this Manual. It should be noted that the intervals quoted are those for a vehicle used in normal operating conditions.

Fuel system general: Inspect the fuel system lines, hoses and connections at

HAYNES HiNT *If a vehicle is used in adverse conditions, such as continuous city driving or in a hot dusty climate, then it is advisable to shorten the maintenance intervals.*

regular intervals for security and condition. In addition also check the associated vacuum hoses and connections. Occasionally lubricate the accelerator control linkages.

Fuel system adjustments: Check and if necessary adjust the engine idling speed and

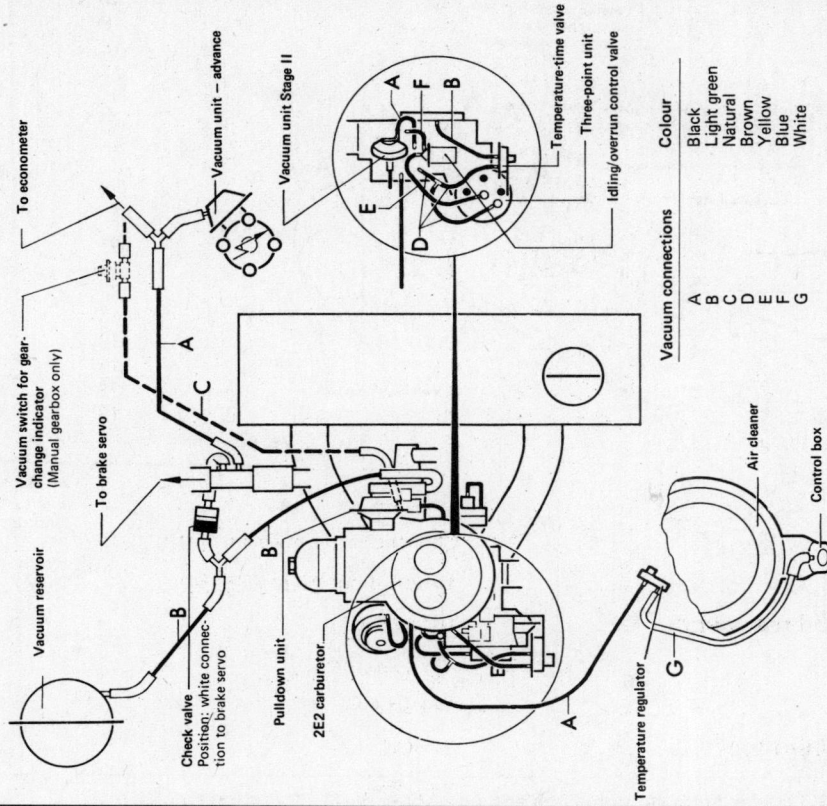

Fig. 3.2 Fuel system vacuum lines and components – 2E2 carburettor models pre August 1983 (Sec 1)

Vacuum connections

	Colour
A	Black
B	Light green
C	Natural
D	Brown
E	Yellow
F	Blue
G	White

Labels: To econometer · Vacuum switch for gear-change indicator (Manual gearbox only) · To brake servo · Vacuum reservoir · Check valve · Position; white connection to brake servo · Pulldown unit · 2E2 carburetor · Temperature regulator · Air cleaner · Control box · Vacuum unit – advance · Vacuum unit Stage II · Temperature-time valve · Three-point unit · Idling/overrun control valve

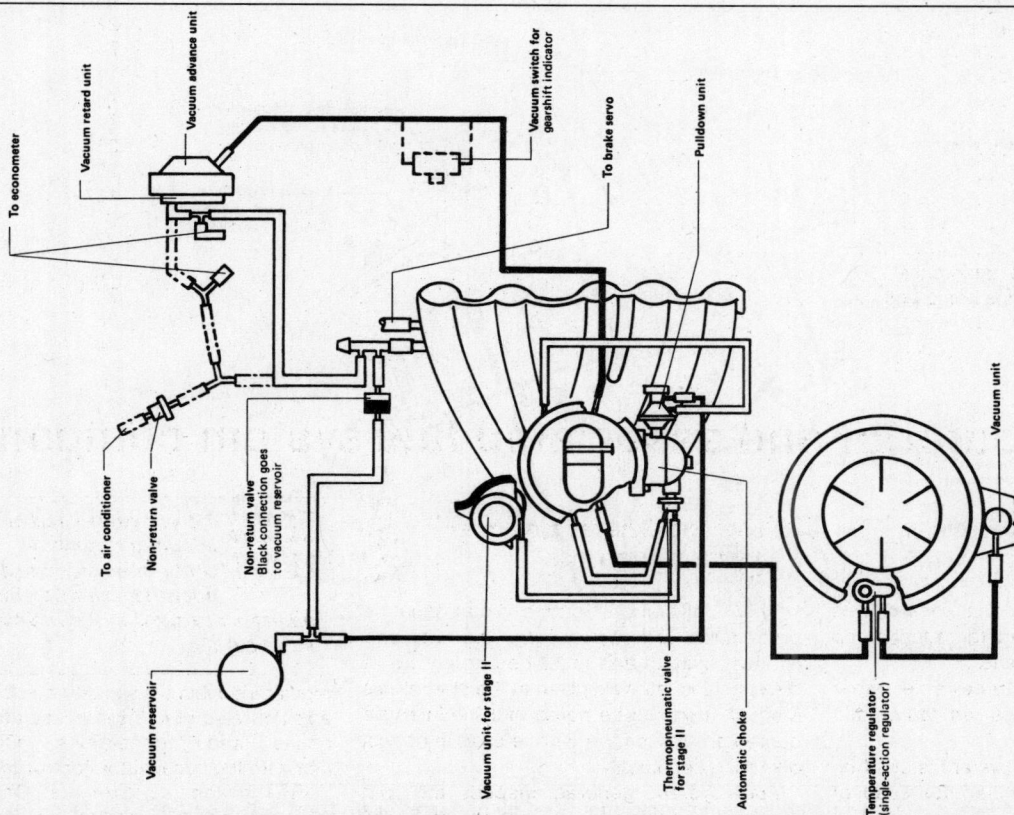

Fig. 3.1 Fuel system vacuum lines and components – 2B5 carburettor on the 1.9 litre engine (Sec 1)

Broken outline components for air conditioner only

Labels: To econometer · Vacuum retard unit · Vacuum advance unit · Vacuum switch for gearshift indicator · To brake servo · Pulldown unit · Vacuum unit · To air conditioner · Non-return valve · Non-return valve · Black connection goes to vacuum reservoir · Vacuum reservoir · Vacuum unit for stage II · Thermopneumatic valve · Vacuum unit for stage II · Automatic choke · Temperature regulator (single-action regulator)

3

Vacuum switch for
gearchange indicator
(Only with manual gearbox)

Solenoid changeover valve
Only with air conditioner

Air conditioner connection

To econometer

To brake servo

Check valve
Position: white connection
to brake servo

Pulldown unit

Air cleaner
Temperature regulator

1B3 carburetor

Vacuum reservoir

Vacuum unit advance

Vacuum connections	Colour
A	Black
B	Light green
C	Natural
D	Pink
E	White
F	Grey

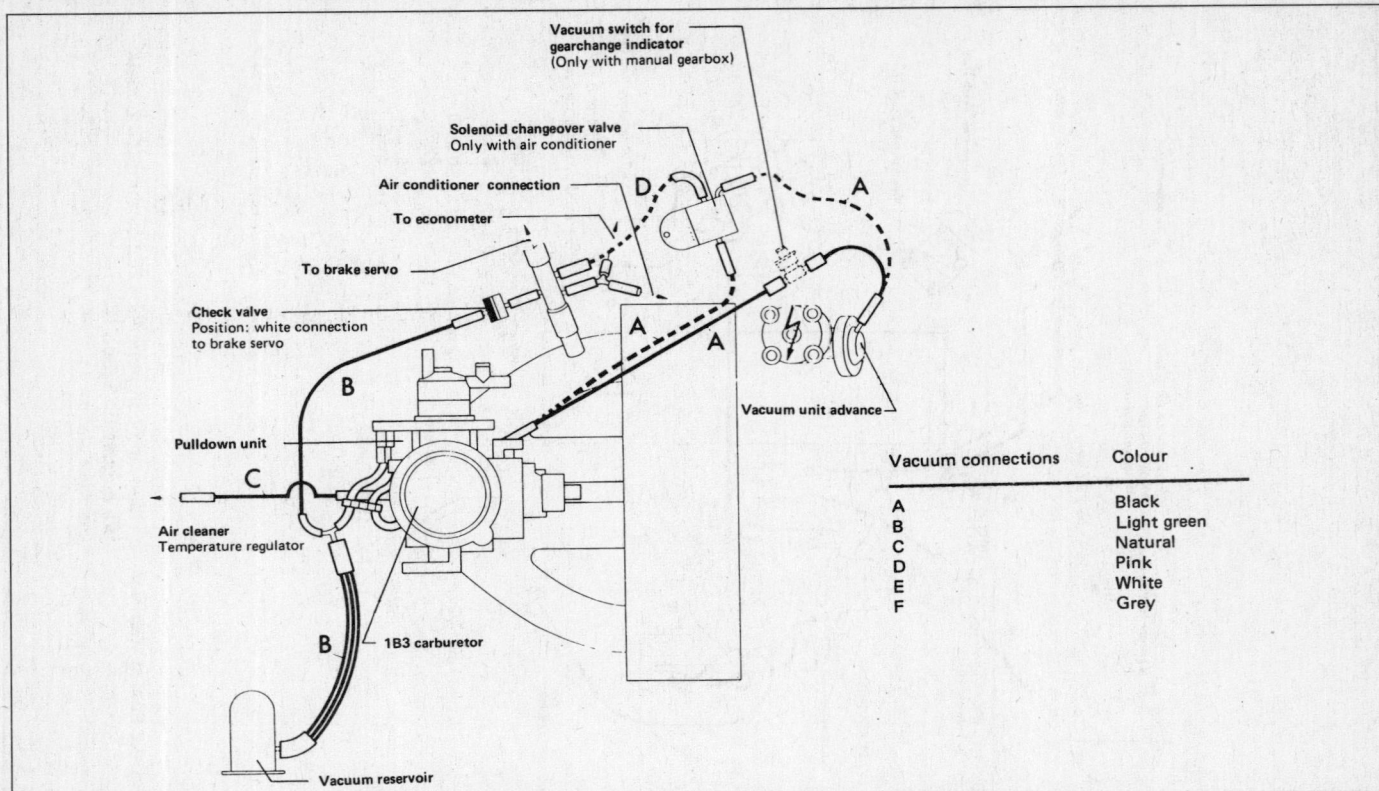

Fig. 3.3 Fuel system vacuum lines and components – 1 B3 carburettor (Sec 1)

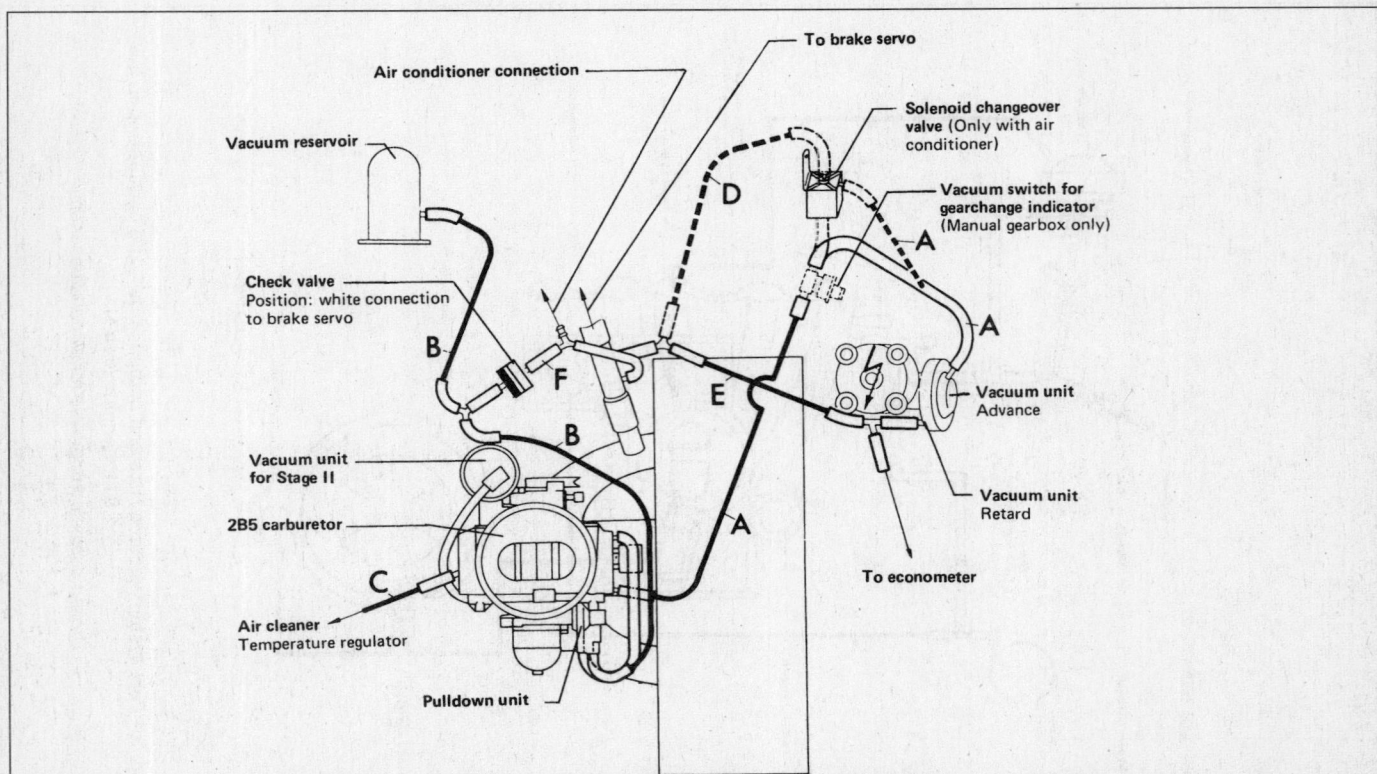

To brake servo

Air conditioner connection

Solenoid changeover
valve (Only with air
conditioner)

Vacuum switch for
gearchange indicator
(Manual gearbox only)

Vacuum reservoir

Check valve
Position: white connection
to brake servo

Vacuum unit
Advance

Vacuum unit
for Stage II

Vacuum unit
Retard

2B5 carburetor

Air cleaner
Temperature regulator

To econometer

Pulldown unit

Fig. 3.4 Fuel system vacuum lines and components – 2B5 carburettor on the 1.6 litre engine (Sec 1)

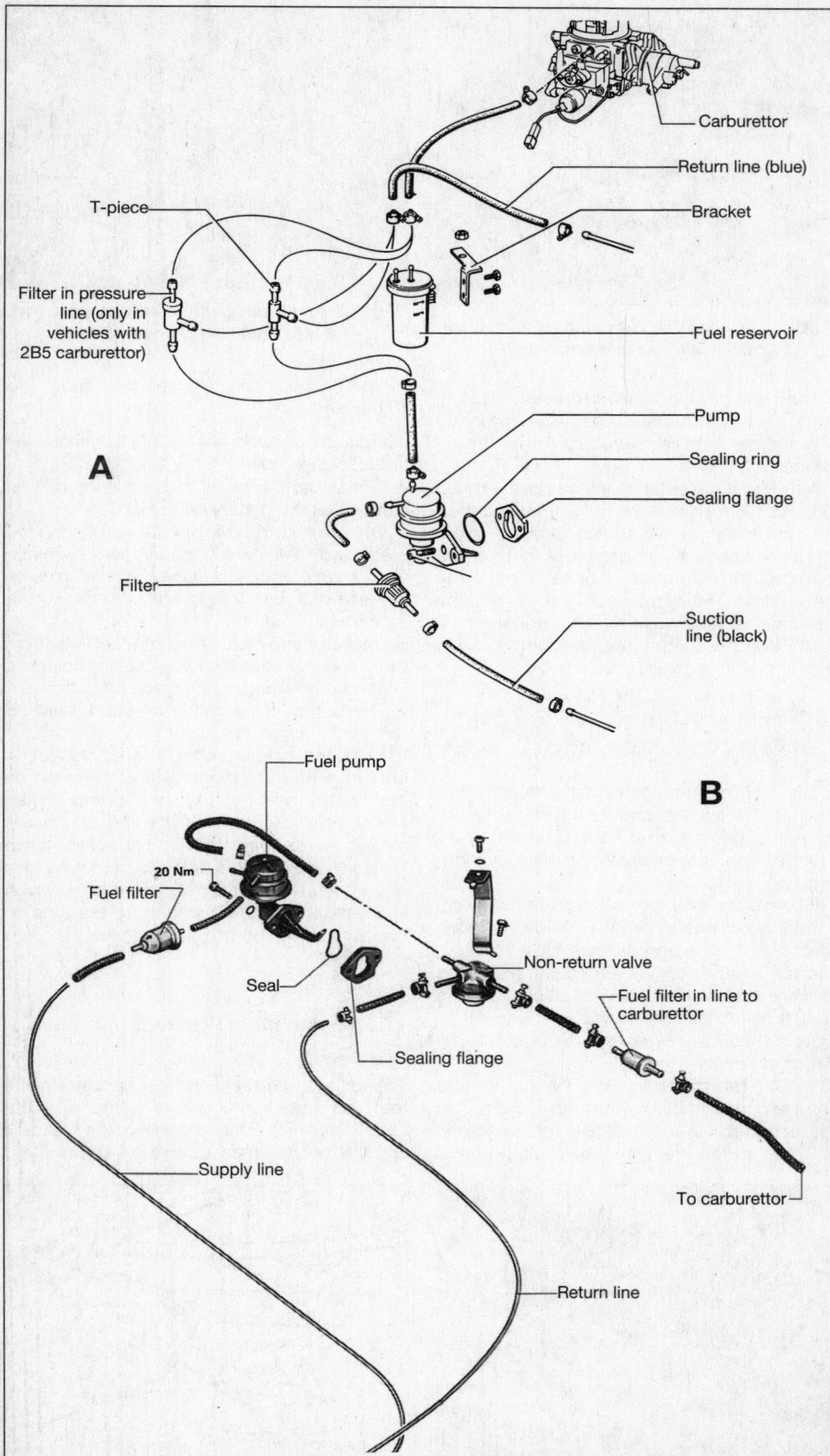

A

B

T-piece

Filter in pressure line (only in vehicles with 2B5 carburettor)

Filter

Carburettor

Return line (blue)

Bracket

Fuel reservoir

Pump

Sealing ring

Sealing flange

Suction line (black)

Fuel pump

20 Nm

Fuel filter

Seal

Sealing flange

Non-return valve

Fuel filter in line to carburettor

Supply line

Return line

To carburettor

Fig. 3.5 Typical fuel system components (Sec 1)

A 1.6 and 1.8 litre   B 1.9 litre

where possible the exhaust CO content (having first checked ignition timing as described in Chapter 4).

Air cleaner: Renew the air cleaner element as described in Section 3.

Fuel filter: Renew the fuel filter as described in Section 7.

Exhaust system: Inspect the exhaust system for signs of joint leaks, excessive corrosion and general security. Repair if necessary (Section 16).

3 Air cleaner element – renewal and cleaning

1 A dirty air cleaner element will cause a loss of performance and an increase in fuel consumption. A new element should be fitted at the specified intervals and the element should be cleaned every twelve months or more frequently under dusty conditions.

2 Release the spring clips securing the air cleaner lid and remove the lid (photo).

3 Cover the carburettor entry port to prevent any dirt entering it when the element is lifted out, and remove the element. Wipe the inside of the air cleaner with a moist rag to remove all dust and dirt and then remove the covering from the entry port, (photo).

4 If cleaning the element, place well away from the vehicle, then tap the air cleaner element to remove dust and dirt. If necessary use a soft brush to clean the outside or blow air at very low pressure from the inside surface towards the outside.

5 Refit the element, clean the cover and put it in place, then clip the cover down, ensuring that the two arrows are aligned.

3

3.2 One of the air cleaner lid retaining clips

3.3 Lifting out the air cleaner element

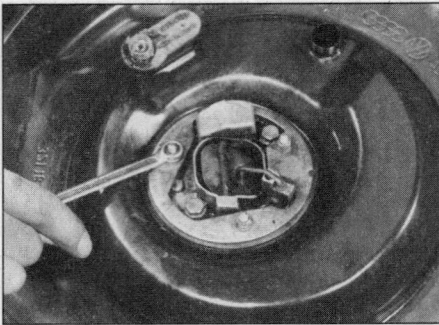

4.2A Unscrew the retaining nuts and . . .

4.2B . . . remove the retaining ring to allow removal of air cleaner body

Fig. 3.6 Air cleaner with element removed (1.6 and 1.8 litre engines) (Sec 5)

A *Thermostat*
B *Intake connection (with regulator flap)*
C *Vacuum unit*

4 Air cleaner – removal and refitting

1 Remove the element, as described in Section 3.
2 Unscrew the nut(s) securing the air cleaner body and remove the adaptor or retaining ring (photos).
3 Note the location of all hoses and tubes, then disconnect them and withdraw the air cleaner body from the carburettor. Remove the sealing ring.
4 Refit in the reverse order of removal ensuring that all hose connections are securely made.

5 Automatic air temperature control – general

1 The air cleaner intake incorporates an automatic temperature control which is operated by vacuum. A flap within the air intake directs warm air from the exhaust manifold, or cold air from the front of the vehicle.
2 To check the control first detach the cold air intake hose from the air cleaner unit and position a mirror at the intake aperture entrance so that the movement of the control flap can be observed.
3 Now detach the hose from the control unit regulator (thermostat) on the underside of the air cleaner unit.
4 By applying suction to the end of this hose, it should operate the control flap. If the control flap fails to operate, check the vacuum hoses for security and condition. Move the regulator flap manually to ensure that it operates freely (Fig. 3.6).
5 To check the temperature regulator (thermostat unit) in the air cleaner body make the above mentioned checks then with the engine cold, start it and allow it to run at idle speed. Detach the hose from the vacuum unit on the intake and check that vacuum is present by placing a finger over the end of the hose. Also check that the regulator flap closes off the warm air intake.
6 Reconnect the vacuum hose and check that the regulator flap fully opens.

7 The position of the regulator flap is controlled by the thermostat in accordance with the ambient temperature within the air cleaner.
8 A further thermostat check can be made on 1.6 and 1.8 models with the engine warmed up and running at idle speed. Check that the regulator flap is fully open, then detach the thermostat/carburettor hose at the carburettor. The regulator flap must resume its initial position within 20 seconds. If defective the thermostat or control valve regulator unit must be renewed.

6 Mechanical fuel pump – testing, removal and refitting

1 The fuel pump is located on the left-hand side of the engine and is operated by an eccentric on the intermediate shaft (four-cylinder engines) or camshaft (five-cylinder engines).
2 If the pump is suspected of malfunctioning, it can be checked in the following way. Disconnect the pump delivery pipe (to filter), position a rag or container beneath the pump delivery pipe connection and disconnect the high tension lead from the ignition coil to prevent the engine firing and the possibility of electrical sparks.
3 With an assistant operating the starter for a few seconds check that the pump is delivering regular spurts of fuel. If satisfactory, refit the pump delivery hose, tighten the

6.6 Fuel pump removal (four-cylinder engine)

clamps. then reconnect the high tension lead to the ignition coil.
4 If the fuel pump is to be removed, first disconnect the battery earth lead.
5 *Ensure that the car is in a well ventilated place and that there is no danger of ignition from sparks or naked flames. Never work on any part of a fuel system when the car is over an inspection pit.*
6 Unclamp and disconnect the fuel lines from the pump. Remove the two securing bolts and lift the pump, flange and gasket off, (photo).
7 The pump is not repairable and must be renewed if defective.
8 Refit the pump using a new gasket in contact with the cylinder block, followed by the flange, sealing ring, and pump. When offering up the pump to the engine, ensure that its lever, or plunger fits correctly against the cam, not under it. Bolt the pump in place and attach the suction (fuel inlet) pipe, which is the pipe from the fuel filter, to the pipe on the cover of the pump.

7 Fuel line filter – renewal

1 A disposable fuel line filter is fitted in the fuel line between the fuel pump and the carburettor on 1.9 litre models or the fuel tank and the fuel pump on 1.6 and 1.8 litre models.

Fig. 3.7 Fuel line filter on 1.6 and 1.8 litre engines (Sec 7)

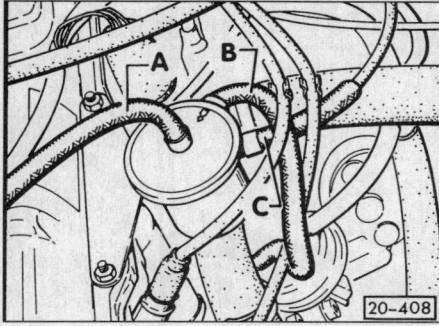

Fig. 3.8 Fuel reservoir (steel type) showing connections (Sec 8)

A To carburettor (not marked)
B From fuel pump (arrow mark)
C To fuel tank (marked F)

2 The fuel filter on the 1.9 litre models does not require renewal during routine maintenance as it is only fitted to collect any particles in the system that may have been left during manufacture. If renewing the fuel line at any time, this filter can then be omitted.

3 To renew the filter unit on the 1.6 and 1.8 litre models disconnect the fuel lines, which are secured with clips. When detaching the fuel lines have the new filter unit ready for immediate fitting to minimise fuel leakage from the disconnected lines. The arrow marking on the filter must face towards the pump.

4 On completion check that the fuel line hoses are secure and that there are no signs of fuel leakage from the connections on restarting the vehicle.

8 Fuel reservoir – removal and refitting

1 The fuel reservoir unit is fitted on 1.6 and 1.8 litre models only and is located between the fuel pump and the carburettor.
2 To remove the reservoir, disconnect the three line hoses and plug them to prevent leakage.
3 Remove the support bracket retaining screws and lift away the reservoir.
4 Refit in the reverse order to removal and then check for any signs of leakage on completion. It should be noted that the reservoir on models fitted with the 1B3 carburettor is manufactured in plastic, whilst on 2E2 carburettor models it is manufactured in metal.

9 Fuel tank – removal, servicing and refitting

Note: *For safety reasons the fuel tank must always be removed in a well ventilated area.*
1 Disconnect the battery negative lead.
2 Remove the fuel tank filler cap.
3 Remove the luggage compartment floor covering and the circular cover from the floor

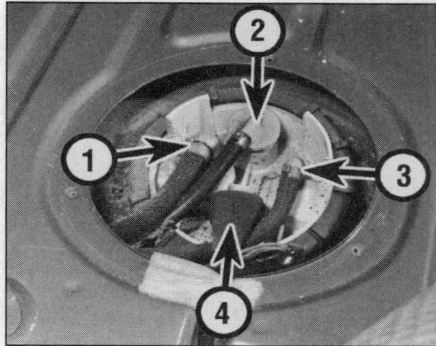

9.4 Fuel tank sender unit and connections (carburettor engine)

1 Suction line *3 Return line*
2 Breather line *4 Wiring connector*

which protects the fuel tank sender unit and connections.
4 Detach the fuel suction and return lines and also the breather line from the sender unit (photo).
5 Pull free and disconnect the sender unit wiring connector.

9.7 Fuel filler pipe connecting hose

6 Jack up the rear of the vehicle and support it on safety stands. Chock the front wheels.
7 Working under the right-hand side wheel arch at the rear, loosen off the connecting hose (filler pipe to tank) retaining clips and detach the connecting hose (photo).
8 Support the tank underneath, making allowance for any fuel remaining in it, and then unscrew the tank retaining strap nuts (Fig. 3.9).

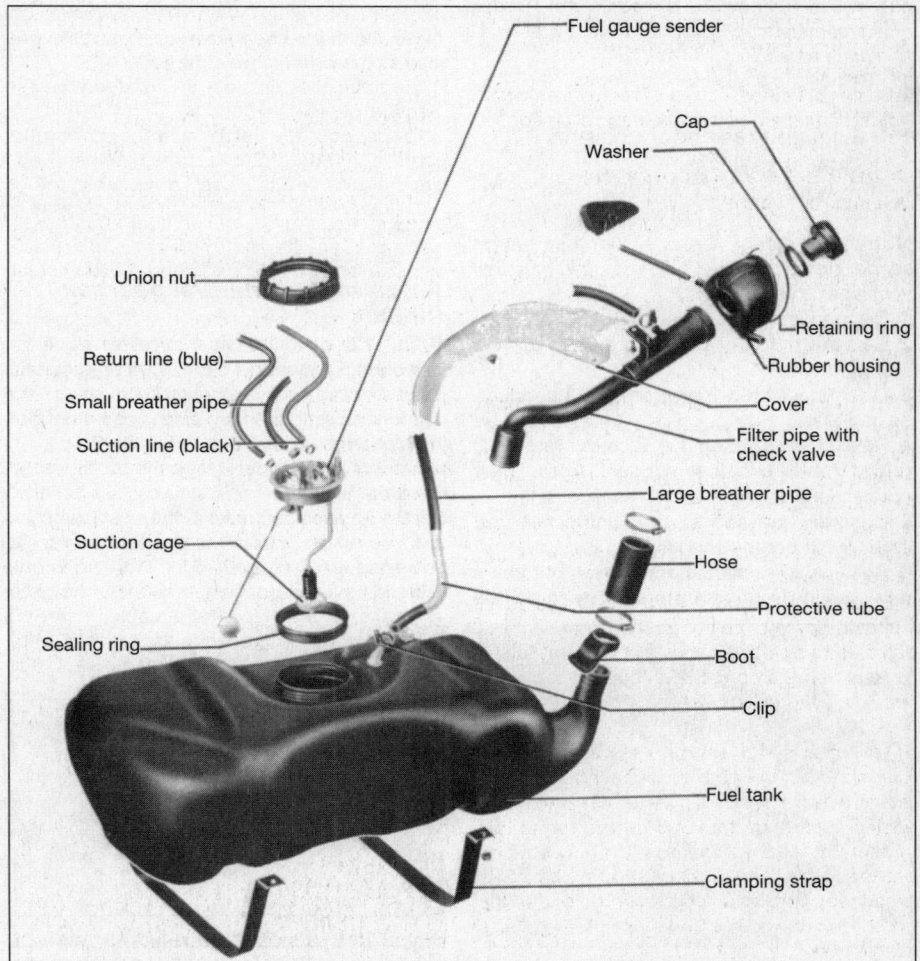

Fig. 3.9 Fuel tank and associated components (Sec 9)

11.2 Disconnecting the throttle cable at the carburettor

Fig. 3.10 Accelerator cable adjustment clip – manual gearbox models (Sec 11)

Fig. 3.11 Disengage the push/pull rods from the rocker linkage (Sec 11)

9 With the straps removed, carefully lower the tank to the point where the large breather line can be disconnected from its connector at the top of the tank. Lower and remove the tank.

10 If the tank is contaminated with sediment or water, remove the gauge sender unit, as described in Section 10 and swill the tank out with clean fuel. If the tank is damaged or leaks, it should be repaired by specialists, or alternatively renewed. **Note:** *Do not, under any circumstances, solder or weld a fuel tank, for safety reasons.*

11 Refitting is a reversal of removal.

10 Fuel gauge sender unit and gravity valve – removal and refitting

Note: *For safety reasons the fuel gauge sender unit must always be removed in a well ventilated area.*

1 Disconnect the battery negative lead.

2 Remove the floor covering in the luggage compartment.

3 Disconnect the wiring from the fuel gauge sender unit.

4 Release the clips and detach the fuel feed, return and breather hoses from their connections on the top of the sender unit.

5 Unscrew the sender unit union nut and carefully lift out the sender unit.

6 Refitting is a reversal of removal, but use a new sealing ring and smear it with a little

glycerine. Ensure that the fuel lines and breather hose connections are correctly made, and are secure. When tightening the union nut ensure that the fuel/breather hose connecting unions face towards the front as shown in photo 9.4.

11 Accelerator cable – removal, refitting and adjustment

Note: *On automatic transmission models refer also to Chapter 7, Figs 7.14 and 7.15.*

1 Remove the air cleaner, as described in Section 14.

2 Disconnect the cable from the carburettor (photo), automatic transmission and accelerator pedal, as applicable, then withdraw the cable. Refitting is a reversal of removal, but make sure that the cable is not kinked. Adjust as follows.

Adjustment – manual gearbox models

3 On 1.6 and 1.8 litre models have an assistant fully depress the accelerator pedal then check that the throttle lever on the carburettor is in the fully open position. If not, reposition the clip shown in Fig. 3.10.

4 On 1.9 litre models, disconnect the cable from the throttle lever at the carburettor then set the accelerator pedal to the idle position so that the distance from the stop on the floor to the pedal is 60.0 mm (2.36 in). With the throttle lever in its idling position, reconnect the cable and take up the slack in the cable. Insert the clip in the nearest groove to the bracket.

5 To check the adjustment on 1.9 litre

models, fully depress the accelerator pedal and check that the distance between the throttle lever and stop is no more than 1.0 mm (0.04 in).

Adjustment – automatic transmission models

1.9 litre engine

6 Run the engine to normal operating temperature to ensure that the choke is fully open, then remove the air cleaner (Section 4).

7 Extract the lock clips from the pull and pushrod balljoints and detach them from the rocker linkage (Fig. 3.11).

8 Actuate the throttle and rotate the automatic choke fast idle cam so that it is clear of the limit screw (Fig. 3.12).

9 Loosen off the ball socket locknut then press the rocker lever against the stop. Rotate the balljoint on the rod and position it so that, when pressed down onto the balljoint, it is not strained in any way, then retighten the locknut.

10 Working inside the car, unscrew and remove the accelerator pedal stop and spacer.

11 Make up a distance piece as shown in Fig. 3.13 using a 135 mm (5.32 in) M8 bolt, and screw it into the pedal stop nut, but do not allow the bolt to touch the accelerator pedal rod.

12 Unscrew and loosen off the locknut on the pushrod (Fig. 3.14).

Fig. 3.12 Automatic choke showing fast idle cam (a) and limit screw (b) (Sec 11)

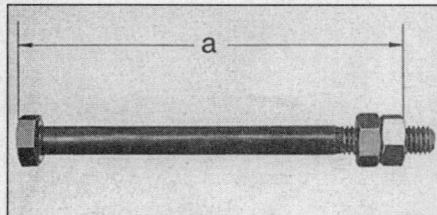

Fig. 3.13 Distance piece required, set nuts at distance (a) (Sec 11)

a = 119 mm (4.68 in)

Fig. 3.14 Pushrod and clamp screw (Sec 11)

Fig. 3.15 Accelerator cable adjuster and locknut (Sec 11)

Fig. 3.16 Gearbox operating lever – move in direction of arrow to idle position (Sec 11)

Fig. 3.17 Hold lever in idle position and tighten pushrod clamp screw (Sec 11)

13 Loosen off the accelerator pedal cable adjuster nut and locknut, (Fig. 3.15).

14 Fit the pushrod ball onto the rocker lever and lock the balljoint.

15 Pivot the gearbox operating lever rearwards to the idle position, then rotate the adjustment nut on the pedal cable so that the cable is engaged without strain whilst leaving the accelerator pedal plate in contact with the bolt (Fig. 3.16).

16 Support the gearbox operating lever in the idle position and retighten the pushrod clamp nut. The bolt can then be removed from underneath the accelerator pedal and the pedal stop and spacer refitted (Fig. 3.17).

17 Check that the throttle opens fully when the accelerator pedal is pressed down to its point of resistance without operating the kickdown. Hold in this position and check that the compensator spring on the pushrod is not compressed and clearance (a) in Fig. 3.18 is as specified.

18 Fully depress the accelerator pedal so that it contacts the pedal stop to compress the compensator spring.

19 When adjusted correctly the throttle must be closed at idle and the gearbox operating lever must be on the stop.

1.6 litre

20 Proceed as described in paragraph 6.

21 Check that the throttle is fully released (closed for normal idle) and that the fast idle speed adjusting screw does not contact with the automatic choke notched disc, but if it does rotate the disc accordingly to correct it.

22 Loosen off the knurled adjuster nut (Fig. 3.19).

23 Loosen off the cable to support bracket locknuts at the inlet manifold and pull the outer cable towards the centre of the engine to take up any play. Do not allow the throttle to open (or the gearbox operating lever).

24 Hand tighten the carburettor side nut against the support bracket and then tighten the engine side lock nut against the opposing side of the support bracket to secure.

25 At the gearbox end, detach the operating lever return spring then get an assistant to fully depress the accelerator pedal against the stop (kickdown position). At the gearbox depress the operating lever in the direction of the arrow (Fig. 3.19) to its stop kickdown position and simultaneously rotate the knurled adjuster nut to take up any play in the cable, then tighten the locknut against the underside of the bracket. Refit the operating lever return spring.

26 Check that the throttle lever is at its idling position and the gearbox operating lever is also at its idle (no throttle) position when the accelerator pedal is fully released.

27 Depress the accelerator pedal to the point where pressure is felt and the throttle is fully open, but without operating the kickdown, then check that the throttle lever is contacting the throttle stop with the kickdown spring uncompressed.

28 Now fully depress the accelerator pedal to the kickdown stop position and check that the gearbox operating lever is in contact with its stop and that the kickdown spring is

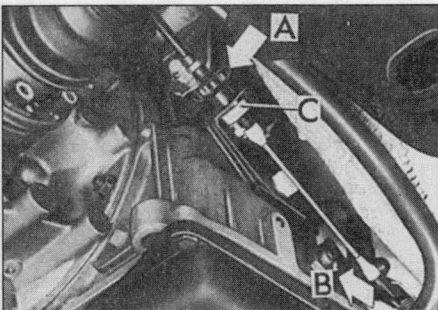

Fig. 3.18 Check clearance (a) as given below (Sec 11)

Full throttle (a) must equal 13.5 mm (0.53 in) Past full throttle (a) must equal 10.5 mm (0.41 in) – compensator spring compressed

compressed the specified amount shown in Fig. 3.20.

1.6 litre with 2E2 carburettor (with overrun cut-off) and 1.8 litre engine

29 Proceed as described in paragraph 6.

30 Loosen off the knurled cable adjuster nut on the side of the transmission.

31 Loosen off the cable nut at the cable locating bracket at the carburettor end.

32 Referring to Fig. 3.21, pivot the warm up lever to the point where the throttle valve operating pin is no longer in contact. Retain

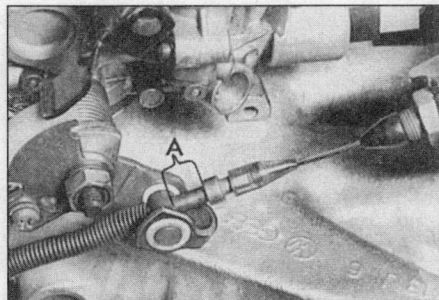

Fig. 3.19 Accelerator pedal cable adjusting nut (A), locknut (C) and operating lever kickdown position direction (B) (Sec 11)

Fig. 3.20 Accelerator cable adjustment check at kickdown position (Sec 11)

A = 10 to 11 mm (0.40 to 0.43 in)

Fig. 3.21 1.6 and 1.8 litre models with automatic transmission – accelerator cable showing 2E2 carburettor connections (Sec 11)

A Warm-up lever
B Throttle valve operating pin
C Connecting lever
D Screwdriver position

Fig. 3.22 Connect up vacuum pump as shown to (A) and seal off connection (B) (Sec 11)

Fig. 3.23 Kickdown spring should compress 8 mm (0.314 in) when measured at gap (a) (Sec 11)

13.2A The 1B3 carburettor cleaned for dismantling

the lever in this position by wedging the connecting lever with a screwdriver as shown.
33 Detach the vacuum hoses from the three-way vacuum connector and then connect up a vacuum pump as shown and seal off and plug connection B (Fig. 3.22).
34 If available use the special Volkswagen vacuum pump shown together with its special tool reference number (VAG 1390).
35 Apply vacuum with the pump so that the diaphragm pushrod is retained in the overrun condition with a gap between the cold idle speed adjusting screw and the diaphragm pushrod.
36 Pull the outer cable away from the carburettor to take up the play ensuring that both the throttle and gearbox operating lever remain in the overrun (closed) position.

37 Hand tighten the adjuster nut (1) then tighten the cable locknut on the opposing (carburettor) side of the support bracket to secure.
38 Proceed as described in paragraphs 25, 26, 27 and 28, but referring to Fig. 3.23 to check the kickdown spring adjustment.

All models
39 Refit the air cleaner as described in Section 4.

12 Carburettor – removal and refitting

1 Disconnect the battery negative lead.
2 Remove the air cleaner, as described in Section 3.

3 Drain half of the coolant from the cooling system, with reference to Chapter 2.
4 Disconnect the coolant hoses from the automatic choke.
5 As applicable, disconnect the wiring from the automatic choke and fuel cut-off solenoid.
6 Disconnect the accelerator cable.
7 Disconnect the fuel and vacuum hoses.
8 Unscrew the through-bolts or nuts, and lift the carburettor from the inlet manifold. Remove the insulating flange gasket.
9 Refitting is a reversal of removal, but clean the mating faces of the carburettor and inlet manifold and always fit a new gasket. Tighten the mounting bolts evenly to the specified torque.

13 Carburettor – dismantling, servicing and reassembly

1 With the carburettor removed from the inlet manifold as described in the previous Section, wash it externally with a suitable solvent and allow to dry.
2 Dismantle and reassemble the carburettor, with reference to Figs. 3.24, 3.25 and 3.26, having first obtained a repair set of gaskets. Before dismantling the automatic choke, note the position of the cover in relation to the carburettor cover. See accompanying photos for dismantling of the 1B3 carburettor.
3 Before removing the respective jets note their locations and note that the air correction

13.2B Unscrew the retaining bolts . . .

13.2C . . . and remove the upper body

13.2D Upper section inverted for inspection/further dismantling as necessary

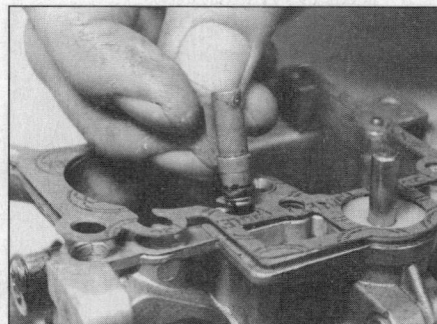

13.2E Injection tube removal

13.2F Always renew the gasket on reassembly

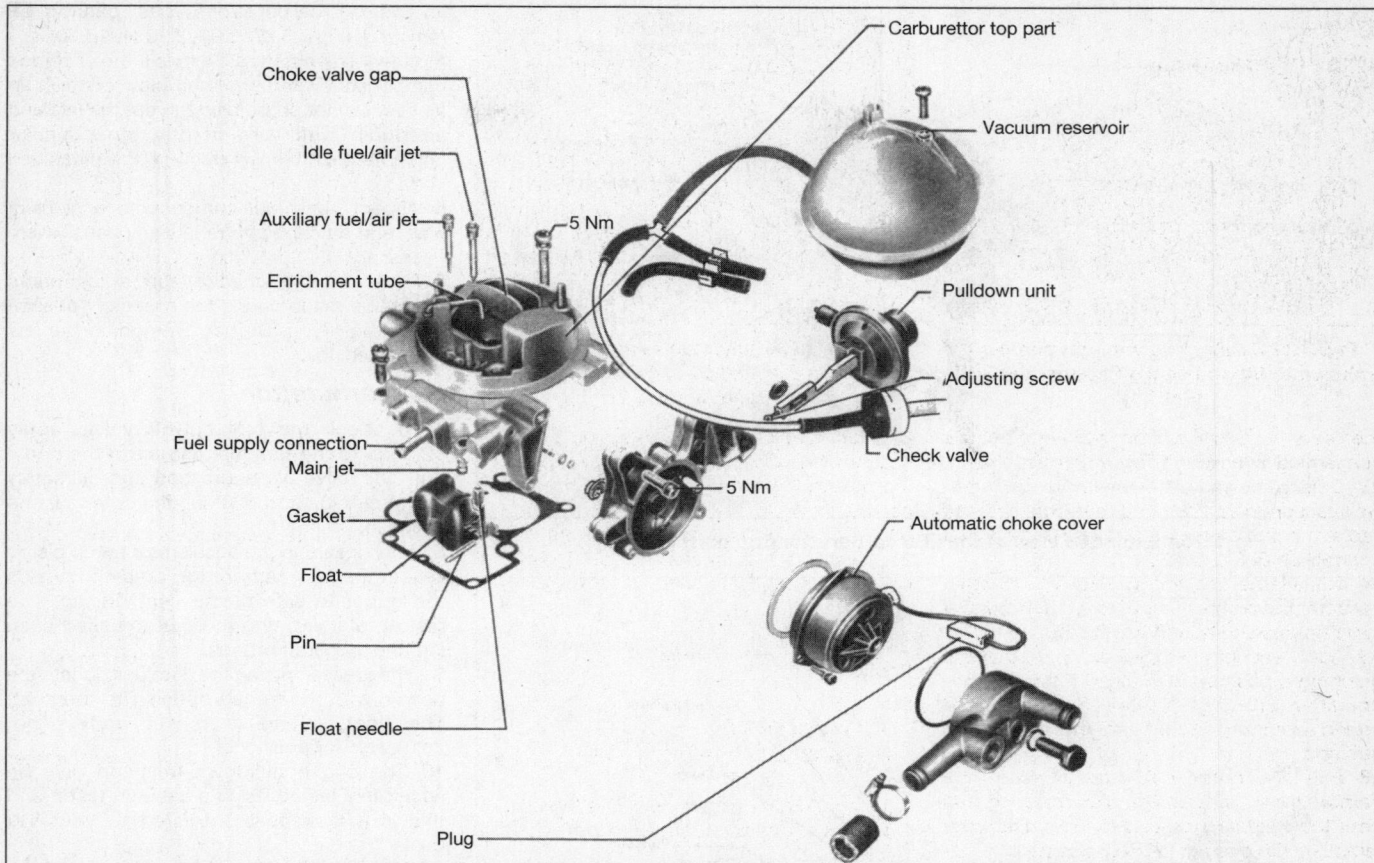

Fig. 3.24A Exploded view of the 1B3 carburettor (top part) (Sec 13)

Choke valve gap
Idle fuel/air jet
Auxiliary fuel/air jet
Enrichment tube
Fuel supply connection
Main jet
Gasket
Float
Pin
Float needle
Plug
Carburettor top part
Vacuum reservoir
5 Nm
Pulldown unit
Adjusting screw
Check valve
5 Nm
Automatic choke cover

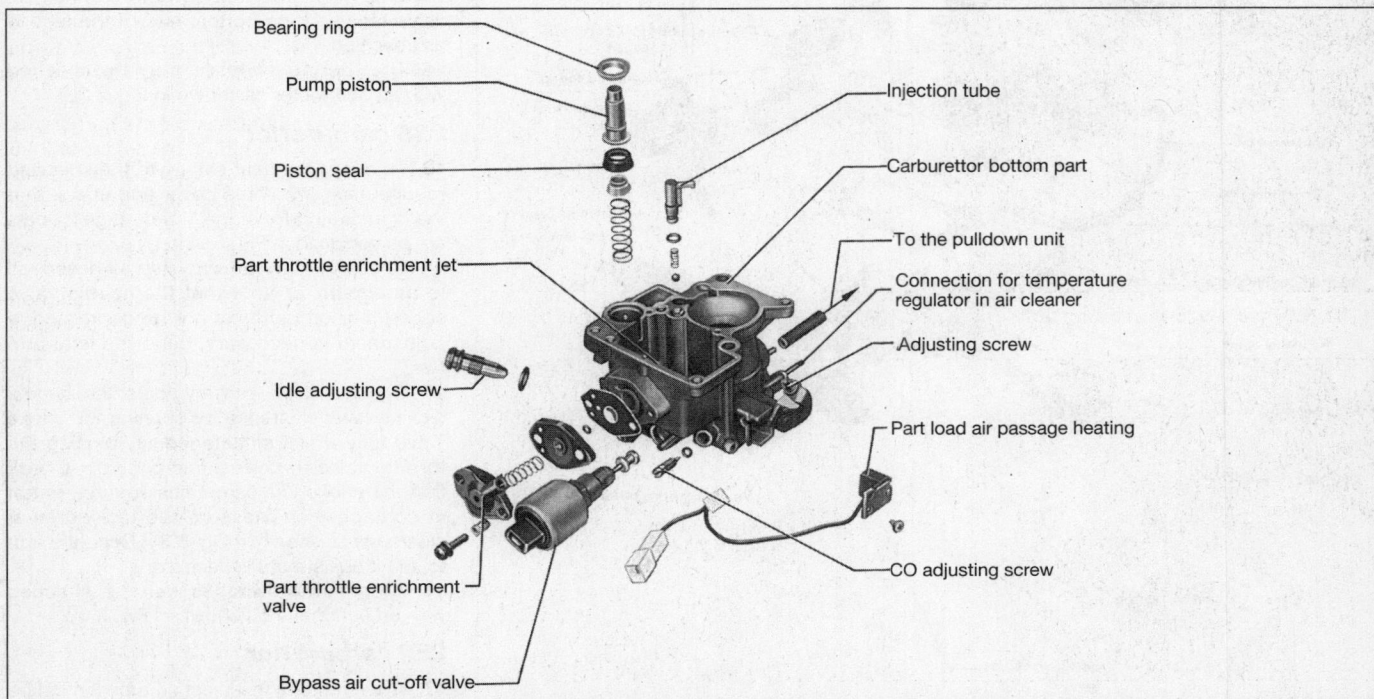

Fig. 3.24B Exploded view of the 1B3 carburettor (bottom part) (Sec 13)

Bearing ring
Pump piston
Piston seal
Part throttle enrichment jet
Idle adjusting screw
Part throttle enrichment valve
Bypass air cut-off valve
Injection tube
Carburettor bottom part
To the pulldown unit
Connection for temperature regulator in air cleaner
Adjusting screw
Part load air passage heating
CO adjusting screw

3

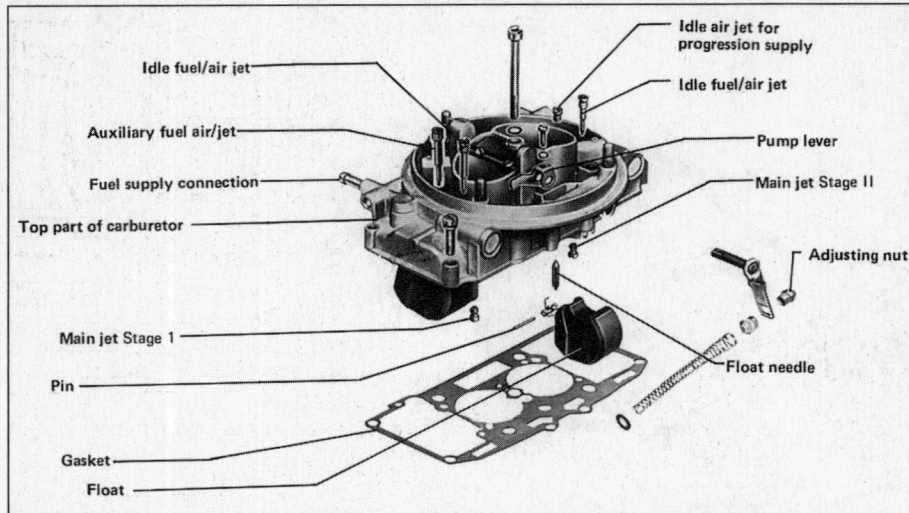

Fig. 3.25A Exploded view of the 2B5 carburettor (top part) (Sec 13)

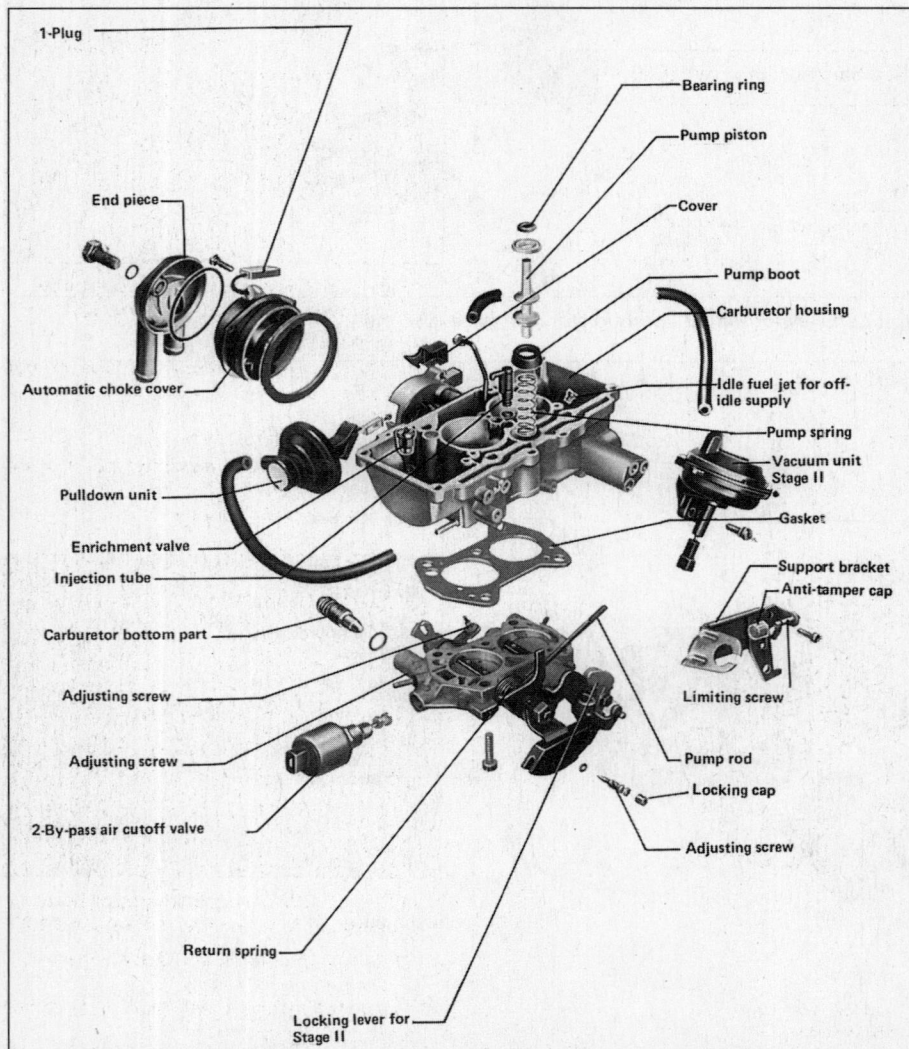

Fig. 3.25B Exploded view of the 2B5 carburettor (bottom part) (Sec 13)

jet on all carburettor types cannot be removed, (Figs. 3.27, 3.28, 3.29 and 3.30).

4 When dismantled, clean the various components with petrol and blow dry with an air line. Do not probe or clean out the jets and apertures with wire or any other similar implement as this will damage the machined surfaces.

5 Inspect the various components for signs of wear and damage and renew any parts where necessary.

6 The following checks and adjustments should be made during the assembly of each carburettor type. Do not overtighten the jets and fastenings.

1B3 carburettor

7 To check the fuel cut-off valve, apply 12 volts to the terminal and earth the body. With the valve pin depressed approximately 3 to 4 mm (0.12 to 0.16 in), the core must be pulled in.

8 When inserting the accelerator pump piston seal in the 1B3 carburettor, press it towards the opposite side of the vent drilling. The piston retaining ring must be pressed flush into the carburettor body.

9 When reassembling the float unit, locate the needle with the forced opening to the float. The float setting is preset and is not adjustable for position.

10 The choke pulldown unit can only be accurately tested using a vacuum tester and this is a task best entrusted to your VW dealer.

11 When refitting the enrichment tube check that its setting clearance, (a) in Fig. 3.31, which is measured between the upperchoke valve face and the bottom end of the tube is as specified.

12 The thermo-switches may be checked with an ohmmeter as shown in Fig. 3.32.

2B5 carburettor

13 To adjust the float setting to the specified requirement, invert the cover and check that the float setting for stage 1 and stage 2 floats are as specified in Fig. 3.33.

14 Check the fuel cut-off valve as described in paragraph 7. Note that the needle valve spring tension pin must not be depressed. If adjustment is necessary, bend the float arm tab.

15 Check and if necessary adjust the stage 2 locking lever clearance by opening the choke valve fully whilst simultaneously closing the throttle valve to normal idle position. Check that the choke idle cam adjuster screw is not in contact, then measure the locking lever clearance as shown in Fig. 3.34. Bend the arm to suit if adjustment is necessary.

16 The thermo-switches can be checked with an ohmmeter as shown in Fig. 3.32.

2E2 carburettor

17 When refitting the injection tube it must be correctly positioned so that fuel is sprayed in line with the recess, shown in Fig. 3.35.

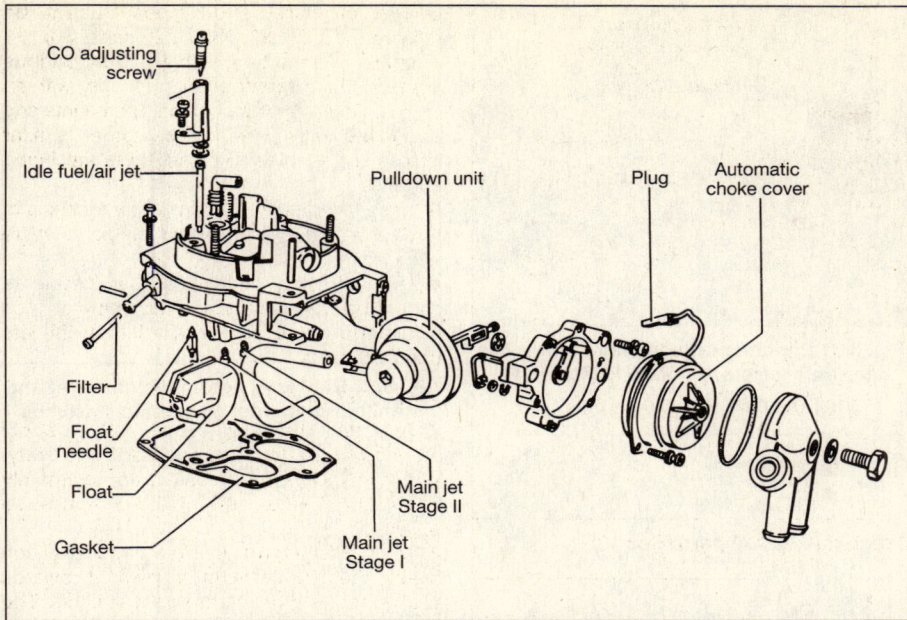

Fig. 3.26A Exploded view of the 2E2 carburettor (top part) (Sec 13)

Labels: CO adjusting screw, Idle fuel/air jet, Pulldown unit, Plug, Automatic choke cover, Filter, Float needle, Float, Gasket, Main jet Stage II, Main jet Stage I

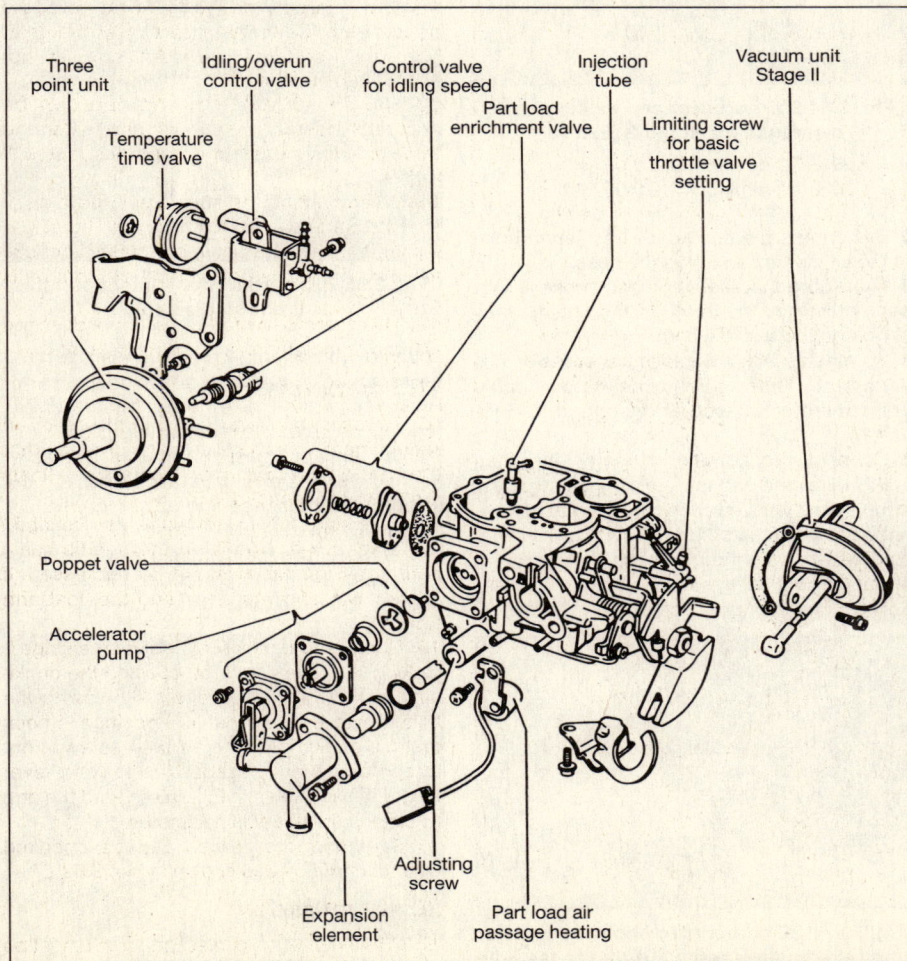

Fig. 3.26B Exploded view of the 2E2 carburettor (bottom part) (Sec 13)

Labels: Three point unit, Idling/overrun control valve, Control valve for idling speed, Injection tube, Vacuum unit Stage II, Temperature time valve, Part load enrichment valve, Limiting screw for basic throttle valve setting, Poppet valve, Accelerator pump, Adjusting screw, Expansion element, Part load air passage heating

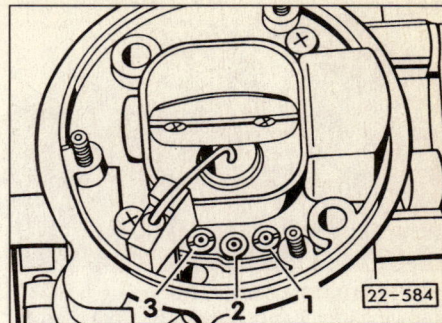

Fig. 3.27 1B3 carburettor upper body showing jet locations (Sec 13)

1 Idle fuel/air jet
2 Air correction jet and emulsion tube (do not remove)
3 Auxiliary fuel/air jet

Fig. 3.28 2B5 carburettor upper body showing jet locations (Sec 13)

1/2 Auxiliary fuel/air jet
3 Air correction jet and emulsion tube (do not remove) – Stage 1
4 Idle fuel/air jet – Stage 1
5 Idle fuel/air jet – Stage 2
6 Air correction jet and emulsion tube (do not remove) - Stage 2
7 Idle air jet (off/idle supply)

Fig. 3.29 2E2 carburettor upper body showing jet locations (top side) (Sec 13)

1 Idle fuel/air jet (beneath CO adjustment screw guide tube)
2 Air correction jet and emulsion tube (do not remove) – Stage 1
3 Air correction jet and emulsion tube (do not remove) – Stage 2

3

Fig. 3.30 2E2 carburettor upper body showing jet locations (bottom side) (Sec 13)

1 Main jet (stage 1)
2 Main jet (Stage 2)
3 Full load enrichment feed pipe
4 Progression feed pipe (Stage 2)

Fig. 3.33 2B5 carburettor – float settings (Sec 13)

Stage I a = 28 ± 1 mm (1.10 ± 0.04 in)
Stage II a = 30 ± 1 mm (1.18 ± 0.04 in)

All models

18 When the carburettor is reassembled and refitted, refer to Section 14 for the necessary adjustments.

14 Carburettor adjustments

1B3 carburettor

Idling speed

1 Run the engine to normal operating temperature and switch off all electrical components.

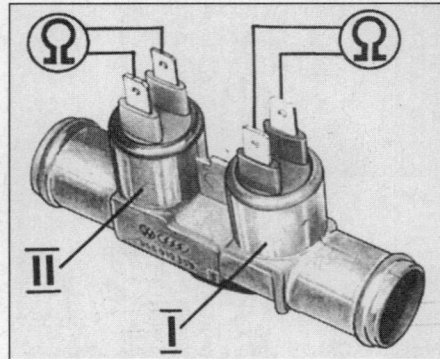

Fig. 3.36 1B3 carburettor – idle mixture adjusting screw location (Sec 14)

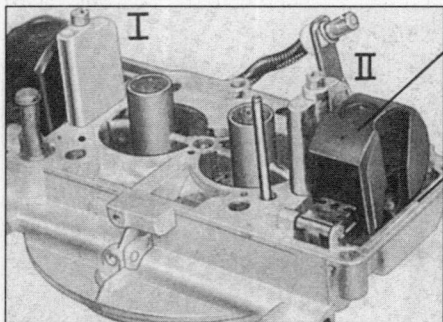

Fig. 3.31 Enrichment tube alignment – 2B3 carburettor from October 1981 (Sec 13)

a = 1.0 ± 0.03 mm (0.04 ± 0.001 in)

Fig. 3.34 2B5 carburettor – locking lever clearance for Stage 2 (Sec 13)

A = 1.00 ± 0.2 mm (0.039 ± 0.007 in)
B = 1.20 ± 0.2 mm (0.047 ± 0.007 in)

2 Disconnect the crankcase ventilation hose at the air cleaner and plug the hose.
3 Make sure that the automatic choke is fully open otherwise the throttle valve linkage may still be on the fast idle cam.
4 On models with automatic transmission it is important that the accelerator cable adjustment is correct as described in Section 11.
5 Connect a tachometer to the engine, then start the engine and let it idle. Check that the idling speed is as given in the Specifications – note that the radiator fan must not be running. If necessary, turn the idling adjusting screw in or out until the idling speed is correct (photo).
6 The idle mixture adjustment screw is

Fig. 3.37 1B3 carburettor showing the idle cam (A) adjusting screw (B) for the fast idle setting (Sec 14)

Fig. 3.32 Checking automatic choke (I) and manifold preheater (II) thermo-switches (Sec 13)

I Below 30°C (86°F) – zero ohm
II Below 50°C (122°F) – zero ohm
Above 40°C (104°F) – infinity
Above 55°C (131°F) – infinity
Note: Where the thermo-switches are individual red the automatic choke thermo-switch is coloured

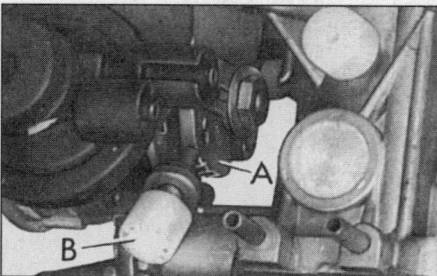

Fig. 3.35 2E2 carburettor – injection tube direction (Sec 13)

covered with a tamperproof cap which must be removed in order to adjust the mixture. However first make sure that current regulations permit its removal (Fig. 3.36).
7 If an exhaust gas analyser is available, connect it to the exhaust system, then run the engine at idling speed and adjust the mixture screw to give the specified CO content percentage. Turn the mixture screw clockwise to weaken the mixture and anti-clockwise to enrich the mixture. Alternatively as a

14.5 Idle speed adjustment – 1B3 carburettor

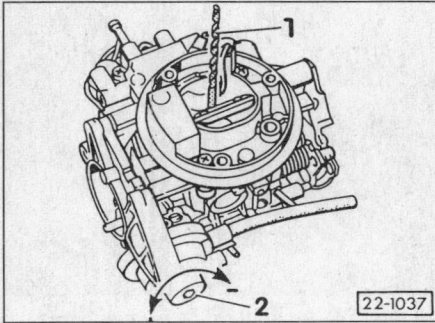

Fig. 3.38 Choke valve gap check using drill shank (1) adjust using socket-head bolt (2) (Sec 14)

Fig. 3.39 Idle limiting screw (C) – 1B3 carburettor (Sec 14)

Fig. 3.40 Accelerator pump adjustment on the 1B3 carburettor (Sec 14)

a Locking screw   b Cam plate

temporary measure adjust the mixture screw to give the highest engine speed, then readjust the idling speed if necessary.

8 After making the adjustment, fit a new tamperproof cap, and reconnect the crankcase ventilation hose.

Fast idling speed

9 With the engine at normal operating temperature and switched off, connect a tachometer and remove the air cleaner.

10 Fully open the throttle valve, then turn the fast idle cam and release the throttle valve so that the adjustment screw is positioned on the highest part of the cam (Fig. 3.37).

11 Without touching the accelerator pedal, start the engine and check that the fast idling speed is as given in the Specifications. If not, turn the adjustment screw on the linkage as necessary. If a tamperproof cap is fitted, renew it after making the adjustment.

Choke pull-down system

12 Remove the air cleaner cover, as described in Section 3.

13 Half open the throttle valve then completely close the choke valve.

14 Without touching the accelerator pedal, start the engine.

15 Close the choke valve by hand and check that resistance is felt over the final 4 mm (0.16 in) of travel. If no resistance is felt there may be a leak in the vacuum connections or the pull-down diaphragm may be broken.

16 Further checking of the system requires the use of a vacuum pump and a gauge, therefore this work should be entrusted to your VW dealer.

Choke valve gap

17 The choke valve gap measurement and adjustment points are shown in Fig. 3.38 for reference purposes only; the use of a vacuum tester and gauge is required, so this task is best left to a VW dealer.

Throttle valve basic setting

18 This setting is made during manufacture and will not normally require adjustment. However, if the setting has been disturbed proceed as follows.

19 First run the engine to normal operating temperature.

20 Remove the air cleaner, as described in Section 3.

21 Disconnect the vacuum advance hose at the carburettor and connect a vacuum gauge.

22 Run the engine at idling speed, then turn the idle limiting screw on the lever until vacuum is indicated on the gauge. Turn the screw out until the vacuum drops to zero, then turn it out a further quarter turn, (Fig. 3.39).

23 After making the adjustment, adjust the idling speed as described in paragraphs 1 to 8.

Electric bypass air heating element

24 Disconnect the wiring from the fuel cut-off solenoid and thermoswitch, and connect a test lamp to the heating element wire and the battery positive terminal.

25 If the lamp lights up, the heater element is in good working order.

Accelerator pump

26 Hold the carburettor over a funnel and measuring glass.

27 Turn the fast idle cam so that the adjusting screw is off the cam. Hold the cam in this position during the following procedure.

28 Fully open the throttle ten times, allowing at least three seconds per stroke. Divide the total quantity by ten and check that the resultant injection capacity is as given in the Specifications. If not, refer to Fig. 3.40 and loosen the cross-head screw, turn the cam plate as required, and tighten the screw.

29 If difficulty is experienced in making the adjustment, check the pump seal and make sure that the return check valve and injection tube are clear.

Automatic choke

30 The line on the cover must be in alignment with the dot on the automatic choke body.

Inlet manifold preheater

31 Using an ohmmeter as shown in Fig. 3.41 check that the resistance of the preheater is

between 0.25 and 0.50 ohms. If not, renew the unit.

2B5 carburettor

Automatic choke

32 The line on the cover must be in alignment with the dot on the automatic choke body.

Idling speed

33 The procedure is identical to that described in paragraphs 1 to 8 inclusive.

Fast idling speed

34 The procedure is as given in paragraphs 9 to 11 inclusive. However, access to the adjustment screw is difficult when it is on the highest part of the cam, so it is necessary to move the cam in order to turn the screw. After making the adjustment. seal the screw with a spot of paint or fit a new tamperproof cap, as necessary.

Choke valve gap

35 To check and adjust the choke valve gap necessitates the use of a vacuum tester and gauge, and in view of this it is a task best entrusted to your VW dealer.

Choke pull-down system

36 The procedure is identical to that described in paragraphs 12 to 16 inclusive.

Accelerator pump

37 The procedure is identical to that

Fig. 3.41 Checking the inlet manifold preheater resistance (Sec 14)

3

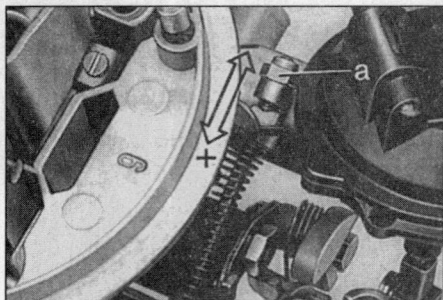

Fig. 3.42 Accelerator pump adjustment nut (a) on the 2B5 carburettor (Sec 14)

Fig. 3.43 Stage I throttle valve basic setting limit screw (A) and anti-tamper cap (B) – 2B5 carburettor (Sec 14)

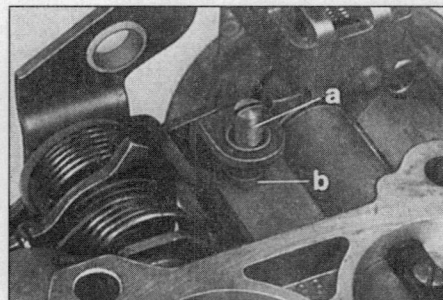

Fig. 3.44 Stage II throttle valve basic setting limit screw (a) and stop (b) – 2B5 carburettor (Sec 14)

described in paragraphs 26 to 29 inclusive. Refer to Fig. 3.42 for adjustment.

Stage 1 throttle valve basic setting

38 This setting is made during manufacture and will not normally require adjustment. However, if the setting has been disturbed proceed as follows. First run the engine to normal operating temperature.

39 Remove the air cleaner, as described in Section 3.

40 Make sure that the fast idle adjusting screw is not contacting the fast idle linkage.

41 Unscrew the basic throttle setting screw until it no longer contacts the linkage, (Fig. 3.43).

42 Open and close the throttle quickly, then turn the setting screw until it just contacts the

Fig. 3.45 Diaphragm pushrod (A) and fast idle adjustment screw (B) – 2E2 carburettor (Sec 14)

Fig. 3.46 Mixture (CO) adjustment screw (B) and control valve idle speed adjustment screw (A) – 2E2 carburettor (Sec 14)

linkage. From this position screw it in a further quarter turn.

43 Fit a tamperproof cap if necessary, then adjust the idling speed, as described in paragraph 33.

Stage 2 throttle valve basic setting

44 This setting is made during manufacture and will not normally require adjustment. However, if the setting has been disturbed proceed as follows. First remove the carburettor.

45 With the Stage 1 throttle valve in the idling position, unscrew the Stage 2 basic throttle setting screw until it no longer contacts the stop (Fig. 3.44).

46 Disconnect the vacuum unit pullrod then turn the setting screw until it just contacts the stop while lightly pressing the throttle lever shut. From this position screw it in a further quarter turn.

47 Fit a new tamperproof cap if necessary, then reconnect the vacuum unit pullrod.

48 Adjust the idling speed, as described in paragraph 33.

2E2 carburettor

Idling speed

49 The procedure is basically the same as that described in paragraphs 1 to 8 inclusive, but note the following.

(a) Check that the fast idle adjustment screw (B in Fig. 3.45) is just making contact with the diaphragm pushrod (A)

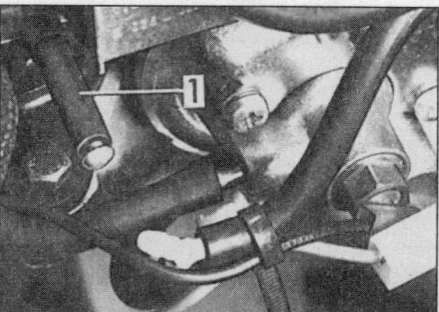

Fig. 3.47 Disconnect and plug vacuum hose (1) for fast idle speed adjustment – 2E2 carburettor (Sec 14)

(b) The idle speed and mixture (CO) adjustment screws are indicated in Fig. 3.46.

Fast idling speed

50 With the engine at normal operating temperature and switched off, connect a tachometer and remove the air cleaner. Plug off the temperature regulator vacuum hose.

51 Detach the Y-piece from the vacuum hose and plug the hose, (Fig. 3.47).

52 Connect up a tachometer to the engine. Start and run the engine and check that the fast idle speed is as given in the Specifications, if not turn the adjustment screw on the linkage as necessary (Fig. 3.45). On completion apply sealant to the screw threads to lock it in position.

Choke pull-down system

53 The procedure is identical to that described in paragraphs 12 to 16 inclusive.

Choke valve gap

54 To check and adjust the choke valve gap necessitates the use of a vacuum tester and gauge and in view of this it is a task best entrusted to your VW dealer.

Stage 1 throttle valve basic setting

55 Proceed as described in Section 11, paragraphs 32 to 35 inclusive. Loosen off the limiting screw to give a clearance between it and the stop, then screw it in so that it just touches the stop. From this point screw it in a further quarter turn. Secure screw with locking compound (Fig. 3.48).

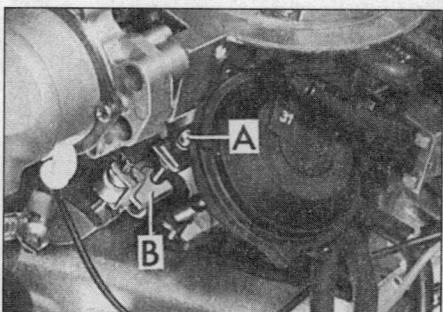

Fig. 3.48 Stage I throttle valve setting limiting screw (A) and stop (B) – 2E2 carburettor (Sec 14)

Fig. 3.49 Throttle valve basic setting showing rod to hold valve open (arrowed), lock lever (1), limiting screw (2) and stop (3) (Sec 14)

Stage 2 throttle valve basic setting

56 This is made during manufacture and will not normally require adjustment. However if the setting has been disturbed proceed as follows. First remove the carburettor (Section 12).

57 Referring to Fig. 3.49, open the throttle valve and hold in this position by inserting a wooden rod or similar implement between the valve and the venturi.

58 Using a rubber band as shown, pre-tension the Stage 2 throttle valve locking lever, then unscrew the limiting screw to provide a clearance between the stop and limiting screw.

59 Now turn the limiting screw in so that it is just in contact with the stop. The limiting screw stop point can be assessed by inserting a piece of thin paper between the screw and stop, moving the paper as the limiting screw is tightened. With the stop point reached turn the screw in a further quarter of a turn then secure it with locking compound. Close both throttle valves then measure the locking lever clearances, A and B in Fig. 3.50. If the clearances are not as specified, bend the arm as necessary.

Accelerator pump

60 To make this check the carburettor must be removed and you will need a vacuum pump and an M8 x 20 mm bolt.

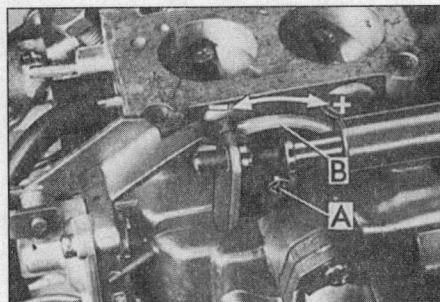

Fig. 3.52 2E2 carburettor showing accelerator pump adjustment points (Sec 14)

A Screw   B Cam plate

Fig. 3.50 Lock lever clearance with throttle valves closed – 2E2 carburettor (Sec 14)

A 0.40±0.1mm (0.015±0.004 in)
B 0.15±0.1mm (0.006±0.004 in)

61 Detach the vacuum hoses from the three point unit then connect up the vacuum pump to the three point unit at A shown in Fig. 3.22 and plug connection B. Apply vacuum with the pump to hold the diaphragm pushrod in the overrun/cut-off position and give a clearance between the fast idle speed and diaphragm pushrod.

62 Pivot up the warm up lever to the point where the throttle valve control pin has clearance and insert the M8 x 20 mm bolt as shown to hold the warm up lever in this position (Fig. 3.51).

63 Hold the carburettor over a funnel and measuring glass then slowly open the throttle valve lever fully five times allowing at least three seconds per stroke. Divide the total quantity by five and check the resultant injection capacity against that given in the Specifications.

64 If adjustment is necessary, refer to Fig. 3.52, loosen off screw A and rotate the cam plate B in the required direction to increase or decrease the injection capacity. On completion, retighten the screw and seal in position with locking compound.

65 The accelerator pump injection capacity can also be checked with the carburettor in position in the vehicle, but as specialised equipment is required this is a task best entrusted to your VW dealer.

Fig. 3.53 2E2 carburettor ready for three-point unit check (Sec 14)

Pushrod to idle point a = 15 mm (0.60 in)
For item 1 refer to text

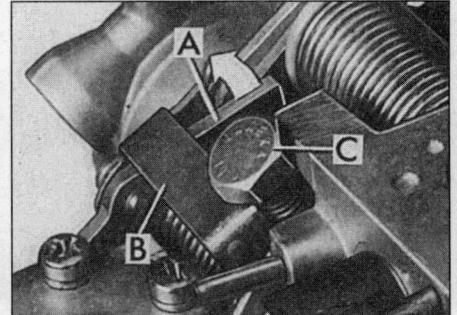

Fig. 3.51 Accelerator pump adjustment check showing warm-up lever (A), lever (B) and bolt (C) – 2E2 carburettor (Sec 14)

Three point unit

66 To check the unit for satisfactory operation you will need a vacuum pump. Detach the vacuum hoses from the unit and attach the vacuum pump to connection 1 in Fig. 3.53. Apply vacuum to pull the diaphragm pushrod to the idle point and then measure the amount of rod protrusion, (a) in Fig. 3.53. which must be as specified.

67 To check the overrun cut-off point, plug off the vacuum connection, B in Fig. 3.22, then apply increased vacuum with the vacuum pump. This should cause the diaphragm pushrod to move to the overrun/cut-off point.

68 Measure the rod protrusion as in paragraph 66. This should be about 10 mm. Hold in this position for a minimum period of one minute. If the rod protrusion is incorrect, or it will not hold for the specified period, then the diaphragm or three point unit probably leak. They are therefore in need of renewal.

Idle/overrun control valve

69 To check this, specialised test equipment is required and it should therefore be entrusted to your VW dealer.

Temperature time valve

70 To check this, specialised test equipment is required and it should therefore be entrusted to your VW dealer.

Automatic choke

71 The line on the cover must be in alignment with the dot on the automatic choke body (photo).

14.71 Automatic choke cover alignment on the 2E2 carburettor

15 Inlet and exhaust manifolds – removal and refitting

1 Remove the air cleaner unit as described in Section 4.
2 Remove the carburettor as described in Section 12.
3 Disconnect the servo unit vacuum hose.
4 Disconnect and detach the coolant hose at the inlet manifold (photo).
5 Disconnect the inlet manifold preheater lead from the base of the inlet manifold (Fig. 3.54).
6 Unbolt and detach the exhaust stabilizer bracket.

7 Unscrew and remove the six Allen screws and detach the manifold.
8 Undo the six nuts securing the exhaust manifold to the down pipe and separate the two.
9 Remove the exhaust manifold to cylinder head nuts (with flat washers) and withdraw the manifold.
10 Refit in the reverse order to removal. Ensure that the mating faces are clean and use new gaskets when assembling.
11 Where given, tighten the fastenings to the specified torque wrench setting.
12 On completion check that the inlet manifold preheater cable is securely connected and located so that it will not fray against adjacent components.

16 Exhaust system – removal and refitting

1 The exhaust systems are shown in Figs. 3.55 and 3.56.
2 Before doing any dismantling work on the exhaust system, wait until the system has cooled down and then saturate the fixing bolts and joints with a proprietary anti-corrosion fluid.
3 When refitting the system, new nuts and bolts should be used, and it may be found easier to cut through the old bolts with a hacksaw, rather than unscrew them.
4 When renewing any part of the exhaust system, it is usually easier to undo the manifold-to-front pipe joint and remove the complete system from the car, then separate the various pieces of the system, or cut out the defective part, using a hacksaw.
5 Refit the system a piece at a time, starting with the front pipe. Use a new joint gasket and note that it has a flanged side, the flanged side should face towards the exhaust pipe (photo).

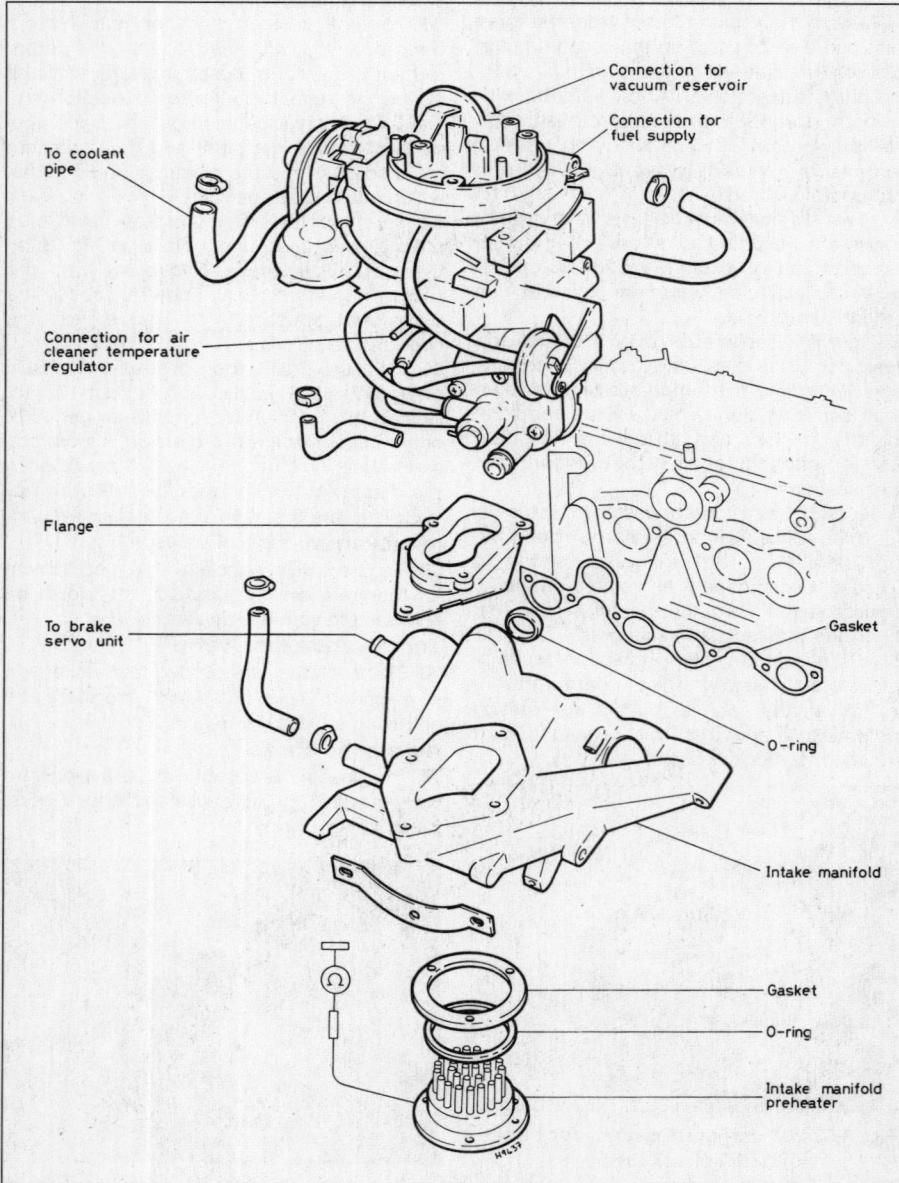

15.4 Inlet and exhaust manifolds shown with carburettor removed. Note coolant hose connections to inlet manifold

16.5 Exhaust front pipe to manifold connection – always renew gasket

Connection for vacuum reservoir

Connection for fuel supply

To coolant pipe

Connection for air cleaner temperature regulator

Flange

To brake servo unit

Gasket

O-ring

Intake manifold

Gasket

O-ring

Intake manifold preheater

Fig. 3.54 Inlet manifold components (as used with the 2E2 carburettor) (Sec 15)

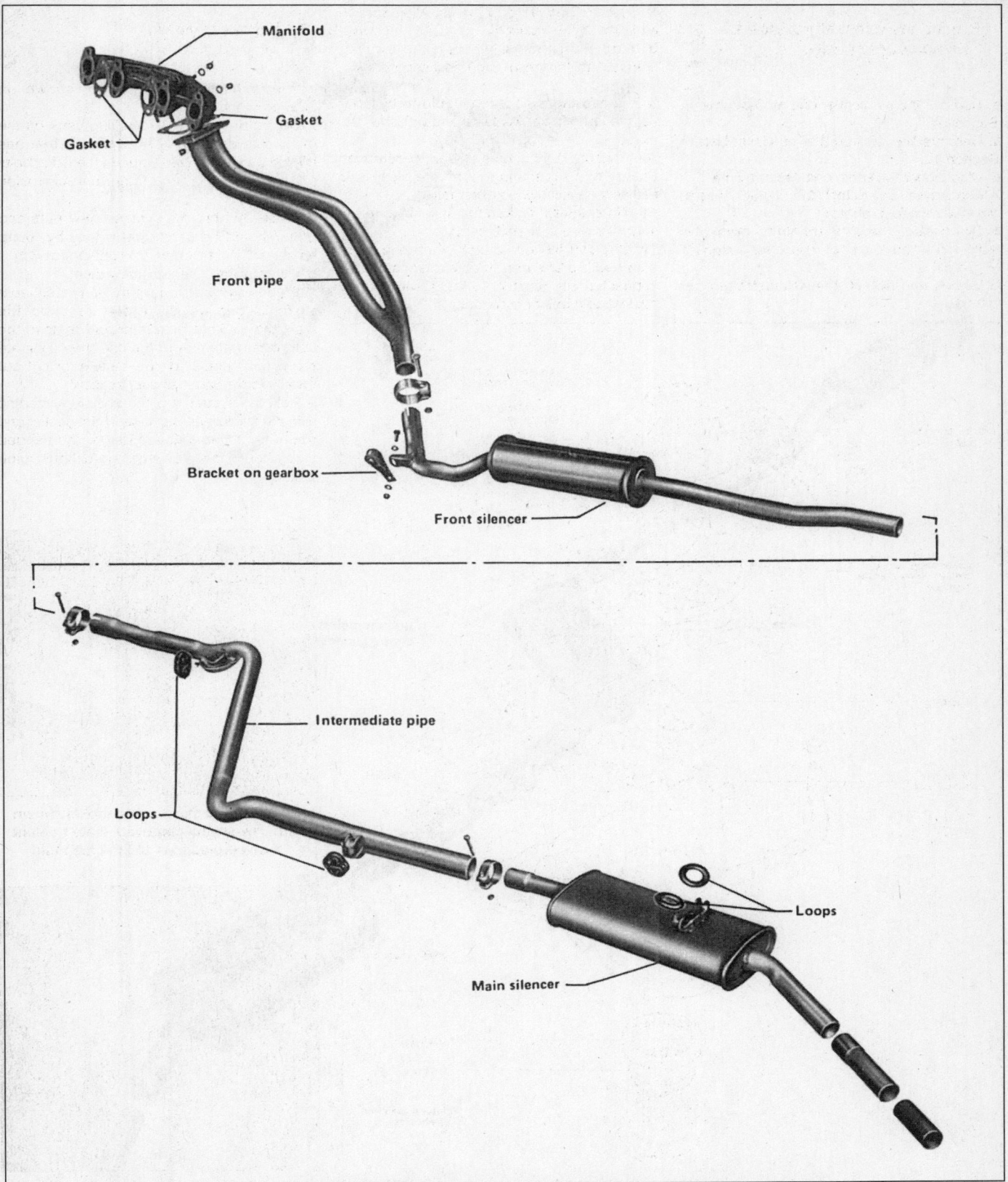

Manifold

Gasket

Gasket

Front pipe

Bracket on gearbox

Front silencer

Intermediate pipe

Loops

Main silencer

Loops

Fig. 3.55 Exhaust system components – four-cylinder engines (Sec 16)

3

Gasket

EGR valve

Warm air deflector plate

Gasket
The beaded edges must face
towards the front exhaust pipe

Gasket

Exhaust manifold

Front exhaust pipe

Front silencer

Seal

Centre silencer

Rubber mount –
centre silencer

Main silencer

Rubber mount –
main silencer

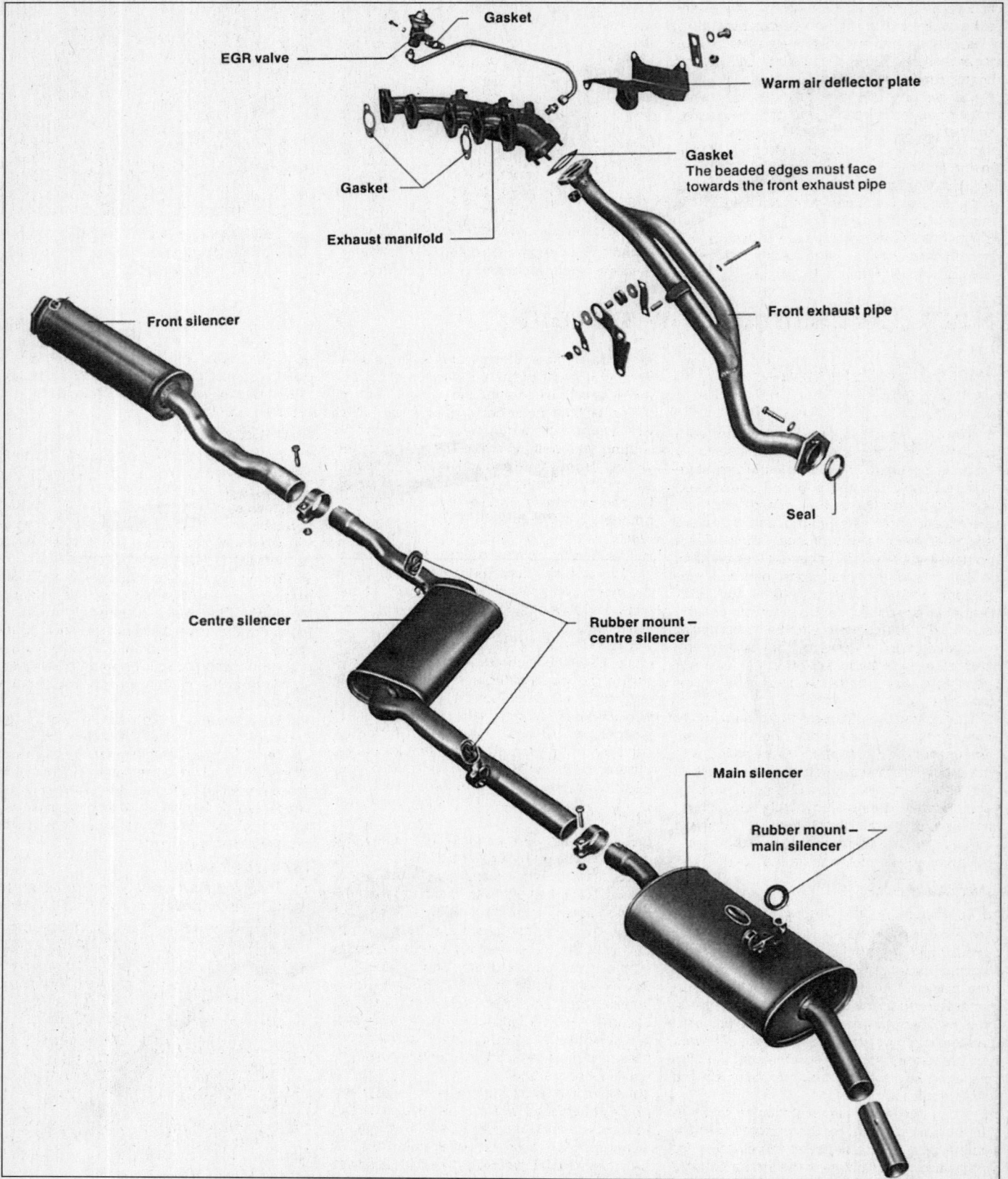

Fig. 3.56 Exhaust system components – five-cylinder engines (Sec 16)

6 Smear all the joints with a proprietary exhaust sealing compound before assembly. This makes it easier to slide the pieces to align them and ensures that the joints will be gas tight. Leave all bolts loose (photo) .

7 Run the engine until the exhaust system is at normal temperature and then, with the engine running at idling speed, tighten all the mounting bolts and clips. starting at the manifold and working towards the rear silencer. *Take care to avoid touching any part of the system with bare hands because of the danger of painful burns.*

8 When the bolts and clips are tightened, it is important to ensure that there is no strain on any part of the system.

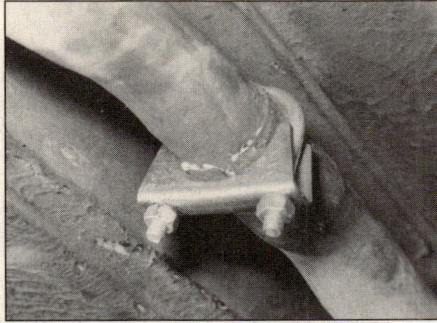

16.6A Typical pipe joint and clamp – use sealing compound when assembling joint

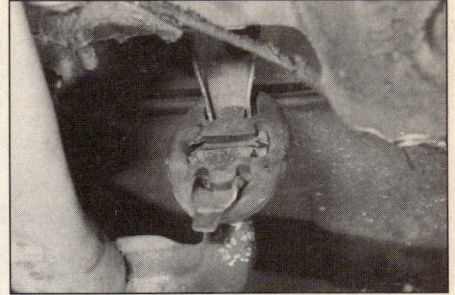

16.6B Exhaust system mounting strap must be in good condition and correctly located

Part B: Fuel injection system

17 Fuel injection system – general description

1 Although the fuel injection system is known as the 'K-Jetronic', it is not an electronically controlled system. The principle of the system is very simple and there are no specialised electronic components. There is an electrically driven fuel pump and there are electrical sensor and switches, but these are no different from those in general use on cars.

2 The following paragraphs describe the system and its various elements. Later Sections describe the tests which can be carried out to ascertain whether a particular unit is functioning correctly, but dismantling and repair procedures of units are not generally given because repairs are not possible.

3 The system measures the amount of air entering the engine and determines the amount of fuel which needs to be mixed with the air to give the correct combustion mixture for the particular conditions of engine operation. The fuel is sprayed continuously by an injection nozzle to the inlet port of each cylinder (photo). This fuel and air is drawn into the cylinder when the inlet valves open.

Airflow meter

4 This measures the volume of air entering the engine and relies on the principle that a circular disc, when placed in a funnel through which a current of air is passing, will rise until the weight of the disc is equal to the force on its lower surface which the air creates. If the volume of air is increased, and the plate were to remain in the same place, the rate of flow of air through the gap between the cone and the plate would increase and the force on the plate would increase.

5 If the plate is free to move, then as the force on the plate increases, the plate rises in the cone and the area between the edge of the plate and the edge of the cone increases, until the rate of air flow and hence the force on the plate, becomes the same as it was at the

former lower flow rate and smaller cone area. Thus the height of the plate is a measure of the volume of air entering the engine.

6 The air flow meter consists of an air funnel with a sensor plate mounted on a lever which is supported at its fulcrum. The weight of the air flow sensor plate and its lever are balanced by a counterweight and the upward force on the sensor plate is opposed by a plunger. The plunger, which moves up and down as a result of the variations in air flow, is surrounded by a sleeve having vertical slots in it. The vertical movement of the plunger uncovers a greater or lesser length of the slots, which meters the fuel to the injection valves.

7 The sides of the air funnel are not a pure cone because optimum operation of the engine requires a different air/fuel ratio under different conditions such as idling, part load and full load. By making parts of the funnel steeper than the basic shape, a richer mixture can be provided for, at idling and full load. By making the funnel flatter than the basic shape, a leaner mixture can be provided.

Fuel supply

8 Fuel is pumped continuously while the engine is running by a roller cell pump running at constant speed: excess fuel is returned to the tank. The fuel pump is operated when the ignition switch is in the START position, but once the starter is released a switch connected to the air plate prevents the pump from operating unless the engine is running.

9 The fuel line to the fuel supply valve incorporates a filter and also a fuel accumulator. The function of the accumulator is to maintain pressure in the fuel system after the engine has been switched off and so give good hot re-starting.

10 Associated with the fuel accumulator is a pressure regulator which is an integral part of the fuel metering device. When the engine is switched off, the pressure regulator lets the pressure to the injection valves fall rapidly to cut off the fuel flow through them and so prevent the engine from 'dieseling' or 'running

on'. The valve closes at just below the opening pressure of the injector valves and this pressure is then maintained by the pressure accumulator.

Fuel distributor

11 The fuel distributor is mounted on the air metering device and is controlled by the vertical movement of the air flow sensor plate. It consists of a spool valve which moves vertically in a sleeve, the sleeve having as many vertical slots around its circumference as there are cylinders on the engine.

12 The spool valve is adjusted to hydraulic pressure on the upper end and this balances the pressure on the air plate which is applied to the bottom of the valve by a plunger. As the spool valve rises and falls, it uncovers a greater or lesser length of metering slot and so controls the volume of fuel fed to each injector.

13 Each metering slot has a differential pressure valve, which ensures that the difference in pressure between the two sides of the slot is always the same. Because the drop in pressure across the metering slot is unaffected by the length of slot exposed, the amount of fuel flowing depends only on the exposed area of the slots.

Cold start valve

14 The cold start valve is mounted in the intake manifold and sprays additional fuel into

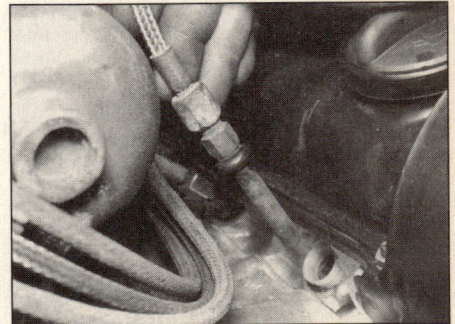

17.3 Injection nozzle, fuel line connection and seal

3

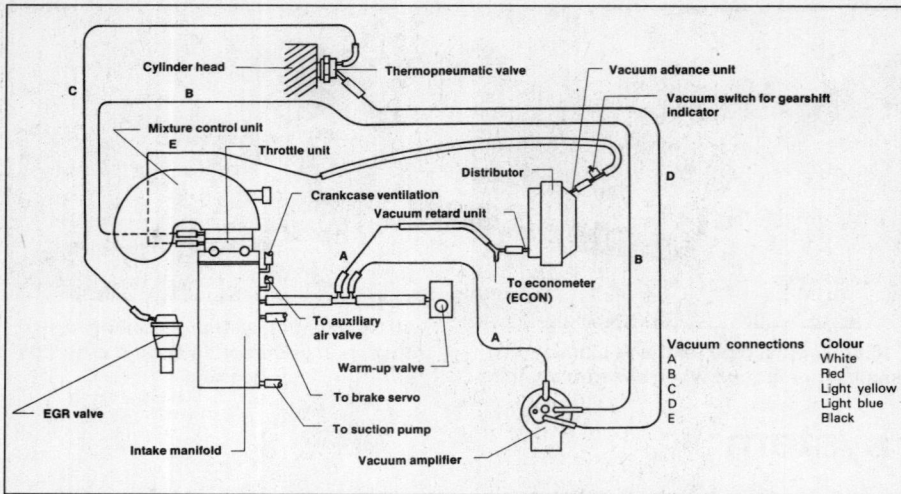

Fig. 3.57 Fuel injection system vacuum circuit and components (Sec 17)

the manifold during cold starting. The valve is solenoid operated and is controlled by a thermo-time switch in the engine cooling system. The thermo-time switch is actuated for a period which depends upon coolant temperature, the period decreasing with rise in coolant temperature. If the coolant temperature is high enough for the engine not to need additional fuel for starting, the switch does not operate.

Pressure surge switch

15 The pressure surge switch is fitted to later models and aids throttle response at low engine temperatures (below 35°C) by causing a momentary injection of fuel from the cold start valve when the throttle is opened. The pressure surge switch operates in conjunction with throttle switch and the thermo-time switch.

Compensation units

16 For cold starting and during warming up, additional devices are required to adjust the fuel supply to the different fuel requirements of the engine under these conditions.

Warm-up regulator (valve)

17 While warming up, the engine needs a richer mixture to compensate for fuel which condenses on the cold walls of the inlet manifold and cylinder walls. It also needs more fuel to compensate for power lost because of increased friction losses and increased oil drag in a cold engine. The mixture is made richer during warming up by the warm-up regulator. This is a pressure regulator which lowers the pressure applied to the control plunger of the fuel regulator during warm-up. This reduced pressure causes the air flow plate to rise higher than it would do otherwise, thus uncovering a greater length of metering slot and making the mixture richer.
18 The valve is operated by a bi-metallic strip which is heated by an electric heater. When

the engine is cold the bi-metallic strip presses against the delivery valve spring to reduce the pressure on the diaphragm and enlarge the discharge cross-section. This increase in cross-section results in a lowering of the pressure fed to the control plunger.
19 When the engine is started, the electrical heater of the bi-metallic strip is switched ON. As the strip warms it rises gradually until it ultimately rises free of the control spring plate and the valve spring becomes fully effective to give normal control pressure.

Auxiliary air device

20 Compensation for power lost by greater friction is compensated by feeding a larger volume of fuel/air mixture to the engine than is supplied by the normal opening of the throttle. The auxiliary air device bypasses the throttle with a channel having a variable aperture valve in it. The aperture is varied by a pivoted plate controlled by a spring and a bi-metallic strip.
21 During cold starting the channel is open and increases the volume of air passing to the engine, but as the bi-metallic strip bends it allows a control spring to pull the plate over the aperture until at normal operating temperature the aperture is closed. The heating of the bimetallic strip is similar to that of the warm-up regulator described above.

18 Maintenance, adjustments and precautions

1 Due to the complexity of the fuel injection system, any work should be limited to the operations described in this Chapter. Other adjustments and system checks are beyond the scope of most readers and should be left to your VW dealer.
2 The mixture setting is pre-set during production of the car and should not normally require adjustment. If new components of the

system have been fitted however, the mixture can be adjusted after reference to Section 25.
3 The only adjustment which may be needed is to vary the engine idle speed by means of the screw mounted in the throttle housing. Use the screw to set the engine speed to that specified when the engine is at the normal operating temperature.
4 Routine servicing of the fuel injection system consists of checking the system components for condition and security, and renewing the air cleaner element at the specified intervals (see Routine Maintenance).
5 Check the system vacuum components for condition and security, these being shown in Fig. 3.57.
6 In the event of a malfunction in the system, reference should be made to the Fault Diagnosis Section at the end of this Chapter, but first make a basic check of the system hoses, connections, fuses and relays for any obvious and immediately visible defects.
7 If any part of the system has to be disconnected or removed for any reason, particular care must be taken to ensure that no dirt is allowed to enter the system.
8 The system is normally pressurised, irrespective of engine temperature, and care must therefore be taken when disconnecting fuel lines: the ignition must also be off and the battery disconnected.
9 Removal, refitting and adjustment of the accelerator cable is similar to that for the 1.9 litre manual gearbox models described in Section 11, except that the throttle lever is located in the throttle housing instead of the carburettor.
10 Removal and refitting of the fuel tank is similar to that for carburettor engine models, as described in Section 9, but it is suggested that the fuel pump and fuel level sender unit are first removed, as described in Section 23.
11 Maintenance and servicing of the exhaust system is as described in Sections 2 and 16. Removal and refitting of the exhaust manifold is similar to that described in Section 15, after removal of the inlet manifold (Section 22).

19 Air cleaner element – removal, cleaning/renewal and refitting

1 Release the spring clips securing the air cleaner cover and separate the cover from the air flow meter (Fig. 3.58).
2 Withdraw the element from the housing.
3 If cleaning the element, place well away from the vehicle then tap the air cleaner element to remove dust and dirt. If necessary use a soft brush to clean the outside or blow air at a very low pressure from the inside surface towards the outside.
4 Wipe clean the inside of the cover.
5 Refit the element and secure the cover by pressing the clips.

20 Air cleaner –
removal and refitting

1 Remove the element as described in the previous Section.
2 Disconnect the cold air and warm air intake hoses from the air cleaner body then detach and remove the air cleaner body (Fig. 3.58).
3 Refit in the reverse order of removal.

21 Air cleaner air intake thermostat –
removal, testing and refitting

1 Unscrew the three retaining screws and withdraw the thermostat unit from the air cleaner body.
2 To test the thermostat, immerse it in water, the temperature of which must initially be below 30°C (86°F). At this temperature the valve flap must pivot to the left (A in Fig. 3.59).
3 When the water temperature is increased to above 38°C (100°F), the intake flap should move to the closed position (B in Fig. 3.59).
4 If defective the thermostat can be removed and renewed. Referring to Fig. 3.60, detach the flap from the retainer at A then press the flap upwards away from retainer B and withdraw the thermostat.
5 Refit in the reverse order to removal.

22 Inlet manifold –
removal and refitting

1 Disconnnect the battery earth lead.
2 Unbolt and detach the throttle valve housing from the manifold (Fig. 3.61).
3 Unbolt and detach the cold start valve unit together with its gasket.
4 Detach the auxiliary air valve, the brake servo hose, suction pump hose, warm-up valve hose and crankcase ventilation hose from the manifold.
5 Remove the retaining bolts and withdraw the manifold from the cylinder head.
6 Refitting is a reversal of the removal procedure, but make sure that the mating faces are clean and always fit new gaskets.

23 Fuel pump –
removal, testing and refitting

1 The fuel pump is located in the fuel tank and access to it (and the fuel level sender unit) is gained after removing the luggage compartment floor covering and cover (Fig. 3.62)
2 It is a sealed unit and the brushes and commutator are not accessible for servicing. The only replacement part available is the gravity check valve on the fuel outlet.
3 Before removing the pump, *first ensure that the car is in a well ventilated place and that there is no danger of ignition from sparks or naked flames. Never work on any part of a fuel*

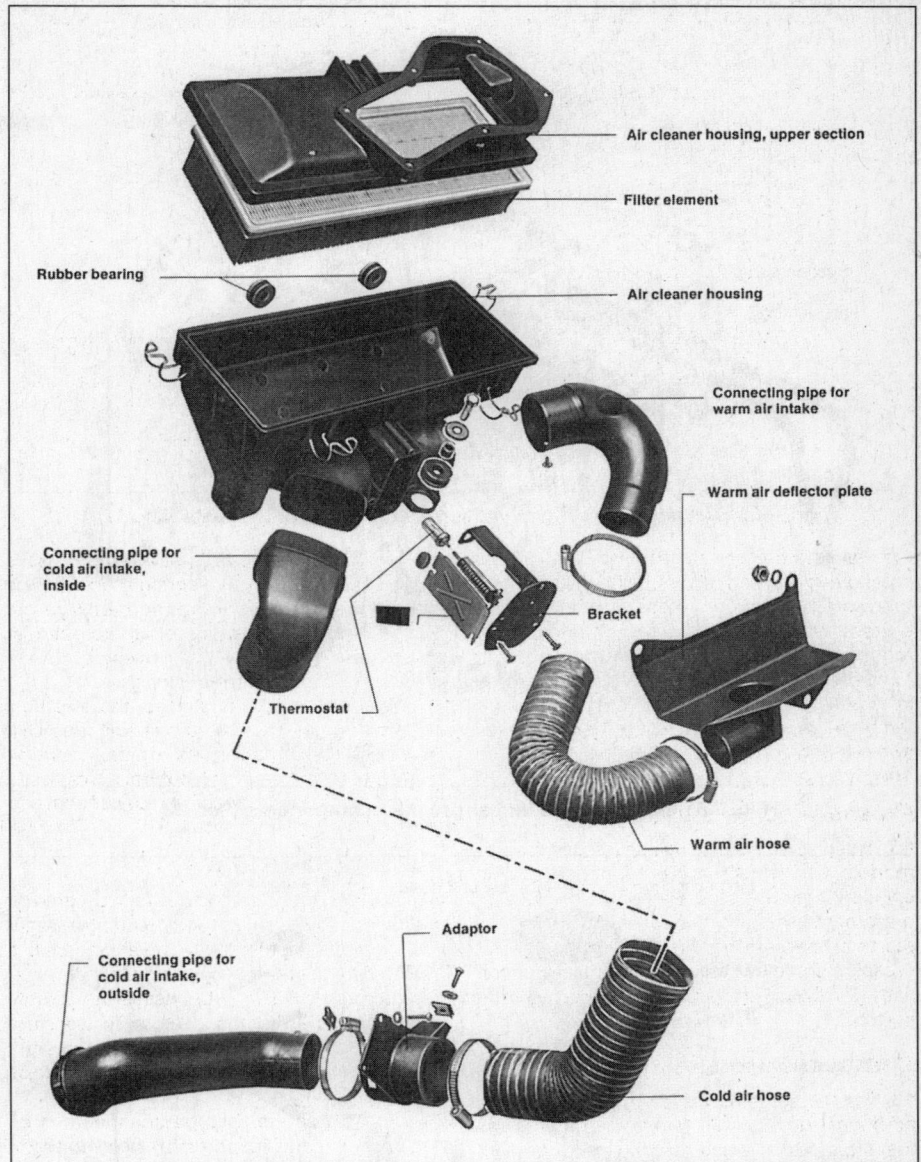

Fig. 3.58 Air cleaner and associated components (Sec 19)

Air cleaner housing, upper section
Filter element
Air cleaner housing
Connecting pipe for warm air intake
Warm air deflector plate
Bracket
Warm air hose
Connecting pipe for cold air intake, inside
Thermostat
Rubber bearing
Connecting pipe for cold air intake, outside
Adaptor
Cold air hose

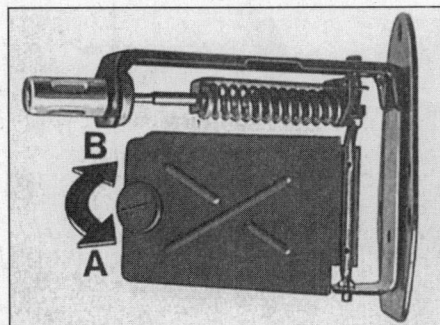

Fig. 3.59 Air intake thermostat (Sec 21)

Position A below 30°C (86°F)
Position B above 38°C (100°F)

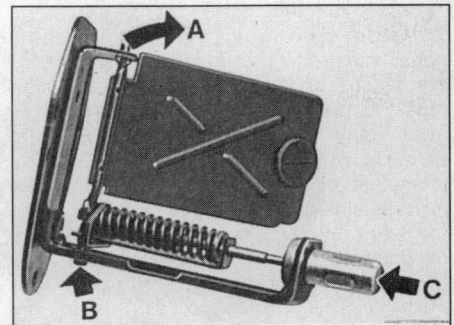

Fig. 3.60 Dismantling air intake thermostat – refer to text for procedure (Sec 21)

3

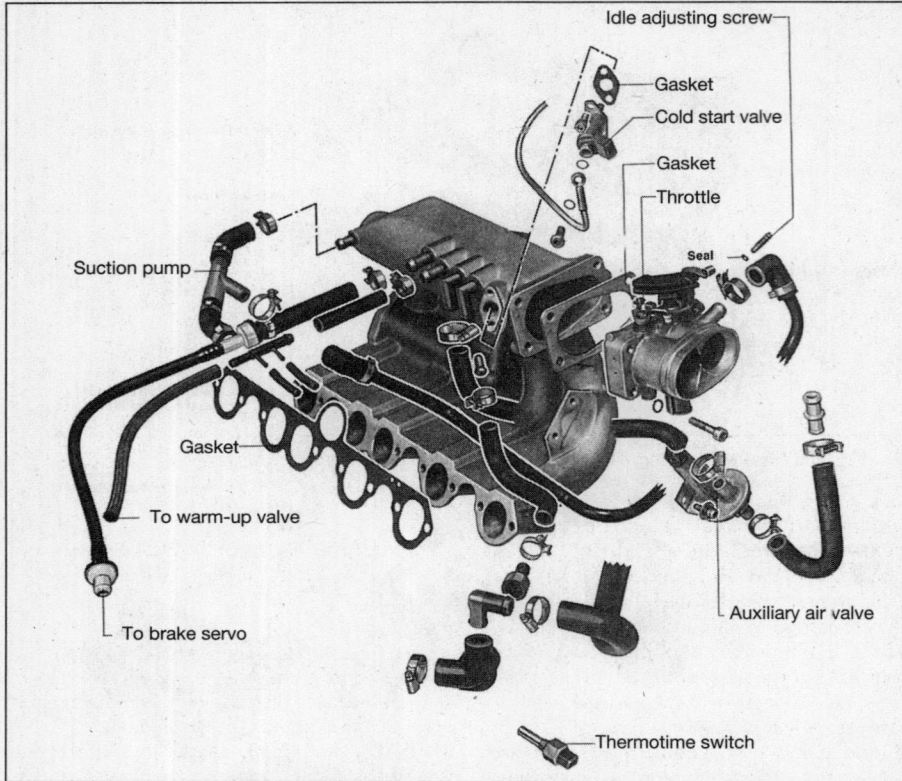

Idle adjusting screw
Gasket
Cold start valve
Gasket
Throttle
Seal
Suction pump
Gasket
To warm-up valve
To brake servo
Auxiliary air valve
Thermotime switch

Fig. 3.61 Inlet manifold and associated components (Sec 22)

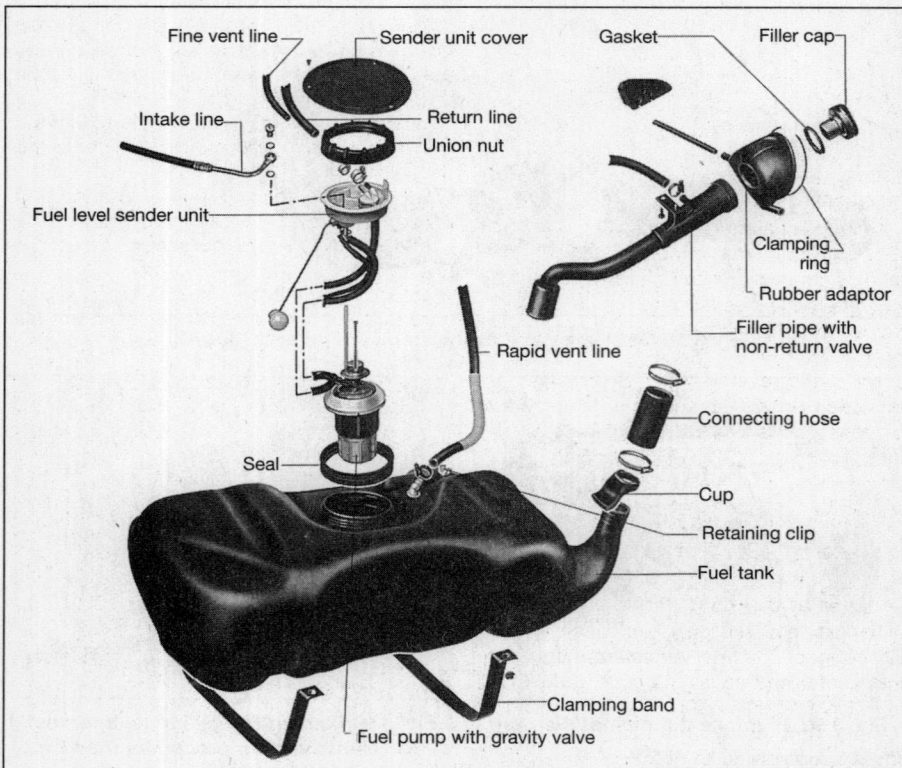

Fine vent line
Sender unit cover
Gasket
Filler cap
Intake line
Return line
Union nut
Fuel level sender unit
Clamping ring
Rubber adaptor
Filler pipe with non-return valve
Rapid vent line
Connecting hose
Seal
Cup
Retaining clip
Fuel tank
Clamping band
Fuel pump with gravity valve

Fig. 3.62 Fuel tank and associated components (Sec 23)

system when the car is over an inspection pit.

4 Disconnect the battery leads and then disconnect the wires multiconnector from the pump.

5 Thoroughly clean all dirt away from the fuel pipe unions on the pump.

6 Detach the vent line and attach a suitable spare hose to its connector. Remove the fuel filler cap then blow into the temporary vent hose. Little or no resistance must be felt.

7 To remove the fuel pump together with the sender and gravity valve units, first detach the fuel supply hose, return hose and vent hose the positions of which are shown in Fig. 3.63.

8 Loosen off the large union nut and withdraw the sender unit from the tank.

9 Undo the pump bracket in the tank and withdraw the fuel pump unit from the tank.

10 Refitting is a reversal of the removal procedure. When refitting the sender unit align its reference mark with the corresponding mark to the front (Fig. 3.64).

24 Idle speed – adjustment

1 Run the engine until the oil temperature is at least 60°C (140°F).

2 Check the ignition timing (Chapter 4), adjusting it if necessary. Leave the wiring disconnected from the idle stabilizer, where applicable.

Fig. 3.63 Fuel tank sender unit connections (Sec 23)

1 Supply (suction) line   *3 Vent line*
2 Return line   *4 Wiring connector*

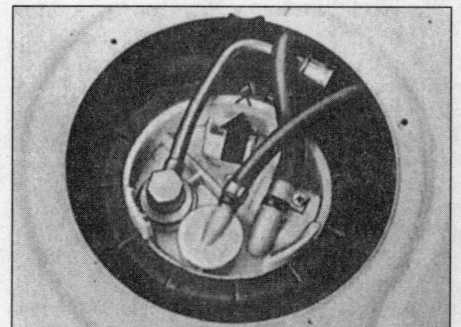

Fig. 3.64 Sender unit orientation – align arrow marks (to front) (Sec 23)

Fig. 3.65 Idle speed adjustment screw (Sec 24)

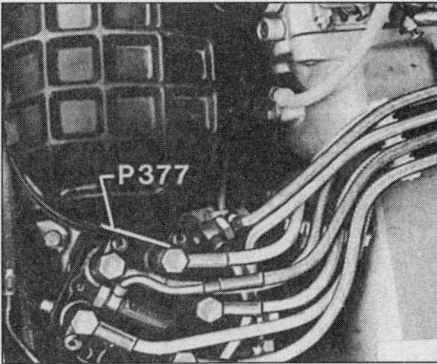

Fig. 3.66 Mixture adjustment screw with special tool number shown (Sec 24)

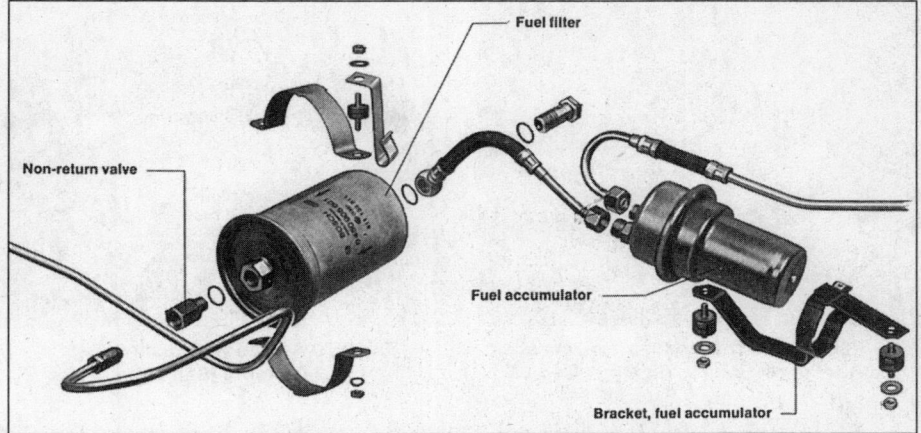

Fig. 3.67 Fuel filter and fuel accumulator with associated components (Sec 26)

3 The main headlights should be turned ON. Disconnect and plug the crankcase breather hose from the valve cover.

4 Remove the locking cap from the adjustment screw on the throttle assembly and turn the screw to achieve the idle speed given in the Specifications. The adjustments should be made only when the electric radiator fan is stationary (Fig. 3.65).

5 After making the adjustment, refit the locking cap over the adjustment screw, and reconnect the wiring to the idle stabilizer, where applicable.

25 Idle mixture – adjustment

1 The idle CO adjustment screw alters the height of the fuel metering distributor plunger relative to the air control plate of the air flow meter.

2 The screw is accessible by removing the locking plug from between the air duct scoop and the fuel metering distributor on the air flow meter casing.

3 Although a special tool is recommended for this adjustment, it can be made using a long, thin screwdriver, (Fig. 3.66).

4 Ensure that the engine is running under the same conditions as those necessary for adjusting the idling speed (see previous Section) and that the idling speed is correct.

5 Connect an exhaust gas analyser to the tailpipe as directed by the equipment manufacturer, and read the CO level.

6 Turn the adjusting screw clockwise to raise the percentage of CO and anti-clockwise to lower it. It is important that the adjustment is made without pressing down on the adjusting screw, because this will move the air flow sensor plate and affect the adjustment.

7 Remove the tool, accelerate the engine briefly and re-check. If the tool is not removed before the engine is accelerated there is a danger of the tool becoming jammed and getting bent.

26 Fuel filter – removal and refitting

1 Disconnect the battery earth lead.

2 Relieve the system pressure as described in Section 28, paragraphs 2 and 3.

3 Position a suitable container beneath the filter pipe connections and disconnect the fuel inlet and outlet pipes from the filter (Fig. 3.67).

4 Loosen the filter clamp bracket screw and withdraw the filter.

5 Refit in the reverse order of removal. On completion restart the engine and check the pipe connections for any signs of leaks.

27 Fuel accumulator – removal and refitting

1 Proceed as described in paragraphs 1 and 2 in the previous Section.

2 Disconnect the fuel pipes from the fuel accumulator and catch the small amount of fuel which will be released.

3 Detach and remove the fuel accumulator from its support bracket.

4 Refitting is a reversal of removal. Check for leaks on completion with the engine restarted.

28 Fuel metering distributor – removal and refitting

1 Disconnect the battery terminals.

2 Ensure that the vehicle is in a well ventilated space and that there are no naked flames or other possible sources of ignition.

3 While holding a rag over the joint to prevent fuel from being sprayed out, loosen the control pressure line from the warm-up valve. The control pressure line is the one connected to the large union of the valve.

4 Mark each fuel line, and its port on the fuel distributor. Carefully clean all dirt from around the fuel unions and distributor ports and then disconnect the fuel lines.

5 Unscrew and remove the connection of the pressure control line to the fuel metering distributor.

6 Remove the locking plug from the CO adjusting screw, then remove the screws securing the fuel metering distributor (Fig. 3.68).

7 Lift off the fuel metering distributor, taking care that the metering plunger does not fall out. If the plunger does fall out accidentally,

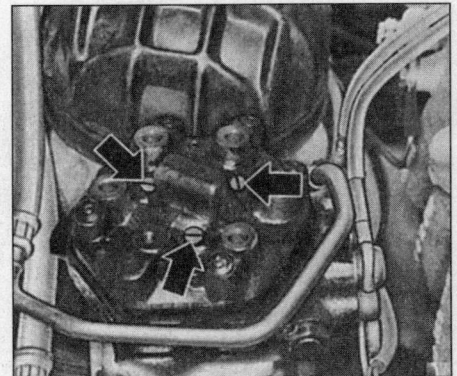

Fig. 3.68 Fuel metering distributor retaining screws (Sec 28)

3

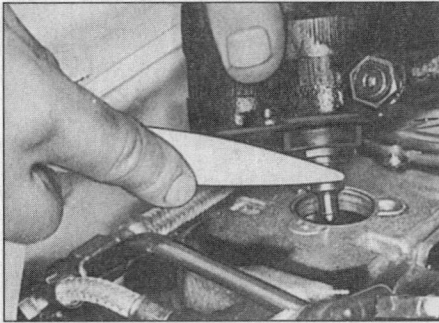

Fig. 3.69 Fuel distributor removal ensuring plunger is retained in position (Sec 28)

Fig. 3.70 Air scoop retaining clips (arrowed) (Sec 29)

Fig. 3.71 Air flow meter retaining bolts (arrowed) (Sec 29)

clean it in fuel and then re-insert it with its chamfered end downwards (Fig. 3.69).

8 Before refitting the metering distributor, ensure that the plunger moves up and down freely. If the plunger sticks, the distributor must be renewed, because the plunger cannot be repaired or replaced separately.

9 Refit the distributor, using a new sealing ring, and after tightening the screws, lock them with paint.

10 Refit the fuel lines and the cap of the CO adjusting screw and tighten the union on the warm-up valve.

29 Air flow meter – removal and refitting

1 Remove the fuel lines from the distributor, as described in paragraphs 1 to 5 of the previous Section.

2 Loosen the clamps at the air cleaner and throttle assembly ends of the air scoop and take off the air scoop (Fig. 3.70).

3 Remove the bolts securing the air flow meter to the air cleaner and lift off the air flow meter and fuel metering distributor (Fig. 3.71 and photo) .

4 The fuel metering plunger should be prevented from falling out when the fuel metering distributor is removed from the air flow meter (see previous Section).

5 Refitting is the reverse of removing, but it is

necessary to use a new gasket between the air flow meter and the air cleaner.

30 Pressure relief valve – removal, servicing and refitting

1 Release the pressure in the fuel system, as described in paragraphs 1 to 3 of Section 28.

2 Unscrew the non-return valve plug and remove the plug and its sealing washer.

3 Take out the O-ring, shims, plunger and O-ring in that order (Fig. 3.72).

4 When refitting the assembly, use new O-rings and ensure that all the shims which were removed are refitted.

31 Thermo-time switch – checking

1 The thermo-time switch energises the cold start valve for a short time on starting and the time for which the valve is switched on depends upon the engine temperature.

2 This check must only be carried out when the coolant temperature is below 30°C (86°F).

3 Pull the connector off the cold start valve and connect a test lamp across the contacts of the connector.

4 Pull the high tension lead off the centre of the distributor and connect the lead to earth.

5 Operate the starter for 10 seconds and note

the interval before the test lamp lights and the period for which it remains alight. Reference to the graph will show that at a coolant temperature of 30°C (86°F) the lamp should light immediately and stay on for two seconds (Fig. 3.73).

6 Refit the high tension lead onto the distributor, and reconnect the lead to the cold start valve.

32 Cold start valve – checking

1 Ensure that the coolant temperature is below 30°C (86°F) and that the car battery is fully charged.

2 Pull the high tension lead off the centre of the distributor and connect the lead to earth.

3 Pull the connectors off the warm-up valve and the auxiliary air unit.

4 Remove the two bolts securing the cold start valve to the inlet manifold and remove the valve, taking care not to damage the gasket (photo) .

5 With fuel line and electrical connections connected to the valve, hold the valve over a glass jar and operate the starter for 10 seconds. The cold start valve should produce an even cone of spray during the time the thermo-time switch is ON.

6 After completing the checks refit the valve and reconnect the leads that were disturbed.

29.3 Air flow meter and fuel distributor unit (inverted)

Fig. 3.72 Pressure relief valve components (Sec 30)

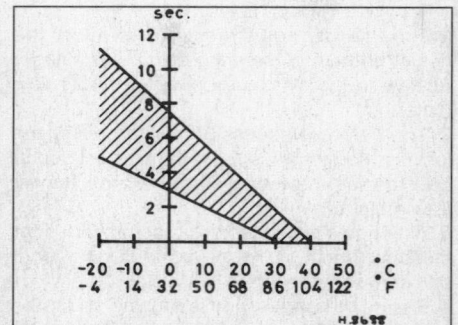

Fig. 3.73 Thermo-time switch graph (switch-on time) (Sec 31)

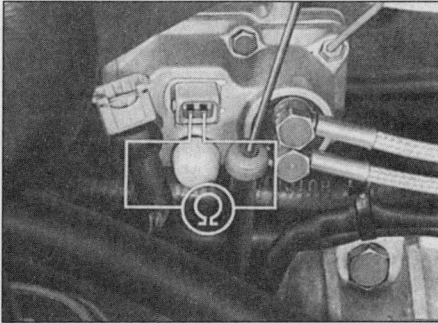

Fig. 3.74 Warm-up regulator heater coil resistance check (Sec 33)

32.4 Cold start valve and attachments

33.1 Warm-up regulator (shown with fuel lines detached)

34.2 Top view of the air flow sensor plate

33 Warm-up regulator – checking

1 With the engine cold, pull the connectors off the warm-up valve and the auxiliary air unit (photo).
2 Connect a voltmeter across the terminals of the warm-up valve connector and operate the starter. The voltage across the terminals should be a minimum of 11.5 volts.
3 Switch the ignition OFF and connect an ohmmeter across the terminals of the warm-up valve. If the meter does not indicate a resistance of about 20 ohms, the heater coil is defective and a new vaive must be fitted (Fig. 3.74).

34 Air flow sensor lever and control plunger – checking

1 For the correct mixture to be supplied to the engine it is essential that the sensor plate is central in the venturi and that its height is correct. First run the engine for a period of about one minute.
2 Loosen the hose clips at each end of the air scoop and remove the scoop. If the sensor plate appears to be off-centre, loosen its centre screw and carefully run a 0.10 mm (0.004 in) feeler gauge round the edge of the plate to centralise it, then re-tighten the bolt (photo).
3 Raise the air flow sensor plate and then quickly move it to its rest position. No resistance should be felt on the downward movement; if there is resistance the air flow meter is defective and a new one must be fitted.
4 If the sensor plate can be moved downwards easily, but has a strong resistance to upward movement, the control plunger is sticking. Remove the fuel distributor (Section 28) and clean the control plunger in fuel. If this does not cure the problem, a new fuel distributor must be fitted.
5 Release the pressure on the fuel distributor, as described in Section 28 and then check the rest position of the air flow sensor plate. The upper edge of the plate should be flush with the bottom edge of the air cone. It is

permissible for the plate to be lower than the edge by not more than 0.5 mm (0.020 in), but if higher, or lower than the permissible limit, the plate must be adjusted (Fig. 3.75).
6 Adjust the height of the plate by lifting it and bending the wire clips attaching the plate to the balance arm, but take care not to scratch or damage the surface of the air cone (photo).
7 After making the adjustment, tighten the warm-up valve union and check the idle speed and CO content.

35 Auxiliary air valve – checking

1 To carry out this test the engine coolant temperature must be below 30°C (86°F).
2 Detach the auxiliary air valve electrical plug and ensure that the contacts in the plug connector are in good condition.
3 Connect up a voltmeter across the contacts as shown (Fig. 3.76), start the engine and run at idle speed. The voltage reading must be a minimum of 11.6V.
4 Leave the engine running at idle speed and pinch the air intake duct to auxiliary air valve hose together. The engine speed should drop.
5 When the engine is warmed up to its normal operating temperature, reconnect the auxiliary valve plug then pinch the hose together again. This time the engine speed should remain unaltered.

34.6 Air flow sensor plate adjustment clip (arrowed)

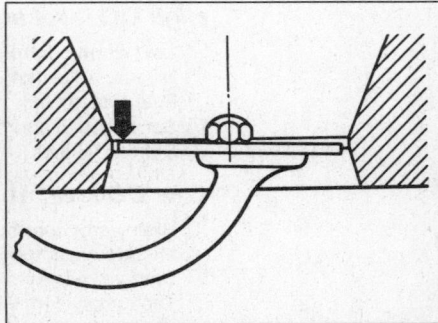

Fig. 3.75 Sensor plate position requirement (Sec 34)

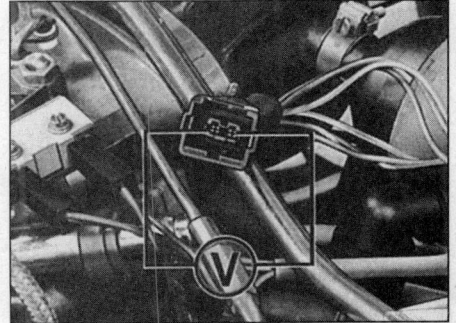

Fig. 3.76 Auxiliary air valve check (Sec 35)

Fault finding – fuel system (carburettor models)

Note: *High fuel consumption and poor performance are not necessarily due to carburettor faults. Make sure that the ignition system is properly adjusted, that the brakes are not binding and that the engine is in good mechanical condition before tampering with the carburettor.*

Fuel consumption excessive

☐ Air cleaner choked, giving rich mixture
☐ Leak from tank, pump or fuel lines
☐ Float chamber flooding due to incorrect level or worn needle valve
☐ Carburettor incorrectly adjusted
☐ Idle speed too high
☐ Choke faulty (sticks on)
☐ Excessively worn carburettor

Lack of power, stalling or difficult starting

☐ Faulty fuel pump
☐ Leak on suction side of pump or in fuel line
☐ Intake manifold or carburettor flange gaskets leaking
☐ Carburettor incorrectly adjusted
☐ Faulty choke
☐ Emission control system defect

Poor or erratic idling

☐ Weak mixture (screw tampered with)
☐ Leak in intake manifold
☐ Leak in distributor vacuum pipe
☐ Leak in crankcase extractor hose
☐ Leak in brake servo hose

Fault finding – fuel system (fuel injection models)

Before assuming that a malfunction is caused by the fuel system, check the items mentioned in the special note at the start of the previous Section.

Engine will not start (cold)

☐ Fuel pump faulty
☐ Auxiliary air device not opening
☐ Start valve not operating
☐ Start valve leak
☐ Sensor plate rest position incorrect
☐ Sensor plate and/or control plunger sticking
☐ Vacuum system leak
☐ Fuel system leak
☐ Thermotime switch remains open

Engine will not start (hot)

☐ Faulty fuel pump
☐ Warm control pressure low
☐ Sensor plate rest position incorrect
☐ Sensor plate and/or control plunger sticking
☐ Vacuum system leak
☐ Fuel system leak
☐ Leaky injector valve(s) or low opening pressure
☐ Incorrect mixture adjustment

Engine difficult to start (cold)

☐ Cold control pressure incorrect
☐ Auxiliary air device not opening
☐ Faulty start valve
☐ Sensor plate rest position faulty
☐ Sensor plate and/or control plunger sticking
☐ Fuel system leak
☐ Thermotime switch not closing

Engine difficult to start (hot)

☐ Warm control pressure too high or too low
☐ Auxiliary air device faulty
☐ Sensor plate/control plunger faulty
☐ Fuel or vacuum leak in system
☐ Leaky injector valve(s) or low opening pressure
☐ Incorrect mixture adjustment

Rough idling (during warm-up period)

☐ Incorrect cold control pressure
☐ Auxiliary air device not closing (or opening)
☐ Start valve leak
☐ Fuel or vacuum leak in system
☐ Leaky injector valve(s), or low opening pressure

Rough idling (engine warm)

☐ Warm control pressure incorrect
☐ Auxiliary air device not closing
☐ Start valve leaking
☐ Sensor plate and/or control plunger sticking
☐ Fuel or vacuum leak in system
☐ Injector(s) leaking or low opening pressure
☐ Incorrect mixture adjustment

Engine backfiring into intake manifold

☐ Warm control pressure high
☐ Vacuum system leak

Engine backfiring into exhaust manifold

☐ Warm control pressure high
☐ Start valve leak
☐ Fuel system leak
☐ Incorrect mixture adjustment

Engine misfires (on road)

☐ Fuel system leak

Engine 'runs on'

☐ Sensor plate and or control plunger sticking
☐ Injector valve(s) leaking or low opening pressure

Excessive petrol consumption

☐ Fuel system leak
☐ Mixture adjustment incorrect
☐ Low warm control pressure

High CO level at idle

☐ Low warm control pressure
☐ Mixture adjustment incorrect
☐ Fuel system leak
☐ Sensor plate and/or control plunger sticking
☐ Start valve leak

Low CO level at idle

☐ High warm control pressure
☐ Mixture adjustment incorrect
☐ Start valve leak
☐ Vacuum system leak

Idle speed adjustment difficult (too high)

☐ Auxiliary air device not closing

Chapter 4 Ignition system

For modifications, and information applicable to later models, see Supplement at end of manual

Contents

Degrees of difficulty

Easy, suitable for novice with little experience	**Fairly easy,** suitable for beginner with some experience	**Fairly difficult,** suitable for competent DIY mechanic	**Difficult,** suitable for experienced DIY mechanic	**Very difficult,** suitable for expert DIY or professional

Specifications

System type . . . . 12 volt battery with transistorized coil ignition (TCI)

Distributor

Rotor rotation . . . . Clockwise
Firing order:
   Four-cylinder . . . . 1-3-4-2 (No 1 at timing belt end)
   Five-cylinder . . . . 1-2-4-5-3 (No 1 at timing belt end)

1.6 litre engine

	WV, YN	YP	DT
Engine type			
Ignition timing – before TDC	9° ± 1° at 900 to 1000 rpm	0 ± 1° at 900 to 1000 rpm	18° ±1° at 700 to 800 rpm
Vacuum hose connection for timing check	Off	On	On

1.8 litre engine

Engine type . . . . **DS**
Ignition timing – before TDC . . . . 18° ±1° at 700 to 800 rpm
Vacuum hose position for timing check . . . . On

1.9 litre engine

Engine type . . . . **WN**
Ignition timing – before TDC . . . . TDC ± 1° at 800 to 850 rpm
Vacuum hose position for timing check . . . . On

2.0 litre engine

Engine type . . . . **JS**
Ignition timing – before TDC . . . . 18° at 800 to 900 rpm
Vacuum hose position for timing check . . . . On

Coil

Primary resistance . . . . 0.52 to 0.76 ohm
Secondary resistance . . . . 2400 to 3500 ohm

Spark plugs

Type:	
1.6 litre engine	Bosch W7D, W7DC, W8D
	Beru 14-7D, 14-8D, 14-7DU/RS 35
	Champion N7YC
	NGK BP6-E
1.8 litre engine	Bosch W6 DO
	Beru 14-6DU
	Champion N79Y
1.9 litre engine	Bosch W5D
	Beru 14-5D
	Champion N7YC
2.0 litre engine	Bosch W6 DO
	Beru 14-6 DU
	Champion N79Y
Spark plug electrode gap:	
1.6 litre engine	0.6 to 0.8 mm (0.023 to 0.031 in)
1.8 litre engine	0.8 to 0.9 mm (0.031 to 0.035 in)
1.9 litre engine	0.6 to 0.8 mm (0.023 to 0.031 in)
2.0 litre engine	0.8 to 0.9 mm (0.031 to 0.035 in)

Torque wrench settings

	Nm	lbf ft
Spark plugs	20	15
Distributor clamp bolt:		
Four-cylinder	25	18
Five-cylinder	15	11

1 General description

The ignition system consists of the battery, coil, distributor and spark plugs. All models covered by this manual have a transistorised ignition system incorporating a magnetic sensor and control unit. On some models a digital idle stabilizer (DIS) unit is incorporated into the system and its function is to stabilise the idle speed by advancing the ignition at low engine speeds (600 to 800 rpm).

The transistorised system functions in a similar manner to a conventional system, but the contact points and condenser are replaced by a magnetic sensor in the distributor and a control unit mounted separately. As the distributor driveshaft rotates, the magnetic impulses are fed to the control unit which switches the primary circuit on and off. No condenser is necessary as the circuit is switched electronically with semiconductor components.

The ignition advance is controlled mechanically by centrifugal weights and by a vacuum capsule mounted on the side of the distributor.

To prevent damage to the system or personal injury observe the precautions given in Section 6.

2 Ignition system – routine maintenance

The following routine maintenance procedures must be carried out at the specified intervals given at the start of this manual.

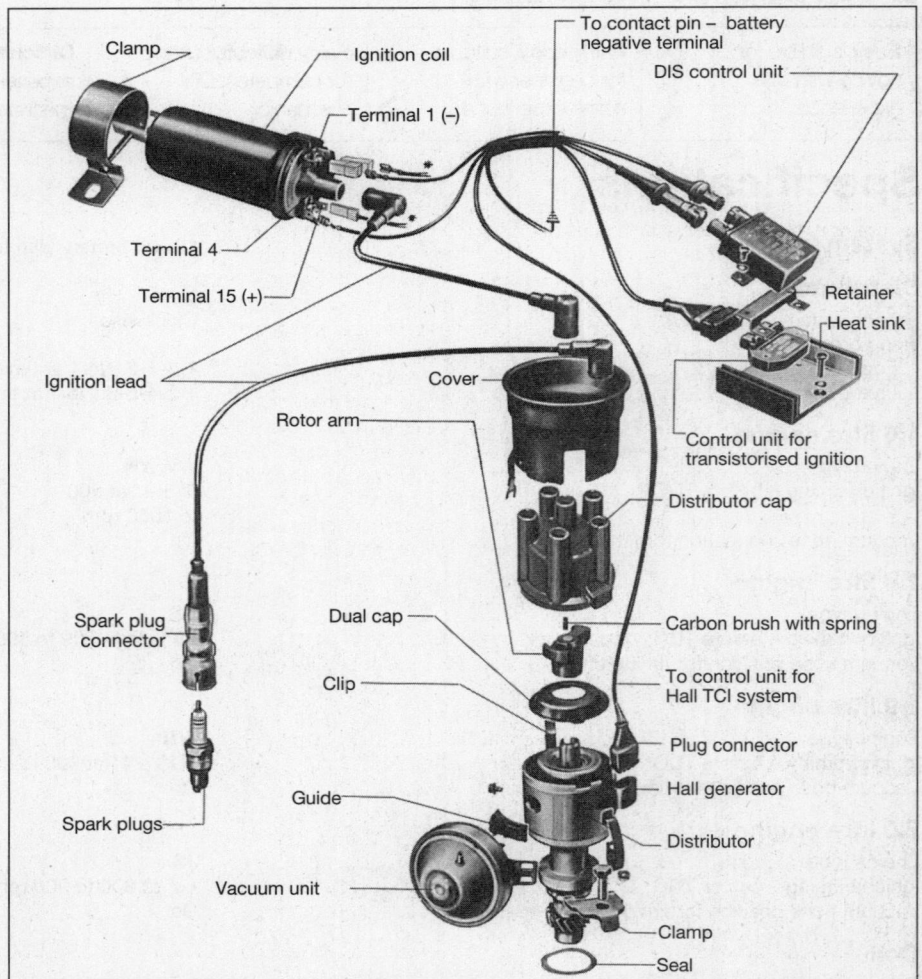

Fig. 4.1 Main components of the ignition system (Sec 1)

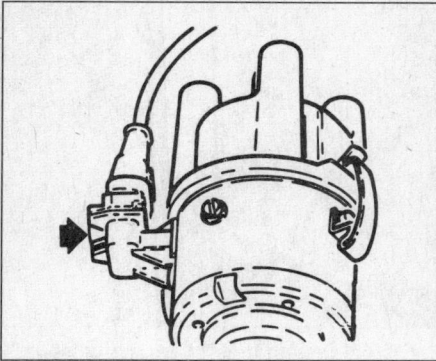

Fig. 4.2 Press wire retaining clip (arrowed) to release wire connector – later models (Sec 3)

1 Remove, clean and reset the spark plugs. Reset the electrode gap before refitting (see Section 11).
2 Clean and inspect the ignition system HT and LT lead connections. Renew if defective in any way.
3 Remove the spark plugs and renew them at the specified intervals with a set of the recommended types.
4 Check and if necessary adjust the ignition timing as described in Section 9.

3 Distributor – removal and refitting

1 Pull the high tension connection from the centre of the ignition coil and remove the caps from the spark plugs.
2 Release the two spring clips securing the distributor cap, then remove the distributor cap with the ignition harness attached. On models with metal screening round the top of the distributor, it is necessary to remove the screw from the bonding strap before the cap can be taken off. Do not allow the metal cover clip to fall inwards as the rotor or trigger wheel may be damaged (photo).
3 Disconnect the control unit lead multi-plug. On some later models it is necessary to release the wire retaining clip to detach the multiplug (Fig. 4.2).
4 Note the exact position of the rotor arm, so

3.2 Distributor cap screen bonding (earth) strap – arrowed (five-cylinder engine)

that the distributor can be fitted with the rotor arm in the same position, and also put mating marks on the distributor mounting flange and base. By marking these positions and also ensuring that the crankshaft is not moved while the distributor is off, the distributor can be refitted without upsetting the ignition timing.
5 Pull the vacuum pipe(s) from the vacuum control unit, marking the position of the pipes, if there is more than one.
6 Remove the bolt and washer from the distributor clamp plate and take the clamp plate off. Remove the distributor and gasket (photos).
7 When refitting the distributor always renew the gasket (photo). Provided the crankshaft has not been moved, turn the rotor arm to such a position that when the distributor is fully installed and the gears mesh, the rotor will turn and take up the position which it held before removal. On the four-cylinder engine if the distributor will not seat fully, withdraw the unit and use pliers to turn the oil pump driveshaft slightly, then try again.
8 Fit the clamp plate and washer, then fit the bolt and tighten it.
9 Fit the distributor cap and clip it in place, then reconnect the high and low tension wires.
10 If the engine has been the subject of overhaul, the crankshaft has been rotated or a new distributor is being fitted, then the procedure for installing the distributor will

3.6A Distributor clamp and bolt (four-cylinder engine)

3.6B Distributor removal (four-cylinder engine)

differ according to engine capacity; refer to Section 8 of this Chapter.

4 Distributor – dismantling, inspection and reassembly

Note: *Before commencing work check that spare parts are individually available for the distributor.*
1 Wipe clean the exterior of the distributor.
2 Pull the rotor arm off the driveshaft, then lift off the dust cap (photo). Do not allow the cap retaining clips to touch the trigger wheel.
3 Mark the trigger wheel in relation to the driveshaft, then prise out the retainer and withdraw the wheel, together with the locating pin (Fig. 4.4 and photo).

3.7 Distributor gasket (four-cylinder engine)

4.2 Removing the dust cap (four-cylinder engine)

4.3 Distributor driveshaft and trigger wheel (four-cylinder engine)

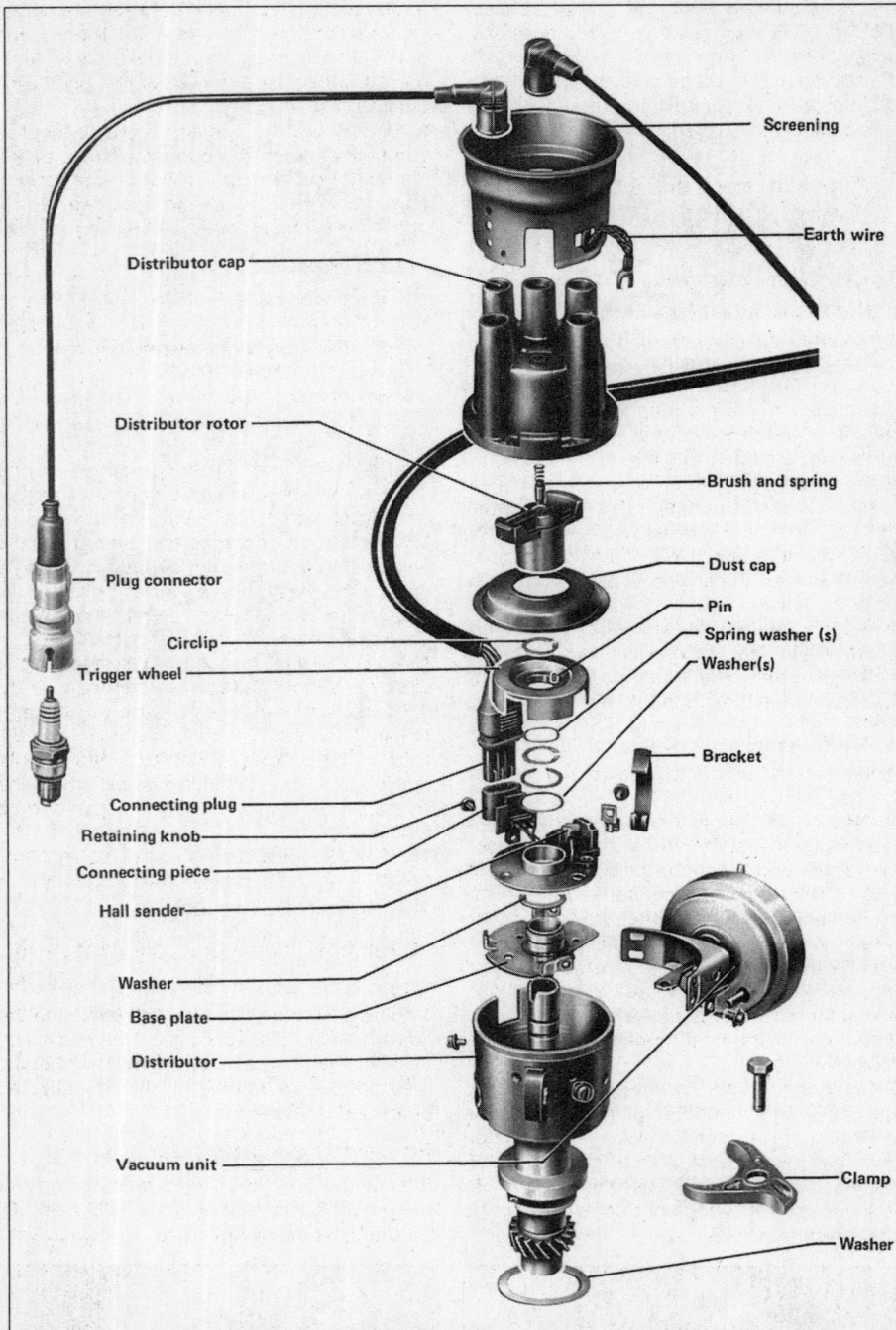

Fig. 4.3 Exploded view of the distributor (four-cylinder type shown) (Sec 4)

Labels (top to bottom, left then right):

Distributor cap
Distributor rotor
Plug connector
Circlip
Trigger wheel
Connecting plug
Retaining knob
Connecting piece
Hall sender
Washer
Base plate
Distributor
Vacuum unit

Screening
Earth wire
Brush and spring
Dust cap
Pin
Spring washer (s)
Washer(s)
Bracket
Clamp
Washer

Fig. 4.4 Prising free the trigger wheel. If distorted it must be renewed (Sec 4)

Fig. 4.5 Thermo-pneumatic valve location (1) (later four-cylinder models) (Sec 5)

2 Check valve

4 Remove the retainer and washers, noting their location.

5 Remove the screws and withdraw the vacuum unit and packing after disconnecting the operating arm.

6 Remove the retainer and washer from the baseplate and also remove the screw securing the socket to the side of distributor. Withdraw the magnetic pick-up and socket together.

7 Remove the screws and lift out the baseplate followed by the washer.

8 Clean all the components and examine them for wear and damage.

9 Inspect the inside of the distributor cap for signs of burning, or tracking. Make sure that the small carbon brush in the centre of the distributor cap is in good condition and can move up and down freely under the influence of its spring.

10 Check that the rotor arm is not damaged. Use an ohmmeter to measure the resistance between the brass contact in the centre of the rotor arm and the brass contact at the edge of the arm. The measured value of resistance should be between 600 and 1400 ohms on the 1.6 and 1.8 models or 1000 ohms on 1.9 and 2.0 models.

11 Suck on the pipe connection to the vacuum diaphragm and check that the operating rod of the diaphragm unit moves. Retain the diaphragm under vacuum to check that the diaphragm is not perforated.

12 Reassemble the distributor in reverse order of dismantling, but smear a little grease on the bearing surface of the baseplate.

5 Thermo-pneumatic valve and check valve – checking

1 Later four-cylinder models fitted with automatic transmission have a thermo-pneumatic valve and check valve fitted. They are located in the vacuum line between the distributor and the carburettor. Their function is to improve acceleration when the engine coolant temperature is below 46°C (113°F) by maintaining the vacuum within the vacuum unit.

2 The thermo-pneumatic valve can be tested for satisfactory operation by blowing through it. At temperatures below 30°C (86°F) the

valve must be closed while at temperatures over 46°C (118°F) it must be open.

3 To check the efficiency of the check valve, blow through the white connection first to see if the valve is open. Blow through the valve from the black connection and check that the valve is shut.

4 Renew the valve(s) if found to be defective. When reconnecting, the white connection of the check valve must face towards the vacuum advance unit on the distributor.

6 Transistorised coil ignition system – precautions

1 To prevent personal injury and damage to the ignition system, the following precautions must be observed when working on the ignition system.

2 Do not attempt to disconnect any plug lead or touch any of the high tension cables when the engine is running, or being turned by the starter motor.

3 Ensure that the ignition is turned OFF before disconnecting any of the ignition wiring.

4 Ensure that the ignition is switched OFF before connecting or disconnecting any ignition testing equipment such as a timing light.

5 Do not connect a suppression condenser or test lamp to the coil negative terminal (1).

6 Do not connect any test appliance or stroboscopic lamp requiring a 12 volt supply to the coil positive terminal (15).

7 If the HT cable is disconnected from the distributor (terminal 4), the cable must immediately be connected to earth and remain earthed if the engine is to be rotated by the starter motor, for example if a compression test is to be done.

8 If a high current boost charger is used, the charger output voltage must not exceed 16.5 volts and the time must not exceed one minute.

9 The ignition coil of a transistorised system must never be replaced by the ignition coil from a contact breaker type ignition system.

10 If an electric arc welder is to be used on any part of the vehicle, the car battery must be disconnected while welding is being done.

11 If a stationary engine is heated to above 80°C (176°F) such as may happen after paint drying, or steam cleaning, the engine must not be started until it has cooled.

12 Ensure that the ignition is switched OFF when the car is washed.

7 Transistorised coil ignition system – testing

Digital Idle Stabilizer (DIS)

1 The DIS is fitted in the circuit between the Hall generator in the distributor and the control unit. It operates at engine speeds between 600 and 840 rpm and effectively advances the ignition in order to increase the engine speed to the correct idling speed. Not all models are fitted with the DIS.

2 To check the DIS, VW garages use an electronic tester connected to the flywheel TDC sensor unit. However, a timing light may be used just as effectively. Connect the timing light then start the engine and allow it to idle. Note the ignition timing.

3 Switch on all the electrical equipment including the heated rear window and main headlights, then check that the ignition timing has advanced. If not, the DIS unit is faulty.

Control unit

4 First check that the ignition coil and DIS are in order.

5 Disconnect the multi-plug from the control unit and measure the voltage between contacts 4 and 2 with the ignition switched on (Fig. 4.6). Note that on early models the control unit is located on the bulkhead, but on later models it is behind the glove compartment. If the voltage is not approximately the same as battery voltage, check the wiring for possible breakage.

6 Switch off the ignition and reconnect the multi-plug.

7 Disconnect the multi-plug from the distributor and connect a voltmeter across the coil primary terminals. With the ignition switched on at least 2 volts must register, falling to zero after approximately 1 to 2 seconds. If this does not occur, renew the control unit and coil.

8 Connect a wire briefly between the centre terminal of the distributor multi-plug and earth. The voltage should rise to at least 5 volts, otherwise the control unit should be renewed (Fig. 4.7).

9 Switch off the ignition and connect a voltmeter across the outer terminals of the distributor multi-plug. Switch on the ignition and check that 5 volts is registered. If not, check the wiring for possible breakage.

Hall Generator

10 First check the ignition coil, DIS, and control unit. The following test should be made within ambient temperature extremes of 0° and 40°C (32° and 104°F).

11 Disconnect the central HT lead from the distributor cap and earth it with a bridging wire.

12 Pull back the rubber grommet on the control unit and connect a voltmeter across terminals 6 and 3 (Fig. 4.8).

13 Switch on the ignition, then turn the engine slowly in the normal direction using a spanner on the crankshaft pulley bolt.

14 On 1.6 and 1.8 litre models the voltage should alternate between 0 and 2 volts.

15 On 1.9 litre and 2.0 litre models the voltage should alternate between 0 to 0.7 volts, and 1.8 volts to battery voltage; if required, the distributor cap can be removed – with a full air gap 0 to 0.7 volts should register, but when the trigger wheel covers the air gap 1.8 volts to battery voltage should register.

16 If the results are not as given in paragraph 14 or 15 the Hall generator is faulty.

8 Ignition timing – basic setting

1 If the distributor has been removed and the ignition timing disturbed, it will be necessary to reset the timing using the following static method before setting it dynamically, as described in Section 9.

Four-cylinder engines

2 Turn the engine using a spanner on the crankshaft pulley bolt until the TDC mark (O) on the flywheel or driveplate is aligned with

Fig. 4.6 Checking the voltage at the control unit multi-plug (Sec 7)

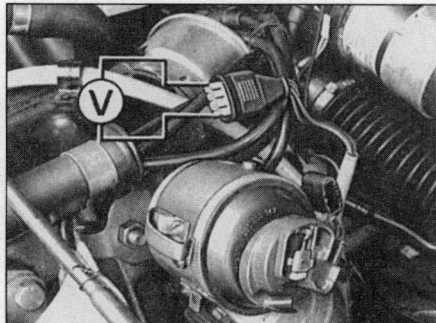
Fig. 4.7 Checking the voltage at the distributor multi-plug (Sec 7)

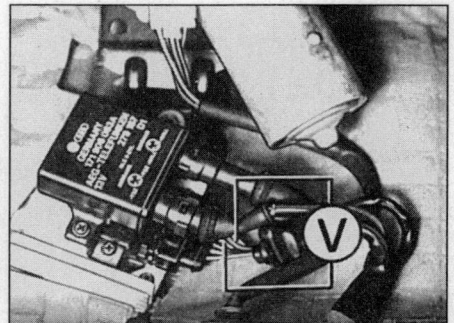
Fig. 4.8 Checking the voltage at the control unit multi-plug terminals 6 and 3 (Sec 7)

Fig. 4.9 TDC timing mark location on the rear of the camshaft gear on four-cylinder engines (Sec 8)

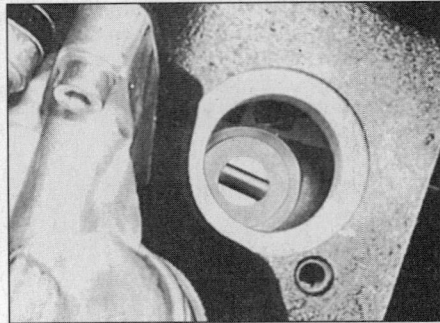

Fig. 4.10 Oil pump shaft lug position when fitting distributor (Sec 8)

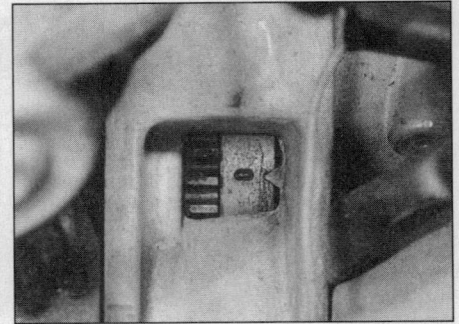

8.2 TDC (top dead centre) timing marks aligned on the four-cylinder engine

the pointer in the timing aperture (photo), which is located next to the distributor.

3 Check that the mark on the rear of the camshaft gear is aligned with the cylinder head cover, as shown in Fig. 4.9. If the mark is on the opposite side of the cylinder head, turn the crankshaft forward one complete revolution, and again align the TDC mark.

4 Turn the oil pump shaft so that its lug is parallel with the crankshaft (Fig. 4.10).

5 With the distributor cap removed turn the rotor arm so that the centre of the metal contact is aligned with the mark on the rim of the distributor body (photo).

6 Hold the distributor over its recess in the cylinder block with the vacuum unit slightly clockwise of the position shown in Fig. 4.11.

7 Insert the distributor fully. As the drivegear meshes with the intermediate shaft, the rotor will turn slightly anti-clockwise and the body can be realigned with the rotor arm to take up the final positions shown in Fig. 4.1 1. A certain amount of trial and error may be required for this. If the distributor cannot be fully inserted into its mounting hole, slightly reposition the oil pump driveshaft lug. Turn the distributor body until the original body-to-cylinder head marks are in alignment then clamp the distributor in position.

Five-cylinder engines

8 Turn the engine using a spanner on the crankshaft pulley bolt until the TDC mark (O) is aligned with the lug in the timing aperture in the flywheel or driveplate housing (Fig. 4.12).

9 Check that the mark on the rear of the camshaft gear is aligned with the upper edge of the valve cover gasket. If the mark is on the opposite side of the cylinder head, turn the crankshaft forward one complete revolution and again align the TDC mark, (Fig. 4.13).

10 With the distributor cap removed, turn the rotor arm so that the centre of the metal contact is aligned with the mark on the rim of the distributor body (Fig. 4.14).

11 Hold the distributor over the location aperture in the rear of the cylinder head with the vacuum unit facing downward.

12 Insert the distributor fully. As the drivegear meshes with the camshaft, the rotor will turn slightly anti-clockwise, and the body can then be realigned with the rotor arm so that the vacuum unit faces slightly to the right (Fig. 4.15).

13 Fit the clamp and tighten the bolt.

8.5 Align the rotor arm metal contact with the TDC mark (arrowed) on the distributor body rim (four-cylinder engine)

Fig. 4.11 Distributor is fitted with rotor arm pointing towards cylinder No 1 firing point (Sec 8)

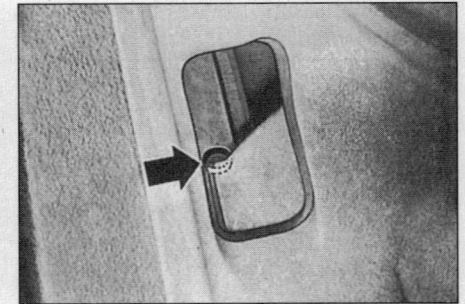

Fig. 4.12 Aligning TDC timing marks on the five-cylinder engine (Sec 8)

Fig. 4.13 TDC timing mark location on the rear of the camshaft gear on five-cylinder engines (Sec 8)

Fig. 4.14 Set distributor arm contact in line with mark on distributor body prior to fitting (five-cylinder engines) (Sec 8)

Fig. 4.15 Distributor fitted position with rotor arm at TDC on No 1 cylinder (five-cylinder engines) (Sec 8)

9.7 Flywheel dynamic timing mark aligned with pointer on bellhousing (four-cylinder engine)

Fig. 4.16 Plug connections to the digital idle stabilizer (DIS) (Sec 9)

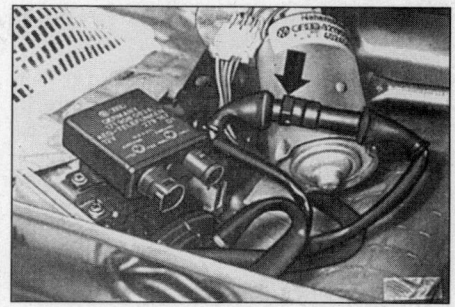

Fig. 4.17 Connect the DIS plugs together (arrowed) when making ignition timing checks (Sec 9)

9 Ignition timing – dynamic setting

1 Check and if necessary adjust the idle speed and idle mixture setting as described in Chapter 3.

2 Where a digital idle stabilizer is fitted, both plugs must be disconnected from the unit and the plugs connected together. Switch the engine off. (Figs. 4.16 and 4.17).

3 Before making the dynamic timing check, the distributor vacuum unit hose on some 1.6 litre engine models must be detached, see Specifications.

4 When checking the ignition timing, the engine must be at normal operating temperature with the choke fully open, but with the radiator cooling fan stationary.

5 The ignition timing may be checked using a digital tester connected to the TDC sensor in the flywheel or driveplate housing. However, as this equipment is not normally available to the home mechanic, the following method describes the use of a timing light.

6 Connect the timing light to the engine in accordance with the manufacturer's instructions.

7 Run the engine at idling speed and direct the timing light through the timing aperture in the flywheel or driveplate housing. The mark on the flywheel or driveplate should appear in line with the pointer or reference edge on the housing. If adjustment is necessary, loosen the clamp bolt and turn the distributor body until the correct position is achieved, then tighten the bolt (photo).

8 Gradually increase the engine speed and check that the ignition advances – the centrifugal advance can be checked by pinching the vacuum hoses, and an indication of the vacuum advance can be obtained by releasing the hoses and noting that a different advance occurs.

9 Switch off the engine and remove the timing light. Reconnect the vacuum hose and digital idle stabilizer plugs, as applicable.

10 Ignition coil – testing

1 It is rare for an ignition coil to fail, but if there is reason to suspect it, use an ohmmeter to measure the resistance of the primary and secondary circuits.

2 Measure the primary resistance between the terminals 1 and 15 and the secondary resistance between the centre HT terminal and terminal 1. The correct resistance values are given in the Specifications (Figs. 4.18 and 4.19).

11 Spark plugs and HT leads – general

1 The correct functioning of the spark plugs is vital for the correct running and efficiency of the engine. The spark plugs should be renewed every 10 000 miles (15 000 km). However, if misfiring or bad starting is experienced in the service period. they must be removed, cleaned and regapped.

2 The condition of the spark plugs will also tell much about the overall condition of the engine.

3 If the insulator nose of the spark plug is clean and white, with no deposits, this is indicative of a weak mixture, or too hot a plug. (A hot plug transfers heat away from the electrode slowly – a cold plug transfers it away quickly).

4 If the tip and insulator nose is covered with hard black-looking deposits, then this is indicative that the mixture is too rich. Should the plug be black and oily, then it is likely that the engine is fairly worn, as well as the mixture being too rich.

5 If the insulator nose is covered with light tan to greyish brown deposits, then the mixture is correct and it is likely that the engine is in good condition.

6 If there are any traces of long brown tapering stains on the outside of the white portion of the plug, then the plug will have to be renewed, as this shows that there is a faulty joint between the plug body and the insulator, and compression is being lost.

7 Plugs should be cleaned by a sand blasting machine, which will free them from carbon more thoroughly than cleaning by hand. The machine will also test the condition of the plugs under compression. Any plug that fails to spark at the recommended pressure should be renewed.

8 The spark plug gap is of considerable importantance, as, if it is too large or too small, the size of the spark and its efficiency will be seriously impaired. The spark plug gap should be set to the figure given in the Specifications at the beginning of this Chapter.

9 To set it, measure the gap with a feeler gauge, and then bend open, or close the outer plug electrode until the correct gap is achieved. The centre electrode should never be bent as this may crack the insulation and cause plug failure, if nothing worse.

10 Always tighten the spark plugs to the specified torque.

11 Periodically the spark plug leads should be wiped clean and checked for security to the spark plugs.

4

Fig. 4.18 Coil terminals for primary resistance check (Sec10)

Fig. 4.19 Coil terminals for secondary resistance check (Sec 10)

Fault finding – ignition system

1 There are two distinct symptoms of ignition faults. Either the engine will not start or fire, or it starts with difficulty and does not run normally.

2 If the starter motor spins the engine satisfactorily, there is adequate fuel and yet the engine will not start, the fault is likely to be on the LT side.

3 If the engine starts, but does not run satisfactorily, it is more likely to be an HT fault.

Engine fails to start

4 If the starter motor spins the engine satisfactorily, but the engine does not start, first check that the fuel supply to the engine is in order, with reference to Chapter 3.

5 Check for broken or disconnected wires to the coil and distributor and for damp distributor cap and HT leads.

6 Refer to Section 7 and carry out the test procedures described.

Engine starts, but misfires

7 Bad starting and intermittent misfiring can be an LT fault, such as poor connection of the coil or the distributor LT leads.

8 If these are satisfactory look for signs of tracking and burning inside the distributor cap, then check the plug leads, plug caps and plug insulators for signs of damage.

9 If the engine misfires regularly, it indicates that the fault is on one particular cylinder.

Chapter 5 Clutch

For modifications, and information applicable to later models, see Supplement at end of manual

Contents

Degrees of difficulty

Easy, suitable for novice with little experience	**Fairly easy,** suitable for beginner with some experience	**Fairly difficult,** suitable for competent DIY mechanic	**Difficult,** suitable for experienced DIY mechanic	**Very difficult,** suitable for expert DIY or professional

Specifications

Type .... Single dry plate and diaphragm spring cover, cable operation

Driven plate
Diameter .... 190 mm (7.48 in), 200 mm (7.87 in), 210 mm (8.26 in) or 215 mm(8.46 in)

Pressure plate
Maximum distortion (inner edge to outer edge) .... 0.3 mm (0.012 in)
Diaphragm spring finger maximum scoring depth .... 0.3 mm (0.012 in)

Cable free play (at clutch pedal) .... 15 mm (0.6 in)

Torque wrench settings	Nm	lbf ft
Pressure plate (clutch cover) bolts	25	18
Clutch release lever pinch bolt	25	18
Release shaft retaining bolt	15	11
Release bearing guide sleeve	15	11

1 General description

The clutch is of single dry plate type with a diaphragm spring cover, and actuation is by cable.

The clutch cover is bolted to the rear face of the flywheel, and the driven plate is located between the cover pressure plate and the flywheel friction surface. The driven plate hub is splined to the gearbox input shaft and is free to slide along the splines. Friction lining material is riveted to each side of the driven plate, and the driven plate hub incorporates cushioning springs to absorb transmission shocks and ensure a smooth take-up of drive.

When the clutch pedal is depressed, the cable moves the release lever, and the release bearing is forced onto the diaphragm spring fingers. As the centre of the spring is pushed in, the outer part of the spring moves out and releases the pressure plate from the driven plate. Drive then ceases to be transmitted to the gearbox.

When the clutch pedal is released, the diaphragm spring forces the pressure plate into contact with the friction linings on the driven plate and at the same time pushes the

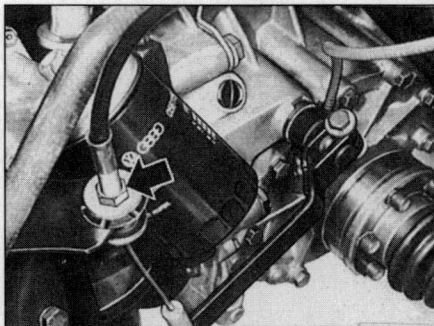
Fig. 5.1 Clutch cable adjuster location (arrowed) – type 014/1 (four-speed) and 013 (five-speed) gearboxes (Sec 2)

driven plate along the input shaft splines into engagement with the flywheel. The driven plate is now firmly sandwiched between the pressure plate and the flywheel and so the drive is taken up.

As the friction linings wear, the pressure plate moves closer to the flywheel and the release cable free play is reduced. Periodic adjustment must therefore be carried out, as described in Section 2. No other routine maintenance is required on the clutch system.

2 Clutch – adjustment

1 The clutch should be checked and, if necessary, adjusted every 10 000 miles (15 000 km).
2 To check the adjustment measure the free play at the clutch pedal. If it is not as given in the Specifications, the cable must be adjusted as follows.

Fig. 5.2 Clutch cable adjuster location (arrowed) – type 093 (five-speed) gearbox (Sec 2)

2.3 Clutch cable and release arm (four-cylinder models)

3.3 Clutch cable ferrule end fitting (014 and 013 type gearbox)

3 On models fitted with the 014 type gearbox (four-speed) and 013 type gearbox (five-speed), loosen off the locknuts at the gearbox end of the cable, reposition the outer cable to give the required free play, then retighten the locknuts (photo).

4 On models fitted with the 093 gearbox (five-speed), extract the adjuster clip from the cable guide and reposition the outer cable accordingly to provide the correct free play, then refit the adjuster clip into position in the appropriate cable groove (Fig. 5.2).

5 On all models fully depress the clutch pedal several times, then recheck the adjustment. Where applicable, lightly grease the adjuster threads to prevent rusting.

3 Clutch cable – renewal

1 Disconnect the battery negative lead.
2 Slacken off the cable adjustment by loosening the locknut(s) and turning the ferrule or extracting the spring clip (as applicable).
3 Release the outer cable from the gearbox bracket and unhook the inner cable from the release arm (photo).
4 Working inside the car, unhook the inner cable from the top of the clutch pedal.
5 Withdraw the clutch cable through the bulkhead into the engine compartment.
6 Fit the new cable using a reversal of the removal procedure, and finally adjust the cable as described in Section 2.

4 Clutch pedal – removal and refitting

1 The clutch and brake pedals share a common bracket assembly and pivot shaft.
2 Disconnect the battery negative lead.
3 Unhook the inner cable from the top of the clutch pedal, with reference to Section 3.
4 On five-cylinder engine models an over-centre return spring is fitted at the top of the pedal. Disconnect it by extracting the circlip and withdrawing the pivot pin (Fig. 5.3).
5 Extract the clip from the end of the pivot shaft and slide off the clutch pedal.
6 Clean the pedal and shaft, then temporarily refit the pedal and check the bushes for wear. If necessary, the bushes can be pressed or driven out and new bushes fitted.
7 The inner bush is of plastic and the outer bush is rubber. To fit the bushes just dip the rubber bush in soapy water and press it into the pedal, then similarly press the plastic bush in from the shouldered side of the pedal flush.
8 Refitting is a reversal of removal, but first lubricate the pivot shaft with a little grease. Finally check the cable free play and, if necessary, adjust it as described in Section 2.

5 Clutch – removal, inspection and refitting

1 Access to the clutch is obtained either by removing the engine (Chapter 1) or by removing the gearbox (Chapter 6). If the clutch requires attention and the engine is not in need of a major overhaul, it is preferable to gain access to the clutch by removing the gearbox, provided that either a pit is available, or the car can be put on ramps to give a good ground clearance (photo).
2 Put a mark on the rim of the clutch pressure plate cover and a corresponding mark on the flywheel so that the clutch can be refitted in exactly the same position.
3 Slacken the clutch cover retaining bolts a

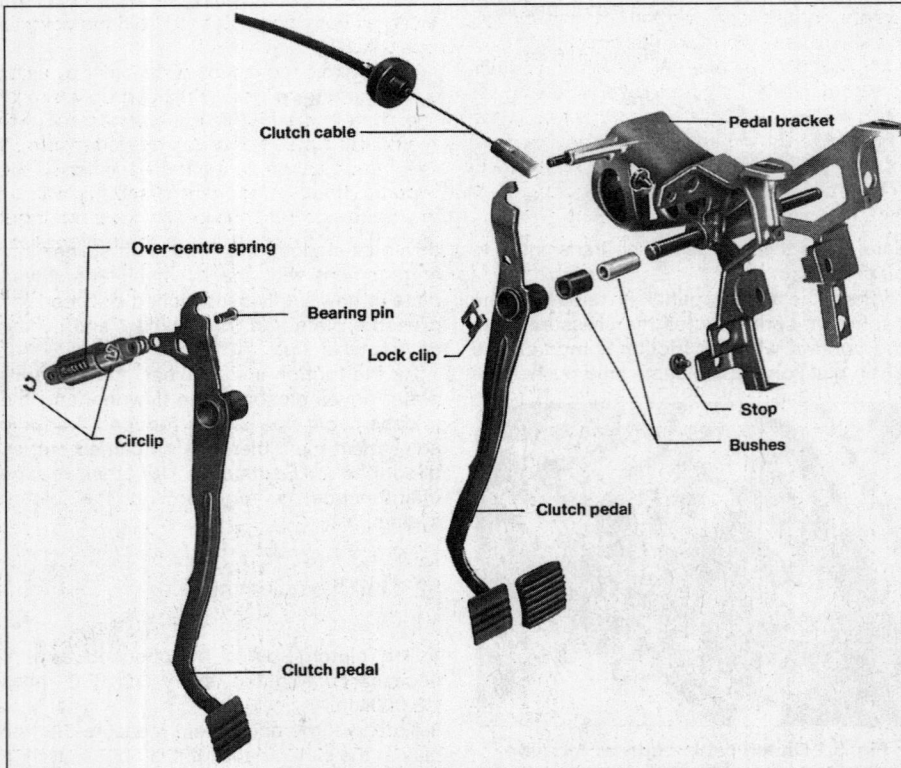

Fig. 5.3 Clutch pedal and support bracket components (Sec 4)

5.1 Engine and gearbox separation gives access to the clutch unit

5.3 Removing the clutch cover and friction plate

Fig. 5.4 Examine the clutch diaphragm spring ends at points arrowed (Sec 5)

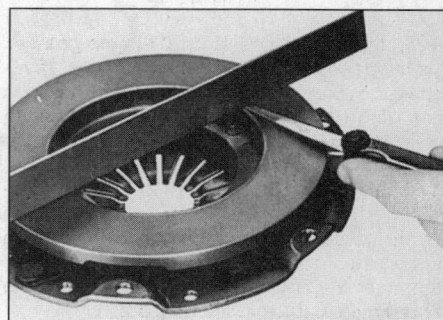

Fig. 5.5 Examine pressure plate for wear and distortion (Sec 5)

turn at a time, working in diagonal pairs round the casing. When all the bolts have been loosened enough to release the tension of the diaphragm spring, remove the bolts and lift off the clutch cover and the friction plate (photo) .

4 Clean the parts with a damp cloth, ensuring that the dust is not inhaled. *Because the dust produced by the wearing of the clutch facing may contain asbestos, which is dangerous to health, parts should not be blown clean or brushed to remove dust.*

5 Examine the fingers of the diaphragm spring for signs of wear, or scoring. If the depth of any scoring exceeds 0.3 mm (0.012 in), a new cover assembly must be fitted (Fig. 5.4).

6 Lay the clutch cover on its diaphragm spring end, place a steel straight edge diagonally across the pressure plate and test for distortion of the plate (Fig. 5.5). If a 0.3 mm (0.012 in) feeler gauge can be inserted in any gap beneath the straight edge, the clutch cover must be discarded and a new one fitted. The check for distortion should be made at several points round the plate.

7 Check that the pressure plate is not badly scored, and shows no signs of cracking, or

burning. Note that the retaining bolts should be renewed on reassembly.

8 Inspect the friction plate and fit a new plate if the surface of the friction material left is approaching the level of the rivets. Discard the plate if the friction material has become impregnated with oil, or shows signs of breaking into shreds (Fig. 5.6).

9 Examine the friction plate splined hub for signs of damage, or wear. Check that when the hub is on the gearbox input shaft, the hub slides smoothly along the shaft and that the radial clearance between the gearbox shaft and clutch hub is small.

10 If there is reason to suspect that the clutch hub is not running true, it should be checked by mounting the hub between centres and checking it with a dial gauge. Unless you have the proper equipment, get your local dealer to make this check.

11 Do not re-use any part which is suspect. Having gone to the trouble of dismantling the clutch, it is well worth ensuring that when reassembled it will operate satisfactorily for a long time. Check the flywheel for scoring and tiny cracks caused by overheating; refinish or renew as necessary (see Chapter 1).

12 Ensure that all the parts are clean, free of oil and grease and are in a satisfactory condition before reassembling.

13 Fit the friction plate so that the torsion spring cages are towards the pressure plate, (photo).

14 Fit the clutch cover to the flywheel

ensuring (where applicable) that the marks made before dismantling are lined up, and insert all bolts finger tight to hold the cover in position.

15 Centralise the friction plate either by using a proprietary tool, or by making up a similar tool to hold the friction plate concentric with the hole in the end of the crankshaft (photo). If the clutch cover is tightened without the friction plate being centralised, it will be impossible to refit the gearbox as the input shaft will not pass through the driven plate hub and engage the spigot bearing.

16 With the centraliser holding the clutch friction plate in position, tighten all the clutch cover bolts a turn at a time in diagonal sequence until the specified torque is achieved. Remove the centring tool and

5

Fig. 5.6 Clutch friction plate run-out check method (Sec 5)

5.13 Locating a new friction plate into position

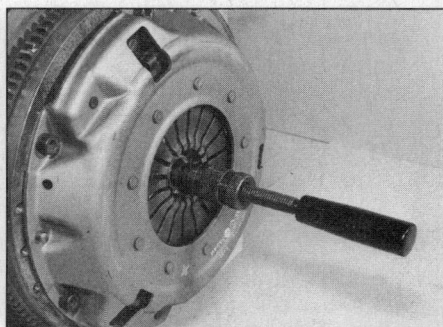

5.15 Centralising the clutch friction plate using a proprietary tool

Bolt
(Secures release shaft,
gearbox 013 only)

Return spring

Rubber bush

Axial guide
(for release shaft,
gearbox 093 only)

**Plastic guide
sleeve**
Do not grease

Circlip

Clutch lever

Clutch cable

Bush

Release shaft

**Retaining
spring**

Clutch cable

Retaining clip

Release bearing

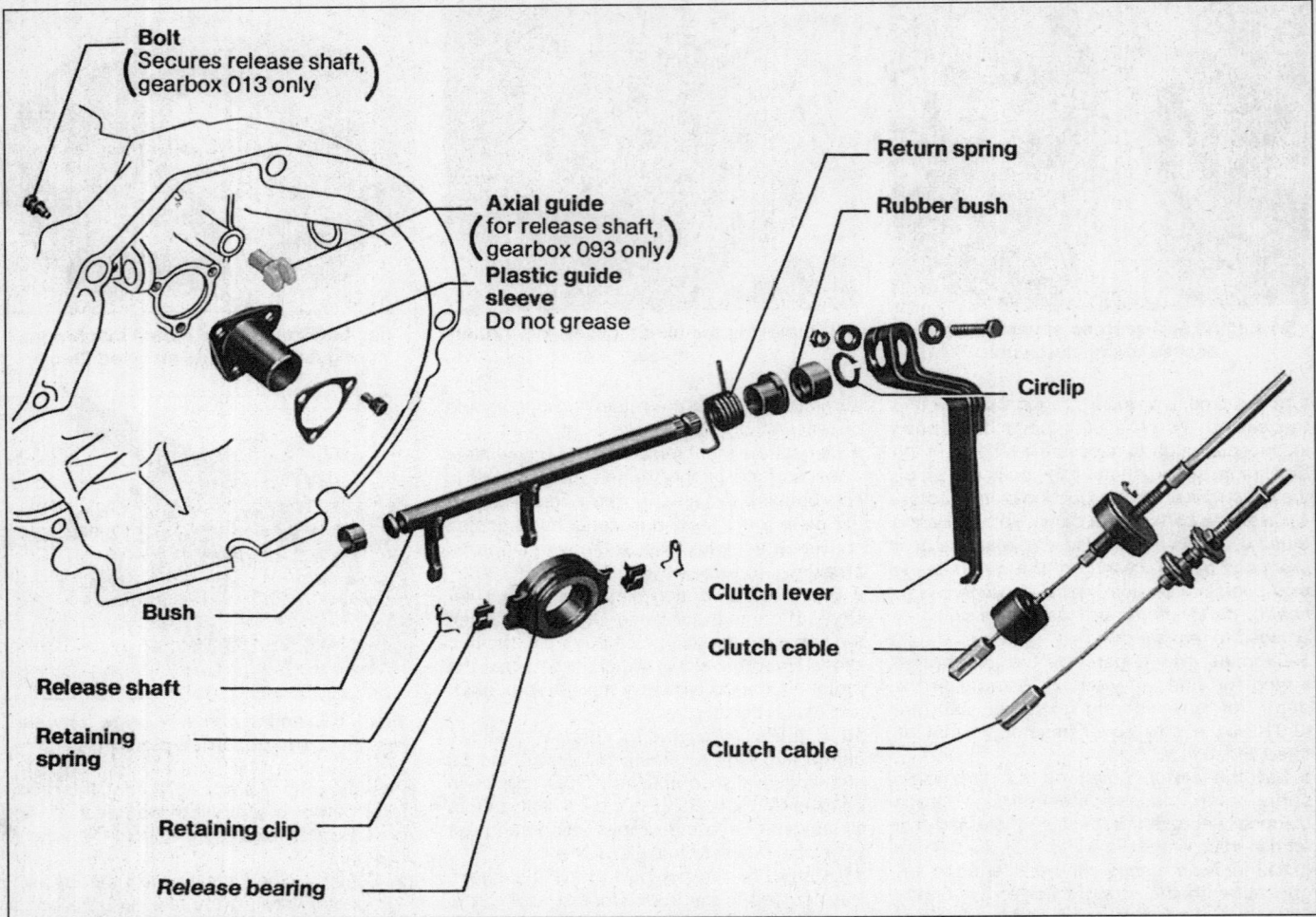

Fig. 5.7 Clutch release mechanism components shown with the two operating cable types available according to gearbox type (Sec 6)

smear the hub splines with molybdenum disulphide grease.

17 Check the release bearing in the front of the gearbox for wear and smooth operation, and if necessary renew it, with reference to Section 6.

18 Refit the engine or gearbox with reference to Chapters 1 or 6.

6 Release bearing and mechanism – removal and refitting

1 To remove the release bearing, either lift the release arm to disengage the forks from the spring clips, or extract the clips from each

side of the bearing, noting how they are fitted (photos). The bearing can then be withdrawn from the guide sleeve.

2 Mark the release lever in relation to the shaft then unscrew the clamp bolt and withdraw the lever (photo).

3 On four-speed (014) and five-speed 013 gearbox models, unscrew and remove the

6.1A Clutch release bearing and release shaft

6.1B Showing the release bearing clips location

6.2 Clutch release lever

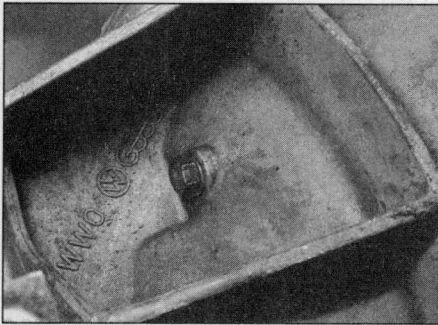
6.3 Release shaft retaining bolt location on four-cylinder models

6.4 Unhooking the release shaft return spring

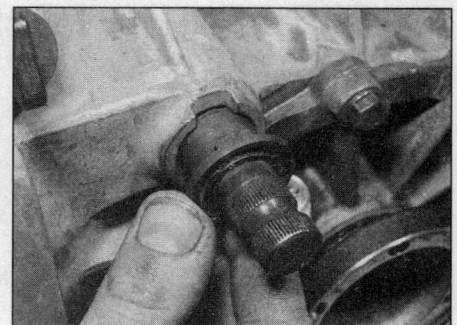
6.5 Removing the release shaft flanged bush

release shaft bolt from the rear of the bellhousing. The bolt engages in the end of the release shaft (photo).
4 Note the position of the return spring on the release shaft then unhook it from the release shaft fork, (photo and Fig. 5.8).
5 Extract the circlip from the splined end of the release shaft and prise out the rubber bush (where fitted) and the flanged bush (photo).
6 Turn the release shaft so that the forks are free of the guide sleeve, then remove the inner end from the bush and withdraw

the shaft from inside the bellhousing (photo).
7 Clean the release bearing with a dry cloth. Do not wash the bearing in solvent, because this will cause its lubricant to be washed out. If the bearing is noisy, or has excessive wear, discard it and obtain a new one.
8 Inspect the release shaft and its bushes for wear. Do not remove the inner bush unless a new one has to be fitted. If a new inner bush is required, the old one will need a special extractor to remove it.
9 Before refitting the release shaft, coat the bearing surfaces with molybdenum disulphide

grease and ensure that the return spring is fitted to the shaft.
10 Refitting is a reversal of removal. However, on four-speed 014 and five-speed models fitted with the 013 gearbox, press in the release shaft until the rubber bush is compressed to approximately 18 mm (0.7 in) (photo) before inserting and tightening the dowel bolt. If a new release lever is being fitted, position it on the splined shaft as shown in Fig. 5.10 (photo). Coat all bearing surfaces with high melting point grease, except for the plastic guide sleeve.

6.6 Removing the release shaft

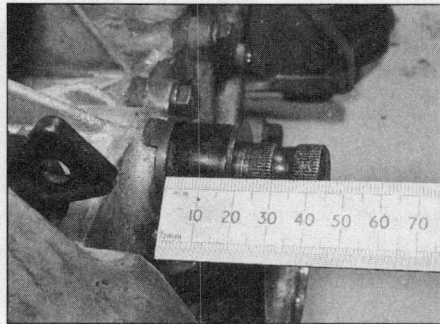
6.10A Checking the release shaft rubber bush dimension

6.10B Checking the release lever fitted dimension

Fig. 5.8 Clutch release mechanism return spring location on the 093 gearbox type (Sec 6)

Fig. 5.9 Insert the dowel bolt and press in the release shaft to compress the rubber bush to 18 mm (0.7 in) – 013 gearbox only (Sec 6)

Fig. 5.10 Release lever fitting dimension (Sec 6)

013 gearbox type – 169 mm (6.65 in)
093 gearbox type – 193 mm (7.6 in)
014 gearbox type – 169 mm (6.65 in)

Fault finding - clutch

Judder when taking up drive

☐ Oil or grease contamination of friction linings
☐ Loose or worn engine/gearbox mountings
☐ Worn friction linings
☐ Worn driven plate or input shaft splines

Clutch slip

☐ Incorrect adjustment
☐ Friction linings worn or contaminated with oil
☐ Weak diaphragm spring

Clutch drag (failure to disengage)

☐ Incorrect adjustment
☐ Driven plate sticking on input shaft splines
☐ Driven plate sticking to flywheel
☐ Input shaft spigot bearing seized

Noise evident on depressing clutch pedal

☐ Dry or worn release bearing
☐ Input shaft spigot bearing dry or worn

Chapter 6 Manual gearbox and final drive

For modifications, and information applicable to later models, see Supplement at end of manual

Contents

Degrees of difficulty

Easy, suitable for novice with little experience	Fairly easy, suitable for beginner with some experience	Fairly difficult, suitable for competent DIY mechanic	Difficult, suitable for experienced DIY mechanic	Very difficult, suitable for expert DIY or professional

Specifications

Type . . . . Four or five forward speeds and reverse. Synchromesh on all forward speeds with integral final drive

Identification

Gearbox code number:

014/1	Four-speed gearbox fitted up to July 1983
014/11	Four-speed gearbox fitted from August 1983
013	Five-speed gearbox fitted to four-cylinder models
093	Five-speed gearbox fitted to five-cylinder models

Ratios

Gearbox 014/1 and 014/11

Gearbox code letters

	MY/IM	2T	4V	5V	8M	UT
Final drive	4.11:1	4.11:1	4.11:1	4.11:1	4.11:1	4.11:1
1st	3.45:1	3.45:1	3.45:1	3.45:1	3.45:1	3.45:1
2nd	1.94:1	1.79:1	1.94:1	1.94:1	1.78:1	1.79:1
3rd	1.29:1	1.13:1	1.28:1	1.28:1	1.28:1	1.065:1
4th	0.91:1	0.83:1	0.91:1	0.88:1	0.90:1	0.77:1
Reverse	3.17:1	3.17:1	3.17:1	3.17:1	3.17:1	3.17:1

Gearbox 013

	QJ	QJ*	3M	2P	9G	2N	8G
Final drive	4.11:1	4.11:1	4.11:1	4.11:1	4.11:1	4.11:1	4.11:1
lst	3.45:1	3.45:1	3.45:1	3.45:1	3.45:1	3.45:1	3.45:1
2nd	1.94:1	1.94:1	1.94:1	1.94:1	1.94:1	1.78:1	1.78:1
3rd	1.29:1	1.28:1	1.28:1	1.28:1	1.28:1	1.13:1	1.13:1
4th	0.90:1	0.90:1	0.90:1	0.96:1	0.90:1	0.82:1	0.82:1
5th	0.68:1	0.73:1	0.73:1	0.80:1	0.73:1	0.68:1	0.68:1
Reverse	3.17:1	3.16:1	3.16:1	3.16:1	3.16:1	3.16:1	3.16:1

*From September 1983

Gearbox 093

	VW and 2V
Final drive	4.90:1
1st	2.85:1
2nd	1.52:1
3rd	0.96:1
4th	0.70:1
5th	0.53:1
Reverse	3.17: 1

Gearbox lubrication

Lubricant type . . . . Hypoid gear oil, viscosity SAE 80 or SAE 80W/90, to API-GL4
Oil capacity:
    Gearbox 014 . . . . 1.7 litre (3.0 Imp pint)
    Gearbox 013 . . . . 2.0 litre (3.5 Imp pint)
    Gearbox 093 . . . . 2.4 litre (4.1 Imp pint)

Synchro ring to cone clearance

	New clearance	Wear limit
1st and 2nd	1.1 to 1.7 mm (0.04 to 0.06 in)	0.5 mm (0.02 in)
3rd and 4th	1.35 to 1.9 mm (0.05 to 0.07 in)	0.5 mm (0.02 in)
5th	1.1 to 1.7 mm (0.04 to 0.06 in)	0.5 mm (0.02 in)

4th gear axial clearance (maximum) . . . . 0.10 to 0.40 mm (0.004 to 0.015 in)
4th gear thrust washers available – all models except 014/11 gearbox . . 3.47 mm (0.136 in)
    3.57 mm (0.141 in)
    3.67 mm (0.144 in)

4th gear circlips available – 014/11 gearbox, and internal
spline gear circlips available – 013 and 093 gearbox) . . . . 2.35 mm (0.0925 in)
    2.38 mm (0.0937 in)
    2.41 mm (0.0948 in)
    2.44 mm (0.0960 in)
    2.47 mm (0.0972 in)

Torque wrench settings

	Nm	lbf ft
Gearbox to engine bolts	55	41
Driveshafts to gearbox	45	33
Drive flange to gearbox:		
014	20	15
013 and 093	25	18
Subframe to body	70	51
Front gearbox support to gearbox	25	18
Bonded rubber mount to body	110	80
Bonded rubber mount to gearbox	25	18
Gearbox mounting to gearbox (093)	40	30
Coupling to steering (093)	45	33
Filler and drain plugs	25	18
Reverse relay lever bolt	35	26
Reverse light switch (014)	30	22
Pinion nut	100	74
Input shaft bolt	45	33

1 General description

Either a four- or five-speed gearbox is fitted according to model, and there are two variations of each gearbox, as listed in the Specifications (Figs. 6.1 and 6.2).

The gearbox is bolted to the rear of the engine in conventional manner, but because of the front wheel drive configuration, drive is transmitted to a differential unit located at the front of the gearbox and then through the driveshafts to the front wheels. All forward gears incorporate synchromesh engagement, and reverse gear is obtained by engaging a spur type idler gear with the 1st/2nd synchro sleeve on the output shaft and a spur gear on the input shaft.

Gearshift is by means of a floor mounted lever, and a single rod and linkage clamped to the selector rod which protrudes from the rear of the gearbox. The selector rod incorporates a finger which engages the other selector rod in the bearing carrier.

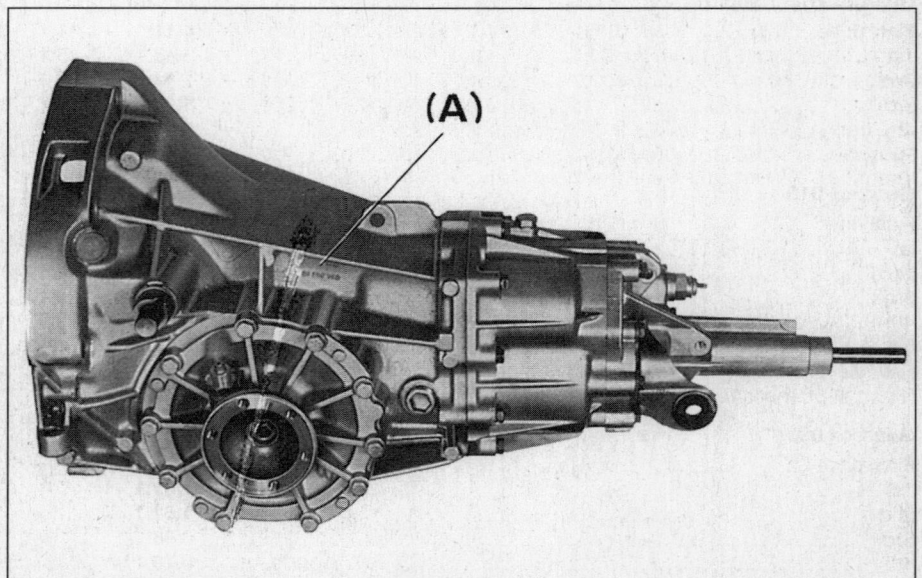

Fig. 6.1 Four-speed manual gearbox showing identification mark location (A) (Sec 1)

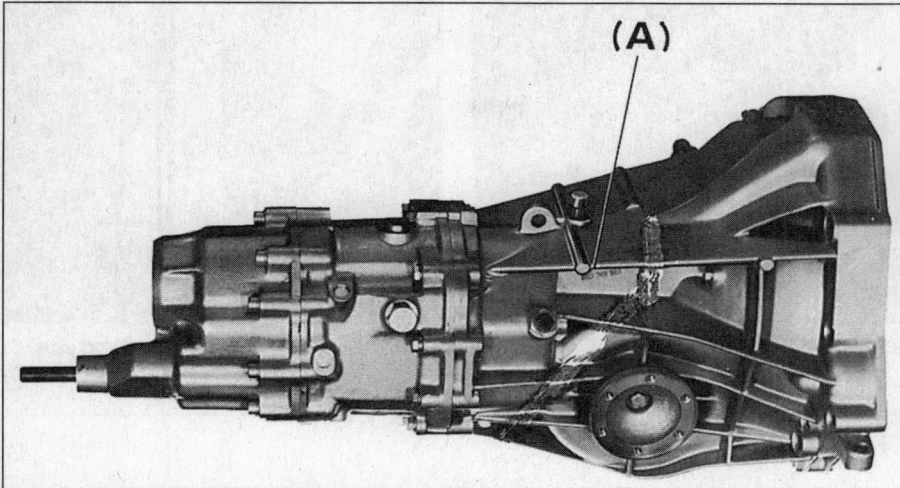

Fig. 6.2 Five-speed manual gearbox showing identification mark location (A) (Sec 1)

2.2 Topping up the gearbox oil level

2 Routine maintenance – gearbox

The manual gearbox requires the minimum amount of maintenance, only the following checks need be made at the specified intervals given at the front of this manual (see Routine Maintenance).

1 Check gearbox for signs of oil leaks: If possible run the vehicle over an inspection pit or raise and support it on safety stands to make this (and the following) check. Inspect the gearbox casing for any signs of serious oil leaks. Oil leakage from the transmission will necessitate further investigation and if serious must be remedied without delay. A very minor leak may be permissible providing regular checks are made to ensure that the leak does not get any worse and to ensure that the gearbox oil level is maintained. Do not confuse gearbox oil leaks with engine oil leaks which may have sprayed onto the gearbox casing.

2 Check the gearbox oil level. The vehicle must be parked level for this check. Remove the oil level/filler plug from the side of the gearbox and check that the oil level is up to the base of the filler orifice. If not, top up with the specified grade of oil and refit the plug (photo).

3 Gearbox – removal and refitting

Note: If necessary, the engine and gearbox on four-cylinder models can be removed together as described in Chapter 1, and the gearbox then separated on the bench. However, this Section describes the removal of the gearbox leaving the engine in situ.

1 Position the front of the car over an inspection pit or on car ramps and apply the handbrake firmly.
2 Disconnect the battery negative lead.
3 Unscrew and remove the upper bolts attaching the gearbox to the engine, noting the location of any brackets.
4 Working underneath the vehicle, disconnect the reverse light switch, gear shift and econometer lead wires and release the cable clip(s).
5 Unscrew the nut and disconnect the speedometer cable from the differential cover or gearbox casing (as applicable).

6 Detach the clutch cable from the gearbox with reference to Chapter 5.
7 Remove the front exhaust downpipe from the manifold and exhaust system with reference to Chapter 3 (photo).
8 On five-cylinder models only, remove the air cleaner (Chapter 3), and also support the front of the engine with a hoist or bar similar to that shown in Fig. 6.3.
9 Disconnect the driveshafts from the drive flanges with reference to Chapter 8 and tie them to one side.
10 On four-cylinder models only, unbolt and remove the centre engine mounting bracket (Fig. 6.4).
11 Unbolt and remove the gearbox front cover.
12 Remove the starter motor, as described in Chapter 10.
13 On five-cylinder models only, disconnect the steering tie-rod bracket from the steering gear with reference to Chapter 11 (Fig. 6.5).
14 Support the gearbox with a trolley jack or stand.
15 Unscrew the lockbolt securing the selector adaptor to the selector rod on the rear of the gearbox (photo). On four-speed gearboxes it may be necessary to first remove the locking wire.
16 Press the support rod from the balljoint and withdraw the adaptor from the selector lever (photo).
17 Unscrew and remove the lower bolts attaching the gearbox to the engine.

6

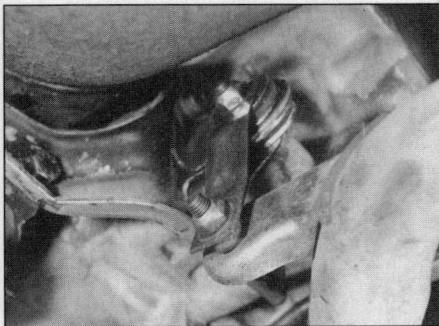

3.7 Exhaust to transmission bracket location

Fig. 6.3 Support bar required for gearbox removal on five-cylinder models – typical (Sec 3)

Fig. 6.4 Unbolt the central engine mounting bracket – four-cylinder models (Sec 3)

Fig. 6.5 Detach the steering tie-rod bracket – five-cylinder models (Sec 3)

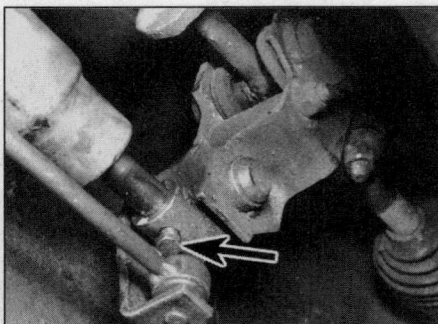

3.15 Remove the adaptor to selector (shift) rod lockbolt (arrowed)

3.16 Disconnect the support rod and selector rod

18 Loosen the bolt securing the gearbox stay to the underbody (photo). Unscrew and remove the stay inner bolt and pivot the stay from the gearbox.

19 Unbolt the bonded rubber mounting from the gearbox.

20 *On four-cylinder models only*, unbolt and remove the front gearbox support bracket.

21 With the help of an assistant withdraw the gearbox from the engine, making sure that the input shaft does not hang on the clutch. Lower the gearbox to the ground.

22 Refitting is a reversal of removal, but lightly lubricate the splines of the input shaft with molybdenum disulphide powder or paste and make sure that the engine/gearbox mountings are free of strain; if necessary refer to Chapter 1 for the correct alignment procedure. Where applicable secure the selector adaptor lockbolt with new locking wire. Adjust the gear lever and linkage as described in Section 11 or 21. Check and, if necessary, top up the oil level.

4 Gearbox – dismantling

Overhauling a manual transmission unit is a difficult and involved job for the DIY home mechanic. In addition to dismantling and reassembling many small parts, clearances must be precisely measured and, if necessary, changed by selecting shims and

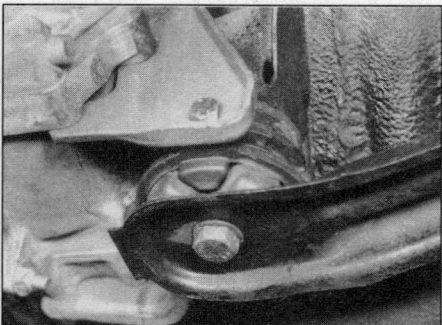

3.18 Gearbox mounting and stay

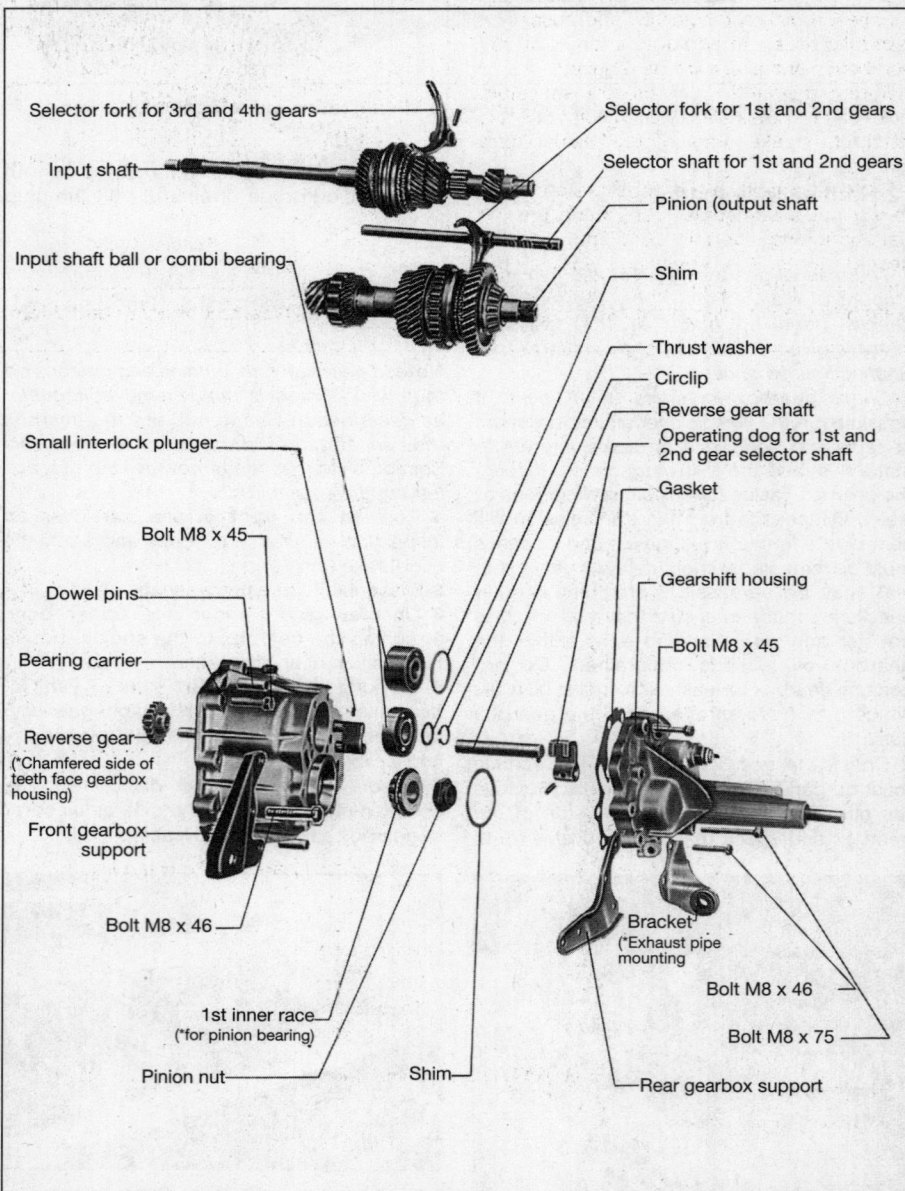

Fig. 6.6 Exploded view of the bearing carrier and gearshift housing and associated components (Sec 4)

spacers. Internal transmission components are also often difficult to obtain, and in many instances, are extremely expensive. Because of this, if the transmission develops a fault or becomes noisy, the best course of action is to have the unit overhauled by a specialist repairer, or to obtain an exchange reconditioned unit.

Nevertheless, it is not impossible for the more experienced mechanic to overhaul the transmission, provided the special tools are available, and that the job is done in a deliberate step-by-step manner so that nothing is overlooked.

The tools necessary for an overhaul may include internal and external circlip pliers, bearing pullers, a slide hammer, a set of pin punches, a dial test indicator, and possibly a hydraulic press. In addition, a large, sturdy workbench and a vice will be required.

During dismantling of the transmission, make careful notes of how each component is fitted, to make reassembly easier and accurate.

Before dismantling the transmission, it will help if you have some idea which area is malfunctioning. Certain problems can be closely related to specific areas in the gearbox, which can make component examination and replacement easier.

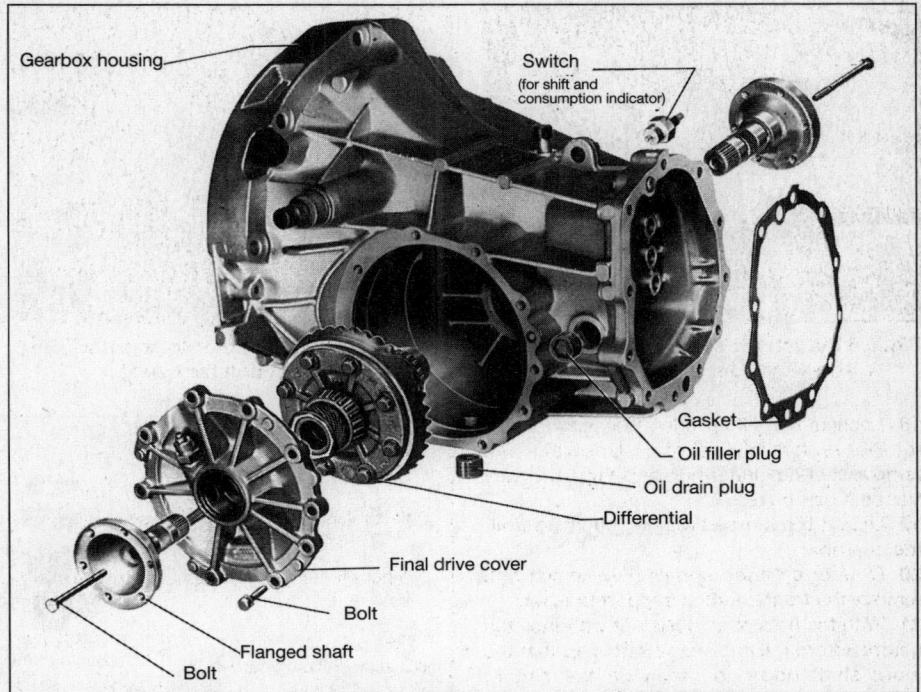

Fig. 6.7 Gearbox housing and differential components 014 gearbox (Sec 4)

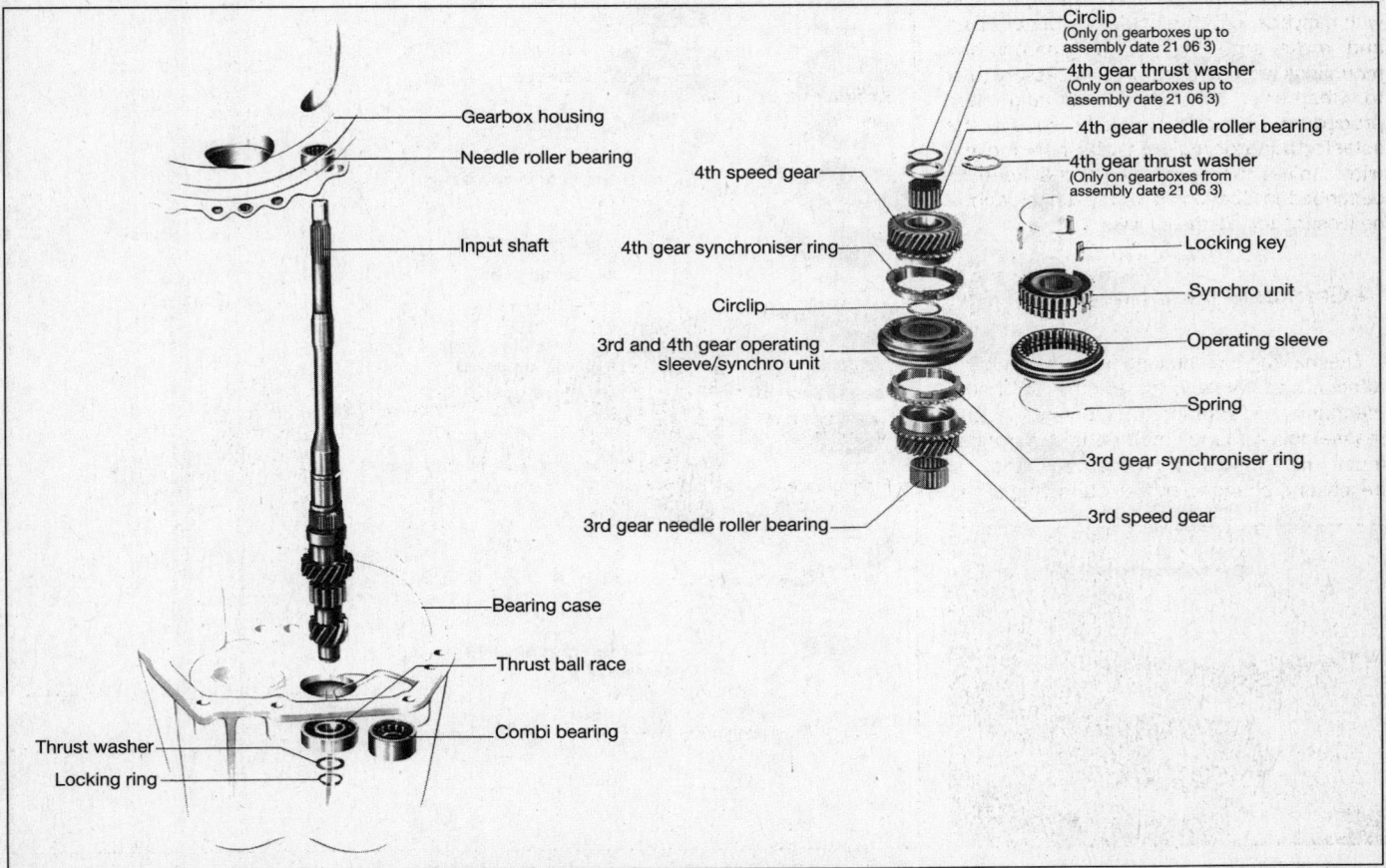

Fig. 6.8 Input shaft components on the 014 gearbox (Sec 4)

Pinion nut

1st inner race/ball bearing

Ball bearing

Pinion nut

Inner race/double taper roller bearing

Outer race/double taper roller bearing*

Adjustment shim S₃

Bearing carrier

2nd inner race/ball bearing

Shim S₃

1st gear needle bearing

Inner race/double taper roller bearing

Needle roller bearing for 1st speed gear*

Locking key

1st speed gear

1st gear synchro ring

Synchro hub

1st/2nd gear synchro sleeve and hub

Sleeve

Spring

2nd gear synchro ring

2nd gear needle bearing

2nd speed gear

3rd speed gear

Circlip

Pinion

(wide shoulder goes towards 4th gear)

4th speed gear

(wide shoulder goes towards pinion head)

Roller bearing

Pin

Roller bearing outer race

Gearbox housing

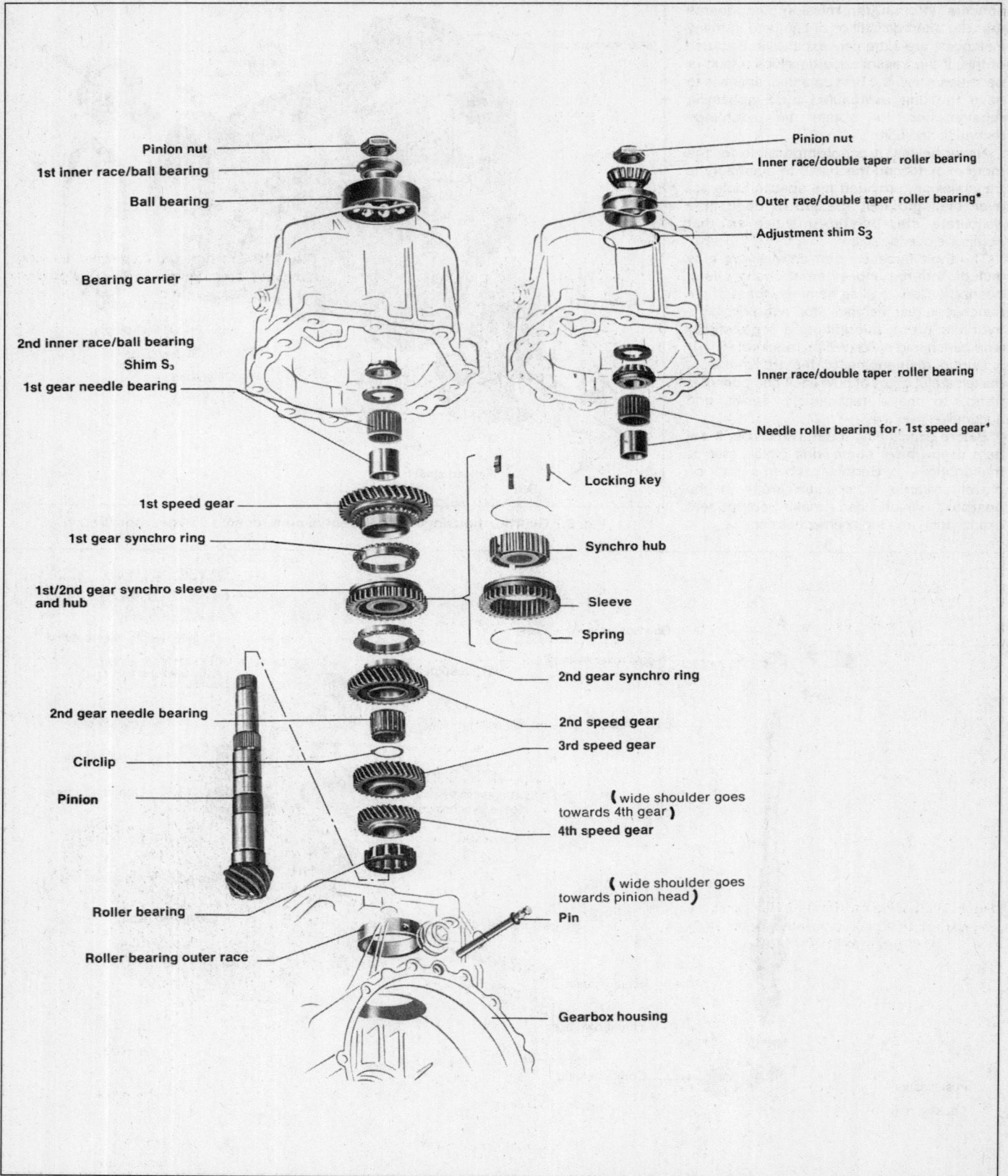

Fig. 6.9 Pinion shaft components on the 014 gearbox with inset showing later (014/II) pinion shaft rear bearing assembly (Sec 4)

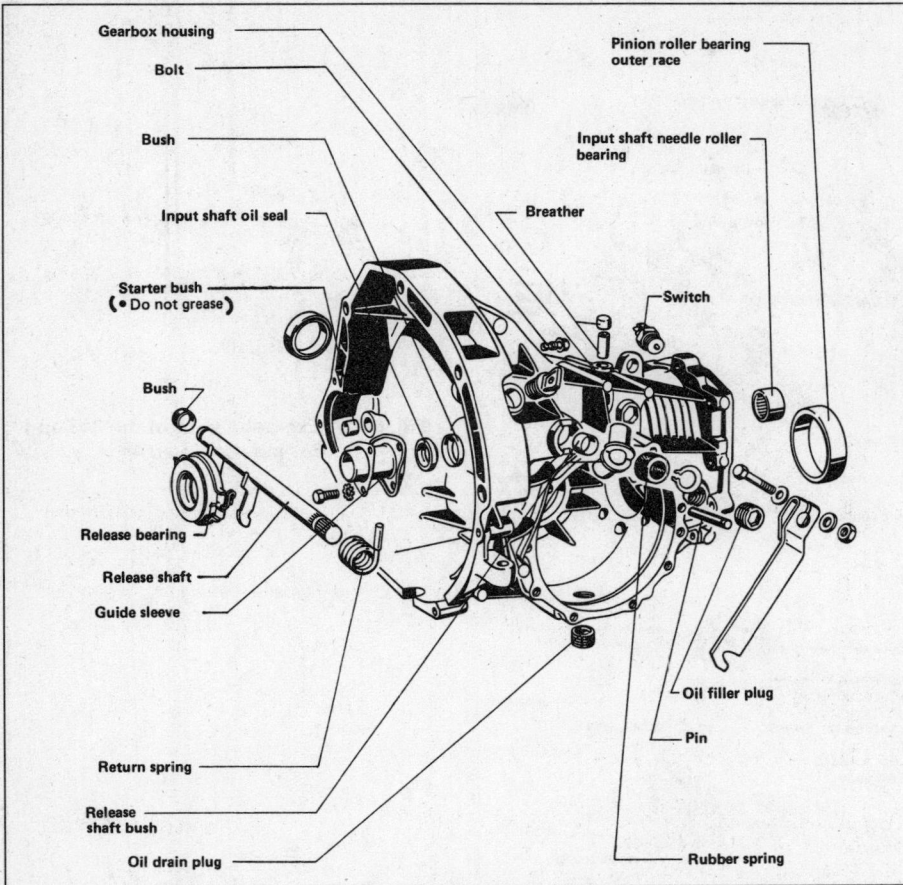

**Fig. 6.10 Gearbox and final drive housing
and associated components – 014 gearbox
(Sec 4)**

Gearbox housing
Bolt
Bush
Input shaft oil seal
Starter bush
(• Do not grease)
Bush
Release bearing
Release shaft
Guide sleeve
Return spring
Release
shaft bush
Oil drain plug

Pinion roller bearing
outer race
Input shaft needle roller
bearing
Breather
Switch
Oil filler plug
Pin
Rubber spring

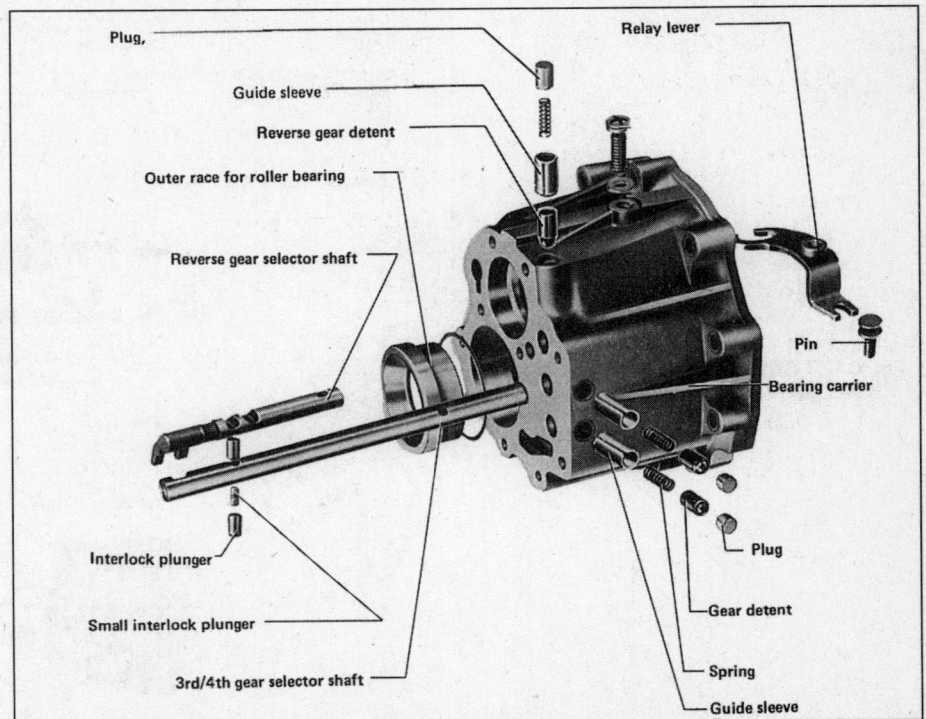

**Fig. 6.11 Bearing carrier on 014/II gearbox
– gearbox 014/I components similar, but
ball-bearing fitted (Sec 4)**

Plug,
Guide sleeve
Reverse gear detent
Outer race for roller bearing
Reverse gear selector shaft
Interlock plunger
Small interlock plunger
3rd/4th gear selector shaft

Relay lever
Pin
Bearing carrier
Plug
Gear detent
Spring
Guide sleeve

6

Gear train

Reversing light switch

Drive flange

Gasket

Dowel pins

Gearbox housing

Final drive cover

Oil filler plug

Dowel sleeve

Hex head bolt,

Oil drain plug

Input shaft

Hex head bolt M 8 x 35 with washer

Differential

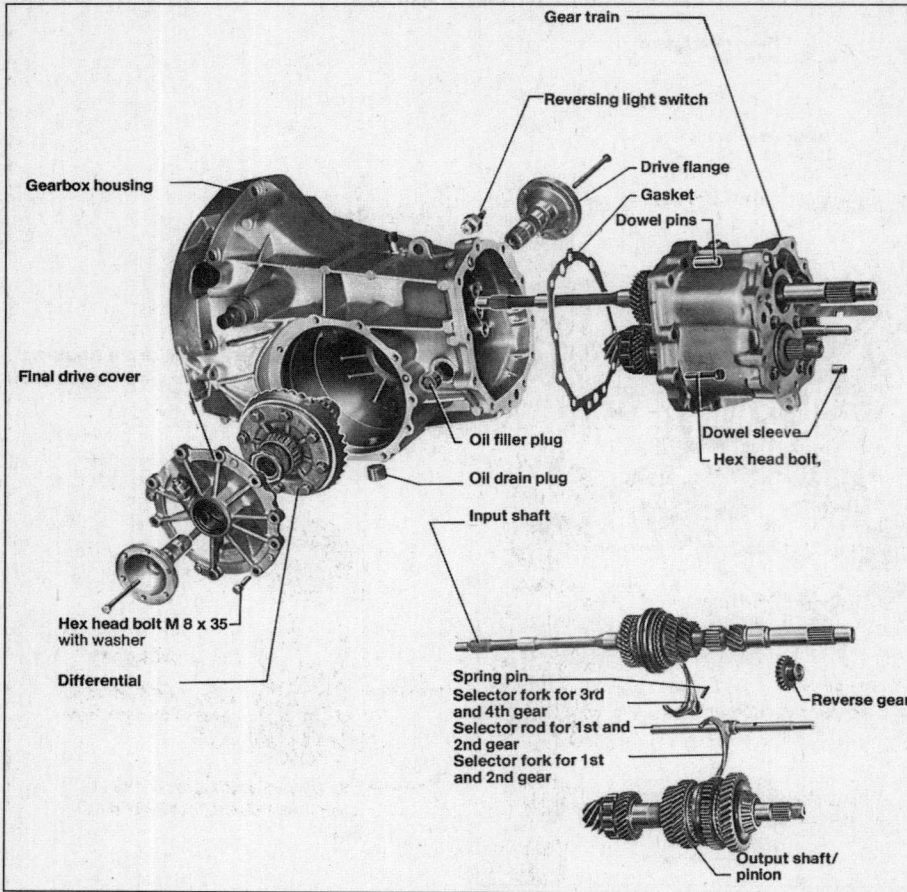

Fig. 6.12A Exploded view of the 013 and 093 gearbox (Sec 4)

Spring pin
Selector fork for 3rd and 4th gear
Selector rod for 1st and 2nd gear
Selector fork for 1st and 2nd gear

Reverse gear

Output shaft/ pinion

Fig. 6.12B Exploded view of the 013 and 093 gearbox (Sec 4) (continued)

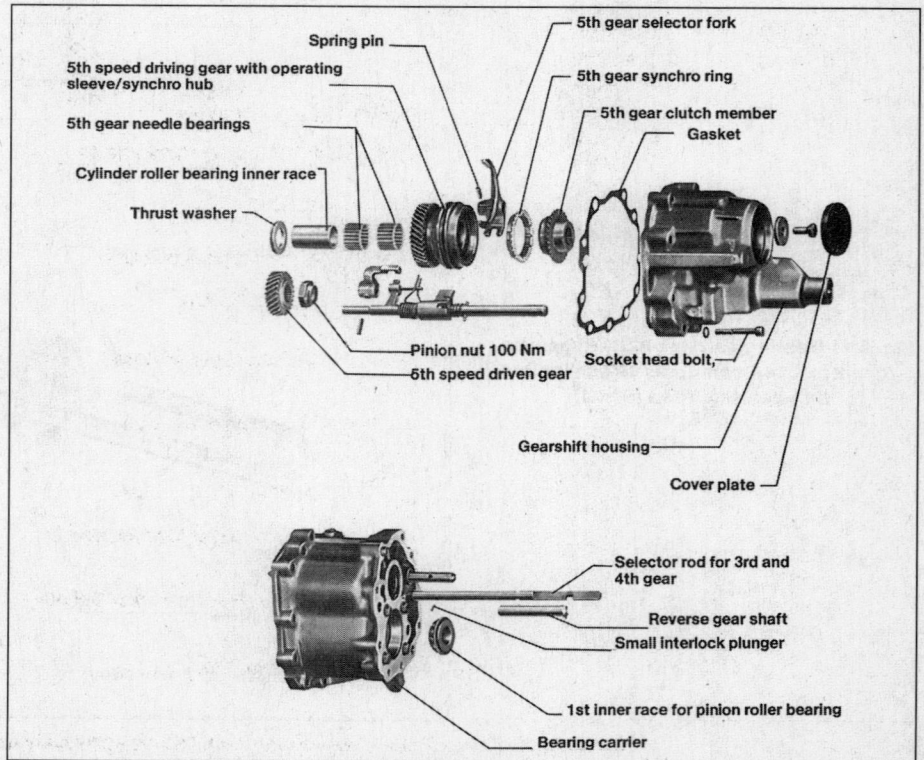

Spring pin

5th gear selector fork

5th speed driving gear with operating sleeve/synchro hub

5th gear synchro ring

5th gear needle bearings

5th gear clutch member

Gasket

Cylinder roller bearing inner race

Thrust washer

Pinion nut 100 Nm

Socket head bolt,

5th speed driven gear

Gearshift housing

Cover plate

Selector rod for 3rd and 4th gear

Reverse gear shaft

Small interlock plunger

1st inner race for pinion roller bearing

Bearing carrier

Circlip

Bearing carrier

Cylinder roller bearing

Gearshift housing

Inner race/cylinder roller bearing

Ball bearing

Input shaft

Circlip

Baffle plate

Spring

Operating sleeve

5th gear synchro hub

5th gear synchro ring

Locking key

3rd gear
3rd gear synchro ring

5th speed driving gear(with synchro hub)

3rd/4th gear operating sleeve/synchro hub

5th speed driving gear with operating sleeve and synchro hub

3rd gear needle bearing

Thrust washer

Operating sleeve

5th gear needle bearing

Locking key

Synchro hub

4th gear needle bearing

Spring

Circlip

Circlip for synchro hub

4th gear synchro ring

4th gear

Thrust washer for 4th gear

Needle bearing for input shaft

Gearbox housing

6

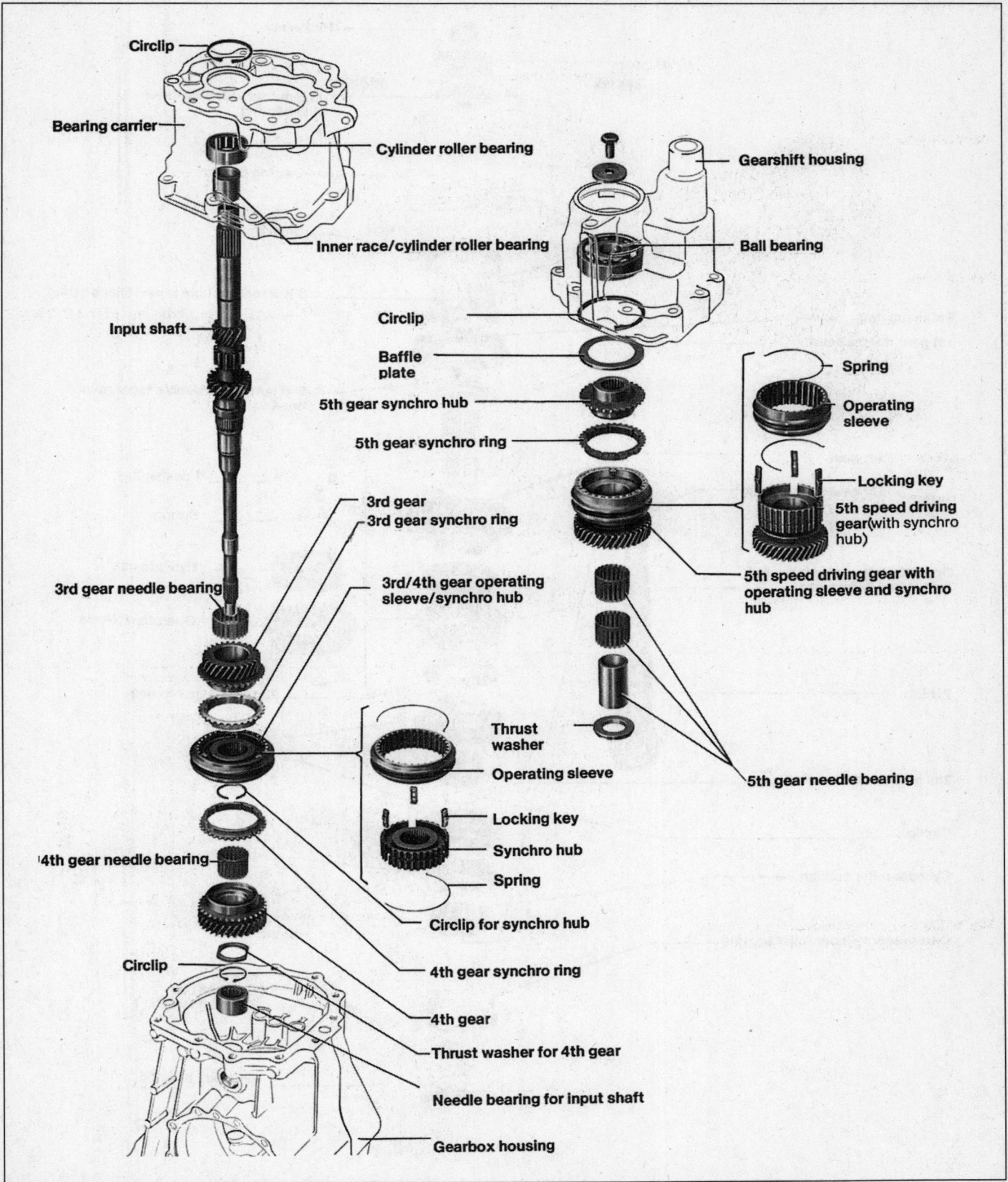

Fig. 6.13 Input shaft components on 013 and 093 gearboxes (Sec 4)

Pinion nut

5th gear

1st inner race/double taper roller bearing

Bearing carrier

Shim S$_3$

Outer race/double taper roller bearing

Retaining ring

1st gear needle bearing

2nd inner race/double taper roller bearing

1st gear

1st gear synchro ring

Locking key

Spring

Synchro hub

Operation sleeve/synchro hub for 1st and 2nd gears

Operating sleeve

Pinion

2nd gear synchro ring

2nd gear needle bearing

2nd gear

3rd gear

Circlip

4th gear

Cylinder roller bearing

Outer race/cylinder roller bearing

Pin
(Gearbox 093:
Socket head bolt)

Gearbox housing

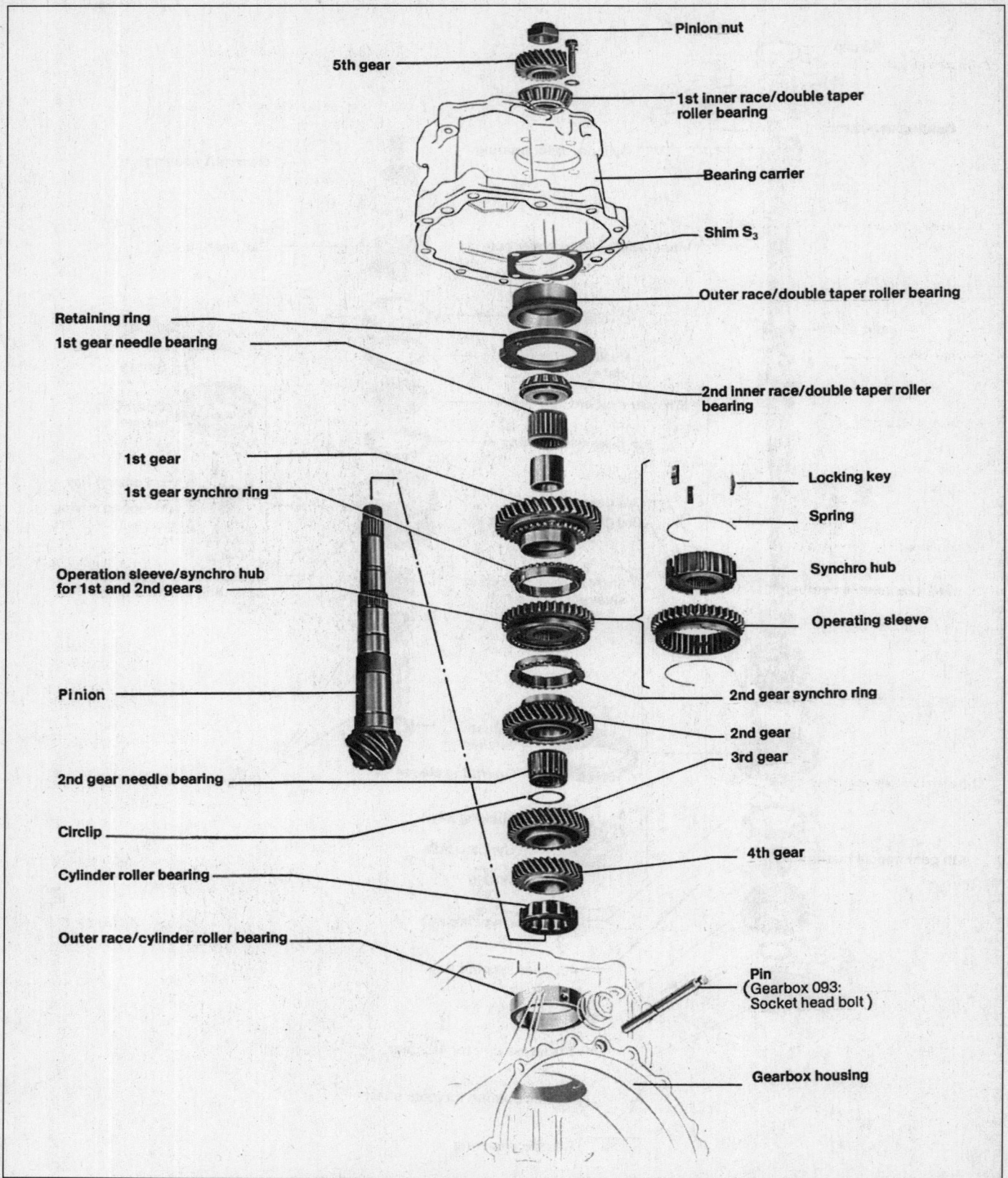

Fig. 6.14 Pinion shaft components on 013 and 093 gearboxes (Sec 4)

Gearbox housing

Input shaft oil seal

Guide sleeve

Starter bush

Release shaft

Release shaft bush

Release bearing

Return spring

Bush (013 only)

Shim S₂

Outer race/taper roller bearing

Outer race/taper roller bearing

Shim S₁

Magnet

Needle bearing for input shaft

Outer race/cylinder roller bearing

Hex head bolt (013 only)

Breather

Oil filler plug

Pin (gearbox 013)
Socket head bolt (gearbox 093)

Rubber bush (013 only)

Clutch lever

Oil drain plug

Speedo drive

Final drive cover

Drive flange oil seal

6

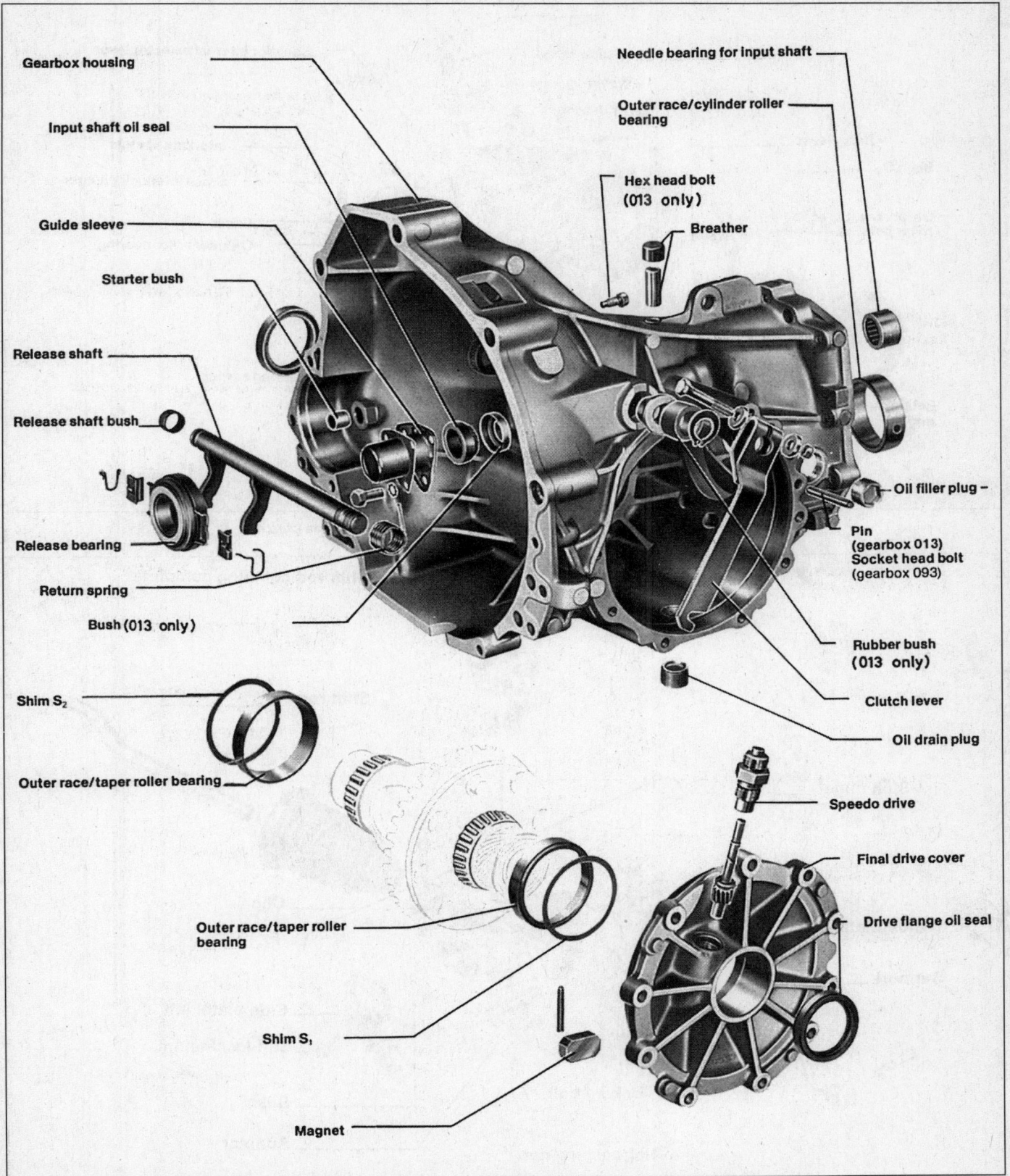

Fig. 6.15 Gearbox/final drive housing – 013 and 093 gearboxes (Sec 4)

Plug

Guide sleeve

Spring

Gear detent

Bush for internal selector lever

Relay lever

Shim S$_3$

Interlock plunger

Small interlock plunger

Outer race/taper roller bearing

Circlip

Cylinder roller bearing

Bearing carrier

Retaining ring

Selector rod for 5th and reverse gears

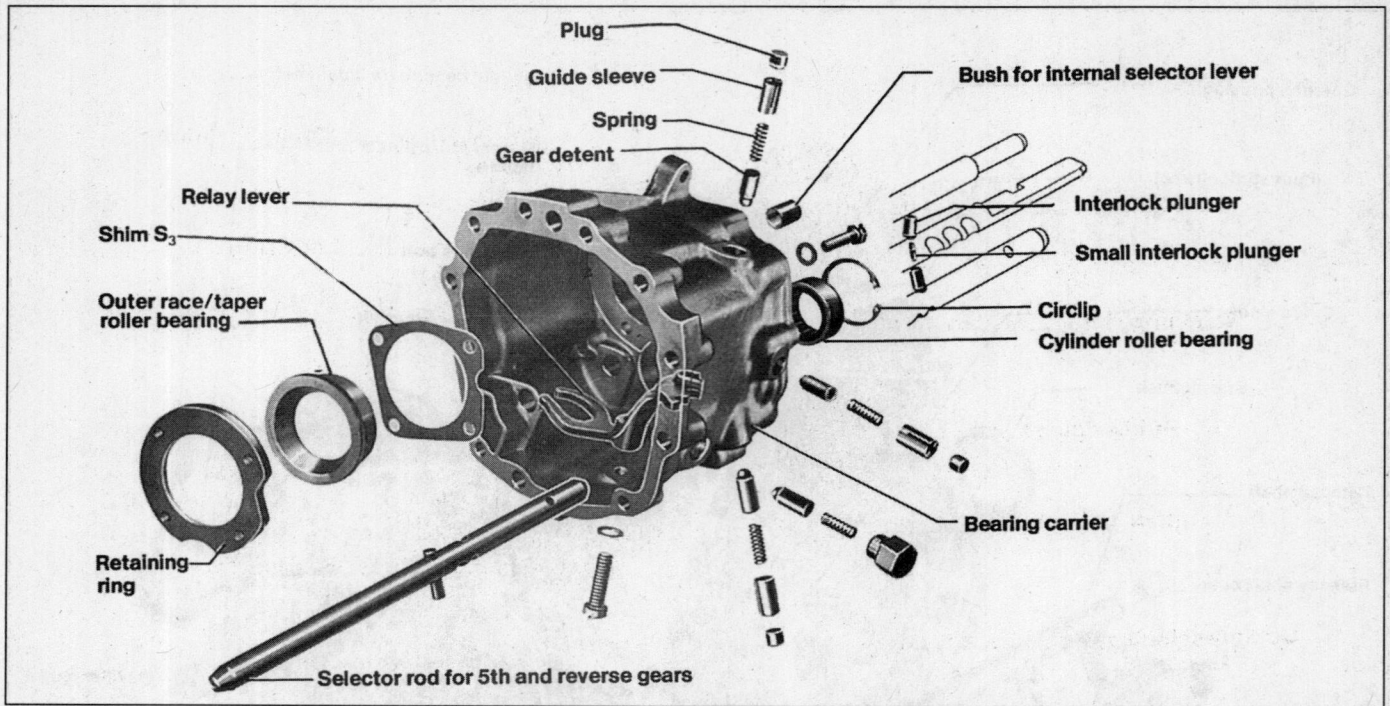

Fig. 6.16 Gear carrier components on 013 and 093 gearboxes (Sec 4)

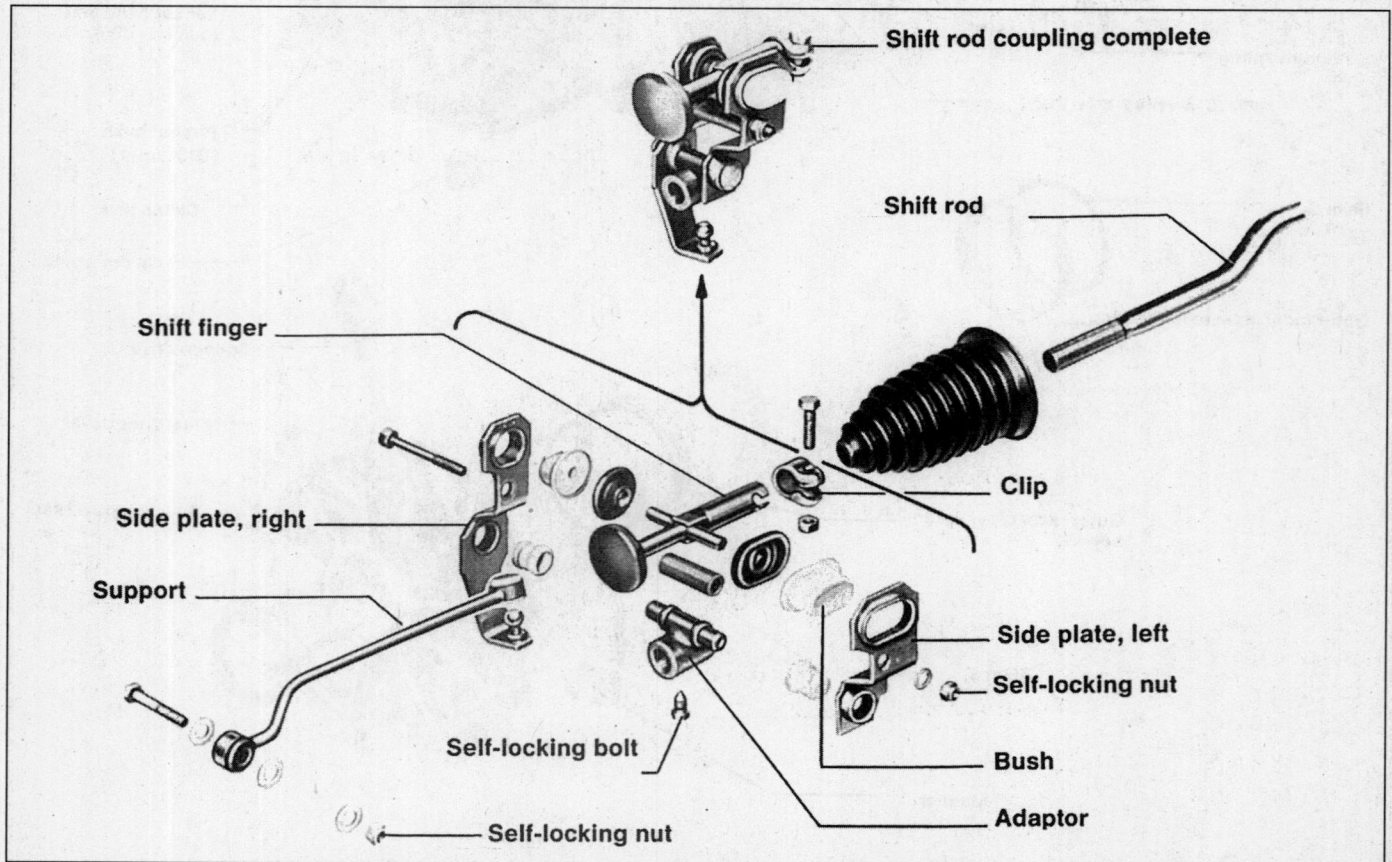

Shift rod coupling complete

Shift rod

Shift finger

Clip

Side plate, right

Support

Side plate, left

Self-locking nut

Self-locking bolt

Bush

Adaptor

Self-locking nut

Fig. 6.17 Exploded view of the 014 gearshift linkage components (Secs 5 and 6)

Housing

Retaining ring

Rubber guide

Upper ball half

Compression spring

Shell

Lower ball half

Gearshift knob

Lock ring

Lever bearing

Gear lever

Lever housing

Reverse stop

Stop pad

6

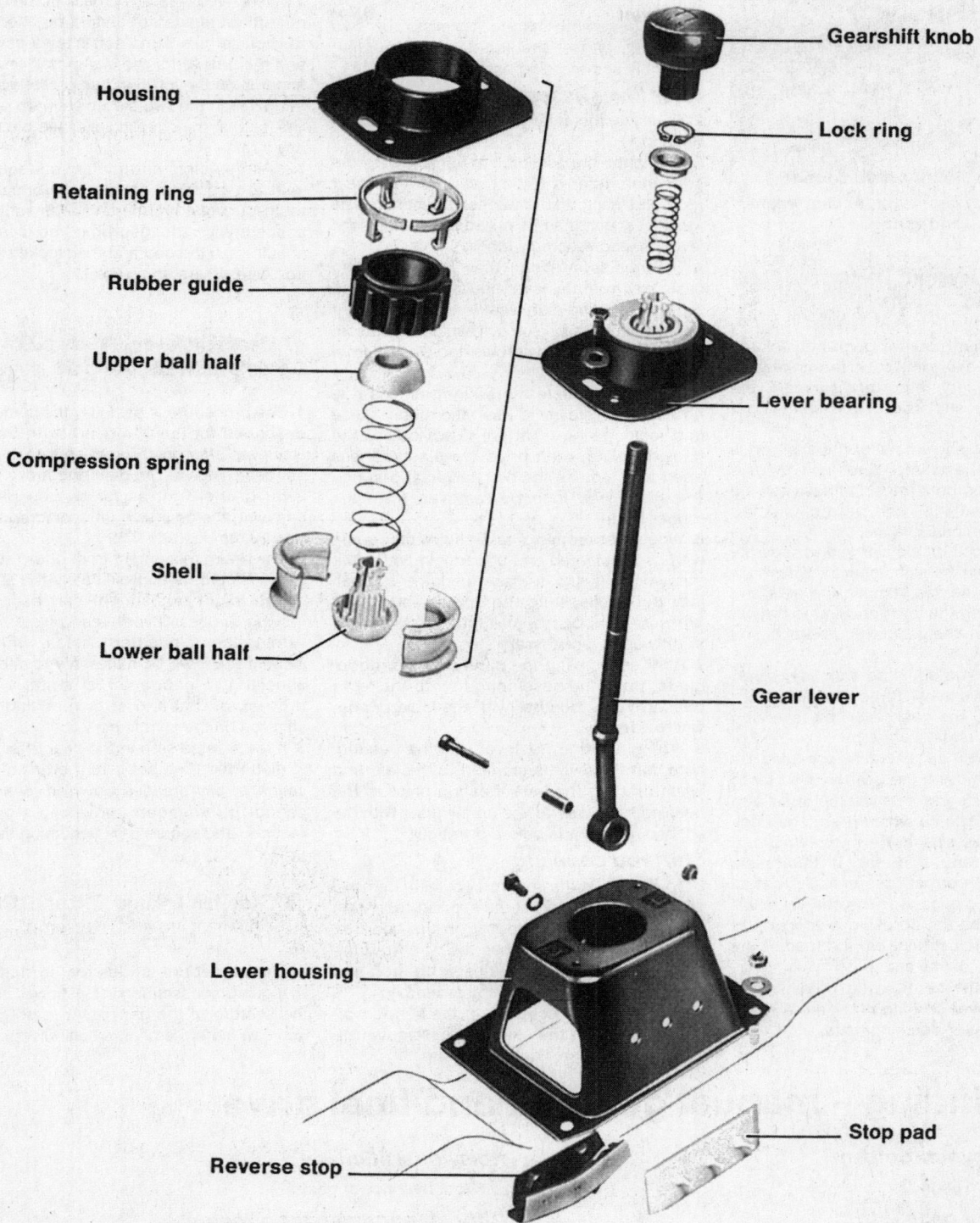

Fig. 6.18 Exploded view of the 014 gearshift lever components (Secs 5 and 6)

6.3 Location of gear lever, circlip, washer and spring

5 Gearshift lever (014) – removal, refitting and adjustment

1 The adjustment of the gearshift linkage requires a special tool, so if the linkage is undone, it is very important to mark the position of the shift rod in the shift finger before separating them.

2 Put a mark to show how far the shift rod is inserted into the clamp, and also mark a horizontal line on both the shift finger and the shift rod so that they can be reconnected without any rotational change.

3 Release the bolt on the clamp and separate the shift rod from the shift finger.

4 From inside the car, remove the nuts and washers securing the lever housing to the car floor and remove the gear lever assembly and shift rod.

5 To separate the shift rod from the lever, undo and remove the shift rod clevis bolt.

6 After refitting the gear lever, by reversing the removal operations, the basic setting of the linkage should be tested by engaging 1st gear and then moving the gear lever as far to the left as it will go. Release the lever and measure the distance which it springs back on its own. This should be between 5 and 10 mm (0.20 and 0.39 in). If this basic adjustment is incorrect, it is unlikely that all the gears can be engaged. The gear lever will either have to be set up using a special VW gauge, or the adjustment can be made by a lengthy process of trial and error.

7 Check that the centralizing holes of the housing and lever bearing plate are aligned when tightening the securing bolts.

6 Gearshift linkage (014) – dismantling and reassembly

1 The gearshift linkage consists of two principal parts, the shift rod coupling assembly and the lever assembly.

Gear lever assembly

2 Remove the gear lever as described in Section 11.

3 Dismantle the assembly by unscrewing the gear knob, removing the circlip from the gear lever and lifting off the washer and spring. The gear lever can then be pulled down out of the lever bearing assembly (photo).

4 Before separating the lever bearing assembly from the lever housing, mark round the lever bearing plate with a scriber so that it can be returned to exactly the same position, then remove the two screws and washer from the plate .

5 Do not dismantle the bearing unless it is necessary to grease it. Push the rubber guide and locking ring (if fitted) down out of the housing plate, then prise the plastic shells apart and remove the ball halves and spring the shells can then be removed from the rubber guide.

6 When reassembling have the rubber guide with its shouldered end uppermost and press the two shells into it. Press the lower ball half into the shells, then the spring and finally press in the upper ball half, pushing the shells slightly apart if necessary.

7 After assembling the parts into the rubber guide, push the assembly up into the lever bearing plate together with the locking ring, where fitted.

8 When inserting the lever into the bearing, note that the lever is cranked to the left, and when refitting the lever bearing plate to the housing, take care to line up the plate with the scribed mark made before dismantling.

Shift rod coupling

9 To dismantle the shift rod coupling, remove the bolt from the end of the support rod. Mark the position of the adaptor on the gearbox selector lever, then remove the wire from the bolt (where applicable), loosen the bolt and remove the shift rod coupling assembly.

10 Prise the ball coupling of the support off its mounting on the side plate. Remove the bolt which clamps the two side plates together and extract the shift finger and its bushes.

11 When reassembling the shift rod coupling, note that the adaptor should be fitted so that the hole for the clamp bolt is towards the front and the groove for the clamp bolt on the shift finger is on the left-hand side. Make sure that the holes in the two side plates are exactly in line, so that the coupling is assembled without any strain.

12 All the joints and friction surfaces of the shift rod coupling should be lubricated with multi-purpose grease and after refitting the assembly to the gearbox, the clamp bolt should be tightened and then locked with soft iron wire (where applicable).

7 Gearshift lever (013 and 093) – removal, refitting and adjustment

1 The procedure is basically the same as that described for the 014 gearbox in Section 5. However, after checking the return distance in 1st gear, engage 5th gear and move the gear lever to the right as far as possible. After releasing the gear lever, it should spring back 5 to 10 mm (0.20 to 0.39 in).

2 The return movement from both 1st and 5th gears should be approximately the same, but if not, slight adjustment can be made by moving the gear lever bearing plate sideways within the elongated holes. If this is insufficient, it will be necessary to either use a special VW gauge, or attempt a lengthy process of trial and error repositioning the shift rod on the shift finger.

3 If the reverse stop in the gear lever housing is disturbed, first set it to the basic position which is fully pressed down. If reverse gear cannot be engaged, move the stop up as required and secure it by tightening the bolts.

8 Gearshift linkage (013 and 093) – dismantling and reassembly

The procedure is identical to that for the 014 gearbox described in Section 6. However, the bottom of the gear lever is not cranked and can therefore be fitted either way round.

Fault finding – manual gearbox and final drive

Ineffective synchromesh

☐ Worn synchro rings

Jumps out of gear

☐ Weak or broken detent spring
☐ Worn selector forks or dogs
☐ Weak synchro springs
☐ Worn synchro unit or gears
☐ Worn bearings or gears

Noisy operation

☐ Worn bearings or gears

Difficult engagement of gears

☐ Worn selector components
☐ Worn synchro units
☐ Clutch fault
☐ Gearbox input shaft spigot bearing seized in end of crankshaft
☐ Incorrect gearshift adjustment

Chapter 7
Automatic transmission and final drive

For modifications, and information applicable to later models, see Supplement at end of manual

Contents

Degrees of difficulty

Easy, suitable for novice with little experience	**Fairly easy,** suitable for beginner with some experience	**Fairly difficult,** suitable for competent DIY mechanic	**Difficult,** suitable for experienced DIY mechanic	**Very difficult,** suitable for expert DIY or professional

Specifications

Type . . . . Three-speed planetary gearbox with hydrodynamic torque converter, final drive differential located between torque converter and gearbox

Identification code number
089 . . . . 1.6 and 1.8 litre engine models
087 . . . . 1.9 and 2.0 litre engine models

Application – code numbers/letters
089/KF (up to October 1982) . . . . Fitted to 1.6 litre engine models
089/KAD (from October 1982) . . . . Fitted to 1.8 litre engine models
089/KAL (from August 1983) . . . . Fitted to 1.6 litre engine models
089/KAH (from August 1983) . . . . Fitted to 1.6 and 1.8 litre engine models
087/RR (up to October 1982) . . . . Fitted to 1.9 litre engine models
087/RBA (October to December 1982) . . . . Fitted to 1.9 litre engine models
087/RBD (from December 1982) . . . . Fitted to 2.0 litre engine models

Ratios (089)

	089/KF	089/KAD	089/KAL	089/KAH
Final drive	3.73 to 1	3.25 to 1	3.42 to 1	3.25 to 1
1st	2.55 to 1	2.71 to 1	2.71 to 1	2.71 to 1
2nd	1.45 to 1	1.50 to 1	1.50 to 1	1.50 to 1
3rd	1.00 to 1	1.00 to 1	1.00 to 1	1.00 to 1
Reverse	2.46 to 1	2.43 to 1	2.43 to 1	2.43 to 1

Ratios (087)

	087/RR	087/RBA	087/RBD	087/RBH
Final drive	3.45 to 1	3.25 to 1	3.25 to 1	3.45 to 1
lst	2.55 to 1	2.71 to 1	2.71 to 1	2.71 to 1
2nd	1.45 to 1	1.50 to 1	1.50 to 1	1.50 to 1
3rd	1.00 to 1	1.00 to 1	1.00 to 1	1.00 to 1
Reverse	2.46 to 1	2.43 to 1	2.45 to 1	2.45 to 1

Lubrication
Transmission fluid type/specification . . . . Dexron type ATF
Final drive oil type/specification . . . . Hypoid gear oil, viscosity SAE 90, to API-GL5, MIL-L-2105B

Lubricant capacity
089 final drive . . . . 0.75 litre (1.3 Imp pt)
087 final drive . . . . 1.0 litre (1.8 Imp pt)
089 and 087 gearbox:
  Total . . . . 6 litre (10.6 Imp pt)
  Service . . . . 3 litre (5.3 Imp pt)

Torque converter

	089	087
Maximum diameter of bush	34.25 mm (1.35 in)	34.12 mm (1.34 in)
Maximum out-of-round of bush	0.03 mm (0.001 in)	0.03 mm (0.001 in)

Torque wrench settings

	Nm	lbf ft
Torque converter to driveplate	30	22
Gearbox to engine	55	41
Mounting to body	40	30
Mounting to gearbox	55	41
Gearbox to final drive housing	30	22
Oil pan	20	15
Strainer cover	3	2
Selector cable clamp nut	8	6
Selector bracket nuts	15	11

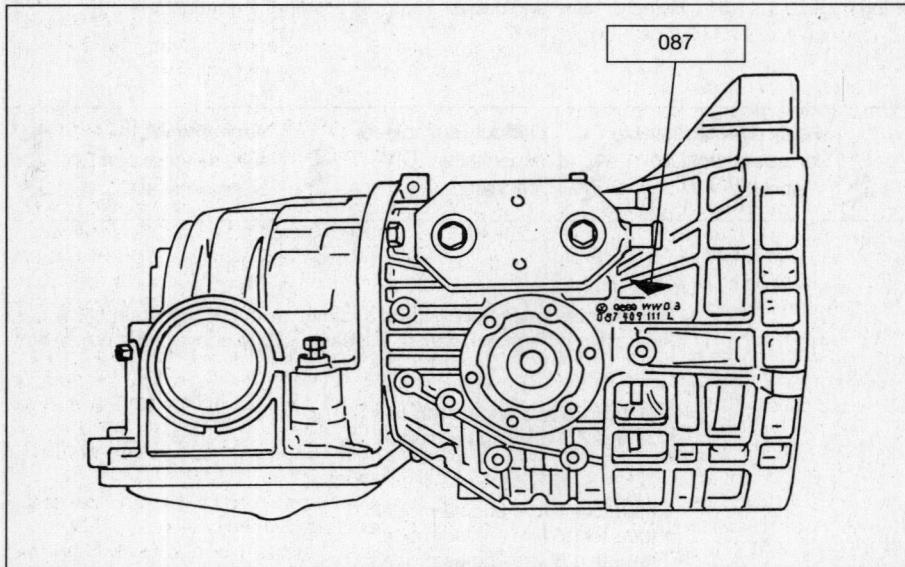

Fig. 7.1 The 087 type automatic transmission unit showing identification mark location (Sec 1)

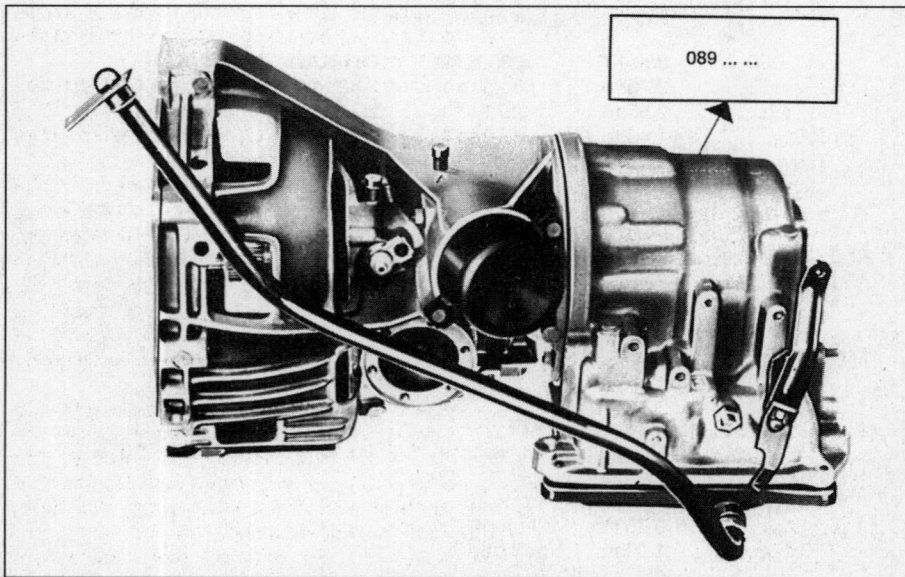

Fig. 7.2 The 089 type automatic transmission unit showing identification mark location (Sec 1)

1 General description

The automatic transmission consists of three main assemblies, these being being the torque converter, which is directly coupled to the engine; the final drive unit which incorporates the differential assembly: and the planetary gearbox with its hydraulically operated multi-disc clutches and brake bands. The gearbox also houses a rear mounted oil pump, which is coupled to the torque converter impeller, and this pump supplies automatic transmission fluid to the planetary gears, hydraulic controls and torque converter. The fluid performs a triple function by lubricating the moving parts, cooling the automatic transmission system and providing a torque transfer medium. The final drive lubrication is separate from the transmission lubrication system, unlike the manual gearbox where the final drive shares a common lubrication system.

The torque converter is a sealed unit which cannot be dismantled. It is bolted to the crankshaft driveplate and replaces the clutch found on an engine with manual transmission.

The gearbox is of the planetary type with epicyclic gear trains operated by brakes and clutches through a hydraulic control system. The correct gear is selected by a combination of three control signals; a manual valve operated by the gearshift cable, a manual valve operated by the accelerator pedal, and a governor to control hydraulic pressure. The gearshift cable and selector lever allow the driver to select a specific gear and override the automatic control, if desired.The accelerator control determines the correct gear for the desired rate of acceleration, and the governor determines the correct gear in relation to engine speed.

Because of the need for special test equipment, the complexity of some of the parts and the need for scrupulous cleanliness when servicing automatic transmissions, the amount which the owner can do is limited, but those operations which can reasonably be carried out are detailed in the following Sections. Repairs to the final drive differential are also not recommended.

The automatic transmission has three forward speeds and one reverse, controlled by a six position lever with the following positions:

P   Park
R   Reverse
N   Neutral
D   Drive
2   Low
1   Low

The selector lever has a push button which must be depressed when selecting the following positions:

From   P to R
           R to P
           N to R
           2 to 1

The selector lever can be moved freely between all other positions. If the lever is set to positions D or 2, the automatic transmission changes gears automatically.

Position D

This position is for normal driving, and once selected the three forward gears engage automatically throughout the speed range from zero to top speed.

Position 2

With the lever in this position, the two lower gears will engage automatically, but the highest gear will not engage. For this reason position 2 should only be selected when the speed of the car is below 70 mph (110 kph). Selecting Position 2 will make use of the engine's braking effect and the actual change can be made without letting up the accelerator pedal.

Position 1

This position is rarely needed but can be used for ascending or descending steep hills. The transmission remains in the lowest gear and Position 1 should only be selected when the car speed is below 40 mph (64 kph).

Reverse

Reverse must only be selected when the car is absolutely stationary and with the engine running at idling speed.

Park

In the park position the transmission is locked mechanically by the engagement of a pawl. This position must only be selected when the car is absolutely stationary, otherwise the transmission will be damaged.

2 Automatic transmission – precautions and maintenance

1 If the car is being towed, the ignition key must be inserted so that the steering wheel is not locked and the gear selector must be in *Neutral*. Because the lubrication of the transmission is limited when the engine is not running, the car must not be towed for more than 30 miles (48 km), or at a speed greater than 30 mph (48 kph) unless the front wheels of the car are lifted clear of the road.

2 Routine maintenance consists of checking the final drive oil level and the automatic transmission fluid level every 10 000 miles (15 000 km) or 12 months. whichever occurs first. When checking the final drive oil level, if the level is found to be too high, fluid may be leaking from the gearbox or hydraulic circuit. It is advisable to check the automatic transmission fluid at periodic intervals between those specified above to ensure that the correct level is maintained.

3 Checking the automatic transmission fluid level should be carried out with the engine warm and running at idling speed with the selector lever in *Neutral* and the handbrake applied. Withdraw the dipstick and wipe it with a piece of clean, lint-free rag. It is important that the rag is both clean and lint-free, because even a tiny speck of dirt can damage or cause a malfunction of the transmission. The level of fluid is satisfactory anywhere between the two marks on the dipstick, but either too high, or too low a level must be avoided. The difference between the marks is equivalent to 0.4 litre (0.7 Imp pt). If necessary, top up the level through the dipstick tube using the specified fluid see *'Recommended lubricants and fluids'* at the beginning of this manual.

4 The automatic transmission fluid must be drained and renewed at 30 000 miles (45 000 km) or every three years, whichever occurs first, see Section 3. Note that the final drive oil does not need changing.

3 Automatic transmission fluid – draining and refilling

1 This job should not be attempted unless clean, dust free conditions can be achieved.

2 With the car standing on level ground, place a container of at least six pints capacity beneath the oil pan of the transmission. For working room beneath the car, jack it up and support it with axle stands, or use car ramps.

3 Unscrew the union nut securing the dipstick tube to the oil pan, pull out the tube and allow the fluid to drain out. **Note:** *Do not drive or tow the vehicle when there is no oil in the transmission!*

4 Remove the retaining screws and withdraw the oil pan. Remove the gasket (Fig. 7.3).

5 Remove the screws and withdraw the cover, strainer, and gasket.

6 Clean the pan and strainer with methylated spirit and allow to dry.

7 Refit the strainer, cover and pan in reverse order using new gaskets and tightening the screws to the specified torque.

8 Insert the dipstick and tighten the union nut.

9 Wipe round the top of the dipstick tube, then remove the dipstick.

10 With the car on level ground, fill the transmission with the correct quantity and grade of fluid, using a clean funnel if necessary.

11 Start the engine and with the handbrake applied, select every gear position once. With the engine idling and the transmission in *Neutral,* check the level of the fluid on the dipstick, and if necessary top up to the lower mark.

12 Road test the vehicle until the engine is at normal temperature, then again check the fluid level and top up if necessary. The amount of fluid which must be added to raise the level from the lower mark to the upper mark on the dipstick is about half a pint (0.4 litre). Do not overfill the transmission, because an excess of fluid will upset its operation and any excess fluid will have to be drained.

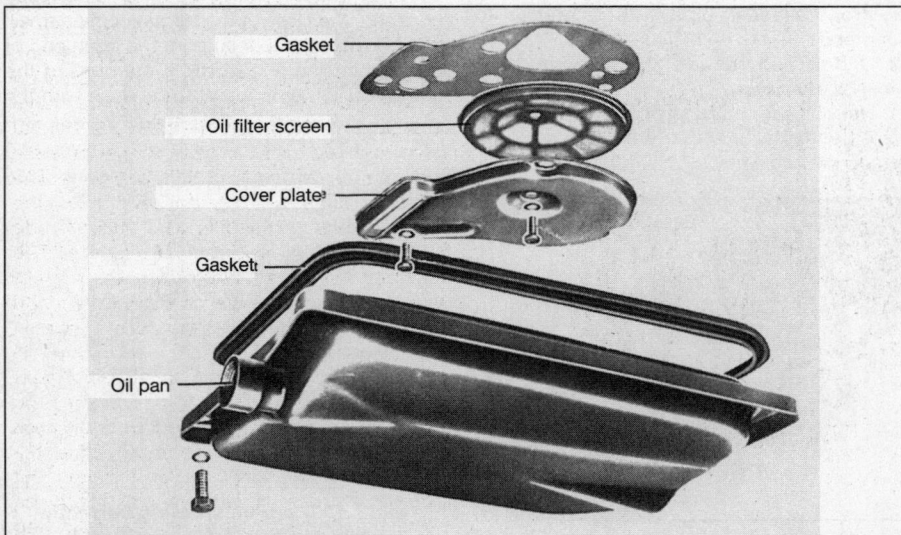

Gasket

Oil filter screen

Cover plate

Gasket

Oil pan

Fig. 7.3 Automatic transmission oil pan and strainer components (Sec 3)

7

Fig. 7.4 Using a special bar to support the front of the engine – typical (Sec 4)

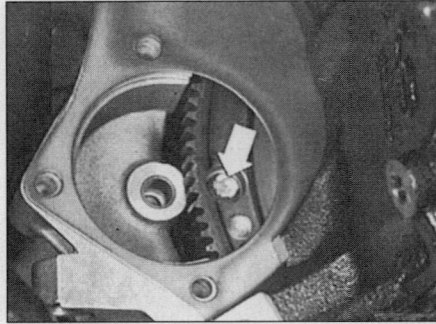

Fig. 7.5 A torque converter to driveplate bolt viewed through the starter motor aperture (Sec 4)

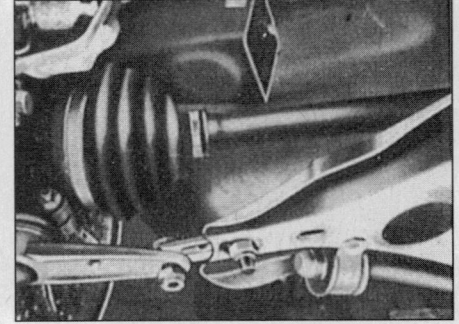

Fig. 7.6 Mark relative positions and detach the balljoint from the suspension wishbone and fit spacer plate (Sec 4)

4 Automatic transmission type 089 – removal and refitting

1 Position the car over an inspection pit or on car ramps. Apply the handbrake.

2 Disconnect the battery negative lead.

3 Support the front of the engine using a hoist or support bar. Fig. 7.4 shows the VW tool specially designed for the job, but it should be relatively easy to make a support bar yourself along similar lines, if you have decided not to use a hoist.

4 Unscrew and remove the upper engine-to-transmission securing bolts.

5 Unscrew and remove the front engine support to the front body member.

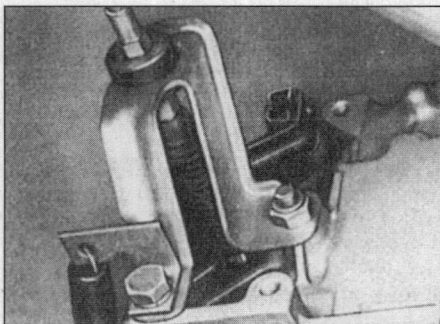

Fig. 7.7 Selector lever cable bracket (Sec 4)

6 Remove the starter motor as described in Chapter 10.

7 Unscrew the three torque converter-to-driveplate bolts while holding the starter ring gear stationary with a screwdriver. It will be necessary to rotate the engine to position the bolts in the starter aperture, using a socket on the crankshaft pulley bolt (Fig. 7.5).

8 Detach and remove the converter cover plate.

9 Unscrew the nut and disconnect the speedometer cable from the transmission. Also release the cable from the clip on the dipstick tube. where applicable.

10 Disconnect the driveshafts from the transmission, with reference to Chapter 8. For additional working room it is necessary to remove one or both driveshafts, or alternatively tie them to one side.

11 Index mark the relative positions of the left-hand steering balljoint and wishbone (suspension arm), then detach the balljoint and fit a suitable spacer plate to them as shown in Fig. 7.6.

12 Unbolt and disconnect the exhaust pipe location bracket from the transmission. For additional working area, unbolt and remove the front exhaust pipe with reference to Chapter 3.

13 Unbolt and remove the selector lever cable bracket, Fig. 7.7.

14 Unhook the return spring, then withdraw the selector lever cable securing clip and detach the cable.

15 Disconnect the pedal cable support bracket from the transmission, then detach the pedal cable.

16 Unscrew the retaining bolt and remove the bonded rubber mounting (Fig. 7.8).

17 Raise the engine using the support bar so that the gearbox is slightly lowered allowing easier removal and subsequent refitting.

18 Unscrew the two retaining bolts, withdraw the throttle bracket and detach the cable (Fig. 7.9).

19 Position a trolley jack or suitable support under the transmission to take its weight. Locate a flat piece of wood between the jack saddle and transmission oil pan to spread the load and prevent damage to the pan.

20 Unscrew and remove the lower engine-to-transmission retaining bolts.

21 With the help of an assistant, withdraw the transmission from the engine, making sure that the torque converter remains fully engaged with the transmission splines.

22 Lower the transmission and remove it from under the car.

23 Refitting is a reversal of removal. but make sure that the torque converter is correctly fitted. If the pump shaft is correctly engaged, the boss of the torque converter will be about 10 mm (0.4 in) from the open end of the bellhousing (Fig. 7.10). If the boss of the torque converter is found to be flush with the

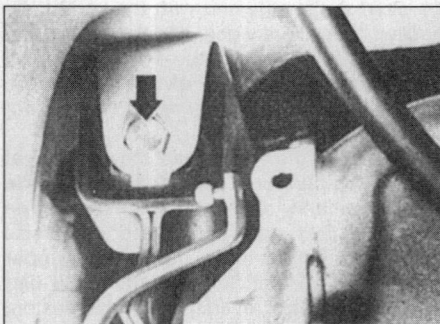

Fig. 7.8 Bonded rubber mounting and retaining bolt (arrowed) (Sec 4)

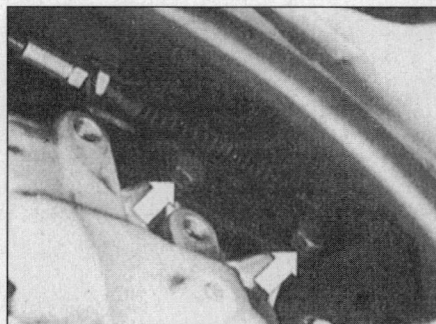

Fig. 7.9 Throttle cable and bracket securing bolts (arrowed) (Sec 4)

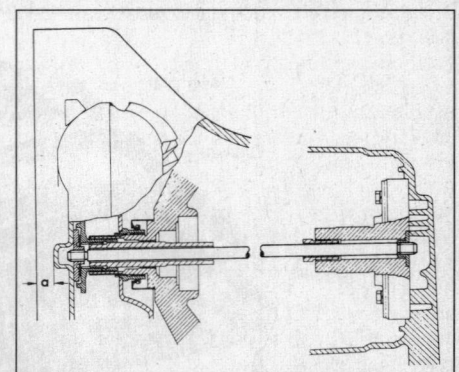

Fig. 7.10 Diagram showing correctly seated torque converter (Sec 4)

a = 10 mm (0.4 in)

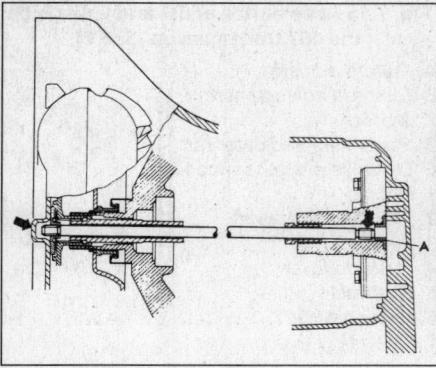

Fig. 7.11 Diagram showing incorrectly seated torque converter with pump shaft disengaged from pump driveplate splines (A) (Sec 4)

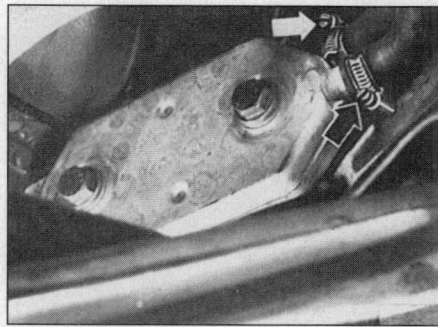

Fig. 7.12 Coolant hoses location on the oil cooler (Sec 5)

Fig. 7.13 Front subframe mounting bolts (Sec 5)

open end of the bellhousing (Fig. 7.11) it is likely that the pump shaft has pulled out of the pump driveplate splines. *The pump driveplate will be destroyed if the gearbox is bolted to the engine with the pump shaft in this position.* Refill the cooling system with reference to Chapter 2. Reconnect the driveshafts with reference to Chapter 8 and the starter with reference to Chapter 10. If necessary adjust the accelerator and selector cables as described in Sections 10 and 11.

5 Automatic transmission type 087 – removal and refitting

1 Proceed as described in paragraphs 1 to 4 in the previous Section.
2 Drain the cooling system as described in Chapter 2 then disconnect the hoses from the oil (ATF) cooler.
3 Unscrew and remove the engine support bolts.
4 Disconnect the accelerator cable support bracket from the transmission then detach the accelerator cable.
5 Disconnect and remove the throttle pushrod.
6 Disconnect the automatic transmission fluid filler tube from the oil pan.
7 Unbolt and disconnect the selector lever cable bracket.
8 Unhook the return spring, detach the selector lever cable retainer and disconnect the cable.
9 Remove the right-hand side deflector plate.
10 Proceed as described in paragraphs 6, 7, 9 and 10 in the previous Section.
11 Unscrew the hose clips then pull free the hoses from the ATF cooler connections shown in Fig. 7.12.
12 Unscrew and remove the front subframe bolts (Fig. 7.13).
13 Remove the retaining bolts and detach the bonded rubber transmission mounting (see Fig. 7.8).
14 Proceed as described in paragraphs 19

to 22 in the previous Section to remove the transmission.
15 Refit the transmission in the reverse order of removal, referring to paragraphs 23 and 24 in the previous Section.
16 In addition, on completion check that the transmission oil cooler hoses are securely located and top up the cooling system as described in Chapter 3.

6 Accelerator pedal and linkage – adjustment

1 The accelerator linkage must be adjusted so that the operating lever on the gearbox is at its idle position when the throttle is closed. If the adjustment is incorrect the shaft speeds will be too high when the throttle is partially open and the main pressure will be too high when the engine is idling.

089 transmission

2 Run the engine to normal operating temperature to ensure that the choke is fully open. Remove the air cleaner unit as described in
Chapter 3.
3 Check that the throttle valve is closed and in its idling position.
4 Loosen both locknuts at the cable bracket on the engine, then pull the cable ferrule away from the engine to eliminate all play and tighten the locknuts in this position.
5 Working inside the car. unscrew and remove the accelerator pedal stop and spacer.
6 Make up a distance piece as shown in Fig. 7.15 using a 135 mm (5.32 in) M8 bolt with two nuts fitted and locked in the position shown in Fig. 7.16.
7 Screw the distance piece bolt into the accelerator pedal stop nut (E in Fig. 7.14).

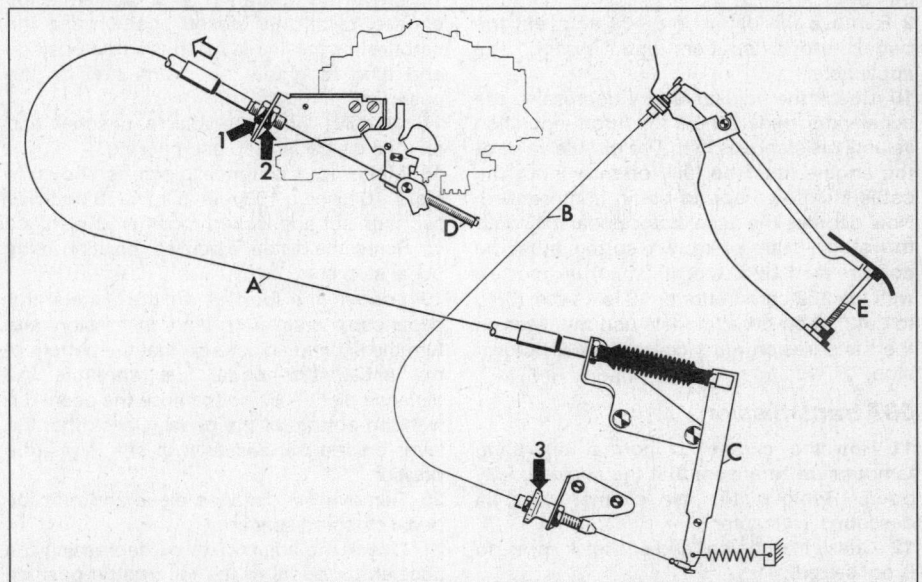

Fig. 7.14 Accelerator pedal and linkage on the 089 transmission (Sec 6)

A *Cable to carburettor linkage and adjuster locknuts (1 and 2)*
B *Accelerator pedal cable and adjuster nut (3)*
C *Operating lever*
D *Spring*
E *Distance piece bolt location*

7

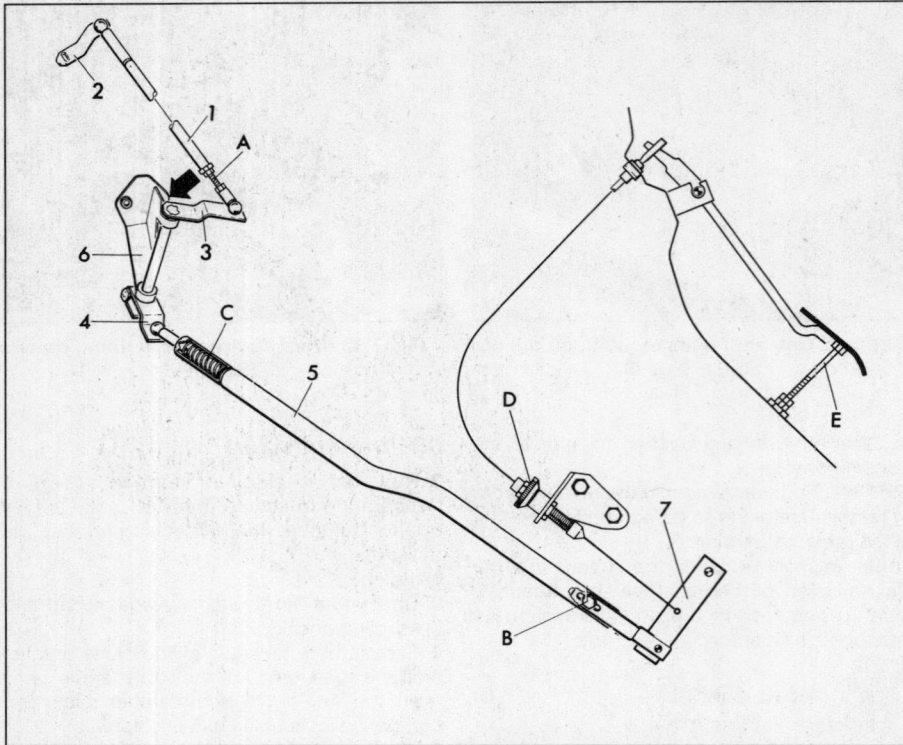

Fig. 7.15 Accelerator pedal and linkage on the 087 transmission (Sec 6)

A Pullrod adjuster
B Pushrod adjuster screw
C Spring
D Pedal cable adjuster nut
E Distance piece bolt location
1 Pullrod
2 Throttle valve lever
3 Lever
4 Pushrod lever
5 Pushrod
6 Relay bracket
7 Operating lever

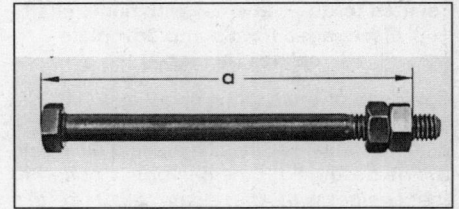

Fig. 7.16 Distance piece made from bolt required for accelerator/pedal linkage adjustment according to transmission type (Sec 6)

089 transmission: bolt length 135 mm (5.3 in) with nuts set at distance (a) to = 124 mm (4.8 in)
087 transmission: bolt length 130 mm (5.1 in) with nuts set at distance (a) to = 119 mm (4.6 in)

8 Loosen the locknut on the accelerator pedal cable bracket on the transmission, and turn the knurled adjuster so that the bottom of the accelerator pedal just contacts the distance piece – do not confuse the pedal rod with the bottom of the pedal. Check that the lever on the transmission is still in its idle position.

9 Remove the distance piece and refit the pedal stop, spacer and switch, as applicable.

10 Check the adjustment by depressing the accelerator pedal to the full throttle position (ie until resistance is felt). The throttle valve at the engine must be fully open without the cable kickdown spring being compressed. Now depress the accelerator pedal fully onto the stop – the kickdown spring must be compressed by 8 mm (0.31 in) on models with the 2E2 carburettor or 10 to 11 mm (0.40 to 0.43 in) on other models and the lever on the transmission must contact the kickdown stop.

087 transmission

11 Run the engine to normal operating temperature to ensure that the choke is fully open. Remove the air cleaner unit as described in Chapter 3.

12 Check that the ball and socket linkage rod is not seized.

13 Referring to Fig. 7.15, remove the pushrod ball socket retainer and disconnect the pushrod from the relay lever.

14 Check that the throttle valve lever is in the idle position (against the stop), then adjust the

threaded end of the pullrod (A) so that the connecting lever touches the stop on the relay lever bracket, (arrowed).

15 Loosen off the clamp bolt on the transmission end of the pushrod (B) then reconnect and secure the pushrod to the relay lever.

16 Move the operating lever (transmission end) to its throttle closed position and the carburettor throttle lever to the idle position and then retighten the clamp bolt on the pushrod.

17 Working within the vehicle, unscrew and remove the pedal stop and packing.

18 Make up a distance piece as shown in Fig. 7.16 using a 130 mm (5.1 in) M8 bolt with two nuts set and locked together at distance (a). Screw the distance bolt into position in the pedal stop nut.

19 Loosen the locknut on the accelerator pedal cable bracket on the transmission, and turn the knurled adjuster so that the bottom of the accelerator pedal just contacts the distance piece – do not confuse the pedal rod with the bottom of the pedal. Check that the lever on the transmission is still in its idle position.

20 Remove the distance piece and refit the pedal stop and spacer.

21 Check the adjustment by depressing the accelerator pedal to the full throttle position (ie until resistance is felt). The throttle valve at the engine must be fully open. Now depress the accelerator pedal fully onto the stop – the transmission lever must contact the kickdown stop.

7 Selector lever and cable – removal, refitting and adjustment

1 Disconnect the battery negative lead.
2 Remove the grub screw and detach the knob from the selector lever (Fig. 7.23).
3 Prise off the cover together with the blanking strip or brushes.
4 Remove the screws and withdraw the console.
5 Disconnect the wiring from the starter inhibitor switch and selector illumination bulb.
6 Unscrew the cable clamp nut, and also unscrew the nut from the floor bracket. Pull the cable clear.
7 Unscrew the nuts and remove the bracket and lever assembly from the floor.
8 Unbolt the cable bracket from the transmission, then extract the circlip and remove the cable end from the lever. Withdraw the cable from under the car.
9 Refitting is a reversal of removal, but before fitting the console adjust the cable as follows. Move the selector lever fully forwards to the P (park) position. Loosen the cable clamp nut, then move the lever on the transmission fully rearwards to the P (park) position. Tighten the cable clamp nut to the specified torque. Lightly lubricate the selector lever and cable pivots with engine oil.

Grub screw

Cover

Console

Grub screw

Contact bridge

Contact plate

Selector cable

Cable clamp

Bracket

O ring

Boot

Circlip

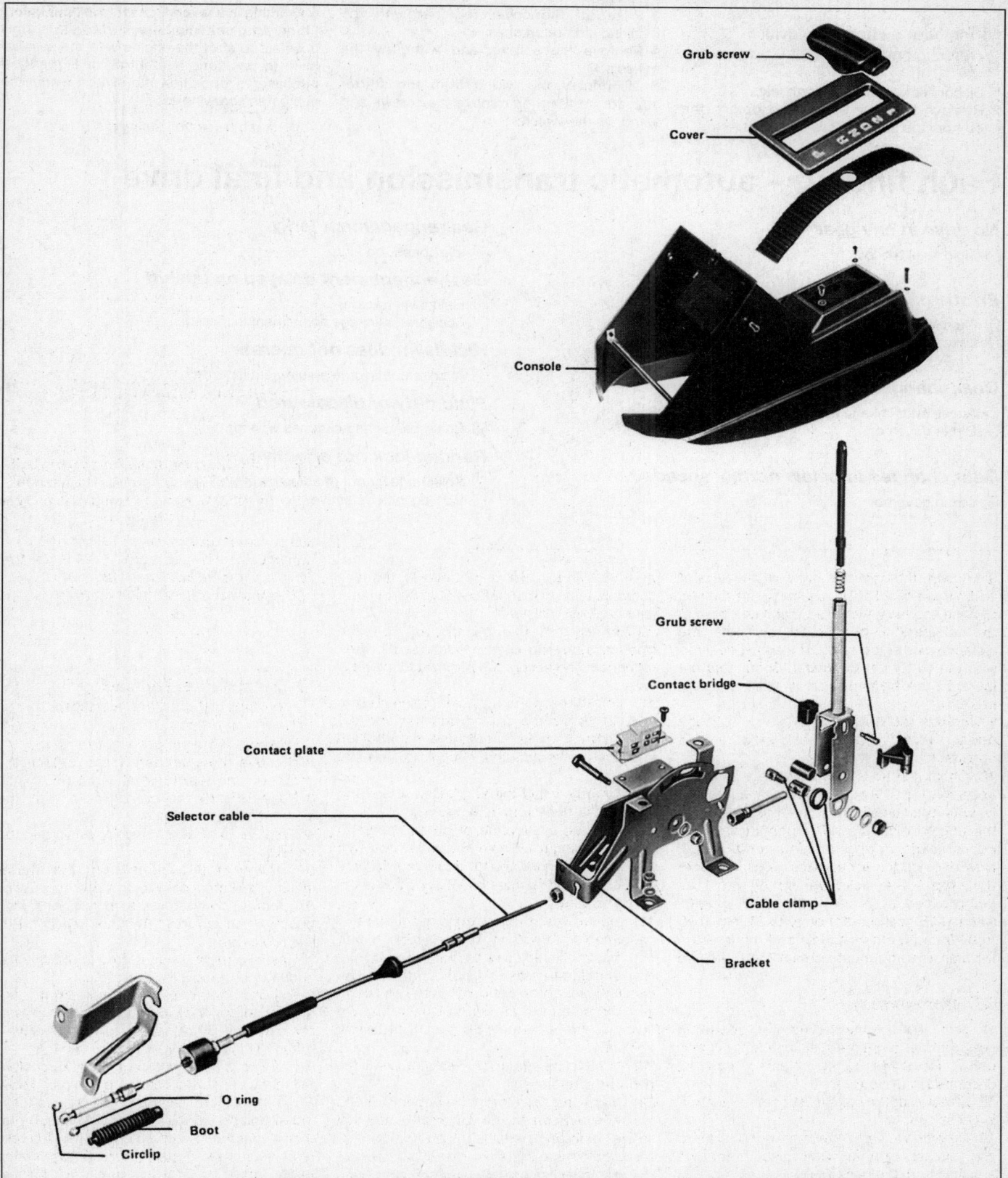

Fig. 7.17 Selector lever and cable components for 087 and 089 automatic transmissions (Sec 7)

7

8 Inhibitor switch – removal, refitting and adjustment

1 Disconnect the battery earth lead.
2 Remove the grub screw and detach the knob from the selector lever.

3 Prise off the cover together with the blanking strip or brushes.
4 Remove the screws and withdraw the console.
5 Disconnect the wiring from the starter inhibitor switch then remove the screws and withdraw the switch.

6 Refitting is a reversal of removal, but before fitting the console check that it is only possible to start the engine with the selector lever in positions N (neutral) or P (park). If necessary, reposition the switch within the elongated screw holes.

Fault finding – automatic transmission and final drive

No drive in any gear

☐ Fluid level too low

Erratic drive in forward gears

☐ Fluid level too low
☐ Dirty filter

Gear changes at above normal speed

☐ Accelerator linkage adjustment incorrect
☐ Dirt in governor

Gear changes at below normal speeds

☐ Dirt in governor

Gear engagement jerky

☐ Idle speed too high

Gear engagement delayed on upshift

☐ Fluid level too low
☐ Accelerator linkage adjustment incorrect

Kickdown does not operate

☐ Accelerator linkage adjustment incorrect

Fluid dirty or discoloured

☐ Brake bands and clutches wearing

Parking lock not effective

☐ Selector lever out of adjustment
☐ Parking lock defective

Chapter 8 Driveshafts

Contents

Degrees of difficulty

Easy, suitable for novice with little experience	**Fairly easy,** suitable for beginner with some experience	**Fairly difficult,** suitable for competent DIY mechanic	**Difficult,** suitable for experienced DIY mechanic	**Very difficult,** suitable for expert DIY or professional

Specifications

Type .... Double constant velocity (CV) joint, tubular driveshafts except on automatic transmission models where the right-hand driveshaft is solid

Shaft length

	Right-hand	Left-hand
Manual gearbox 014 and 013	530 mm (20.866 in)	530 mm (20.866 in)
Manual gearbox 093 (pre January 1983)	498.5 mm (19.625 in)	544.1 mm (21.421 in)
Manual gearbox 093 (from January 1983)	496.0 mm (19.53 in)	541.6 mm (21.32 in)
Automatic gearbox 089	492.2 mm (19.38 in)	589.5 mm (23.21 in)
Automatic gearbox 087	483.9 mm (19.05 in)	583.0 mm (22.95 in)
Automatic gearbox 087	476.7 mm (18.77 in)	574.5 mm (22.62 in)

CV joint lubricant .... VAG G6 grease

Torque wrench settings

	Nm	lbf ft
Inner CV joint to flange	45	33
Driveshaft nut	230	170

1 General description and maintenance

Drive is transmitted from the splined final drive differential gears and drive flanges to the inner constant velocity (CV) joints. The driveshafts are either tubular or solid, as given in the Specifications, and are splined at each end to the CV joint hubs. The outer CV joints incorporate splined shafts which are attached to the front wheel hubs.

The driveshaft joints are sealed and require no maintenance apart from checking the rubber boots at the specified routine maintenance intervals for any sign of leakage or damage, in which case they must be renewed.

If the joints are suspected of excessive wear, noticeable when changing from acceleration to overrun and vice versa, the shafts should be removed and the joints dismantled to inspect for wear or damage, and overhauled or renewed as necessary.

2 Driveshafts – removal and refitting

1 Prise the hub cap from the centre of the roadwheel.
2 Unscrew the driveshaft nut. *This is tightened to a very high torque and no attempt to loosen it must be made unless the full weight of the car is on the roadwheels.*

2.6A Driveshaft inner joint coupling bolts (arrowed)

3 Loosen the four bolts securing the roadwheel.
4 Jack the car and support the car securely on stands, or wood blocks.
5 Remove the roadwheel.
6 Using a socket wrench, remove the bolts from the inner driveshaft coupling and separate the driveshaft from the gearbox drive flange. Note the location of the bolt lock plates (photos).

2.6B Removing the bolts and lock plates from the driveshaft inner joint coupling

8

Fig. 8.1 Disconnect the front suspension balljoint from the suspension wishbone arm (Sec 2)

7 On manual transmission models it is necessary to disconnect the right-hand side front suspension balljoint from the suspension wishbone. On automatic transmission models it will be necessary to disconnect both the right and left-hand side balljoints from the wishbones. To do this first mark the exact position of the balljoint on the wishbone – this is important as the wheel camber setting is adjusted by moving the balljoint. Unscrew the nuts and release the balljoint.

8 Using a hub puller, press the driveshaft out of the front wheel hub, then withdraw the driveshaft.

9 Clean the splined end of the driveshaft and the wheel hub.

10 Refitting is a reversal of the removal procedure.

11 Tighten the nuts and bolts to the specified torque. If a suitable torque wrench is not available to tighten the driveshaft nut, tighten the nut firmly, then take the car to a VW agent for final tightening – the front wheel camber can also be checked at the same time.

3 Driveshafts – overhaul

1 With the driveshaft removed as described in the preceding Section, remove the two clamps on the rubber boot at the inner end of the shaft and pull the boot down the shaft and off the coupling.

2 Remove the circlip from the end of the shaft, and using a suitable drift, remove the protective cap from the outer ring.

3 Support the inner joint, then press the driveshaft out. Remove the dished washer noting its fitted direction.

4 Remove the two clamps from the outer joint boot and pull the boot down the shaft.

5 To release the hub from the circlip, strike the inner face of the hub outer ring with a soft face mallet, (Fig. 8.3).

6 Withdraw the outer joint from the driveshaft together with the spacer and dished washer.

7 Before starting to dismantle the outer joint, mark the position of the hub in relation to the cage and housing. Because the parts are

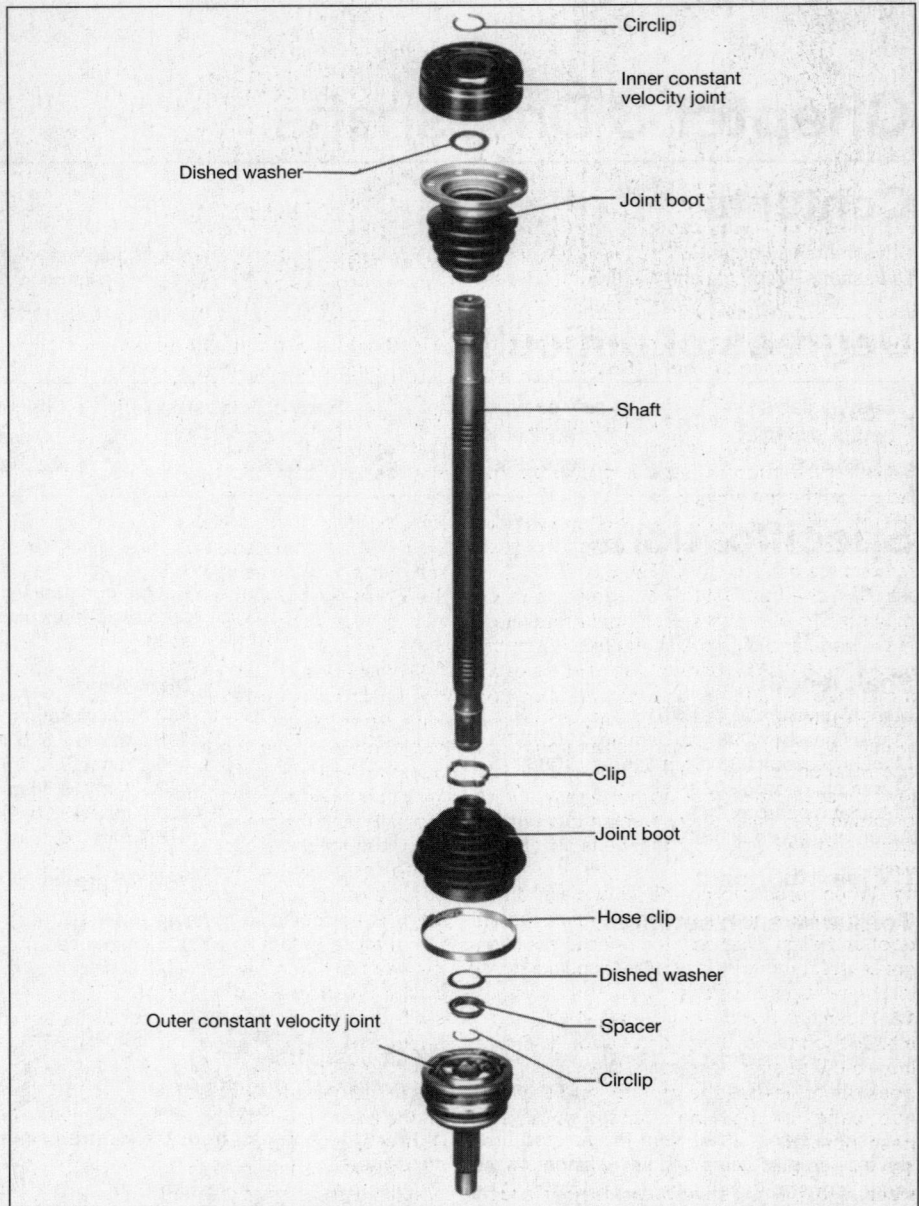

Fig. 8.2 Exploded view of the driveshaft (Sec 3)

- Circlip
- Inner constant velocity joint
- Dished washer
- Joint boot
- Shaft
- Clip
- Joint boot
- Hose clip
- Dished washer
- Spacer
- Circlip
- Outer constant velocity joint

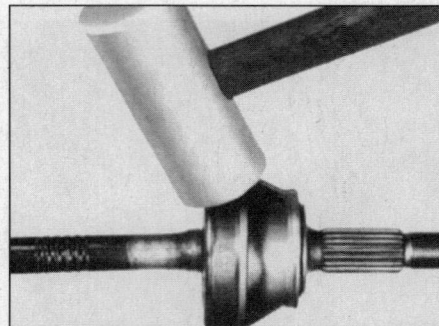

Fig. 8.3 Outer CV joint removal from shaft using soft-faced hammer (Sec 3)

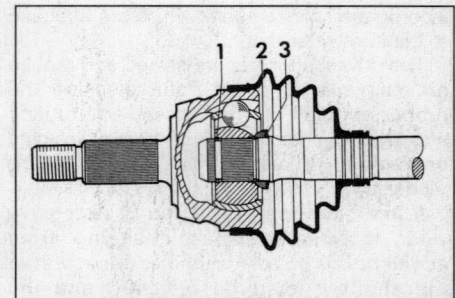

Fig. 8.4 Cross-section view of the outer joint (Sec 3)

1 Circlip   2 Spacer   3 Dished washer

Fig. 8.5 Arrange hub and cage as shown to remove the balls (Sec 3)

Fig. 8.6 Align rectangular openings in cage with housing to remove (Sec 3)

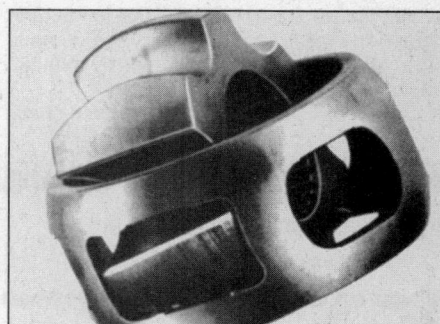
Fig. 8.7 Outer CV joint hub removal from cage (Sec 3)

hardened this mark will either have to be done with a grinding stone, or with paint.

8 Swivel the hub and cage and take out the balls one at a time (Fig. 8.5).

9 Turn the cage until the two rectangular openings align with the housing and then remove the cage and hub (Fig. 8.6).

10 Turn the hub until one segment can be pushed into one of the rectangular openings in the cage and then swivel the hub out of the cage (Fig. 8.7). The parts of the joint make up a matched set and no individual parts can be replaced. If there is excessive play in the joint which is noticeable when changing from acceleration to overrun, or vice versa, a new joint must be fitted, but do not renew a joint because the parts have been polished by wear and the track of the balls is clearly visible.

11 When reassembling the joint, clean off all the old grease and use a new circlip, rubber boot and clips. Use only the special coupling grease, recommended by VW for packing the joint – see Specifications.

12 Press half a sachet of grease (45 g, 1.6 oz) into the joint and then fit the cage and hub into the housing, ensuring that it will be possible to line up the mating marks of the hub, cage, and housing after the balls have been inserted.

13 Press the balls into the hub from alternate sides: when all six have been inserted check that the mating marks on the hub, cage and housing are aligned.

14 Fit a new circlip into the groove of the hub then squeeze the remainder of the grease into the joint so that the total amount is 90 g (3.2 oz).

15 The inner joint is dismantled in a similar way. Pivot the hub and cage and press them out of the housing as shown in Fig. 8.8.

16 Press the balls out of the cage, then align two grooves and remove the hub from the cage (Fig. 8.9).

17 When reassembling the joint, press half of the charge of grease into each side of the joint. Note that the chamfer on the splined hub must be on the larger diameter side of the outer ring. It will be necessary to pivot the joint hub when reassembling in order to align the balls with the grooves (Fig. 8.10).

18 Check the operation of the inner joint unit when assembled by moving the hub through its full axial range by hand.

19 It is advisable to fit new rubber boots to the shaft; a defective boot will soon lead to the need to fit a new joint due to wear caused by grit entering the joint. Fit the boots to the shaft and put any residual grease into the boots.

20 Fit the dished washer to the inner end of the driveshaft and locate the protective cap on the boot. Note that the concave side of the dished washer must face the joint (Fig. 8.11). On solid driveshaft models the dished washer is splined on its inner diameter.

21 Press the inner joint onto the end of the driveshaft and secure it with a new circlip.

Fig. 8.8 Inner CV joint hub and cage removal (Sec 3)

22 Tap the protective cap onto the outer ring.

23 Locate the dished washer and spacer on the outer end of the driveshaft, and check that the retaining circlip is located in the shaft groove.

24 Using a mallet, drive the outer joint onto the driveshaft until the circlip engages the groove.

25 Fit new clamps to each end of the rubber

Fig. 8.9 Removing the inner joint hub from cage (arrows indicate grooves) (Sec 3)

Fig. 8.10 Reassembling the inner joint (Sec 3)

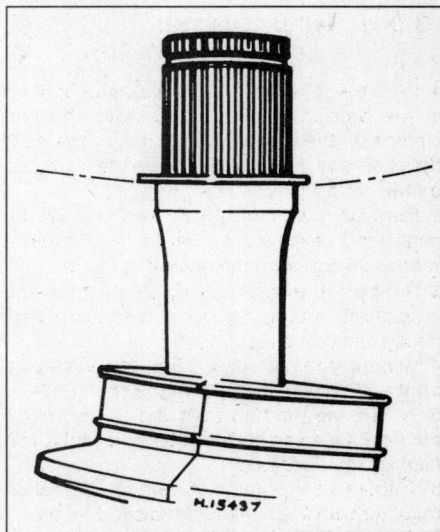
Fig. 8.11 Correct location of the dished washer on the driveshaft inner end (Sec 3)

8

boots, locate the boots over the joints, and tighten the clamps. If necessary the crimped type clamps can be replaced by plastic, ratchet type.

26 When the rubber boots are secured check that they are not deformed. An inward fold of the boot can be caused by a vacuum when fitting and can be relieved by lifting the small diameter (shaft end) of the boot (Fig. 8.12).

Fig. 8.12 Check for deformity in the rubber boot on reassembly (Sec 3)

Fault finding - driveshafts

Vibration and noise on lock

☐ Worn driveshaft joints

Noise on taking up drive or between acceleration and overrun

☐ Worn driveshaft joints
☐ Worn front wheel hub and driveshaft splines
☐ Loose driveshaft bolts or nut

Chapter 9 Braking system

Contents

Degrees of difficulty

Easy, suitable for novice with little experience	**Fairly easy,** suitable for beginner with some experience	**Fairly difficult,** suitable for competent DIY mechanic	**Difficult,** suitable for experienced DIY mechanic	**Very difficult,** suitable for expert DIY or professional

Specifications

System type . . . .   Four wheel hydraulic, with discs on front, and drums on rear. Twin diagonally split hydraulic circuits with vacuum servo assistance. Self-adjusting rear brakes. Load sensitive rear brake pressure regulator on some models. Cable operated handbrake on rear wheels.

Disc brakes

Teves caliper

Disc thickness	10 mm (0.394 in)
Disc wear limit	8 mm (0.314 in)
Maximum run-out allowable	0.06 mm (0.002 in)
Pad thickness	10 mm (0.394 in)
Minimum pad thickness allowable (including backplate)	7 mm (0.276 in)

Mk II caliper

Disc thickness	12 mm (0.472 in)
Disc wear limit	10 mm (0.394 in)
Maximum run-out allowable	0.06 mm (0.002 in)
Minimum pad thickness allowable (including backplate)	7 mm (0.276 in)

Mk II caliper - 85 kW variant

Disc thickness	20 mm (0.787 in)
Disc wear limit	18 mm (0.708 in)
Maximum run-out allowable	0.06 mm (0.002 in)
Minimum pad thickness allowable (including backplate)	7 mm (0.276 in)

Drum brakes

	Four-cylinder manual gearbox models	All other models
Internal diameter	180 mm (7.087 in)	200 mm (7.874 in)
Wear limit	181 mm (7.126 in)	201 mm (7.913 in)
Drum radial run-out	0.05 mm (0.002 in)	0.05 mm (0.002 in)
Wheel cylinder diameter	14.29 mm (0.563 in)*	17.46 mm (0.688 in)
Brake lining minimum thickness	2.5 mm (0.098 in)	2.5 mm (0.098 in)

* 17.46 mm (0.688 in) on models fitted with pressure limiter

Brake master cylinder

Type	ATE (Teves) or FAG (Schafer)
Brake fluid type	Hydraulic fluid to FMVSS 116 DOT 3 or DOT 4

9

Torque wrench settings

	Nm	lbf ft
Caliper or caliper bracket to wheel bearing housing bolts	70	52
Upper and lower caliper guide bolts (Mk II caliper)	40	30
Disc guard (splash) plate	10	7
Wheel bolts	110	81
Rear brake backplate	60	44
Servo to bearing bracket	20	15

1 General description

The braking system is hydraulic with servo-assistance, and there are disc brakes on the front wheels and drum brakes on the rear. The system has a tandem master cylinder which incorporates two completely independent braking circuits, each circuit operating a front wheel and the diagonally opposite rear wheel. This ensures that if one circuit fails, the car can still be brought to rest in a straight line, even though the braking distance will be greater.

Some models incorporate a brake pressure regulator, which limits the pressure applied to the rear brake cylinders to a proportion of that applied to the front, and so prevents the rear wheels from locking.

The rear brakes are self-adjusting and incorporate a wedged key which automatically adjusts the length of the brake shoe upper pushrod on each wheel.

The handbrake operates on the rear wheels only and on some models incorporates a switch which illuminates a warning light on the instrument panel – the same light also warns of low brake fluid level.

2 Maintenance – brake system

1 The brake fluid level should be checked every week – the reservoir is translucent and the fluid level should be between the MIN and MAX marks. If necessary, top up with the specified brake fluid (Fig. 9.1). However, additional fluid will only be necessary if the hydraulic system is leaking, therefore the source of the leak must first be traced and rectified. Note that the level will drop slightly as the front disc pads wear, but in this case it is not necessary to top up the level.

2 Every 10 000 miles (15 000 km) or 12 months, if this occurs sooner, the hydraulic pipes and unions should be checked for chafing, leakage, cracks and corrosion. At the same time check the operation of the brake pressure regulator and check the disc pads and rear brake linings for wear. Also check the servo vacuum hose for condition and security.

3 Check the brake warning device for correct operation by switching the ignition on and releasing the handbrake. Now press the contact on the reservoir filler cap down and get an assistant to check that the handbrake and dual circuit warning lamp light up.

4 Renew the brake fluid every 2 years.

3 Disc pads – inspection and renewal

1 Pad thickness can be checked without removing the roadwheel. Turn the wheel until the brake pad is visible through one of the openings in the wheel rim.

2 With the aid of a torch to increase visibility, measure the thickness of each brake pad including its metal backplate and compare this with the minimum value given in the Specifications. A rough guide to the amount of life remaining in brake pads which are nearing their minimum is that the rate of wear is about 1 mm (0.039 in) for every 1000 km (620 miles) driving, (Fig. 9.2 and 9.3).

3 If the brake pads are to be re-used, they must be refitted to the position from which they were taken. To ensure this, mark the pads before removing them.

4 First apply the handbrake, then jack up the front of the car and support it on axle stands. Remove the roadwheels.

Teves caliper

5 Extract the retaining spring, then use a punch to drive out the pad retaining pins (Fig. 9.4).

6 Pull out the inner pad using a hooked instrument.

7 Press the floating caliper frame outwards to disengage the pad from the projection on the caliper frame and then pull out the outer pad (Fig. 9.5).

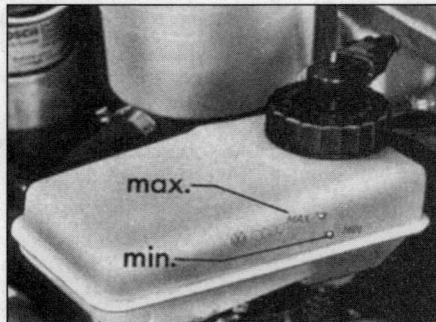

Fig. 9.1 Master cylinder fluid level to be maintained between the MAX and MIN markings (Sec 2)

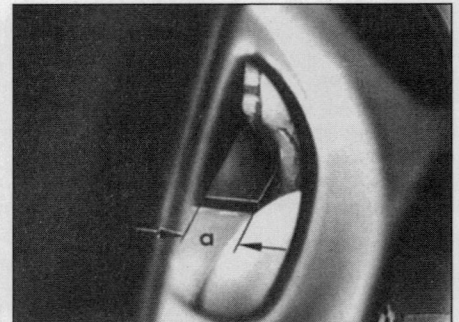

Fig. 9.2 Check the disc pads for wear by measuring thickness (a) through one of the wheel rim apertures – Teves caliper (Sec 3)

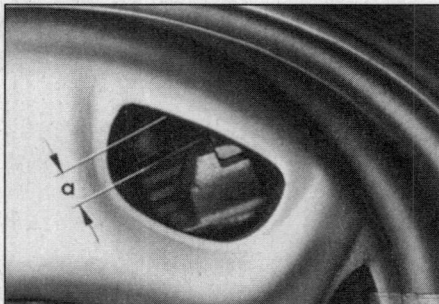

Fig. 9.3 Disc pad thickness check (a) through one of the wheel rim apertures for the Mk II caliper (Sec 3)

Fig. 9.4 Remove the pad retaining pin (Teves caliper) (Sec 3)

Fig. 9.5 Outer pad and locating projection on the Teves caliper (Sec 3)

Fig. 9.6 Checking the piston position with a 20° gauge – arrow indicates forward rotation of disc (Sec 3)

Fig. 9.7 The later type Mk II caliper unit (Sec 3)

Fig. 9.8 The later type Mk II caliper brake pads and retaining springs (Sec 3)

8 Using two pieces of wood, lever the piston back into the cylinder. While doing this check that brake fluid will not overflow from the reservoir, and if necessary use a flexible plastic bottle to extract some fluid. **Note**: *Brake fluid is poisonous, so no attempt should be made to siphon the fluid by mouth.*

9 Check that the piston cutaway is positioned at 20° to the horizontal (see Fig. 9.6) and, if necessary, rotate the piston as required.

Mk II caliper

10 Unhook and remove the pad retaining springs at the top and the bottom of the caliper (photo). Note that a later Mk II caliper is fitted from February 1984 in which the retaining springs are of the flat section type (Figs. 9.7 and 9.8).

11 Unscrew and remove the upper and lower caliper guide pins then withdraw the caliper unit (photos).

12 Remove the brake pads from the carrier noting fitted positions (photo).

13 Compress the piston back into the cylinder as
described in paragraph 8.

14 The repair kit includes new caliper housing guide pins and these will differ according to type. The later type pins fitted to calipers from February 1984 on, consist of a socket head screw and pin guide sleeve (Fig. 9.9).

3.10 Early type Mk II caliper brake pad retaining spring

Teves and Mk II calipers

15 Brush the dust and dirt from the caliper, piston. disc and pads, *but do not inhale it, as it is injurious to health.* Scrape any scale or rust from the disc and pad backing plates.

16 Fitting the brake pads is a reversal of the removal procedure, but on the later Mk II type caliper locate the retaining springs onto the carrier prior to fitting the pads to position. Also on the later Mk II caliper type, as the caliper is fitted into position, simultaneously insert the guide pins (socket head screws) to ensure that the threads are in alignment. Fully fitting the caliper at this stage could mean that the retaining springs will get distorted later when fitting the pins, which in turn will give noisy braking.

Fig. 9.9 Socket-head screws and guide sleeves fitted to later Mk II calipers (Sec 3)

17 On completion the brake pedal should be depressed firmly several times with the car stationary, so that the disc pads take up their normal running position. Also check and if necessary top up the brake fluid level in the reservoir.

4 Disc caliper –
removal and refitting

1 Remove the disc pads, as described in Section 3.

2 Remove the brake fluid reservoir filler cap and tighten it down onto a piece of polythene sheeting placed over the reservoir filler hole in order to prevent the loss of fluid in the following procedure. Alternatively fit a brake

9

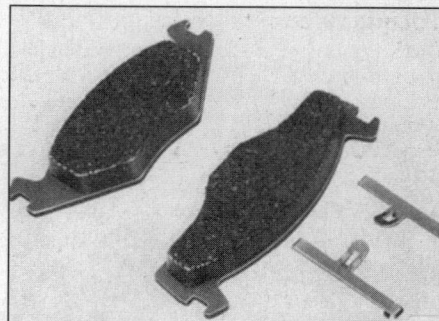

3.11A Unscrew the guide pins (early Mk II type shown) . . .

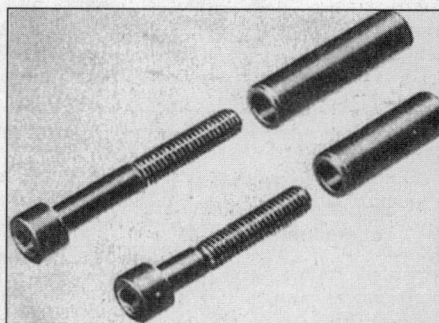

3.11 B . . . and withdraw the caliper unit

3.12 Remove the brake pads from the carrier (Mk II caliper)

Fig. 9.10 Teves caliper and associated components (Sec 4)

Pads

Wheel bearing housing

Driveshaft

Lock washer

Caliper

Brake disc

Fig. 9.11 The Mk II caliper and associated components (Sec 4)

Wheel bearing housing

Splash plate

Driveshaft

Lock washer

Pads

Disc

Sealing ring

Dust cap

Circlip

Bleeder screw

Locating spring

Cylinder

Piston

Mounting frame

Floating frame

Brake pads

Fig. 9.12 Exploded view of the Teves caliper (Sec 5)

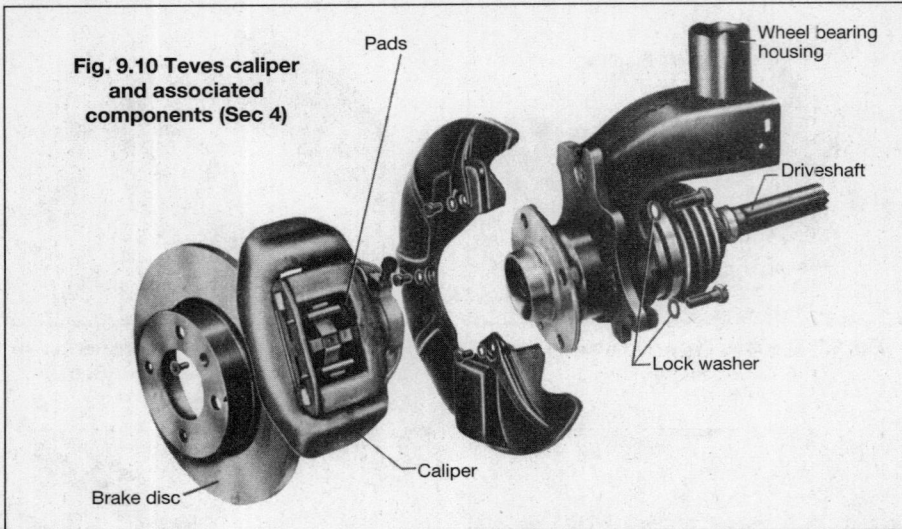

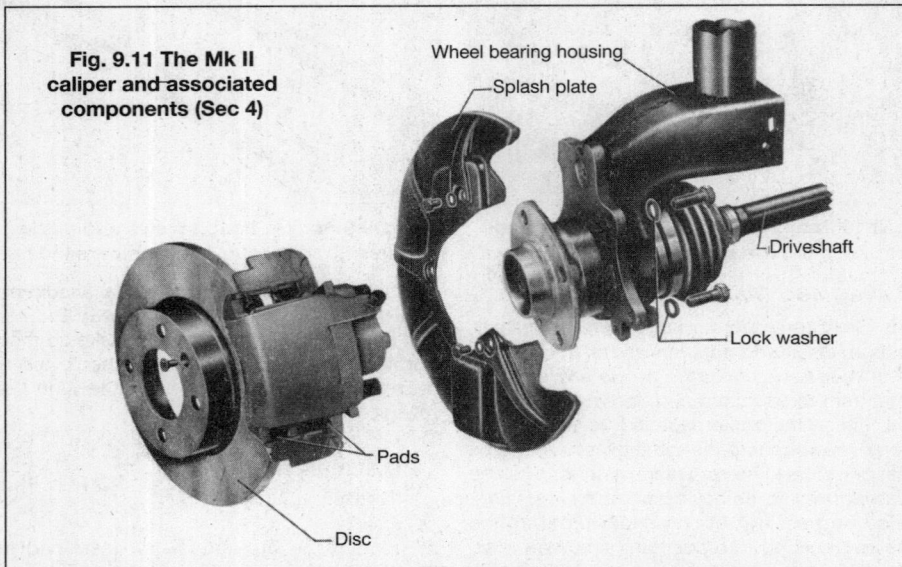

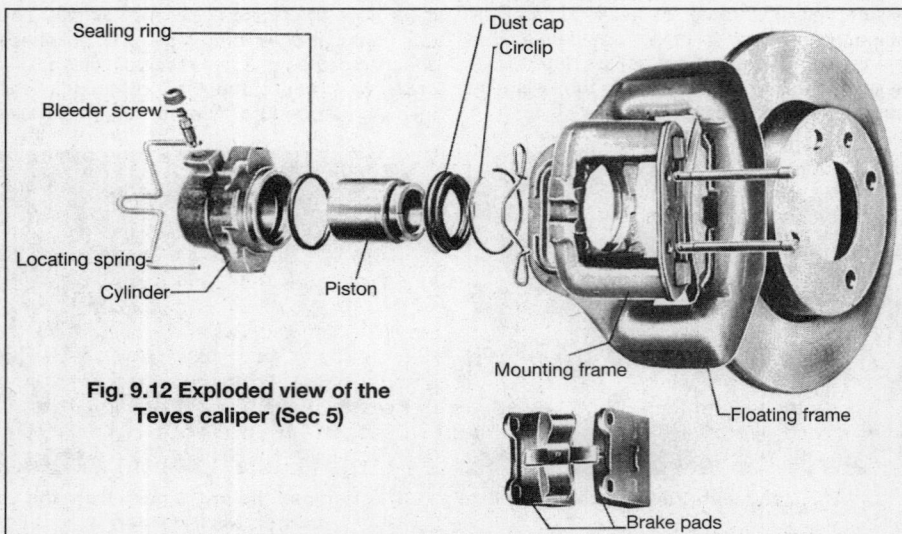

hose clamp on the hydraulic hose attached to the caliper.

3 Loosen the hydraulic hose union, but do not attempt to unscrew it at this stage.

Teves caliper

4 Unscrew the two bolts securing the caliper to the suspension strut while supporting the caliper (Fig. 9.10).

5 Unscrew the caliper from the hydraulic hose and plug the end of the hose.

Mk II caliper

6 The caliper unit can be removed in the same manner as that described in paragraphs 10 and 11 in the previous Section. Do not allow the caliper to hang on the hydraulic hose (Fig. 9.11).

7 Unscrew the caliper from the hydraulic hose and plug the end of the hose. The caliper is now free.

8 If necessary, unbolt the caliper bracket from the suspension strut.

Teves and MK II calipers

9 Refitting is a reversal of removal but the following special points concerning the caliper or caliper bracket retaining bolts should be noted.

10 New retaining bolts must always be used when refitting the caliper or caliper bracket, but note that on later models the hex-headed retaining bolts and lock washers have been superseded by flanged locking bolts (Fig. 9.13). Additionally, the bolt holes in the

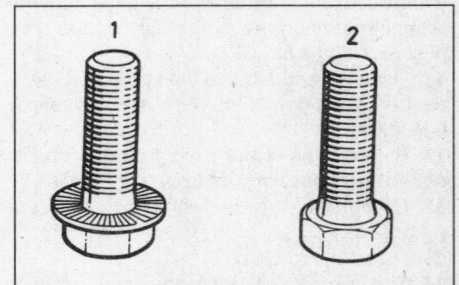

1   2

Fig. 9.13 Brake caliper retaining bolts (Sec 4)

1 Later type flanged locking bolts
2 Early type hex-headed bolts

Fig. 9.14 Caliper chamfered bolt holes (arrowed) to accept later type flanged locking bolts (Sec 4)

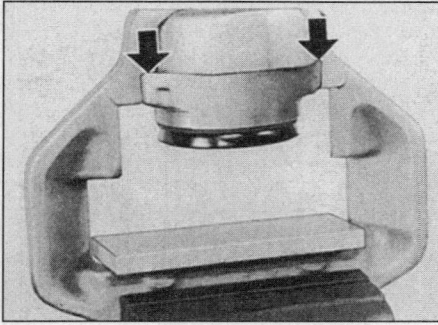

Fig. 9.15 Removing the cylinder on the Teves caliper (Sec 5)

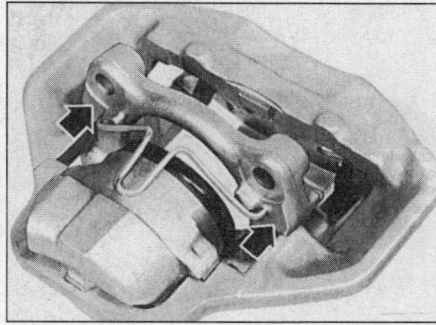

Fig. 9.16 Installing the mounting frame on the Teves caliper (Sec 5)

Fig. 9.17 Dust cap location on Mk II caliper piston (Sec 5)

Top – early type   *Bottom – later type*

caliper have been provided with a 45° chamfer on their internal circumference to accept the new bolts. As only the flanged locking bolts are now supplied by VW parts stockists, it will be necessary to chamfer the bolt holes of early calipers to accept the flanged bolts. This can be done with a countersunk drill bit or suitable scraper. The earlier hex-headed bolts **must not** be used on calipers with chamfered bolt holes.

11 Ensure that all mating faces are clean when refitting and tighten the bolts and unions to the specified torque. Finally bleed the hydraulic system as described in Section 14.

5 Disc caliper – overhaul

1 Remove the caliper as described in the previous Section, then clean it externally taking care not to allow any foreign matter to enter the hydraulic hose aperture.

Teves caliper

2 Press the mounting frame off the floating frame by hand and separate the two frames.
3 Place a block of wood in the floating frame and drive out the cylinder assembly, using a soft metal drift (Fig. 9.15).
4 Use a foot pump or a compressed air supply to blow the piston out of the cylinder, but support the cylinder with the piston facing downwards to avoid the risk of injury when the piston is ejected.
5 Use a blunt screwdriver to prise the seal out of the piston bore, taking great care not to scratch the bore of the cylinder.
6 Clean the caliper components thoroughly with methylated spirit and allow to dry.
7 Fit a new seal into the groove in the cylinder bore. Apply a thin coat of brake cylinder paste to the seal and to the cylinder. Then squeeze the piston into the cylinder using a vice fitted with soft jaws.
8 Rest the floating frame on the bench, then insert the brake cylinder and tap it fully into the frame.

9 With the locating spring fitted to the mounting frame, press the mounting frame onto the floating frame (Fig. 9.16).
10 Set the piston cutaway 20° to the horizontal as described in Section 3. The angle can be checked by making a simple gauge out of card.

Mk II caliper

11 Use a foot pump, or a compressed air supply to blow the piston out of the cylinder, but place a block of wood inside the frame to prevent damage to the piston. Remove the dust cap.
12 Use a blunt screwdriver to prise the seal out of the piston bore, taking great care not to scratch the bore of the cylinder.
13 Clean the caliper components thoroughly with methylated spirit and allow to dry.
14 Fit a new seal into the groove in the cylinder bore. Apply a thin coat of brake cylinder paste to the seal and to the cylinder.
15 Fit the dust cap on to the piston as shown in Fig 9.17, and then offer the piston up to the cylinder and fit the sealing lip of the dust cap into the groove of the cylinder bore, using a screwdriver. On later models, the dust cap inner lip locates in the groove within the cylinder housing.
16 Smear brake cylinder paste over the piston and then press the piston into the cylinder until the outer lip of the dust cap springs into place in the piston groove.
17 Check the guide pins, bolts and dust caps for condition and, if necessary, renew them. The bolts are self-locking and the manufacturers recommend that they are renewed whenever they are removed.

6 Brake disc – examination, removal and refitting

1 Jack up the front of the car and support it on axle stands. Remove the roadwheel.
2 Rotate the disc and examine it for deep scoring or grooving. Light scoring is normal, but if excessive the disc should be removed and renewed, or ground by a suitably qualified

engineering works. Use a micrometer in several positions to check the disc thickness.
3 To remove the brake disc first remove the caliper as described in Section 4. However, do not disconnect the hydraulic hose – suspend the caliper by wire from the coil spring making sure that the hydraulic hose is not strained.
4 Remove the countersunk cross-head screw and withdraw the brake disc from the hub.
5 Refitting is a reversal of removal, but make sure that the mating faces of the hub and disc are clean.

7 Rear brake shoes – inspection and renewal

1 The thickness of the rear brake linings can be checked without removing the brake drums. Remove the rubber plug which is above the handbrake cable entry on the brake backplate. Use a torch to increase visibility and check the thickness of friction material remaining on the brake shoes. If the amount remaining is close to the minimum given in the Specifications, a more thorough examination should be made by removing the brake drum (photo).

9

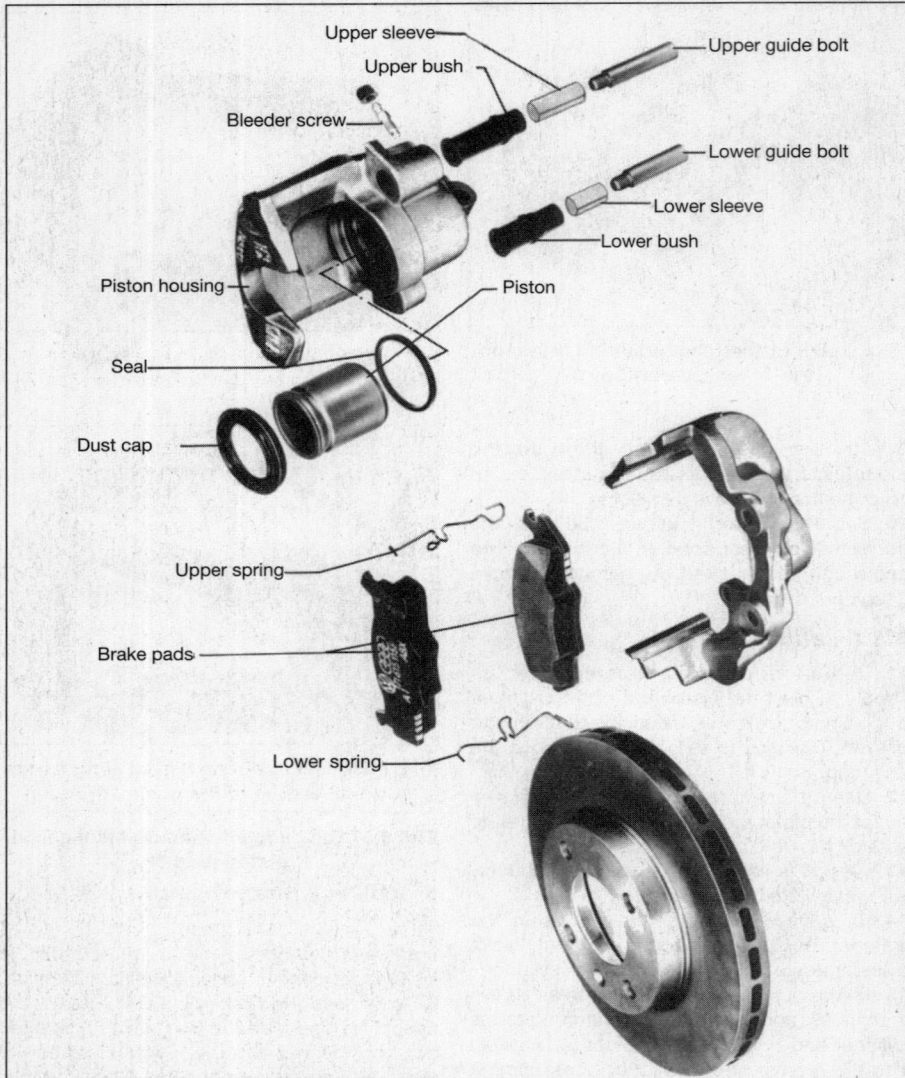

Fig. 9.18 Exploded view of the Mk II caliper (Sec 5)

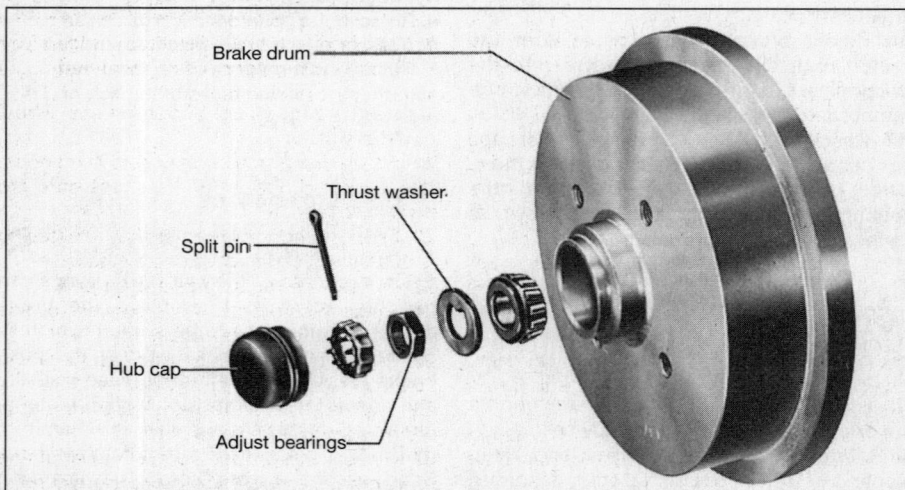

Labels (top to bottom, left to right):
Upper sleeve
Upper bush
Upper guide bolt
Bleeder screw
Lower guide bolt
Lower sleeve
Lower bush
Piston housing
Piston
Seal
Dust cap
Upper spring
Brake pads
Lower spring

Fig. 9.19 Rear brake drum and wheel bearing assembly components (Sec 7)

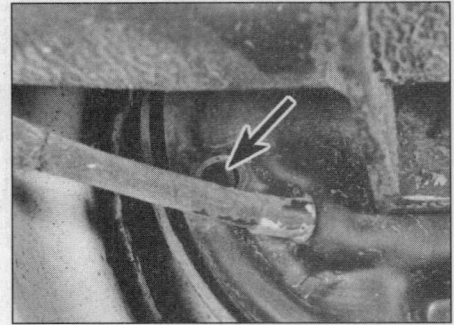

Labels:
Brake drum
Thrust washer
Split pin
Hub cap
Adjust bearings

7.1 Brake shoe inspection aperture in rear brake backplate

7.7 Note position of brake shoes and springs prior to dismantling

7.8 Remove the rear brake shoe retaining springs

2 Chock the front wheels, then jack up the rear of the car and support it on axle stands. Remove the rear wheels and release the handbrake.

3 Remove the cap from the centre of the brake drum by tapping it on alternate sides with a screwdriver or blunt chisel.

4 Extract the split pin and remove the locking ring.

5 Unscrew the nut and remove the thrust washer (Fig. 9.19).

6 Withdraw the brake drum, making sure that the outer wheel bearing does not fall out. If the brake drum binds on the shoes, insert a small screwdriver through a wheel bolt hole and lever up the wedged key in order to release the shoes.

7 Note the position of the brake shoes and

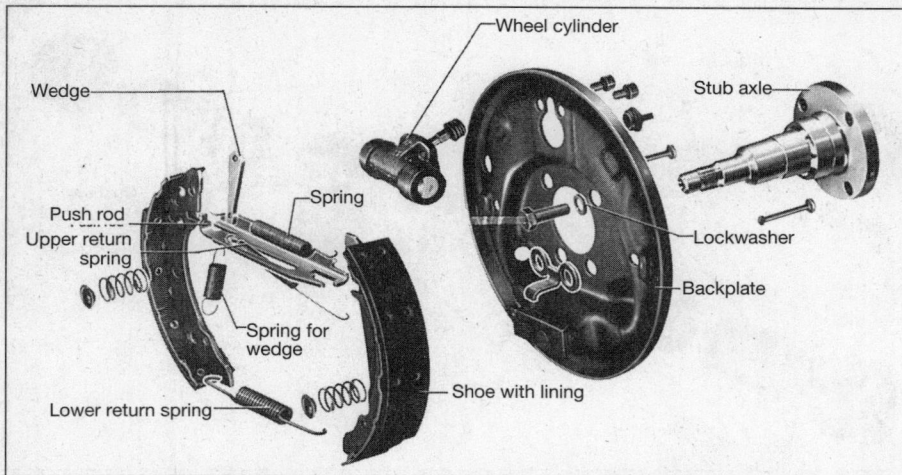

Fig. 9.20 Rear brake assembly components – 180 mm type (Sec 7)

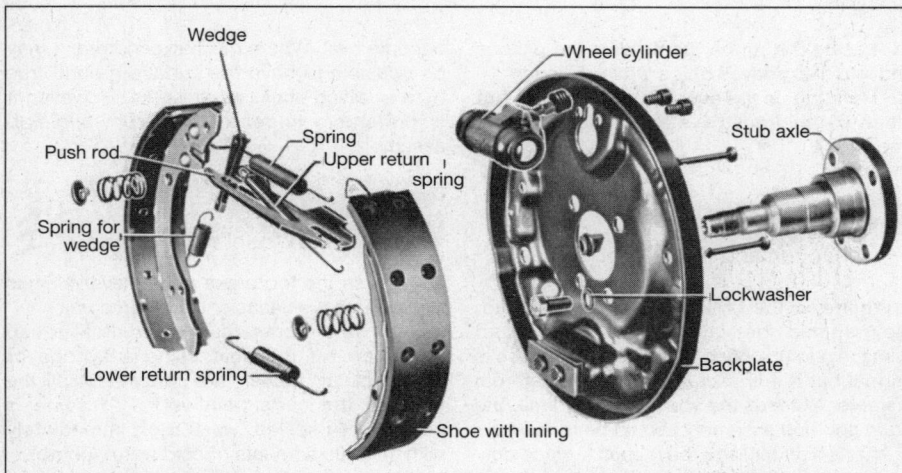

Fig. 9.21 Rear brake assembly components – 200 mm type (Sec 7)

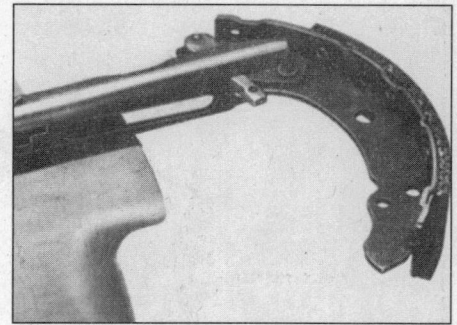

Fig. 9.22 Clamp the pushrod when separating from the brake shoe (Sec 7)

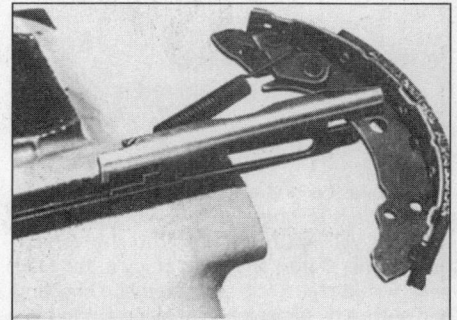

Fig. 9.23 Reconnecting the locating spring and brake shoe to the pushrod (Sec 7)

7.22 Rear brake self-adjusting wedged key and spring connection (arrowed)

springs and mark the webs of the shoes, if necessary, to aid refitting (photo).

8 Using pliers, depress the spring retainer caps, turn them through 90° and remove them together with the springs (photo).

9 Prise the brake shoes away from the lower pivot support and then disengage the lower return spring.

10 Detach the handbrake cable from the lever.

11 Using suitable pliers, unhook the wedge spring and the upper return spring and remove the brake shoes.

12 To separate the pushrod from the brake shoe, clamp the pushrod in a vice and unhook the retaining spring (Fig. 9.22). As they are separated note how the wedged key is located.

13 Brush the dust from the brake drum, brake shoes, and backplate. *but do not inhale it, as it is injurious to health*. Scrape any scale or rust from the drum.

14 Measure the brake shoe lining thickness. If worn down to the specified minimum

amount, renew all four rear brake shoes.

15 Clean the brake backplate. If there are any signs of loss of grease from the rear hub bearings, the oil seal should be renewed with reference to Chapter 11. If hydraulic fluid is leaking from the wheel cylinder, it must be repaired or renewed, as described in Section 8. Do not touch the brake pedal while the shoes are removed. Position an elastic band over the wheel cylinder pistons to retain them.

16 Apply a little brake grease to the contact areas of the pushrod and handbrake lever.

17 Clamp the pushrod in a vice, then hook the tensioning spring on the pushrod and leading shoe and position the shoe slot over the pushrod (Fig. 9.23).

18 Fit the wedged key between the shoe and pushrod with the lug facing towards the backplate.

19 Locate the handbrake lever on the trailing shoe in the pushrod, and fit the upper return spring.

20 Connect the handbrake cable to the handbrake lever, swivel the shoes upward and

locate the top of the shoes on the wheel cylinder pistons.

21 Fit the lower return spring to the shoes, then lever the bottom of the shoes onto the bottom anchor.

22 Fit the spring to the wedged key and leading shoe as noted during dismantling (photo).

23 Fit the retaining springs and caps.

24 Press the wedged key upwards to give the maximum shoe clearance.

25 Fit the brake drum and adjust the wheel bearings, with reference to Chapter 11.

26 Depress the brake pedal once firmly in order to adjust the rear brakes.

27 Repeat the procedure on the remaining rear brake, then lower the car to the ground.

28 Check that the operation of both the foot and handbrake are satisfactory.

9

8.3 Brake pipe union connection to the rear wheel cylinder (arrowed) with mounting bolts above (each side of bleed nipple)

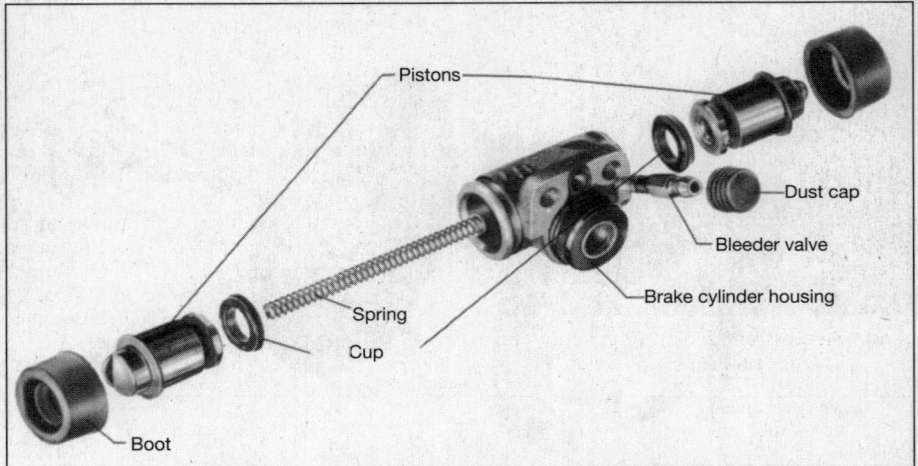

Fig. 9.24 Exploded view of the rear wheel cylinder (Sec 8)

8 Rear wheel cylinder –
removal, overhaul and refitting

1 Remove the brake shoes, as described in Section 7.
2 Remove the brake fluid reservoir filler cap and place it to one side away from dirt and where it will not drip fluid onto the vehicle paintwork. Cover the master cylinder filler neck with a piece of clean polythene sheeting and secure by fastening in position around the filler neck with an elastic band. This will prevent excessive loss of fluid in the following procedure.
3 Loosen the brake pipe union on the rear of the wheel cylinder (photo).
4 Using an Allen key, unscrew the wheel cylinder mounting bolts.
5 Unscrew the brake pipe union and withdraw the wheel cylinder from the backplate. Plug the end of the hydraulic pipe, if necessary.
6 Clean the exterior of the wheel cylinder. taking care not to allow any foreign matter to enter the hydraulic pipe aperture.
7 Remove the rubber boots from the ends of the cylinder and extract the two pistons and the spring between them.
8 Inspect the cylinder bore for signs of scoring and corrosion and the pistons and seals for wear. If the cylinder is satisfactory a repair kit can be used; otherwise the cylinder should be discarded and a new complete assembly fitted. If servicing a cylinder, use all the parts in the repair kit. Clean all the metal parts, using methylated spirit if necessary, *but never petrol or similar solvents*, then leave the parts to dry in the air, or dry them with a lint-free cloth.
9 Apply brake cylinder paste to the seals and fit them so that their large diameter end is nearest to the end of the piston.
10 Smear brake cylinder paste on to the pistons and into the bore of the cylinder. Fit a piston into one end of the cylinder and then the spring and other piston into the other end. Take care not to force the pistons into the cylinders because this can twist the seals.

11 Locate the rubber boots over the pistons and into the grooves of the wheel cylinder.
12 Refitting is a reversal of removal, but bleed the hydraulic system as described in Section 14.

9 Brake drum –
inspection and renovation

1 Whenever the brake drums are removed, they should be checked for wear and damage. Light scoring of the friction surface is normal but if it is excessive, or if the internal diameter exceeds the specified wear limit, the drum and hub assembly should be renewed.
2 After a high mileage the friction surface may become oval. Where this has occurred, it may be possible to have the surface ground true by a qualified engineering works. However, it is preferable to renew the drum and hub assembly.

10 Master cylinder –
removal and refitting

1 Depress the footbrake pedal several times to dissipate the vacuum in the servo unit.
2 Remove the brake fluid reservoir filler cap and draw off the fluid using a syringe or flexible plastic bottle. Take care not to spill the fluid on the car's paintwork – if some is accidentally spilled, wash it off immediately with copious amounts of cold water (photo).

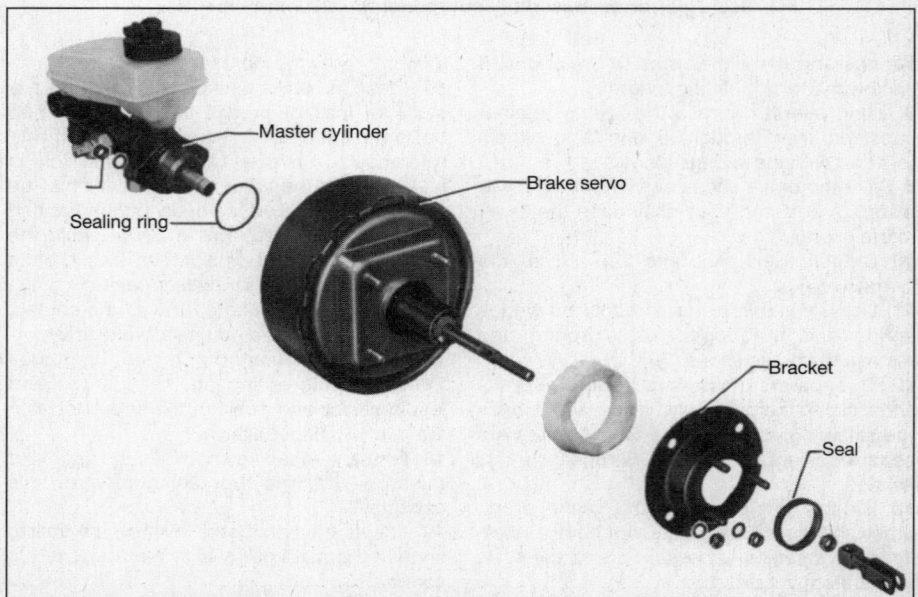

Fig. 9.25 Brake master cylinder and servo unit connecting components (Sec 10)

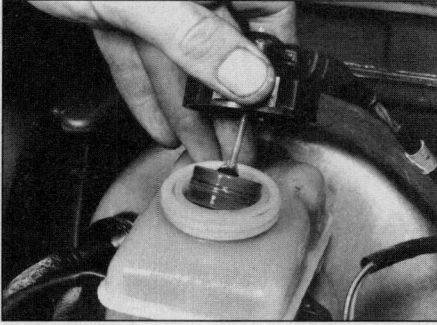

10.2 Removing the master cylinder cap with level warning device

3 Refit the filler cap, but disconnect the wiring for the fluid level warning circuit.
4 Place some rags beneath the master cylinder, then unscrew the brake pipe union nuts and pull the pipes just clear of the master cylinder. Plug the ends of the pipes or cover them with masking tape.
5 Where applicable disconnect the wiring from the brake light switch.
6 Unscrew the mounting nuts and withdraw the master cylinder from the front of the servo unit. Remove the seal (Fig. 9.25).
7 Refitting is a reversal of removal, but always fit a new seal. Finally bleed the hydraulic system as described in Section 14.

11 Master cylinder – overhaul

Note: *A FAG (Schafer) or ATE (Teves) master cylinder will be fitted and it is important to identify which type is on your vehicle so that the correct repair kit is supplied.*
1 Empty the master cylinder and reservoir and discard the fluid.
2 Pull the reservoir off the master cylinder.
3 Remove the stop screw from the cylinder and then remove the circlip from the mouth of the bore (Fig. 9.26).
4 Pull out the piston pushrod assembly, the secondary piston assembly and finally the loose spring.
5 Inspect the bore of the cylinder for signs of wear, damage and corrosion. If the cylinder is in good condition it may be re-used with a service kit, otherwise obtain a new cylinder assembly.
6 Clean all the metal parts, using methylated spirit if necessary, *but never petrol or similar solvents*, then leave the parts to dry in the air or dry them with a clean lint-free cloth. Remove the seals, but note their fitting orientation.
7 Use all the parts supplied in the repair kit and moisten all components with brake fluid before fitting them. Make sure that the correct repair kit is obtained as there are two types of master cylinder – ATE and FAG. If the repair kit supplied is a 'ready to use' kit proceed from paragraph 14.
8 Reassemble the secondary piston assembly in the following order. Fit the two piston seals at the larger diameter end of the piston. The piston seals are identified by their chamfer and groove; the sealing lips of the two seals should face away from each other. Fit the cup washer, the primary cup with its lips towards the closed end of the cylinder, the support ring and finally the conical spring.
9 While holding the master cylinder body vertically with its open end pointing downwards, feed the secondary piston assembly into the cylinder. Guide the lips of the seals into the bore carefully.
10 Fit the cup washer, primary cup, support ring, the cylindrical spring, the stop sleeve and the stroke limiting screw to the short end of the primary piston and tighten the screw firmly. To the longer end of the piston fit the metal washer, the secondary cup with its lip towards the washer, the plastic washer, the other secondary cup, with its lip facing the same way as the first one, and finally the second metal washer.
11 Insert the primary piston assembly into the cylinder bore with the spring end first. Insert the stop screw, and if necessary move the piston so that the stop screw can be screwed in fully. After screwing in the stop screw, fit the circlip to the mouth of the cylinder bore.
12 Fit a new seal over the end of the master cylinder.

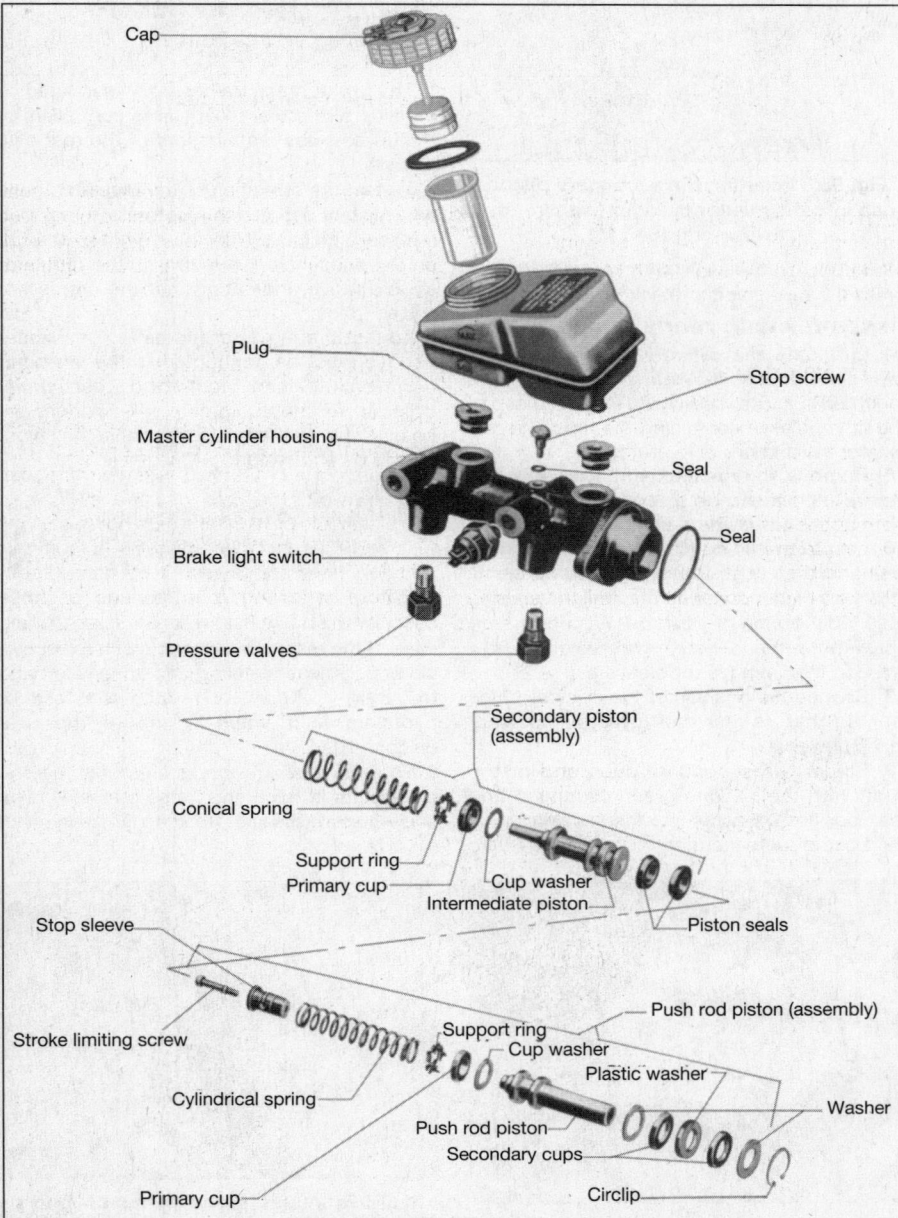

Cap
Plug
Master cylinder housing
Brake light switch
Pressure valves
Stop screw
Seal
Seal
Secondary piston (assembly)
Conical spring
Support ring
Primary cup
Cup washer
Intermediate piston
Piston seals
Stop sleeve
Stroke limiting screw
Cylindrical spring
Push rod piston
Secondary cups
Primary cup
Support ring
Cup washer
Plastic washer
Push rod piston (assembly)
Washer
Circlip

Fig. 9.26 Exploded view of the master cylinder (Sec 11)

9

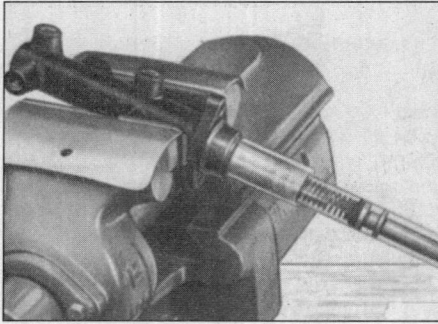

Fig. 9.27 Inserting 'ready-to-use' master cylinder piston kit (Sec 11)

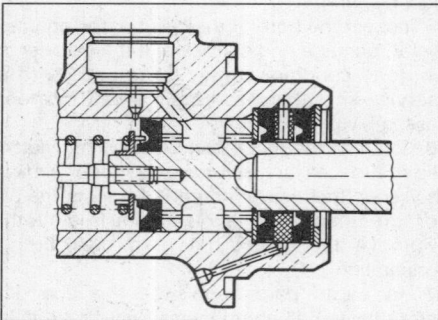

Fig. 9.29 Sectional view showing sealing cup locations in the master cylinder (Sec 11)

12.1 Brake pressure regulator showing pipe connections and pressure spring attachment

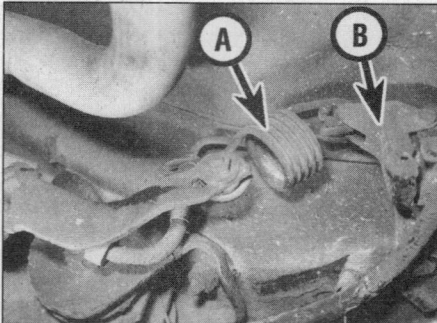

Fig. 9.28 Inserting the secondary piston cap into the master cylinder (with 'ready-to-use- kit) (Sec 11)

13 Lubricate the seals with brake fluid then press the reservoir into them .

Assembly with ready to use kit

14 Lubricate the cylinder bore with brake cylinder paste then with the cylinder unit mounted in a vice, detach the large end cap of the fitting sleeve containing the repair kit and extract the parts.
15 Remove the second end cap from the sleeve then push the sleeve short end in to align both ends of the sleeve.
16 Locate the fitting sleeve into the cylinder wall and then push the cylinder components into the cylinder so far as they will travel using a suitable punch or pushrod. Fit a new stop screw into the cylinder wall using a new washer, then remove the sleeve (Fig. 9.27).
17 Change the position of the cylinder in the vice so that its rear end (cylinder opening) faces upwards.
18 Smear the secondary cups and piston shaft with the silicone grease supplied and then position a washer into the housing.
19 Locate the secondary cup into the fitting

sleeve, then fit the cup in position on the pushrod piston shaft with the sleeve and insert into the cylinder.
20 Push the secondary cup down using the long sleeve section and pull back the short sleeve section the thickness of the cup (Fig. 9.28 and 9.29).
21 Insert the plastic washer, the second cup and then the second washer. Locate circlip to secure, fit the residual pressure valves and the brake light switch.
22 Fit a new seal over the end of the master cylinder.
23 Lubricate the seals with brake fluid then press the reservoir into them to complete.

12 Brake pressure regulator – testing and adjustment

1 The brake pressure regulator is located to the side of the rear axle, and is controlled by the up-and-down movement of the rear axle (photo).
2 To test the operation of the regulator, have an assistant depress the footbrake firmly then release it quickly. With the weight of the car on the suspension, the arm on the regulator should move, indicating that the unit is not seized (photo).
3 To test the regulator for leakage, pressure gauges must be connected to the left-hand front caliper and right-hand rear wheel cylinder. As the equipment will not normally be available to the home mechanic, this work should be entrusted to a VW dealer.

13 Hydraulic brake lines and hoses – removal and refitting

1 Before removing a brake line or hose unscrew the brake fluid reservoir filler cap and seal off the reservoir filler neck with a piece of clean polythene sheeting and an elastic band to retain it. This will reduce the loss of hydraulic fluid when a line air hose is disconnected.
2 To remove a rigid brake line, unscrew the union nuts at each end, prise open the clips and withdraw the line. Refitting is a reversal of removal (photo).

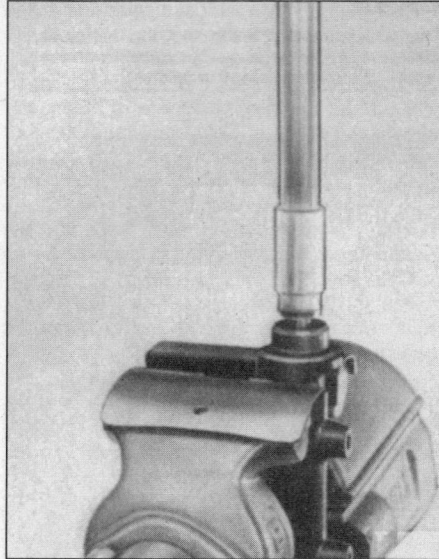

12.2 Brake pressure spring (A) and regulator arm (B) attachments

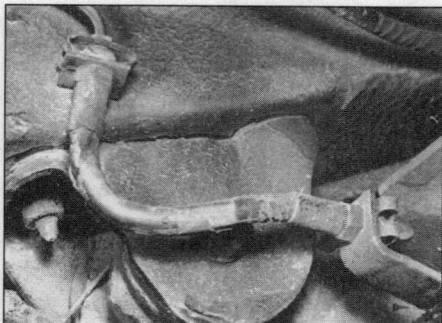

13.2A Typical rigid-to-flexible brake line connections with support brackets

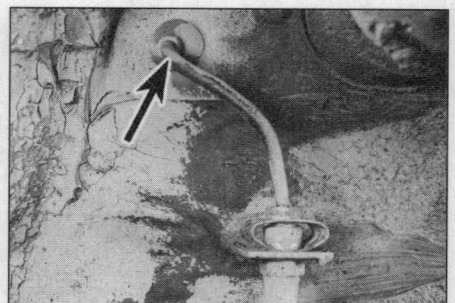

13.2B Grommet (arrowed) must support rigid brake line where passing through body panel

3 To remove a flexible brake hose, unscrew the union nut securing the rigid brake line to the end of the flexible hose while holding the end of the flexible hose stationary. Remove the clip and withdraw the hose from the bracket. Unscrew the remaining end from the component or rigid pipe according to position. Refitting is a reversal of removal, but position flexible hoses clear of adjacent steering and suspension components.

4 Bleed the hydraulic system as described in Section 14 after fitting a rigid brake line or flexible brake hose.

14 Hydraulic system – bleeding

1 If any of the hydraulic components in the braking system have been removed or disconnected, or if the fluid level in the master cylinder has been allowed to fall appreciably, it is inevitable that air will have been introduced into the system. The removal of all this air from the hydraulic system is essential if the brakes are to function correctly and the process of removing it is known as bleeding.

2 There are a number of one-man, do-it-yourself, brake bleeding kits currently available from motor accessory shops. It is recommended that one of these kits should be used wherever possible, as they greatly simplify the bleeding operation, and also reduce the risk of expelled air and fluid being drawn back into the system.

3 If one of these kits is not available then it will be necessary to gather together a clean jar and a suitable length of clear plastic tubing which is a tight fit over the bleed screw, and also to engage the help of an assistant.

4 Before commencing the bleeding operation, check that all rigid pipes and flexible hoses are in good condition and that all hydraulic unions are tight. Take great care not to allow hydraulic fluid to come into contact with the car's paintwork, otherwise the finish will be seriously damaged. Wash off any spilled fluid immediately with cold water.

5 If hydraulic fluid has been lost from the master cylinder, due to a leak in the system, ensure that the cause is traced and rectified before proceeding further, or a serious malfunction of the braking system may occur.

6 To bleed the system, clean the area around the bleed screw at the wheel cylinder to be bled. If the hydraulic system has only been partially disconnected and suitable precautions were taken to prevent further loss of fluid, it should only be necessary to bleed that part of the system. However, if the entire system is to be bled, the following sequence must be adhered to:

   1 *Right-hand rear wheel cylinder*
   2 *Left-hand rear wheel cylinder*
   3 *Right-hand front caliper*
   4 *Left-hand front caliper*

When bleeding the rear wheel cylinders on models with a brake pressure regulator, have

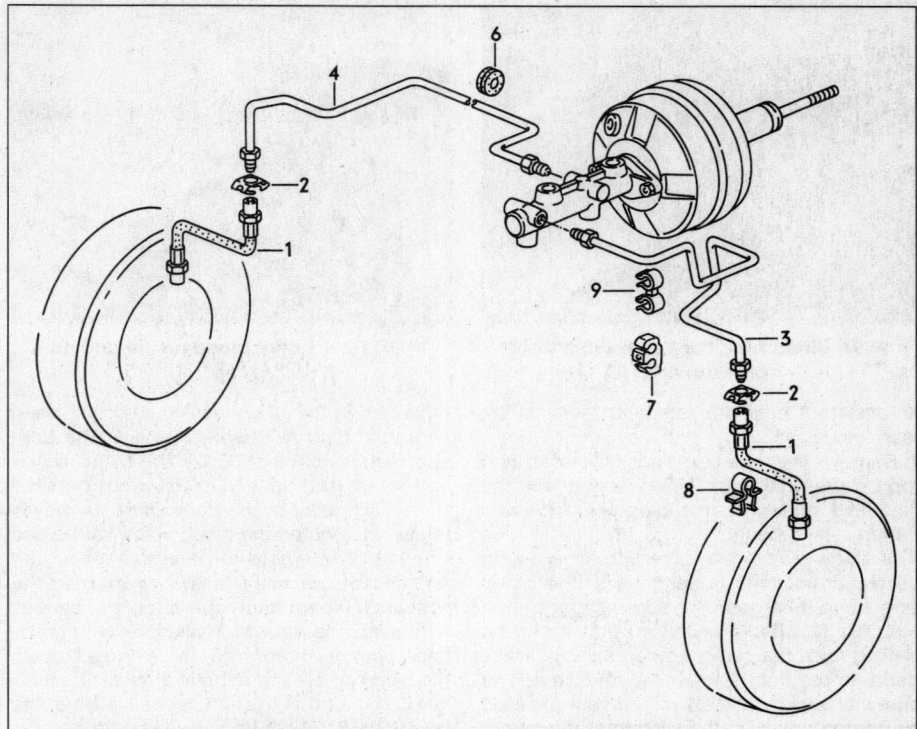

Fig. 9.30 Brake line components to front brakes (Sec 13)

1 *Flexible line to caliper*	4 *Master cylinder rigid line*	8 *Location bracket – hose*
2 *Bracket clip*	6 *Grommet*	9 *Location bracket – hose*
3 *Master cylinder rigid line*	7 *Location bracket – rigid line*	

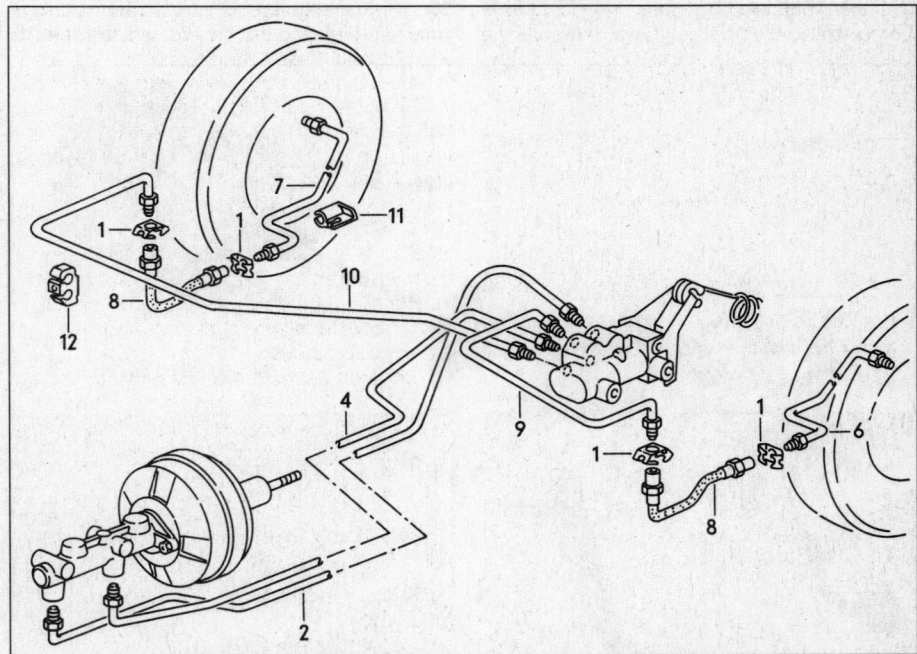

Fig. 9.31 Brake line components to rear brakes (Sec 13)

1 *Hose bracket clip*	8 *Flexible hoses*
2 and 4 *Rigid line – master cylinder to pressure regulator unit*	9 and 10 *Rigid lines – from pressure regulator*
6 and 7 *Rigid line – to wheel cylinder*	11 *Rigid line clip*
	12 *Rigid line location bracket*

9

14.9A Bleed tube connected to a front brake caliper

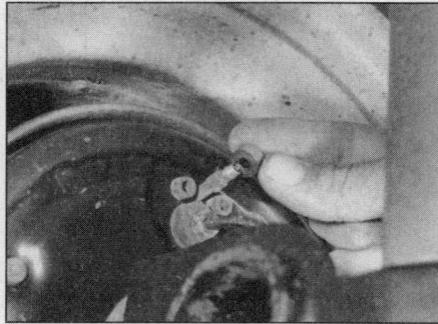

14.9B Rear brake bleed nipple and dust cap

an assistant hold the regulator lever to the rear.

7 Remove the master cylinder filler cap and top up the reservoir. Periodically check the fluid level during the bleeding operation and top up as necessary.

8 if a one-man brake bleeding kit is being used, connect the outlet tube to the bleed screw and then open the screw half a turn. If possible position the unit so that it can be viewed from the car, then depress the brake pedal to the floor and slowly release it. The one-way valve in the kit will prevent expelled air from returning to the system at the end of each stroke. Repeat this operation until clean hydraulic fluid, free from air bubbles, can be seen coming through the tube. Now tighten the bleed screw and remove the outlet pipe.

9 If a one-man brake bleeding kit is not available, connect one end of the plastic tubing to the bleed screw and immerse the other end in the jar containing sufficient clean hydraulic fluid to keep the end of the tube submerged (photos). Open the bleed screw half a turn and have your assistant depress the brake pedal to the floor and then slowly release it. Tighten the bleed screw at the end of each downstroke to prevent expelled air and fluid from being drawn back into the system. Repeat this operation until clean hydraulic fluid, free from air bubbles, can be seen coming through the tube. Now tighten the bleed screw and remove the plastic tube.

10 If the entire system is being bled the procedures described above should now be repeated at each wheel. Do not forget to recheck the fluid level in the master cylinder at regular intervals and top up as necessary.

11 As an alternative to the methods just described, the brake (and clutch, see Chapter 13 Supplement) may be bled using a pressure bleeding kit. The kit, which uses the air pressure from the spare tyre is available from most accessory stores.

12 By connecting a pressurised container to the master cylinder fluid reservoir, bleeding is then carried out by simply opening each bleed screw in turn and allowing the fluid to run out, rather like turning on a tap, until no air is visible in the expelled fluid.

13 By using this method, the large reserve of hydraulic fluid provides a safeguard against air being drawn into the master cylinder during bleeding which often occurs if the fluid level in the reservoir is not maintained.

14 Pressure bleeding is particularly effective when bleeding 'difficult' systems or when bleeding the complete system at time of routine fluid renewal .

15 When completed. recheck the fluid level in the master cylinder, top up if necessary, and refit the cap. Check the 'feel' of the brake pedal; this should be firm and free from any 'sponginess', which would indicate air still present in the system.

16 Discard any expelled hydraulic fluid as it is likely to be contaminated with moisture, air and dirt, which makes it unsuitable for further use.

15 Brake pedal – removal and refitting

1 The brake and clutch pedals share a common bracket assembly and pivot shaft. However, on some models the pivot shaft is fixed to the bracket and on other models it is free to move (Fig. 9.32).

2 Working inside the car, extract the clip and remove the clevis pin securing the servo pushrod or intermediate pushrod to the pedal (photo). Remove the lower panel, if necessary.

3 Extract the clip from the end of the pivot shaft, unhook the return spring (where fitted) and withdraw the pedal from the shaft. On some models it is necessary to loosen the bracket mounting nuts and move the bracket to one side.

4 Check the pedal bushes for wear. If necessary drive them out from each side and press in new bushes using a soft-jawed vice.

5 Refitting is a reversal of removal, but lubricate the pivot shaft with a little multi-purpose grease.

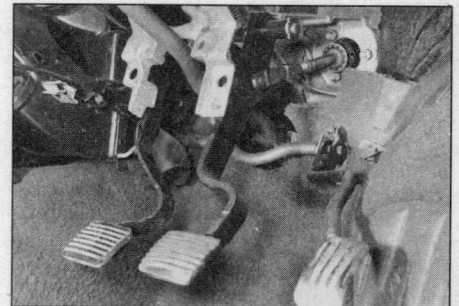

15.2 General view of the brake and clutch pedal assembly (shown with lower trim panel removed)

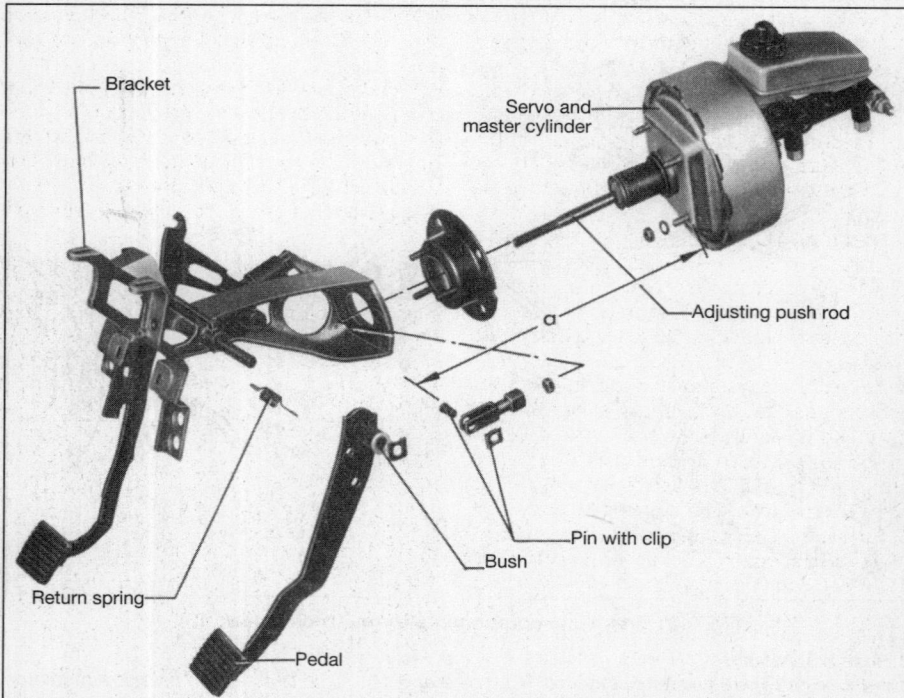

Fig. 9.32 Brake pedal and vacuum servo unit components (Sec 15)

16 Handbrake lever – removal and refitting

1 Chock the front wheels, then jack up the rear of the car and support it on axle stands.
2 Unscrew the nut from the handbrake pullrod and remove the compensator bar (Fig. 9.33).
3 Pull the rubber bellows from the underbody and off the pullrod.
4 Working inside the car, prise out the gaiter and move it up the handbrake lever.
5 Disconnect the wiring from the handbrake warning switch.
6 Unscrew the mounting bolts and withdraw the handbrake lever assembly and switch bracket.
7 Remove the central screws and withdraw the lower cover, then extract the clips, push out the clevis pin and remove the pullrod.
8 If necessary the ratchet may be dismantled by removing the clips and clevis pin or grinding off the rivet head. However, if the ratchet is known to be worn, renew the complete assembly.
9 Refitting is a reversal of removal, but lubricate the pivots and compensator bar with multi-purpose grease. Finally adjust the handbrake, as described in Section 17.

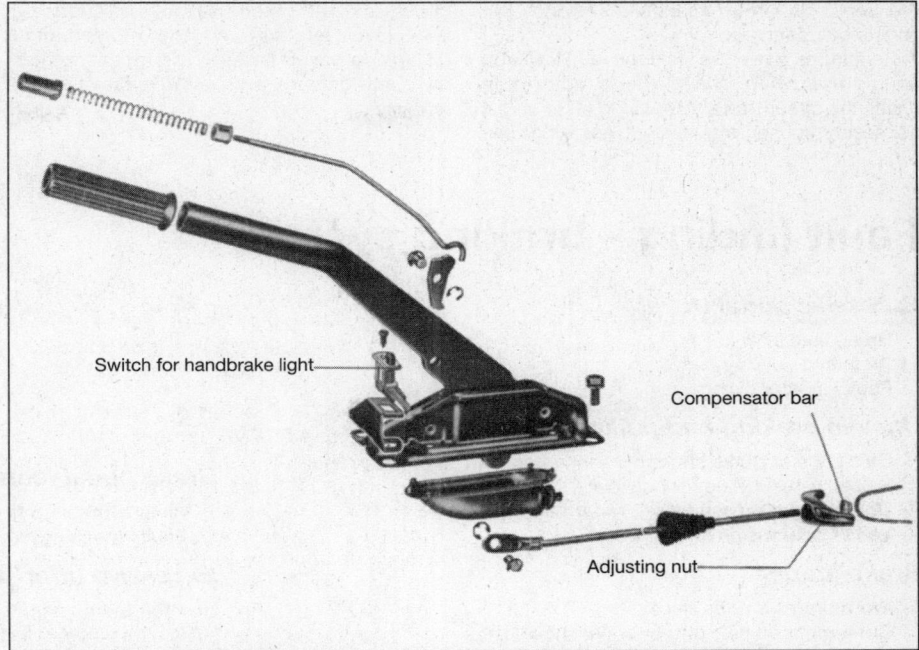

Fig. 9.33 Handbrake lever and associated components (Sec 16)

Switch for handbrake light

Compensator bar

Adjusting nut

17 Handbrake – adjustment

1 Chock the front wheels, then jack up the rear of the car and support it on axle stands.
2 Fully release the handbrake lever, then firmly depress the brake pedal once.
3 Pull the handbrake lever onto the 2nd notch and check that it is just possible to turn the rear wheels by hand. If necessary, adjust the position of the nut on the end of the pullrod beneath the car. The nut is located next to the compensator bar (photo).
4 Fully release the handbrake lever and check that both rear wheels turn freely.
5 Lower the car to the ground.

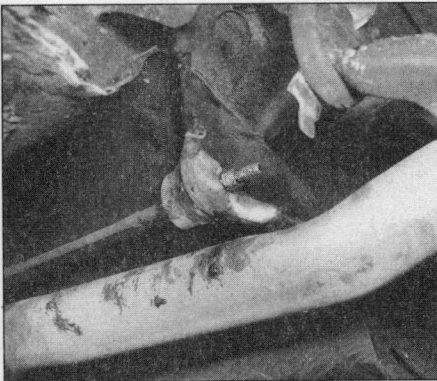

17.3 Handbrake adjustment nut and compensator

18 Vacuum servo unit – description and testing

1 The vacuum servo unit is located between the brake pedal and the master cylinder and provides assistance to the driver when the brake pedal is depressed. The unit operates by vacuum from the inlet manifold.
2 The unit basically consists of a diaphragm and non-return valve. With the brake pedal released, vacuum is channelled to both sides of the diaphragm, but when the pedal is depressed, one side is opened to the atmosphere. The resultant unequal pressures are harnessed to assist in depressing the master cylinder pistons.
3 Normally, the vacuum servo unit is very reliable, but if the unit becomes faulty, it should be renewed. In the event of a failure, the hydraulic system is in no way affected, except that higher pedal pressures will be necessary.
4 To test the vacuum servo unit depress the brake pedal several times with the engine switched off to dissipate the vacuum. Apply moderate pressure to the brake pedal then start the engine. The pedal should move down slightly if the servo unit is operating correctly.
5 To test the check valve in the vacuum hose, disconnect it from the hose then blow through the valve in the direction of the arrow marking. Air should pass through the check valve. However if air is supplied in the reverse direction through the valve it should not, as the valve must be closed to airflow in that direction. Renew the valve unit if found to be defective.

19 Vacuum servo unit – removal and refitting

1 Remove the master cylinder, as described in Section 10.
2 On right-hand drive models remove the glovebox, with reference to Chapter 12.
3 Pull the vacuum hose free from the servo unit connector and where applicable, the non-return valve.
4 Working inside the vehicle detach the lower trim panel on the driver's side.
5 Disconnect the pushrod clevis from the brake pedal by releasing the clip and withdrawing the clevis pin (photo).
6 Unscrew the mounting nuts and withdraw

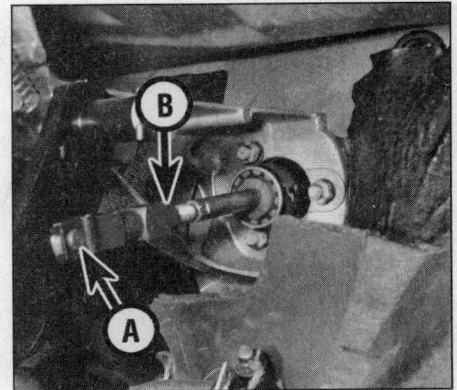

19.5 Servo unit pushrod to pedal attachment

A Clevis, pin and clip   B Adjustment locknut

9

the servo unit from the bulkhead into the engine compartment.

7 Refitting is a reversal of removal. However, on models where the clevis is adjustable check the dimension shown in Fig. 9.34 and, if necessary, loosen the locknut, reposition the clevis, and re-tighten the locknut. Lubricate the clevis pin with a little molybdenum disulphide based grease. The mounting nuts are self-locking and should always be renewed.

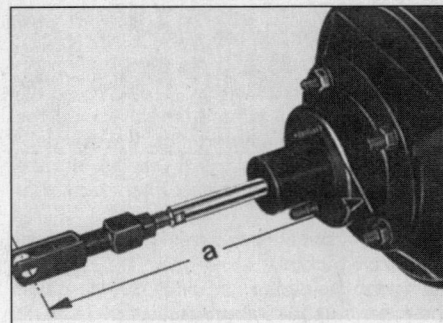

Fig. 9.34 Servo unit pushrod setting dimension (a) (Sec 19)

a = 220 mm (8.66 in)

Fault finding – braking system

Excessive pedal travel

☐ Brake fluid leak
☐ Air in hydraulic system
☐ Faulty master cylinder

Uneven braking and pulling to one side

☐ Disc pads or brake shoes contaminated with oil or brake fluid
☐ Seized wheel cylinder or caliper
☐ Unequal tyre pressures
☐ Loose suspension anchor point

Brake judder

☐ Worn drums and/or discs
☐ Contaminated disc pad or brake shoe linings
☐ Loose suspension anchor point

Brake pedal feels 'spongy'

☐ Air in hydraulic system
☐ Faulty master cylinder

Excessive effort to stop car

☐ Faulty servo unit
☐ Seized wheel cylinder or caliper
☐ Contaminated disc pad or brake shoe linings

Chapter 10 Electrical system

For modifications, and information applicable to later models, see Supplement at end of manual

Contents

Degrees of difficulty

Easy, suitable for novice with little experience	**Fairly easy,** suitable for beginner with some experience	**Fairly difficult,** suitable for competent DIY mechanic	**Difficult,** suitable for experienced DIY mechanic	**Very difficult,** suitable for expert DIY or professional

Specifications

System type .... 12 volt, negative earth

Battery

Capacity .... 36, 45, 54 or 63 Ah

Alternator **Motorola or Bosch**

Output	45, 65, 75 or 90 Amp
Minimum brush length*	5.0 mm (0.2 in)
Drivebelt tension – drivebelt up to 1000 mm (39.3 in) long:	
New belt	2 mm (0.078 in)
Used belt	5 mm (0.196 in)
Drivebelt tension – drivebelt over 1000 mm (39.3 in) long:	
New belt	10 mm (0.4 in)
Used belt	15 mm (0.6 in)
Stator winding resistance:	
Bosch 45 Amp	0.18 to 0.20 ohm
Bosch 65 Amp	0.10 to 0.11 ohm
Bosch 75 and 90 Amp	0.10 ohm maximum
Motorola 65 Amp	0.13 to 0.15 ohm
Motorola 45 Amp with 6 wire ends	0.27 to 0.3 ohm
Motorola 45 Amp with 3 wire ends	0.09 to 0.11 ohm
Diode resistance	50 to 80 ohm

10

Alternator (continued)

Rotor winding resistance:

Bosch 45 Amp	3.4 to 3.7 ohm
Bosch 65 Amp	2.8 to 3.0 ohm
Bosch 75 and 90 Amp	3.0 to 4.0 ohm
Motorola 45 and 65 Amp	3.8 to 4.2 ohm

Starter motor

Type	Pre-engaged
Commutator minimum diameter	33.5 mm (1.32 in)
Commutator maximum run-out	0.03 mm (0.001 in)
Commutator insulation undercut	0.5 to 0.8 mm (0.02 to 0.03 in)
Minimum brush length	13.0 mm (0.5 in)

Fuses (late models)*

No	Function	Rating (amp)
1	Radiator fan	30
2	Brake lights	10
3	Cigarette lighter, radio, clock, central locking and interior lights	15
4	Emergency light system	15
5	Fuel pump (electric)	15
6	Foglights (main current)	15
7	Side and tail lights – left side	10
8	Side and tail lights – right side	10
9	High beam (right) and high beam warning light	10
10	High beam (left)	10
11	Windscreen wipers, washers and headlamp washers	15
12	Coolant warning lamp, exterior mirror adjuster and sliding/ tilting roof. Rear wiper/washer (Passat)	15
13	Rear window and outside mirror heating	15
14	Blower	20
15	Reverse lights, automatic transmission shift illumination	10
16	Horn (single)	15
17	Automatic choke, idle cut-off valve and electric manifold heating	10
18	Horn (dual), front seating heating, stop/start system and brake warning lamp	15
19	Indicators	10
20	Number plate light, foglights (switch current) and glovebox light	10
21	Low beam – left side	10
22	Low beam – right side	10

*For early models, see Fig. 10.26

Separate holder fuses (above fusebox)

Rear foglight	10
Electric window lifters	30

Bulbs

	Wattage
Headlamps:	
Standard	45/40
Halogen	60/55
Sidelight	4
Direction indicators	21
Stop/tail lights	21/5
Reversing lights	21
Foglamp (rear)	21
Foglamp (front)	55
Number plate light	10
Interior light	10
Ashtray and heater illumination	1.2
Instrument panel illumination	3
Instrument panel warning lights:	
All except charge and temperature warning lights	1.2
Charge and temperature warning lights	2
Glove compartment light	4
Luggage compartment light	5

Motorola or Bosch

1 General description

The electrical system is of 12 volt negative earth type. The battery is charged by a belt-driven alternator which incorporates a voltage regulator. The starter is of pre-engaged type incorporating four brushes, and on this type of motor a solenoid moves the drive pinion into engagement with the starter ring gear before the motor is energised.

Although repair procedures are given in this Chapter, it may well be more economical to renew worn components as complete units.

2 Electrical system – maintenance

The following routine maintenance procedures should be undertaken at the specified intervals given at the start of this manual.

1 *Battery:* Refer to Section 4 in this Chapter.
2 *Alternator:* Refer to Section 7 in this Chapter.
3 *Alternator drivebelt:* Check condition and adjustment of the drivebelt as described in Section 9 of this Chapter.
4 *Vehicle lighting:* Periodically check that all of the front and rear lights are functioning correctly. The headlights and (where applicable) the foglights should be checked for alignment as described in Section 27. Renew any defective bulbs.
5 *Windscreen/rear window wipers:* Check their operation (having wet the glass first) and also the condition of the wiper blade rubbers. Renew if necessary. At the same time, check that the windscreen, rear window and headlamp washers (as applicable) operate in a satisfactory manner. Check and when necessary top up the fluid reservoirs.
6 *Wiring:* Periodically check the wiring and connections for condition and security.

3 Battery – removal and refitting

1 The battery is located in the engine compartment just in front of the bulkhead (photo).
2 Loosen the negative terminal clamp and disconnect the lead.
3 Loosen the positive terminal clamp and disconnect the lead.
4 Unscrew the bolt and remove the battery holding clamp.
5 Lift the battery from the platform, taking care not to spill any electrolyte.
6 Refitting is a reversal of removal, but make sure that the polarity is correct before connecting the leads, and do not overtighten the clamp bolts. **Note:** *The curved side of the holding clamp must contact the base of the battery.*

4 Battery – maintenance

1 Where a conventional battery is fitted, the electrolyte level of each cell should be checked every month and, if necessary, topped up with distilled or de-ionized water until the separators are just covered. On some batteries the case is translucent and incorporates minimum and maximum level marks. The check should be made more often if the car is operated in high ambient temperature conditions.
2 Where a low maintenance battery is fitted it is not necessary to check the electrolyte level.
3 Every 15 000 km (10 000 miles) or 12 months, whichever occurs first, disconnect and clean the battery terminals and leads. After refitting them, smear the exposed metal with petroleum jelly.
4 At the same time, inspect the battery clamp and platform for corrosion. If evident, remove the battery and clean the deposits away, then treat the affected metal with a proprietary anti-rust liquid and paint with the original colour.
5 When the battery is removed, for whatever reason, it is worthwhile checking it for cracks and leakage. Cracks can be caused by topping up the cells with distilled water in winter *after* instead of *before* a run. This gives the water no chance to mix with the electrolyte, so the former freezes and splits the battery case. If the battery case is fractured, it may be possible to repair it with a proprietary compound. but this depends on the material used for the case. If electrolyte has been lost from a cell, refer to Section 5 for details of adding a fresh solution.
6 If topping up the battery becomes excessive and the case is not fractured, the battery is being over-charged and the voltage regulator will have to be checked.
7 If the car covers a very small annual mileage, it is worthwhile checking the specific gravity of the electrolyte every three months to determine the state of charge of the battery. Use a hydrometer to make the check, and compare the results with the following table.

	Normal climates	Tropics
Discharged	1.120	1.080
Half charged	1.200	1.160
Fully charged	1.280	1.230

8 If the battery condition is suspect, first check the specific gravity of electrolyte in each cell. A variation of 0.040 or more between any cells indicates loss of electrolyte or deterioration of the internal plates.
9 A further test can be made using a battery heavy discharge meter. The battery should be discharged for a maximum of 15 seconds at a load of three times the ampere-hour capacity (at the 20 hour discharge rate). Alternatively connect a voltmeter across the battery terminals and spin the engine on the starter

3.1 Battery location showing lead connections and retaining clamp and bolt (arrowed) at base

5.2 Replenishing the battery electrolyte

with the ignition disconnected (see Chapter 4), and the headlamps, heated rear window and heater blower switched on. If the voltmeter reading remains above 9.6 volts, the battery condition is satisfactory. If the voltmeter reading drops below 9.6 volts, and the battery has already been charged as described in Section 6, it is faulty and should be renewed.

5 Battery – electrolyte replenishment

1 If after fully charging the battery, one of the cells maintains a specific gravity which is 0.040 or more lower than the others, but the battery also maintains 9.6 volts during the heavy discharge test (Section 4), it is likely that electrolyte has been lost.
2 If a significant quantity of electrolyte has been lost through spillage, it will not suffice merely to refill with distilled water. Top up the cell with a mixture of 2 parts sulphuric acid to 5 parts distilled water (photo).
3 When mixing the electrolyte *never* add water to sulphuric acid – *always* pour the acid slowly onto the water in a glass container. *If water is added to sulphuric acid, it will explode!*
4 Top up each cell as required to a level 5 mm (0.20 in) above the separators or to the acid level marking on the battery casing.
5 After topping up the cell with fresh electrolyte, recharge the battery and check the hydrometer readings again.

10

6 Battery – charging

1 In winter when a heavy demand is placed on the battery, such as when starting from cold and using more electrical equipment, it is a good idea to occasionally have the battery fully charged from an external source at a rate of 10% of the battery capacity (ie 6.3 amp for a 63 Ah battery). It is necessary to disconnect the battery leads for charging.

2 Continue to charge the battery until no further rise in specific gravity is noted over a four hour period.

3 Alternatively, a trickle charger, charging at a rate of 1.5 amp can be safely used overnight.

4 Special rapid 'boost' charges, which are claimed to restore the power of the battery in 1 to 2 hours, can be dangerous unless they are thermostatically controlled, as they can cause serious damage to the battery plates through overheating.

5 While charging the battery, ensure that the temperature of the electrolyte never exceeds 37.8°C (100°F) and loosen the vent caps (where applicable).

6 If a maintenance-free battery is fitted, consult your dealer when it is necessary to recharge it from an external source as specialised procedures are sometimes required depending on battery type.

7 Alternator – maintenance and special precautions

1 Periodically wipe away any dirt which has accumulated on the outside of the unit, and also check that the plug is pushed firmly on the terminals. At the same time, check the tension of the drivebelt and adjust it if necessary as described in Section 9.

2 Take extreme care when making electrical circuit connections on the car, otherwise damage may occur to the alternator or other electrical components employing semi-conductors. Always make sure that the battery leads are connected to the correct terminals. Before using electric-arc welding equipment to repair any part of the car, disconnect the battery leads and alternator multi-plug. Never run the engine with the alternator multi-plug or a battery lead disconnected.

8 Alternator – removal and refitting

1 Disconnect the battery negative lead.

2 Loosen the alternator adjustment nuts or bolts and swivel the alternator toward the engine. If difficulty is experienced, loosen the pivot bolt.

3 Slip the drivebelt from the pulley.

4 Note the location of the wiring then disconnect it from the rear of the alternator (photo). On some models it will first be necessary to remove the duct cover and hose.

8.4 Alternator multi-plug (Bosch)

8.5A Remove upper timing cover plug . . .

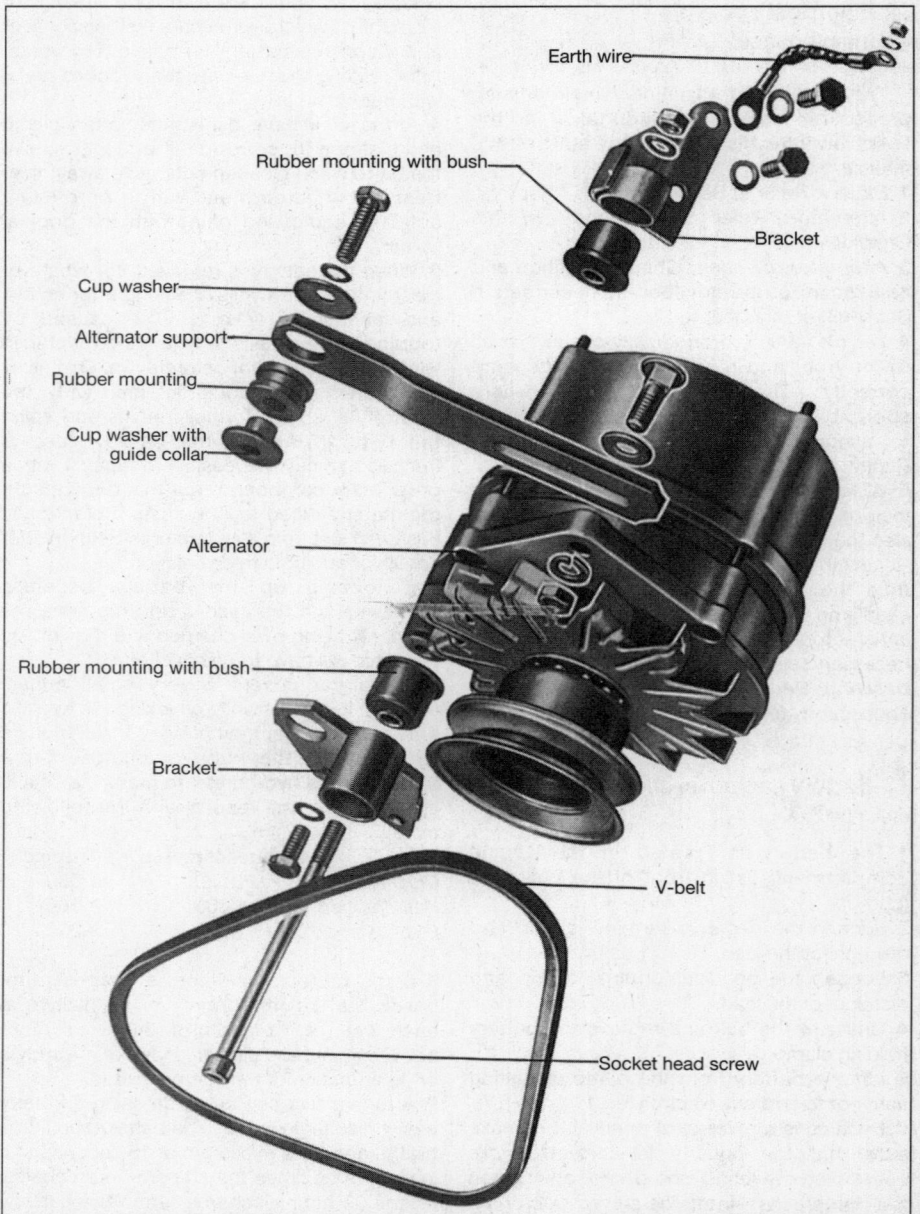

Fig. 10.1 Alternator and mounting/adjustment components – 1.6 litre engine (Sec 8)

8.5B . . . for access to alternator bolt

8.5C Alternator adjuster strap bolt removal – note spacing washer

10.2A Bosch 45 Amp alternator rear end view

5 Remove the link bolt and pivot bolt, and withdraw the alternator from the engine. If necessary, remove the upper timing cover plug in order to remove the pivot bolt (photos).

6 If necessary, the mounting brackets may be unbolted, and on four-cylinder engines the rubber bushes removed (Fig. 10.1). When refitting the bracket on five-cylinder engines, the short bolt must be tightened first.

7 Refitting is a reversal of removal, but adjust the drivebelt as described in Section 9.

9 Alternator drivebelt – renewal and adjustment

1 Every 15 000 km (10 000 miles) the alternator drivebelt should be checked for condition, and re-tensioned. If there are signs of cracking or deterioration, the drivebelt should be renewed.

2 To remove the drivebelt, loosen the adjustment nuts or bolts and swivel the alternator toward the engine. If necessary loosen the pivot bolt.

3 Slip the drivebelt from the pulleys. On models fitted with power steering, it will first be necessary to remove the drivebelt in order to remove the alternator drivebelt.

4 Fit the new drivebelt over the pulleys, then lever the alternator from the engine until the specified tension is achieved using firm thumb pressure mid-way between the pulleys. Lever the alternator at the pulley end to prevent any torsional damage.

5 Tighten the adjustment nuts or bolts followed by the pivot bolts.

6 Run the engine for several minutes, then recheck the tension and adjust if necessary.

10 Alternator – servicing

Note: *The voltage regulator and brushes can be removed without removing the alternator. However, the following complete dismantling procedure assumes that the alternator is on the bench.*

1 Clean the exterior of the alternator.

Bosch 45A and 65A

2 Remove the screws and withdraw the voltage regulator and brushes. If the brushes

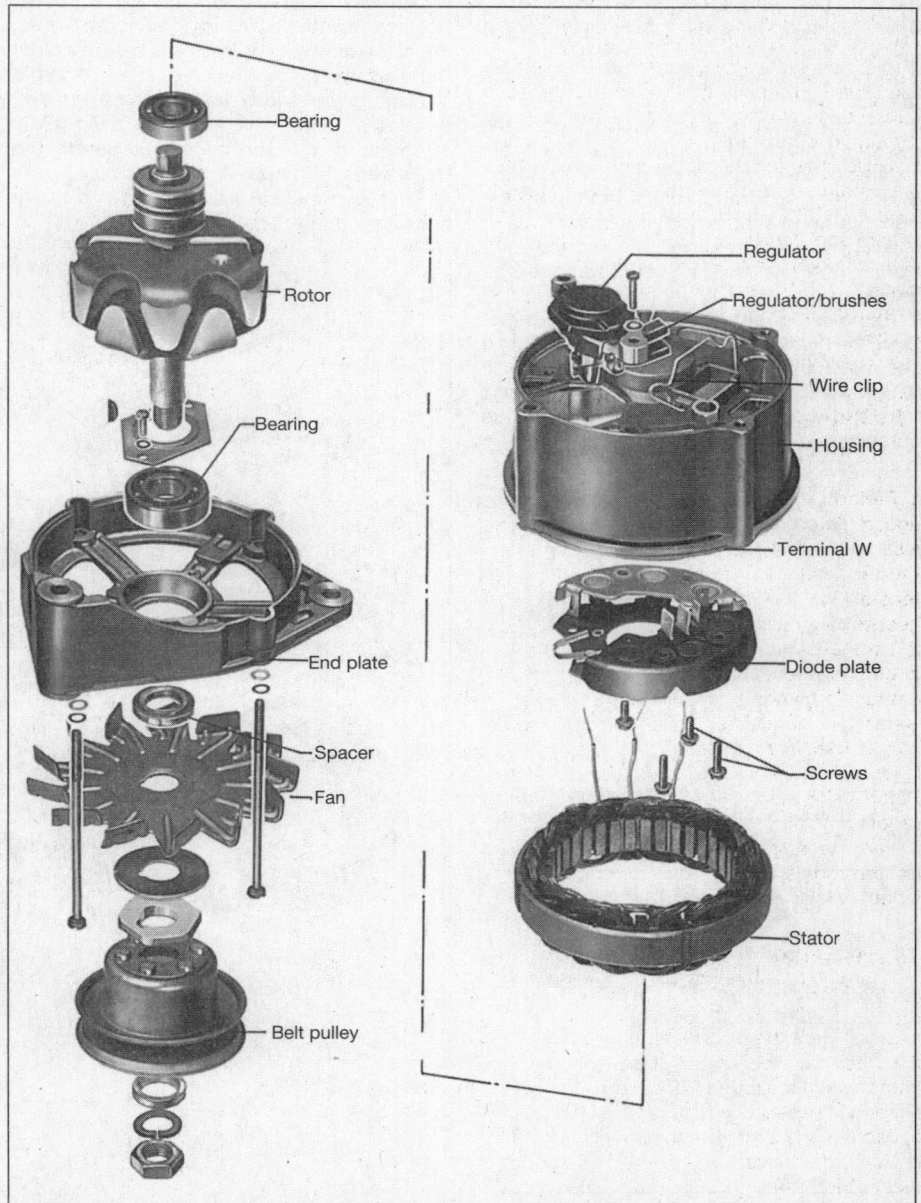

Fig. 10.2 Exploded view of the Bosch 45 Amp and 65 Amp alternator (Sec 10)

Bearing
Rotor
Bearing
End plate
Spacer
Fan
Belt pulley
Regulator
Regulator/brushes
Wire clip
Housing
Terminal W
Diode plate
Screws
Stator

10

10.2B Removing the regulator and brush holder (Bosch 45 amp) . . .

10.2C . . . to check the brushes for excessive wear

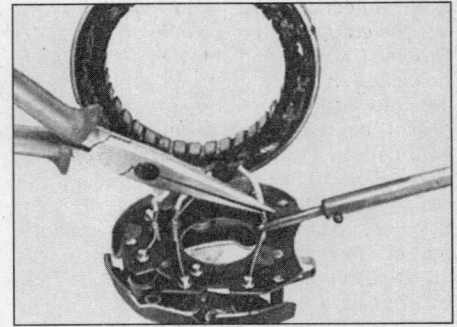

Fig. 10.3 Using pliers as a heat sink when unsoldering the stator wires (Sec 10)

are worn below the specified minimum length, they cannot be renewed individually, only in unit with the regulator unit (photos).

3 Mark the end housings and stator in relation to each other (Fig. 10.2).

4 Grip the pulley in a vice then unscrew the nut and withdraw the washer, pulley and fan, together with any spacers. Prise out the key.

5 Unscrew the through-bolts and tap off the front housing, together with the rotor.

6 Using a mallet or puller, remove the rotor from the front housing and remove the spacers.

7 Remove the screws and retaining plate. Drive the bearing from the front housing with a soft metal drift.

8 Using a suitable puller, remove the bearing from the slip ring end of the rotor, but take care not to damage the slip rings.

9 Carefully separate the stator from the rear housing without straining the three wires. Identify each wire for location, then unsolder them using the minimum of heat to avoid damage to the diodes. Long nosed pliers may be used as a heat sink, as shown in Fig. 10.3.

10 Remove the screws and separate the diode plate from the rear housing.

11 Remove the wave washer from the rear housing (if fitted).

12 Check the stator windings for a short to ground by connecting an ohmmeter or test bulb between each wire and the outer ring. Check that the internal resistance between the wires is as given in the Specifications using an ohmmeter between wires 1 and 2, then 1 and 3, and 2 and 3 – the numbering of the wires is of no importance, (Fig. 10.4).

13 Using the ohmmeter check that the resistance of each diode is as given in the

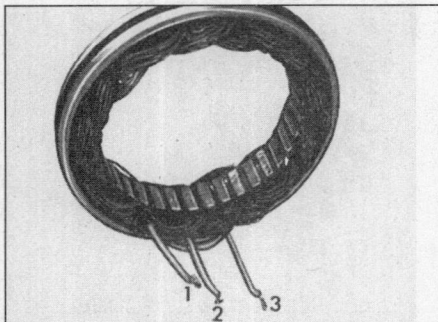

Fig. 10.4 Check the stator winding for a short circuit (Sec 10)

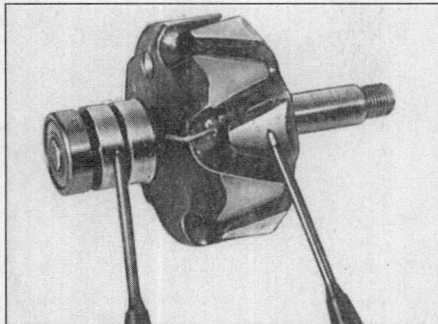

Fig. 10.5 Check the rotor windings for a short circuit (Sec 10)

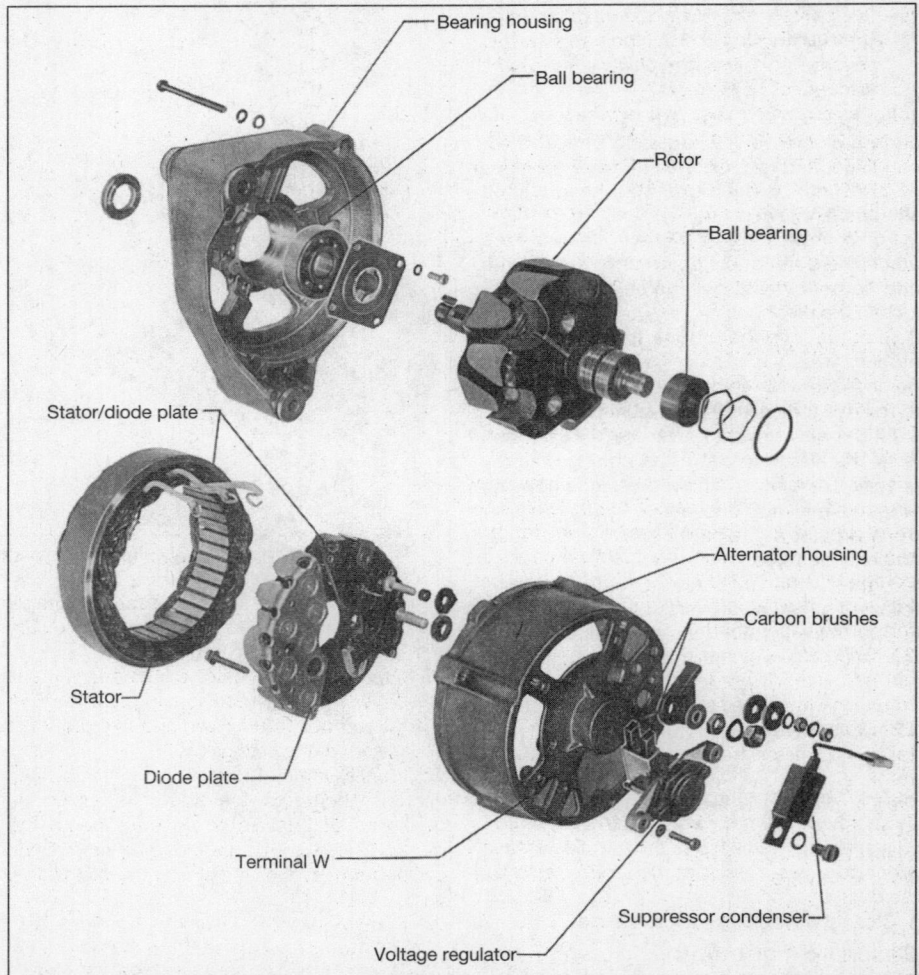

Fig. 10.6 Exploded view of the Bosch 75 and 90 Amp alternators (Sec 10)

Specifications when the ohmmeter is connected across the diode in one direction. Reverse the wires and check that there is now no resistance.

14 Check the rotor windings for a short to ground by connecting an ohmmeter or test bulb between each slip ring and the winding core (Fig. 10.5). Check that the internal resistance of the winding is as given in the Specifications using an ohmmeter between the two slip rings.

15 Clean all the components and obtain new bearings, brushes etc, as required.

16 Reassembly is a reversal of dismantling, but when fitting the bearing to the front housing, drive it in using a metal tube *on the outer race* making sure that the open end of the bearing faces the rotor. When fitting the rear bearing to the rotor, drive it on using a metal tube *on the inner race*, making sure that the open end of the bearing faces the rear housing.

Bosch 75A and 90A

17 The procedure is identical to that described in paragraphs 2 to 16 with the following exceptions.

(a) *A suppressor condenser is attached to the rear housing and can be removed by unscrewing the single retaining screw and disconnecting the lead.*

(b) *The diode plate has two terminals protruding through the rear housing, and there are two wave washers and a plain washer in the rear housing.*

(c) *The front (drive end) ball bearing must be fitted with the casing side towards the bearing housing. The rear bearing may be installed either way round as both sides are closed.*

Motorola 45A and 65A

18 Remove the screws and lift the voltage regulator from the rear housing. Note the location of the wires, then disconnect them and withdraw the regulator.

19 Remove the screws and withdraw the brush holder and brushes. If the brushes are worn below the specified minimum length renew the brush holder and brushes as a complete unit (Fig. 10.8).

20 Mark the end housings and stator in relation to each other.

21 Grip the pulley in a vice, then unscrew the nut and withdraw the washer, pulley and fan, together with any spacers. Prise out the key.

22 Unscrew the through-bolts and nuts, and tap off the front housing, together with the rotor.

23 On the 45A alternator remove the three screws securing the three bearing retaining plates to the front housing.

24 On the 65A alternator, remove the three screws securing the one-piece bearing retaining plate to the front housing.

25 Using a mallet or puller, remove the rotor and bearing from the front housing.

26 Using a suitable puller, remove the

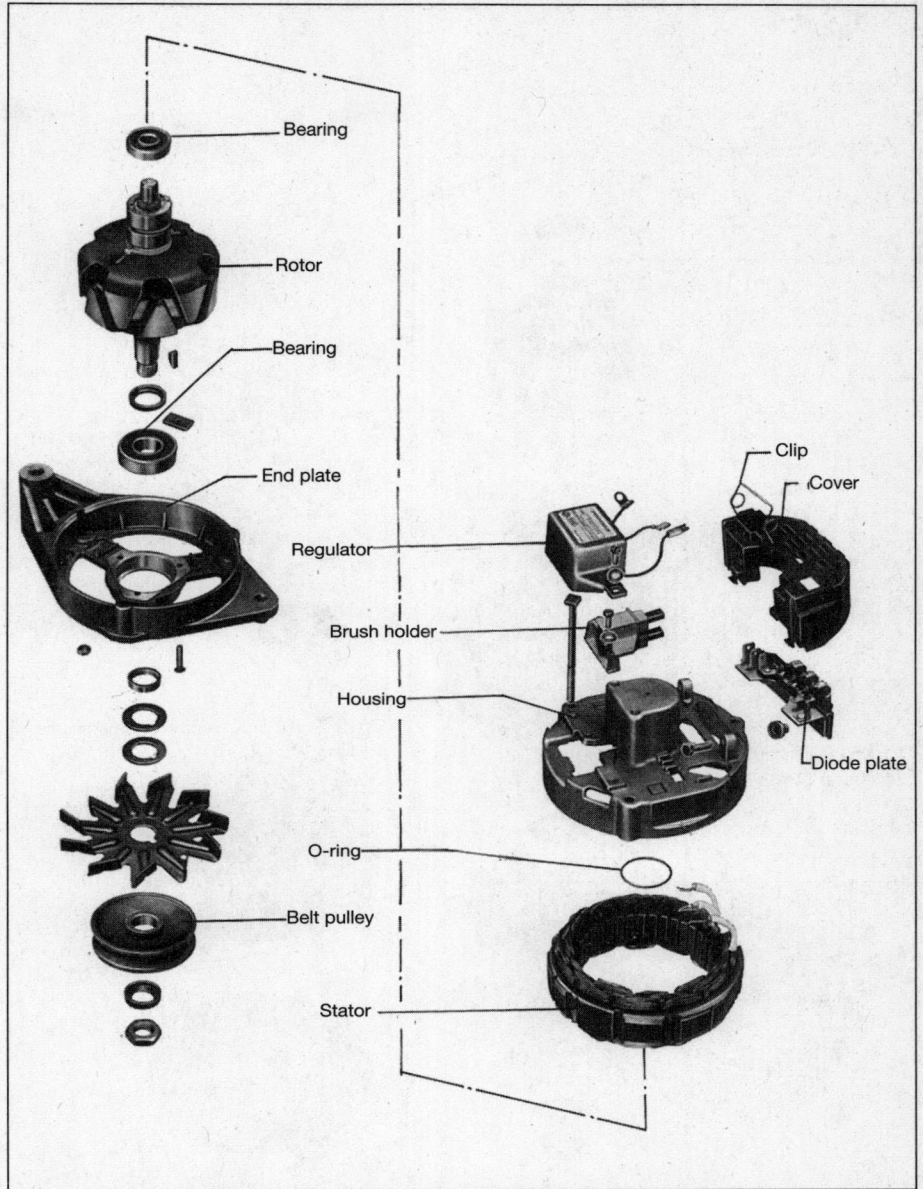

Fig. 10.7 Exploded view of the Motorola 45 Amp alternator (Sec 10)

bearings from each end of the rotor, together with the retaining plate and spacer. Take care not to damage the slip rings.

27 On the 45A alternator unclip and remove the cover from the rear housing. On the 65A model, remove the screws and withdraw the cover from the rear housing.

28 Identify the stator wires for location, then unsolder them from the diode plate using the minimum of heat to avoid damage to the diodes, (Fig. 10.10). Long nosed pliers may be used as a heat sink in a similar manner to that shown in Fig. 10.3.

29 Remove the stator from the rear housing.

30 On the 65A alternator, unsolder the D+ wire from the diode plate using the procedure described in paragraph 28. Note the wire

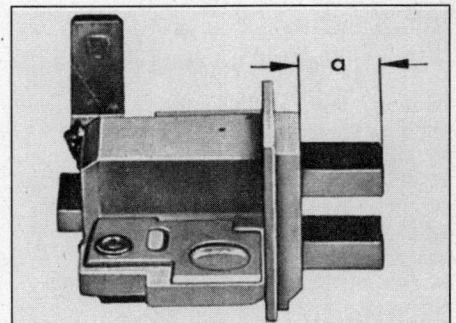

Fig. 10.8 Check dimension (a) for brush minimum length – Motorola alternator (Sec 10)

10

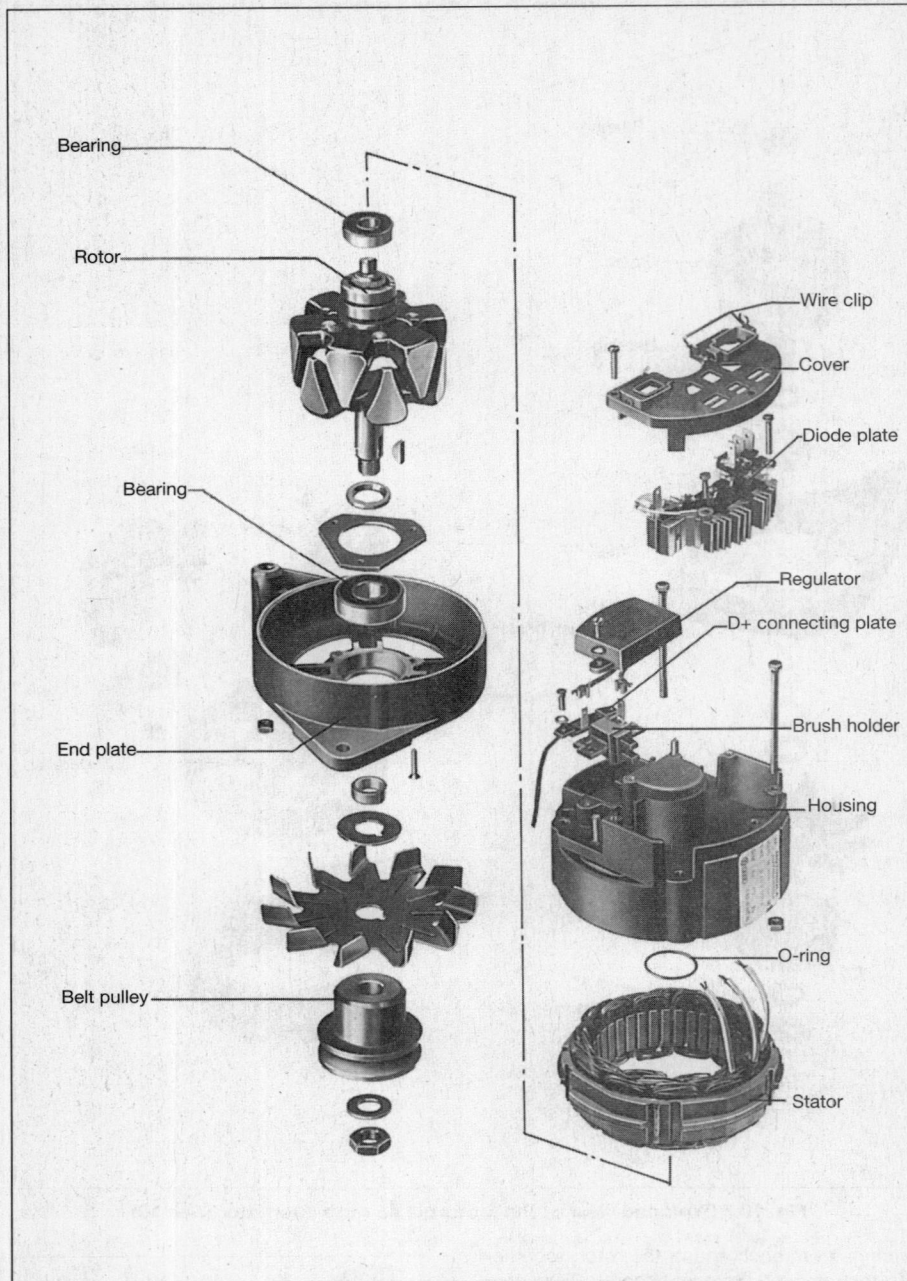

Fig. 10.9 Exploded view of the Motorola 65 Amp alternator (Sec 10)

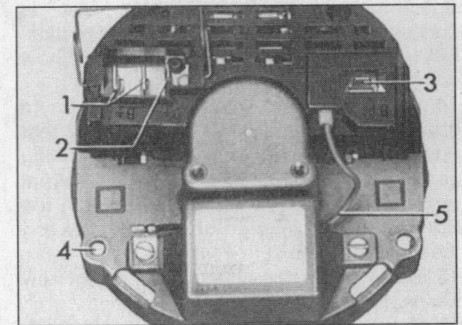

Fig. 10.10 Stator wire connections on the Motorola alternator (Sec 10)

1 *B+*   4 *Earth*
2 *D+ for warning lamp*   5 *To regulator*
3 *B+ suppression condenser*

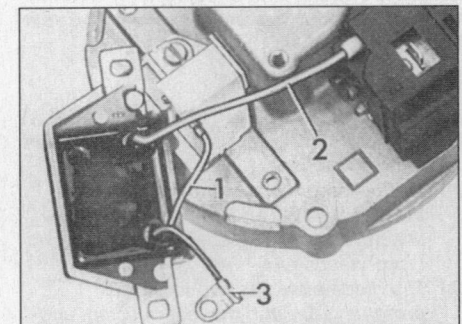

Fig . 10.11 Voltage regulator wire terminals on the Motorola 45 Amp alternator (Sec 10)

1 *Green wire DF*   3 *Black wire Earth*
2 *Red wire D+*

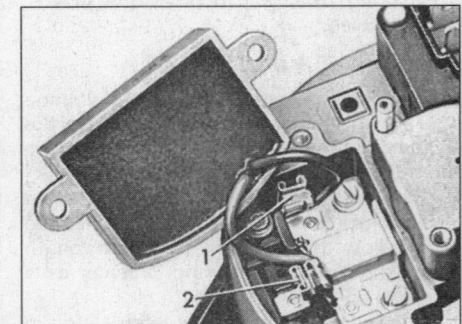

Fig. 10.12 Voltage regulator connections on the Motorola 65 Amp alternator (Sec 10)

1 *Green wire DF*   2 *Red wire D+*

location, then remove the screws and withdraw the diode plate.

31 Remove the O-ring from the rear housing.

32 Check the stator, diodes, and rotor, as described in paragraphs 12, 13 and 14.

33 Clean all the components and obtain new bearings, brushes, etc, as required.

34 Reassembly is a reversal of dismantling, but lubricate the rear housing O-ring with multi-purpose grease before fitting.

35 Check that the voltage regulator wiring connections are correct and securely made (Figs. 10.11 and 10.12).

11 Starter motor – testing in the car

1 If the starter motor fails to respond when the starter switch is operated, first check that the fault is not external to the starter motor.

2 Connect a test lamp between chassis earth and the large terminal on the starter solenoid, terminal 30. This terminal is connected directly to the battery and the test lamp should light whether or not the ignition switch is operated (Fig. 10.13 or 10.14).

3 Remove the test lamp connection from the large terminal (30) and transfer it to the smaller terminal (50) on the solenoid. The lamp should light only when the starter switch is in its Start position.

4 If both these tests are satisfactory the fault is in the starter motor.

5 If the starter motor is heard to operate, but the engine fails to start, check the battery terminals, the starter motor leads and the engine-to-body earth strap for cleanliness and tightness.

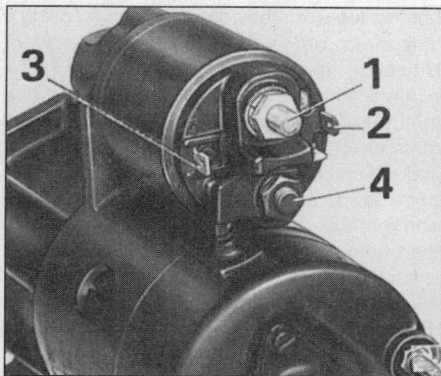

Fig. 10.13 Starter motor solenoid connections (Sec 11)

1 *Terminal 30 (from the battery)*
2 *Terminal 15a (to coil on vehicles with series resistance or resistance wire only)*
3 *Terminal 50 (from ignition/starter switch)*
4 *Field winding connection*

12 Starter motor – removal and refitting

1 Disconnect the battery negative lead.
2 Note the location of the wires on the starter solenoid, then disconnect them by removing the nut and screws, as applicable (photo) .
3 Unscrew the mounting bolts and withdraw the starter motor (photo).
4 Refitting is a reversal of removal, but make sure that the mating faces are clean and

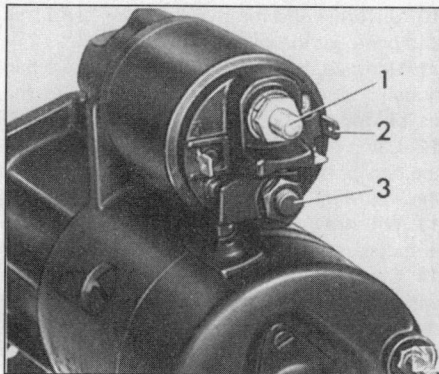

Fig. 10.14 Alternative starter motor solenoid connections on some models (Sec 11)

1 *Terminal 30 (from battery)*
2 *Terminal 50 (from ignition switch)*
3 *Field winding connection*

tighten the mounting bolts to the specified torque.

13 Starter motor – overhaul

1 Mark the housings and mounting bracket (where fitted) in relation to each other.
2 Remove the nuts and washers and withdraw the end mounting bracket (where fitted).
3 Unscrew the through-bolts and remove them (photo).

12.2 Wiring connections to starter motor solenoid

4 Remove the screws and withdraw the small end cover and gasket, then extract the circlip and remove the shims. Note the exact number of shims, as they determine the shaft endfloat. Withdraw the bearing housing (photos).
5 Lift the brush springs and extract the field winding brushes (photo) .
6 Note the position of the brush holder plate in relation to the field winding housing, then withdraw the plate from the armature.
7 Unscrew the nut and disconnect the field wire from the solenoid.
8 Remove the field winding housing from the armature.
9 Unscrew the three bolts and remove the solenoid from the drive end housing. Unhook the core from the lever.

12.3 Starter motor mounting bolts (A) and front mounting bracket (B)

13.3 Unscrew and remove the through-bolts

13.4A Remove the end cover (arrowed) . . .

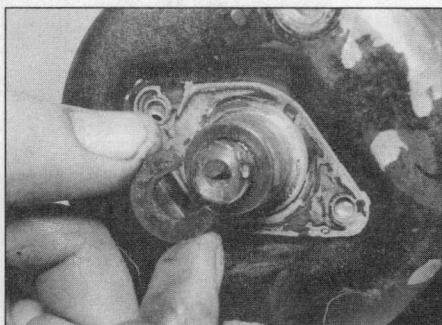

13.4B . . . the circlip . . .

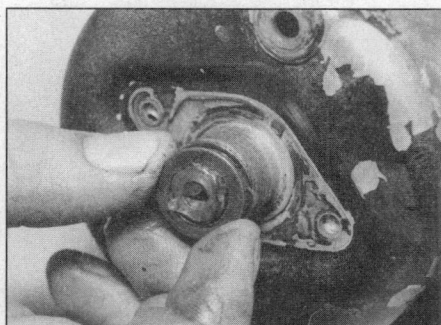

13.4C . . . and shims

13.4D Withdraw the bearing housing

13.5 Lift spring to extract the brushes from the holders

10 Unscrew and remove the lever pivot bolt and prise out the lever cover pad.

11 Remove the armature from the drive end housing and disengage the lever from the pinion drive.

12 Using a metal tube, drive the stop ring off the circlip, then extract the circlip and pull off the stop ring.

13 Withdraw the pinion drive from the armature.

14 Clean all the components in paraffin and wipe dry, then examine them for wear and damage. Check the pinion drive for damaged teeth and make sure that the one-way clutch only rotates in one direction. If the shaft bushes are worn they can be removed using a soft metal drift and new bushes installed. However, the new bushes must first be soaked in hot oil for approximately five minutes. Clean the commutator with a rag moistened with a suitable solvent. Minor scoring can be removed with fine glasspaper, but deep scoring will necessitate the commutator being skimmed in a lathe and then being undercut. Commutator refinishing is a job which is best left to a specialist.

15 Measure the length of the brushes. If less than the minimum given in Specifications, fit new brushes. Crush the old brushes to free the copper braid and file the excess solder off

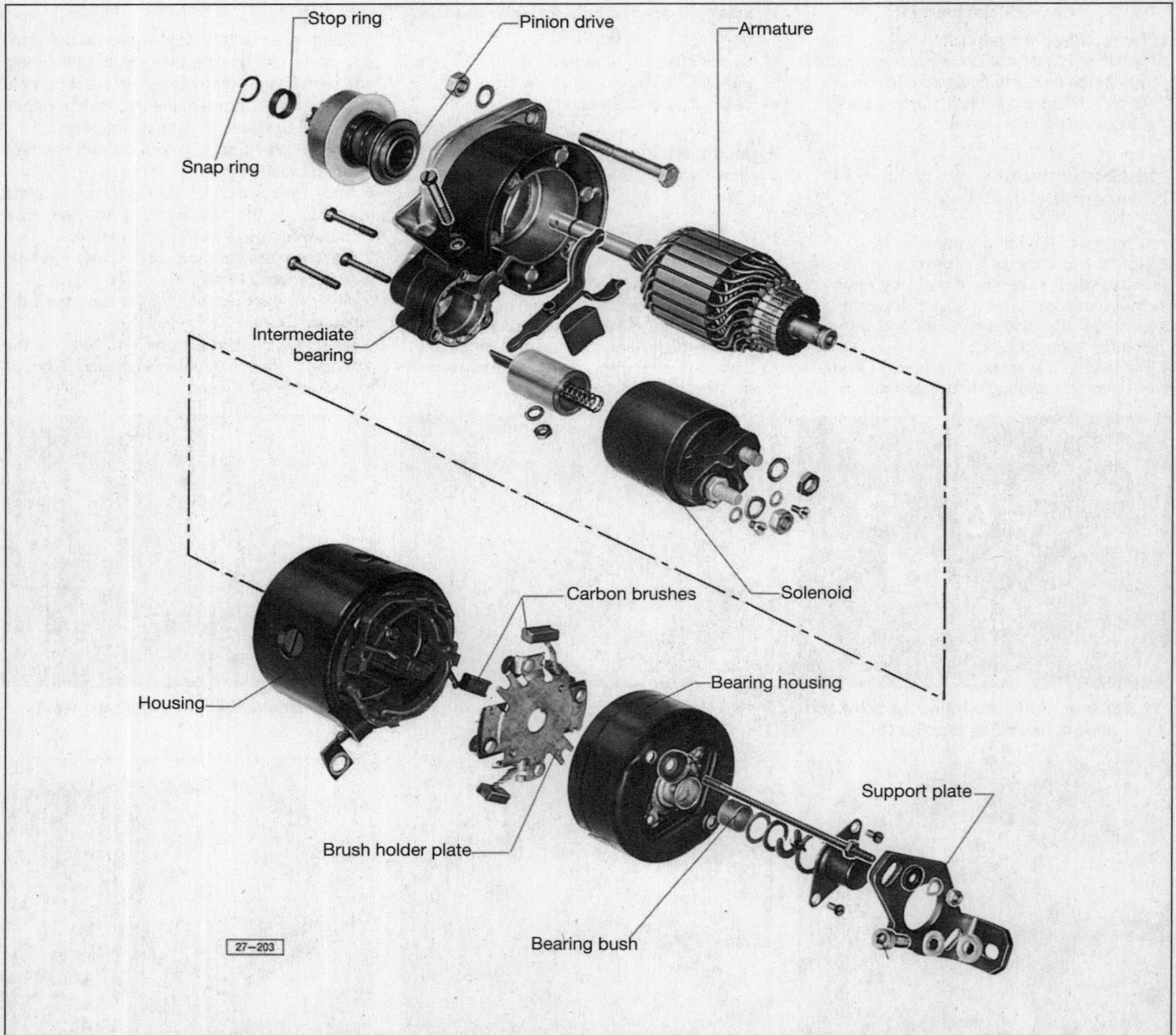

Fig. 10.15 Exploded view of the 0.8 and 1.7 kW starter motor (Sec 13)

Bearing bush

Snap ring

Armature

Stop ring

Pinion drive

Housing

Carbon brushes

Solenoid

Bearing housing

Brush holder plate

Bearing bush

10

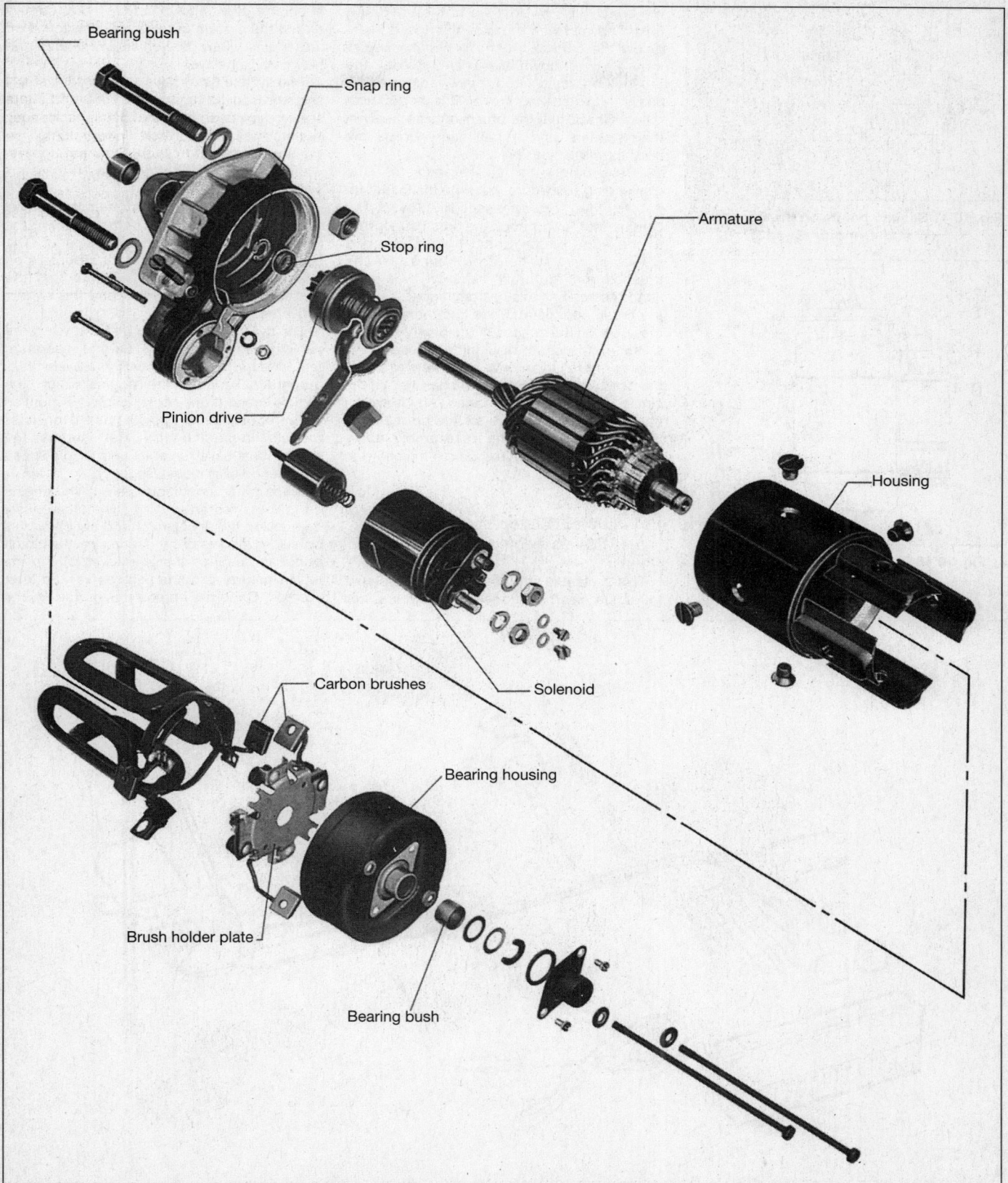

Fig. 10.16 Exploded view of the 1.1 kW starter motor (Sec 13)

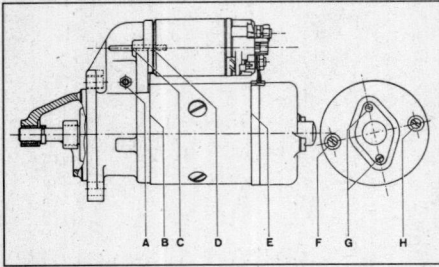

Fig. 10.17 Sealing points on the 0.8 and 1.7 kW starter motor (Sec 13)

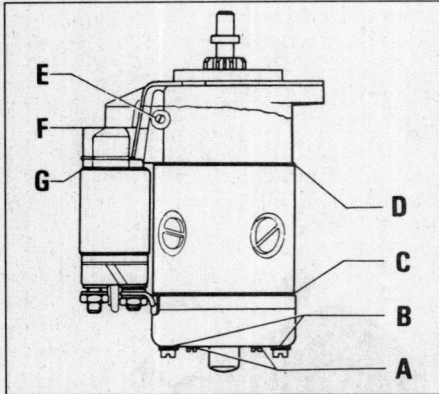

Fig. 10.18 Sealing points on the 1.1 kW starter motor (Sec 13)

the braid. Fit new brushes to the braid and hold the braid with a pair of flat nosed pliers below the carbon brush, to prevent solder from flowing down the braid. Solder the brushes on using radio quality solder and a heavy duty soldering iron of at least 250 watt rating. Check that the brushes move freely in their holders and if necessary dress the brushes with a fine file.

16 Reassembly is a reversal of the dismantling procedure, but note the following. To fit the brush assembly over the commutator, either hook the brush springs on to the edge of the brush holder, or bend pieces of wire to hold the springs off the brushes until the brush assembly has been fitted. As soon as this has been done, release the brush springs and position them so that they bear on the centres of the brushes. Apply a little molybdenum disulphide grease to the splines of the pinion drive. Fit a new circlip to the pinion end of the shaft and ensure that the circlip groove is not damaged. Any burrs on the edges of the groove should be removed with a fine file. During reassembly apply sealing compound to the points indicated in Figs. 10.17 and 10.18.

14 Automatic stop/start system (ASS) – general

This system is fitted as optional equipment to manual gearbox models only and is a fuel economy device. Activated by a control switch the system automatically switches off the engine when the vehicle is stationary in heavy traffic delays.

Two system types have been fitted, an early system used up to October 1983 and a late system type used from that date. On the early system the restart switches are mounted on the accelerator and clutch pedal assemblies, whilst the later model type has the restart switch on the gearbox.

On both early and late models the control switch is mounted in the same place, to the right of the driver on the facia panel. The switch is marked SSA and incorporates an integral warning light to indicate that the switch is in the ON position and the system activated.

The system should only be used when the vehicle has reached its normal operating temperature. When activated, the system will automatically stop the engine when the vehicle speed drops below 2 kph (1.2 mph) on early models or 5 kph (3.1 mph) on later models. In addition on later models, the engine must have run at its normal idle speed for a period of at least 2 seconds.

When traffic conditions permit, the engine can be restarted by simultaneously depressing the accelerator and clutch pedals on earlier models or by depressing the clutch pedal and moving the gear lever fully to the left in neutral or engaging reverse on later models. Once the engine has restarted, the

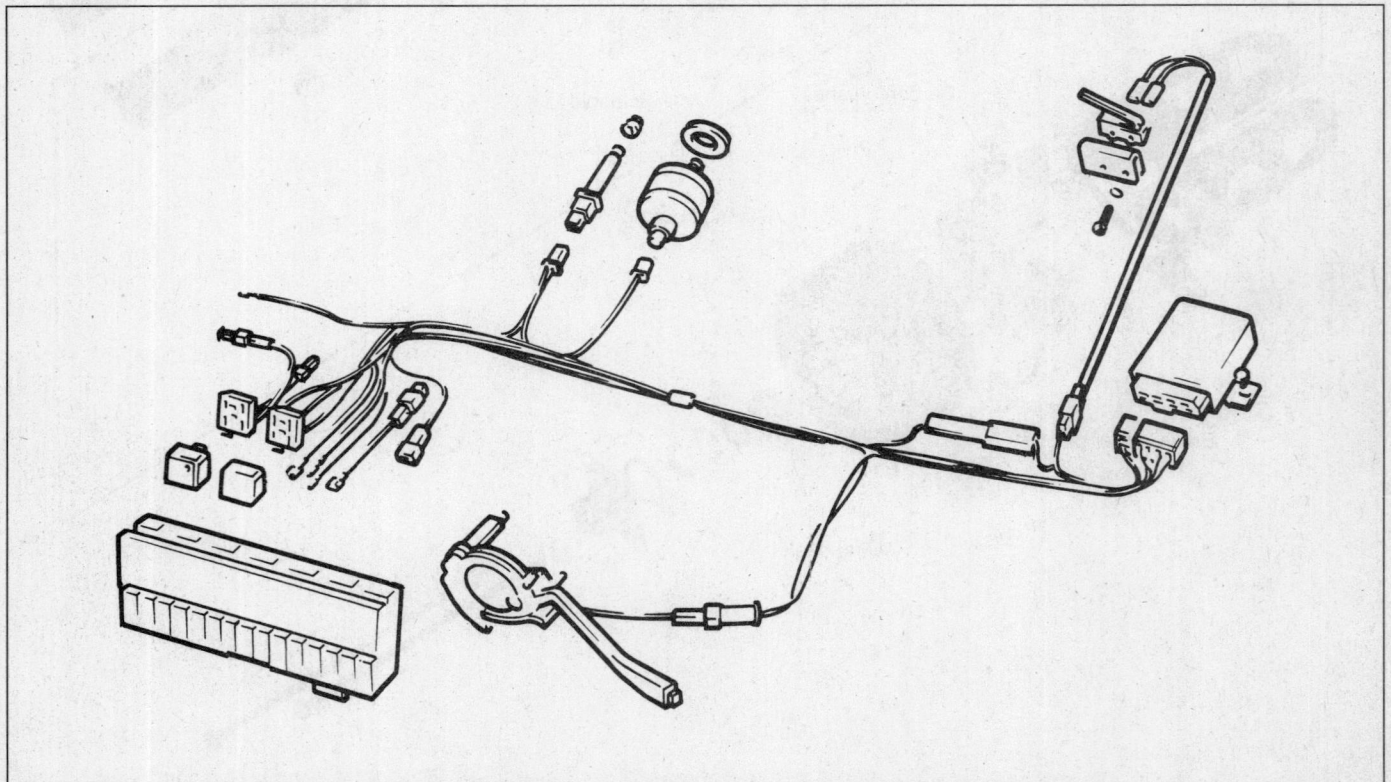

Fig. 10.19 Automatic stop/start system components (up to 1983) (Sec 15)

15.3 Automatic stop/start system control unit (early models)

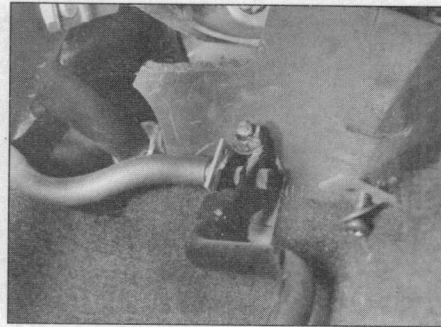

15.6 Accelerator pedal clamp bolt

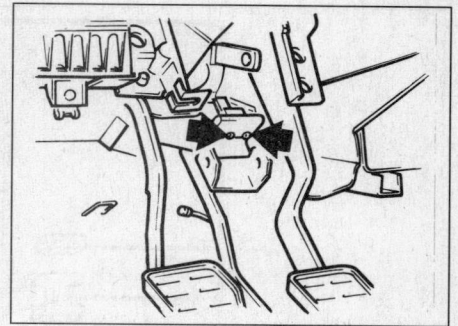

Fig. 10.20 Accelerator pedal switch location and retaining screws (automatic stop/start system up to 1983) (Sec 15)

gear engagement can be made in the normal manner. If when restarted the engine stalls or for any reason stops the restart procedure should be repeated, but on later models the gear lever must be moved back into neutral within 6 seconds.

The following safety precautions should be noted when using the ASS, these being:

(a) *Do not use the system when the engine temperature is below 55°C (131°F) or when the ambient temperature is very low as the engine will take longer to warm up.*
(b) *Do not allow the vehicle to roll when the engine is switched off, check that the handbrake is fully applied.*
(c) *During extended delays switch the engine off in the normal manner with the ignition key as electrical accessories will otherwise be left on and the battery run down.*
(d) *If leaving the vehicle for any length of time, switch off the ASS and always take the ignition key with you.*

15 Automatic stop/start system components (up to 1983) – removal and refitting

The system layout and components are shown in Fig. 10.19. Should the system malfunction in any way, have the fault diagnosed by your VW dealer. The following system components may be removed and refitted.

Control unit

1 Disconnect the battery earth lead.
2 The control unit is located directly behind the radio/heater control panel and these will therefore have to be removed for access as described in Section 43.
3 Disconnect the wiring connector from the control unit then remove the securing screw each side of the unit and withdraw it (photo).
4 Refit in the reverse order of removal.

Accelerator pedal switch

5 Disconnect the battery earth lead.
6 Remove the accelerator pedal. Depending on model, this is achieved by unscrewing the clamp bolt (photo) or removing the securing clip and bush and withdrawing the pedal from or with the pivot shaft as applicable.
7 Detach the wiring connector and withdraw the pedal switch. On some models it is necessary to first remove the switch retaining screws (Fig. 10.20).
8 Refit in the reverse order of removal.

Clutch pedal switch

9 Disconnect the battery earth lead.
10 Detach and remove the lower instrument panel trim on the driver's side.
11 Disconnect the wiring connector from the switch.

12 You will now need VW special tool number 2041 or to fabricate an equivalent tool. Push and locate the tool between the pedal bracket and the relay plate/fusebox. Secure the tool (2041) in front of the clutch pedal switch and pull the switch out of its sleeve (Fig. 10.21).
13 Refit in the reverse order of removal. A new switch must always be used when refitting and must be pushed home into its sleeve by hand with the aid of a hammer shaft. Check that the switch is fully home then reconnect the wire connector.
14 Before refitting the lower trim panel, fully depress the clutch pedal to engage the switch over the retainer.

Speed sender unit

15 Disconnect the battery earth lead.
16 Unscrew and remove the drive cable at each end of the sender unit, detach the wiring connector and withdraw the unit (photos).
17 Refit in the reverse order of removal.

Diode

18 This is located under the dashboard on the driver's side.
19 Disconnect the battery earth lead, remove the lower trim panel for access, disconnect and remove the diode from the control unit wiring harness.
20 Refit in the reverse order to removal ensuring that secure connections are made.

10

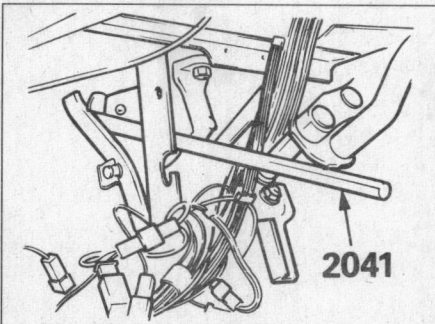

Fig. 10.21 Clutch pedal switch removal showing location of VW tool 2041 (automatic stop/start system) (Sec 15)

15.16A Automatic stop/start system speed sender unit drive cable removal . . .

15.16B . . . and wiring connector removal

Connector for coil

Connector for reversing light switch and gearbox switch for gearshift and consumption indicator

Gearbox switch

Switch-off relay for rear window and oil pressure monitor

Switch unit

Connector for oil pressure switch

Connector for starter (terminal 50)

Speed sensor

Connector for ignition switch

Main switch

Overrun control valve

Connector for thermo-time valve for overrun cutoff

Connector for manifold preheating (hedgehog)

Connector for instrument loom

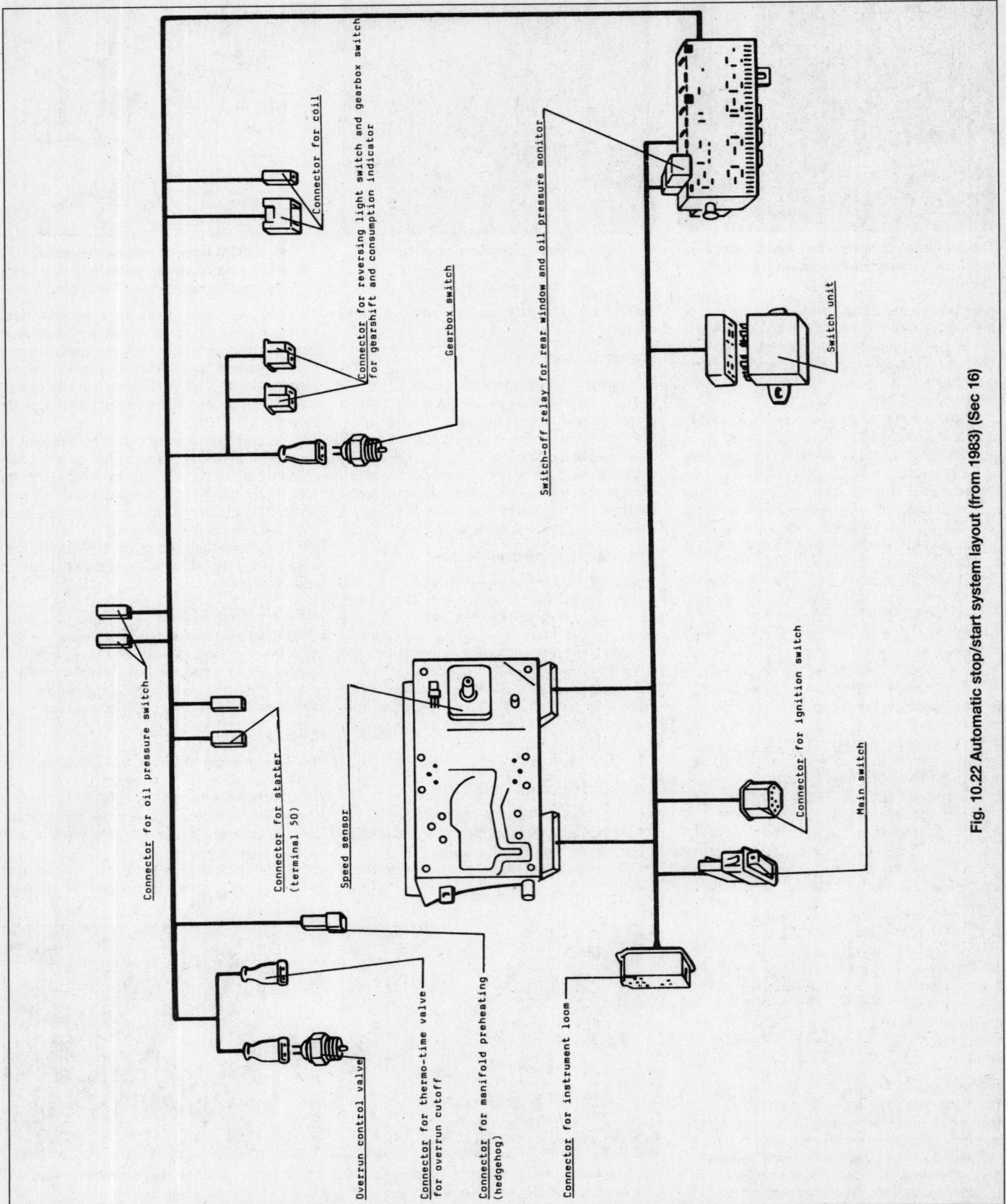

Fig. 10.22 Automatic stop/start system layout (from 1983) (Sec 16)

Fig. 10.23 Switch control unit location – automatic stop/start system from 1983 (Sec 16)

Fig. 10.24 Speed sensor location and securing screws – automatic stop/start system from 1983 (Sec 16)

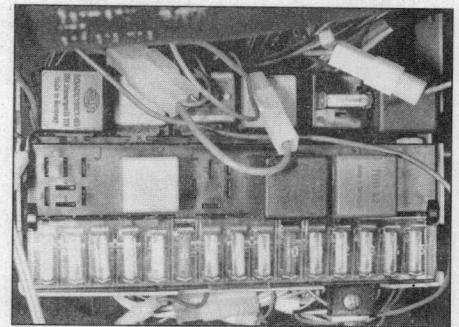

Fig. 10.25 Fuses and relays location – typical (Sec 17)

16 Automatic stop/start system components (from 1983) – removal and refitting

The system layout and components are shown in Fig. 10.22. Should the system malfunction in any way have the fault diagnosed by your VW dealer. The following components may be removed and refitted.

Switch control unit

1 Disconnect the battery earth lead.
2 The switch is located under the dashboard on the driver's side. Unclip and remove the lower trim panel for access (Fig. 10.23).
3 Disconnect the wiring multi-connector from the switch then pull the switch from the relay plate.
4 Refit in the reverse order to removal.

Speed sensor unit

5 Disconnect the battery earth lead.
6 The speed sensor is located on the rear face of the speedometer. Remove the instrument panel as described in Section 19, then undo the two retaining screws and pull the sensor unit free (Fig. 10.24).
7 Refit in the reverse order to removal.

Overrun control valve

8 Disconnect the battery earth lead.
9 Remove the air cleaner unit as described in Chapter 3.
10 Pull free the connector hoses then undo and remove the two retaining screws and remove the unit.
11 It should be noted that the valve contains a diode which is wired in parallel to the solenoid and if checking it for polarity watch the polarity.
12 Refitting is the reverse procedure to removal.

ASS switch

13 Disconnect the battery earth lead.
14 Use a thin flat blade screwdriver and carefully prise free the switch from the facia panel. When released from the panel, withdraw the switch and disconnect the wire connector.
15 Refit in reverse order of removal.

17 Fuses and relays – general

1 The fuses and relays are located behind the left-hand side of the facia. Access to them is gained by removing the curved cover inside the glovebox (right-hand drive models) or storage space (left-hand drive models).
2 The fuse and relay circuits are shown on the back of the cover, together with some spare fuses. In addition, in-line fuses are fitted to

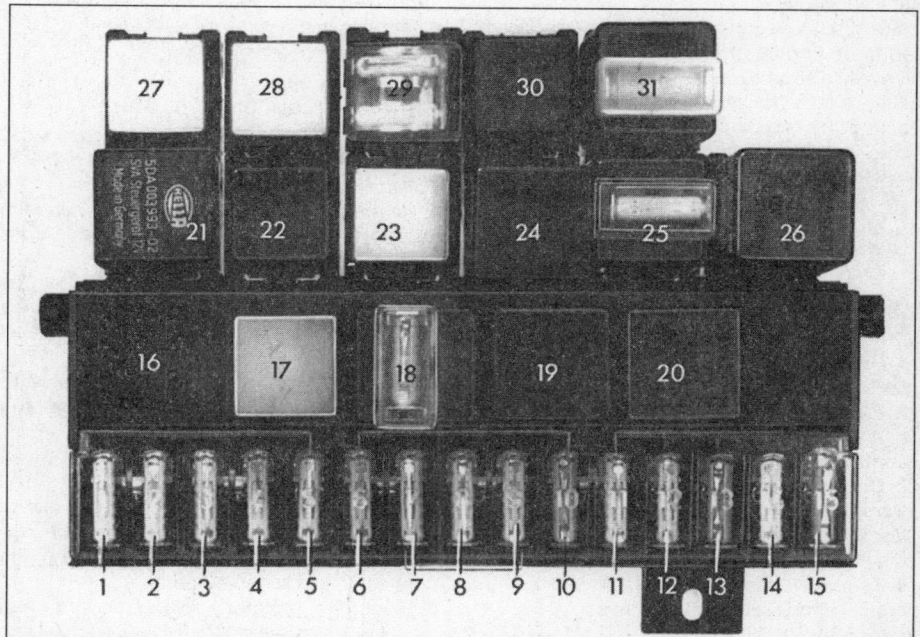

Fig. 10.26 Fuses and relay positions viewed from the passenger seat – early models (Sec 17)

Fuses		Rating	Relays and control units
1	Dipped beam left	8A	16 Vacant
2	Dipped beam right	8A	17 Relief relay (for x contact)
3	High beam left	8A	18 Fuel pump relay
4	High beam right	8A	19 Wash/wipe intermittent facility relay
5	Heated rear window	16A	20 Emergency light relay
6	Brake lights, turn signals	8A	21 Control unit for gearchange indicator
7	Cigarette lighter, glovebox light, radio	8A	22 Dual horn relay
8	Turn signals	8A	23 Manifold pre-heating relay/glow plug relay
9	Reversing lights, horn	8A	24 Rear wiper and washer relay
10	Fresh air blower	16A	25 Fog light relay
11	Wiper motor, gearchange/consumption indicator	8A	26 Headlight washer relay
12	Number plate light, glovebox light	8A	27 Change-over relay for day driving lights
13	Side light right	8A	28 Switch relay for day driving lights
14	Side light left	8A	29 Fuse box
15	Radiatorfan	25A	30 Window lifter relay
			31 Air conditioner relay

Note: *The number and location of the relays fitted may vary according to model and options fitted*

10

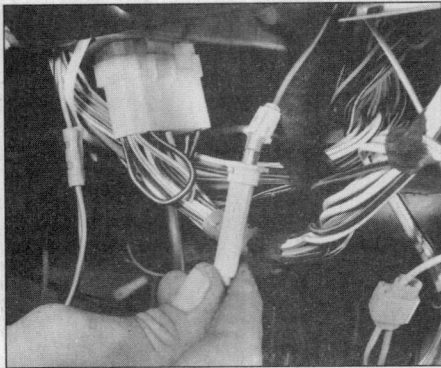

17.2 An in-line fuse

Fig. 10.27 Location for drilling hole in order
to remove the lock cylinder (Sec 18)

a = 12mm (0.47 in)   b = 10 mm (0.39 in)

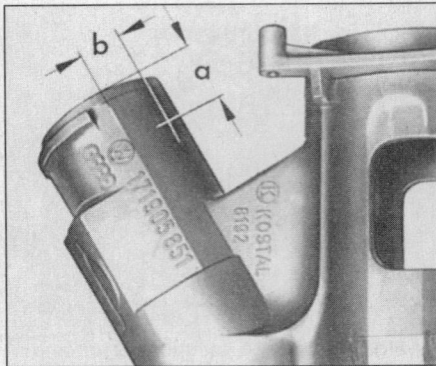

Fig. 10.28 Instrument panel insert trim
retaining screw locations (arrowed)
(Sec 19)

certain models, as given in the Specifications, and their location is shown in the wiring diagrams given at the end of this Manual (photo).

3 Always renew a fuse with one of identical rating, and never renew it more than once without finding the source of the trouble (usually a short circuit).

4 All relays are of the plug-in type. Relays cannot be repaired and if one is suspect, it should be removed and taken to an auto electrical workshop for testing.

18 Ignition switch/steering column lock – removal and refitting

1 Remove the steering column combination switch, as described in Section 21.

2 Prise the lockwasher from the inner column and remove the spring followed by the contact ring.

3 Using an Allen key, unscrew the clamp bolt from the steering lock housing and slide the housing from the outer column. Disconnect the multiplug.

4 To remove the ignition switch, unscrew the cross-head screw inside the housing.

5 To remove the lock cylinder it will be necessary to drill a 3 mm (0.118 in) diameter hole in the housing at the location shown in Fig. 10.27 – the retaining pin can then be depressed and the lock cylinder removed.

6 Refitting is a reversal of removal.

19 Instrument panel cluster – removal and refitting

1 Disconnect the battery earth lead.

2 Remove the steering wheel as described in Chapter 11. Although not essential, this does allow greater freedom of access.

3 Using a flat blade tool, carefully prise free and remove the switch trim panel between the steering column and instrument cluster panel (photo).

4 Unscrew and remove the four screws retaining the dash panel insert trim and withdraw the trim (Fig. 10.28).

5 Unscrew and remove the cluster panel retaining screws.

6 Reach through the side and to the rear of the cluster panel and disconnect the speedometer drive cable by pressing the connector lugs together (photo).

7 Pivot the cluster panel forward and

withdraw it sufficiently to detach the wiring multi-connectors and the vacuum hose, (photos). When fully removing the instrument cluster panel take care not to damage the printed circuit foil.

8 Refitting is a reversal of the removal sequence, but make sure that all connections are securely made and check instruments for satisfactory operation on completion.

19.3 Switch trim panel removal

19.6 Speedometer drive cable connector

19.7A Detaching the vacuum hose from the instrument panel cluster

19.7B Instrument panel cluster removed

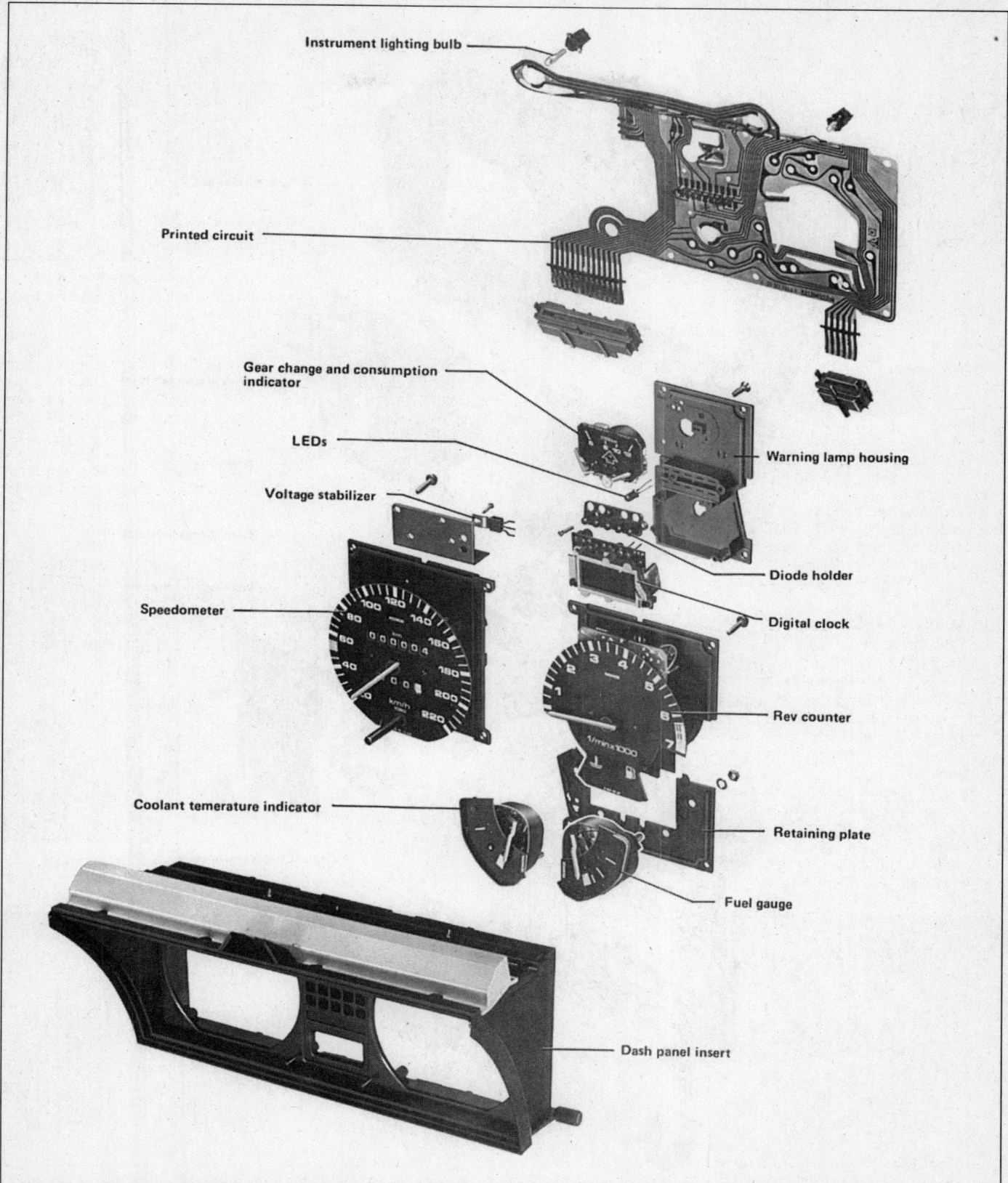

Instrument lighting bulb

Printed circuit

Gear change and consumption indicator

LEDs

Voltage stabilizer

Speedometer

Coolant temerature indicator

Warning lamp housing

Diode holder

Digital clock

Rev counter

Retaining plate

Fuel gauge

Dash panel insert

10

Fig. 10.29 Exploded view of instrument panel cluster unit with digital clock and gearchange/consumption indicator (Sec 20)

Dash panel insert

Voltage stabilizer

Speedometer

Clock

Gear change indicator

Warning lamp housing

Fuel gauge

Retaining plate

Instrument lighting bulb

Diode holder

LEDs

Coolant temperature gauge

Printed circuit

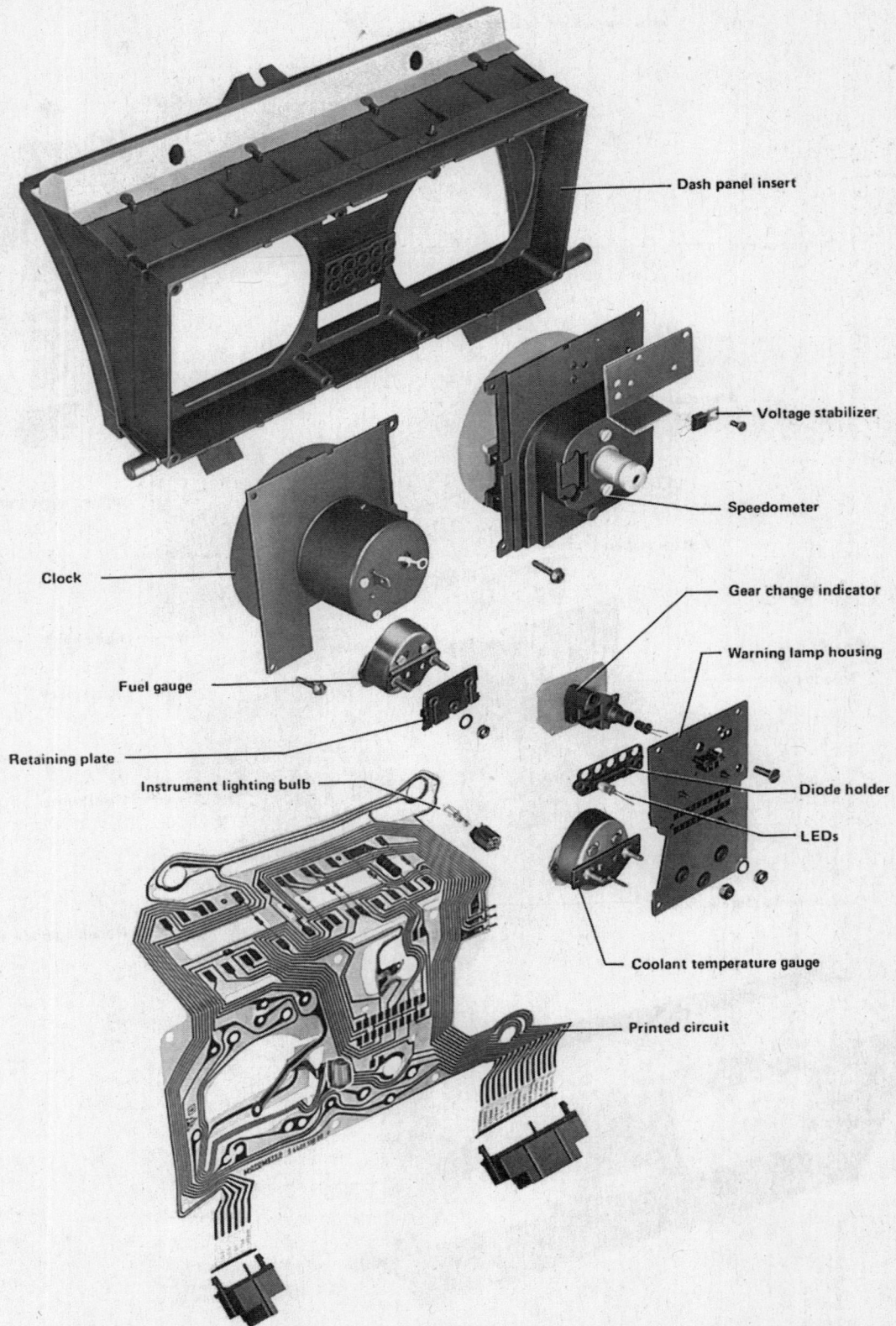

Fig. 10.30 Exploded view of instrument panel cluster unit with normal type clock and gearchange indicator (Sec 20)

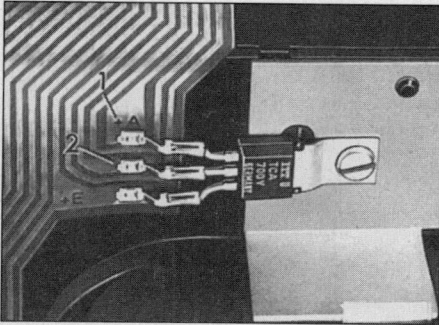
Fig. 10.31 Voltage stabilizer test terminals (Sec 20)

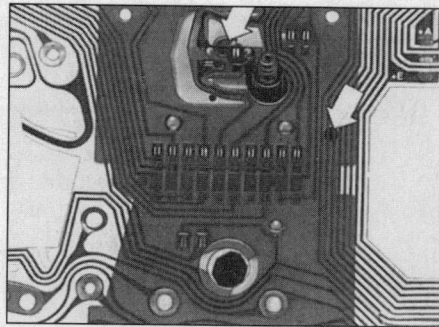
Fig. 10.32 Gearchange/consumption indicator (bottom arrow) and retaining screw (top arrow) (Sec 20)

Fig. 10.33 Gearchange/consumption indicator connections (Sec 20)

1 LED (plus)
2 LED (minus)
3 Solenoid valve (plus)
4 Solenoid valve (earth)
5 Vacuum connection

Fig. 10.34 Normal type clock connections (Sec 20)

1 Earth connection   2 Plus (live) connection

Fig. 10.35 Digital clock retaining screws (arrowed) (Sec 20)

Fig. 10.36 LED locations in the warning lamp housing (Sec 20)

K1 High beam (yellow)
K2 Alternator (red)
K3 Oil pressure (red)
K5 Indicators (green)
K29 Glow time (yellow) – diesel
K48 Gearchange indicator (yellow)

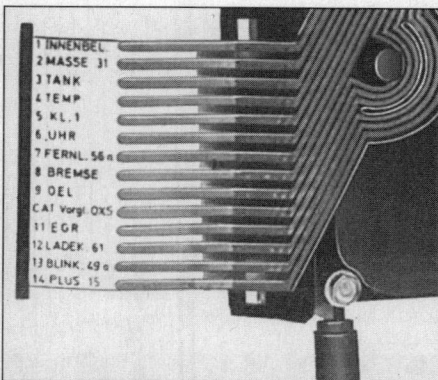
Fig. 10.37 Printed circuit functions with 14 pin connector (Sec 20)

1 Instrument lighting
2 Earth (terminal 31)
3 Fuel gauge (to sender)
4 Coolant temperature gauge (to sender)
5 Tachometer (terminal 1)
6 Clock
7 High beam warning light (terminal 56a)
8 Vacant
9 Oil pressure warning lamp (to switch)
10 Glow time warning lamp – diesel
11 Vacant
12 Alternator warning lamp (terminal 61)
13 Indicator warning lamp (terminal 49a)
14 Plus wire (terminal 5)

20 Instrument panel cluster – dismantling, testing and reassembly

1 Remove the instrument panel cluster as described in Section 19.

2 Remove the relevant instrument with reference to Figs 10.29 or 10.30 but take particular care not to damage the printed circuit foil.

3 To test the voltage stabilizer connect a voltmeter between the terminals shown in Fig. 10.31 with a 12 volt supply to the remaining terminal. A constant voltage of 10 volts must be registered. If the voltage is above 10.5 volts or below 9.5 volts renew the voltage stabilizer.

4 The accuracy of the fuel gauge can be checked by draining the fuel tank and then adding exactly 5 litres of fuel. After leaving the ignition switched on for at least two minutes the fuel gauge needle should be level with the upper edge of the red reserve zone. If not, either the fuel gauge or tank unit is faulty.

5 If renewing the gearchange/consumption indicator avoid touching the back of the gauge. Removal necessitates detaching the printed circuit, the warning lamp housing and removing the retaining screw (arrowed in Fig. 10.32). Renew the diode or consumption indicator as necessary.

6 When renewing the normal type clock (which incorporates the fuel gauge) it is important to ensure the correct printed circuit connections when refitting. The connections are shown in Fig. 10.34.

7 The digital type clock is secured by two retaining screws, Fig. 10.35. When removing the clock take care not to allow the adjuster pins for the hours and minutes to fall out.

8 The warning lamp LED indicators in the lamp housing are positioned as shown in Fig. 10.36. When renewing the LEDs, each diode can be pulled free from the retainer plate, but note how one of the connector prongs is offset. This is the minus connection and it is important that they are correctly refitted. If necessary the diode holder unit can be removed by carefully levering it free from the warning lamp housing.

9 The individual circuits of the printed circuit foil can be checked for continuity using Figs. 10.37, 10.38, and 10.39 for reference.

10 If renewing the printed circuit foil it should be noted that a common type is supplied for all models. If fitting a new printed circuit foil to the dash insert on models with a normal type

10

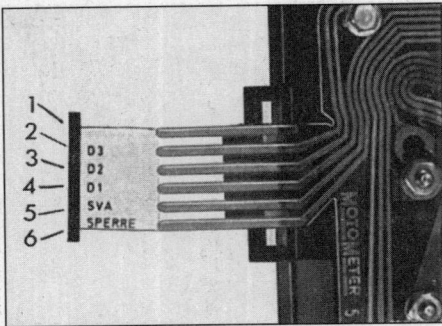

Fig. 10.38 Printed circuit functions with 6 pin connector (Sec 20)

1 to 4 Vacant
5 LED (gearchange indicator)
6 Solenoid valve (consumption indicator)

clock it is necessary to cut off the connector pins used for the digital clock and vice versa for models with the digital clock.

11 The instrument light bulbs can be renewed by withdrawing the holder and pulling free the bulb (photo). Refit in reverse.

12 Reassembly and refitting of the instrument cluster is a reversal of the removal procedure. Check the various functions for satisfactory operation on completion.

21 Steering column combination switches – removal and refitting

1 Disconnect the battery earth lead.

2 Remove the steering wheel as described in Chapter 11.

3 Unscrew and remove the lower to upper steering column shroud screws and detach the lower cover.

4 Disconnect the multi-plug then remove the three screws and withdraw the combination switches (photos).

5 The switches are not repairable or adjustable and if defective must be renewed.

6 Refitting is a reversal of removal. If fitting an

Fig. 10.39 Printed circuit connection designations (Sec 20)

Voltage stabilizer
1 Plus (+) input
2 Earth (–)
3 Output

Coolant temperature gauge
4 To gauge sender
5 Earth (–)
6 Output

Fuel gauge
7 To gauge sender
8 To voltage stabilizer output

Digital clock
9 Plus (+)
10 Earth (–)

Tachometer
11 Earth (–)
12 From ignition coil (terminal 1)
13 To voltage stabilizer input

Gearchange/consumption
14 To 6 pin connector indicator
15 LED earth
16 To 6 pin connector
17 Solenoid valve

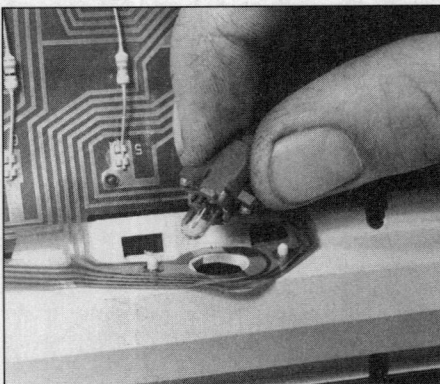

20.11 Instrument panel cluster bulb removal

21.4A Detach the multi-plug lead connector . . .

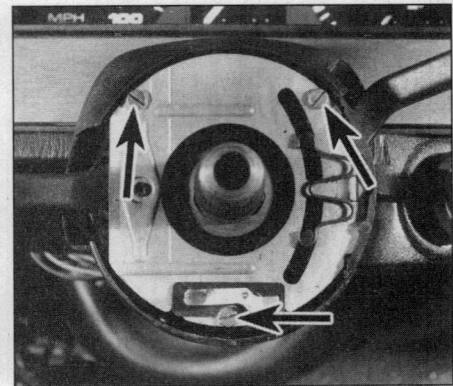

21.4B . . . then remove the three retaining screws (arrowed)

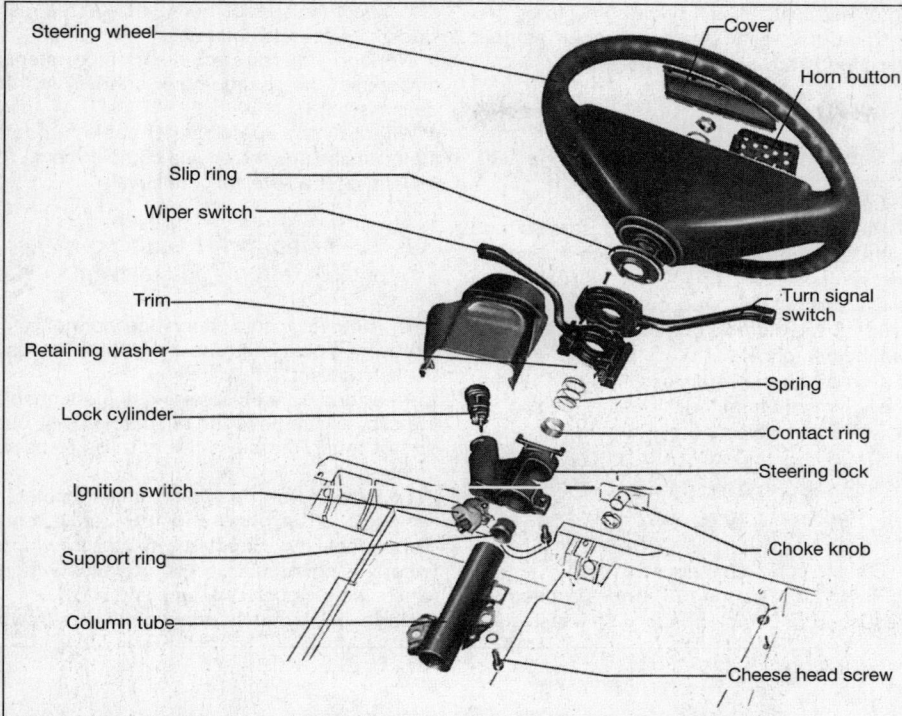

Fig. 10.40 Steering column assembly components (Sec 21)

Steering wheel — Cover — Horn button — Slip ring — Wiper switch — Trim — Retaining washer — Lock cylinder — Ignition switch — Support ring — Column tube — Turn signal switch — Spring — Contact ring — Steering lock — Choke knob — Cheese head screw

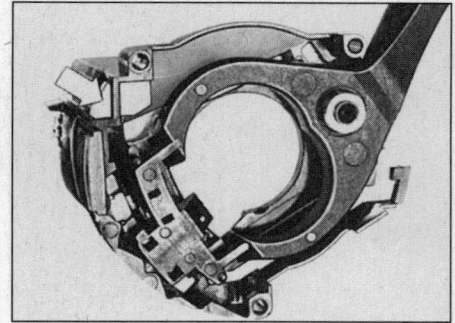

Fig. 10.41 Wedge location in windscreen wiper switch unit (Sec 21)

intermittent device for the windscreen wipers, locate the relay holder in place of the bridge and withdraw the wedge (Fig. 10.41).

7 Tighten the switch securing screws securely to ensure a good earth connection. Check switches for satisfactory operation on completion.

22 Rocker switches – removal and refitting

1 Rocker switches are fitted to the lower facia panel and to the central console, their functions and positions depending on model. The function of each switch is indicated by a symbol on its face, and some switches incorporate a warning light.

2 Disconnect the battery earth lead.

23.2 Location of door courtesy light switch

3 To remove a lower facia panel switch, first carefully prise free the trim panel.

4 Reach behind the switch concerned and compress the retaining lugs whilst simultaneously pushing the switch from its aperture.

5 Disconnect the lead connector to fully remove the switch.

6 To remove the switch bulb, twist free the holder and then pull the bulb out for inspection and if necessary renewal.

7 Refitting is a reversal of removal. Check switch operation on completion.

23 Door and luggage compartment light switches – removal and refitting

1 Disconnect the battery earth lead.

Door switch

2 Open the door, unscrew the switch retaining screw and withdraw the switch (photo).

3 Disconnect the switch wire.

4 Refit in the reverse order of removal.

Luggage compartment switch

5 The luggage compartment switch location is dependent on the model and year, the two main types being shown in Fig. 10.42.

6 In each case the switch is secured by retaining screws and with these removed the switch can be withdrawn and the wiring connector detached.

7 Refit in the reverse order of removal.

Boot light switch – Santana

8 The switch is fitted to the underside of the boot lid adjacent to the nearside number plate

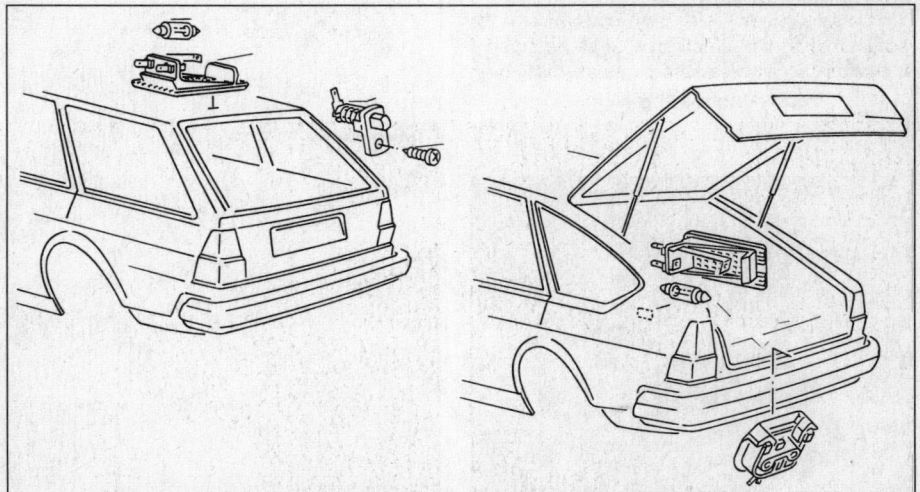

Fig. 10.42 Luggage compartment lights and switches on the Variant and Passat saloon (Sec 23)

10

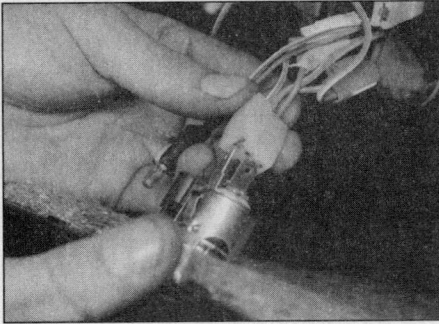

24.3 Detaching the plug connector from the cigar lighter. The illumination bulb is also shown removed

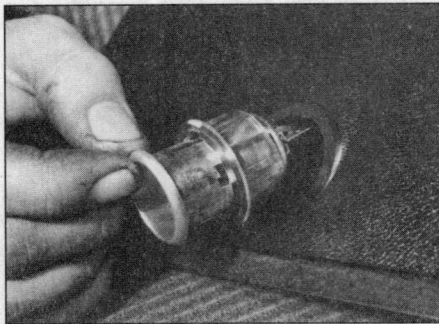

24.4 Cigar lighter unit removal

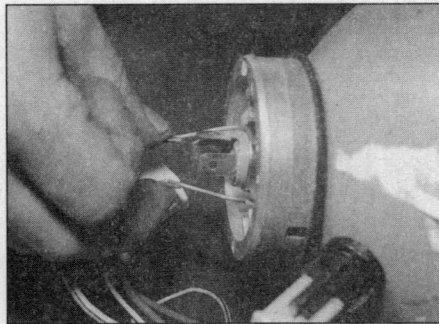

26.2A Unhook the clip . . .

26.2B . . . to remove headlight bulb

light. This switch has a single screw fixing and is removed in the same manner as that described above for other models.

24 Cigarette lighter – removal and refitting

1 Detach the battery earth lead.
2 Remove the two retaining screws and withdraw the trim panel in which the lighter unit is mounted.
3 Disconnect the lead connector from the rear of the lighter unit and pull free the bulbholder (photo).
4 Compress the retaining clips and extract the lighter unit from the trim panel (photo).
5 Refit in the reverse order of removal.

25 Speedometer cable – removal and refitting

1 Disconnect the battery earth lead.
2 Refer to Section 19 and proceed as described in paragraphs 2 to 6 inclusive and

disconnect the speedometer cable at the rear side of the speedometer unit.
3 Working in the engine compartment, disconnect the speedometer cable from the transmission.
4 Withdraw the speedometer cable through the bulkhead into the engine compartment.
5 Refitting is a reversal of removal.

26 Headlamps and headlamp bulbs – removal and refitting

1 To remove a bulb, open the bonnet and remove the plug and rubber cap from the rear of the headlight.
2 Press and twist the retaining ring or unhook the clip, and remove the headlamp bulb, but do not touch the glass if it is to be re-used (photos).
3 To remove the headlamp unit complete, remove the surround and front grille, and disconnect the direction indicator wiring (where wing mounted), or foglamp wiring (where wing mounted) as applicable.
4 Undo and remove the securing screws

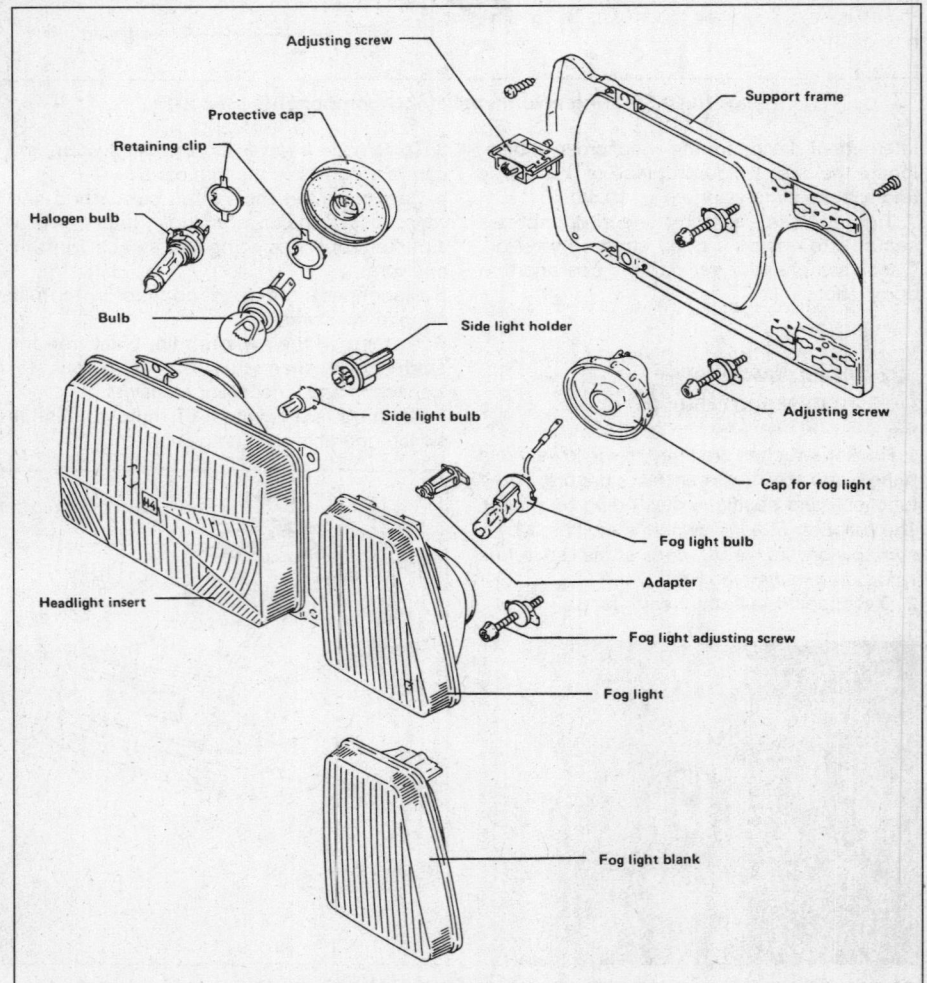

Fig. 10.43 Combined head and foglamp unit components (Sec 26)

Adjusting screw
Protective cap
Support frame
Retaining clip
Halogen bulb
Bulb
Side light holder
Side light bulb
Adjusting screw
Cap for fog light
Fog light bulb
Adapter
Headlight insert
Fog light adjusting screw
Fog light
Fog light blank

Fig. 10.44 Head and foglamp unit securing screw positions (Sec 26)

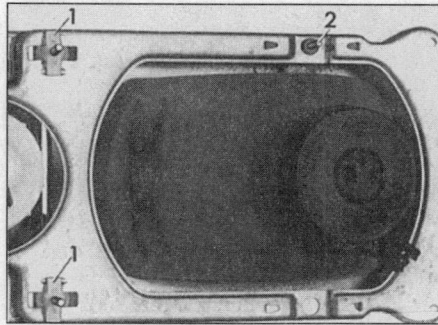

Fig. 10.45 Headlamp insert retainers (1) and adjuster screw (2) (Sec 26)

27.4 Headlamp/foglamp adjuster screw locations

A Headlamp vertical adjuster
B Headlamp horizontal adjuster
C Foglamp adjuster

arrowed in Fig. 10.44, and remove the combined headlamp, fog or indicator lamp unit.

5 To remove the headlamp insert, rotate the plastic retainers 90° and remove the adjustment screw (Fig. 10.45).

6 Refitting is a reversal of the removal procedure. On completion check the lights for satisfactory operation and check the head and spotlamp beam alignments as described in Section 27.

27 Headlamps – alignment

1 It is recommended that the alignment is carried out by a VW dealer using modern beam setting equipment. However, in an emergency, the following procedure will provide an acceptable light pattern.

2 Position the car on a level surface with tyres correctly inflated, approximately 10 metres (33 feet) in front of and at right-angles to a wall or garage door.

3 Draw a horizontal line on the wall or door at headlamp centre height. Draw a vertical line corresponding to the centre-line of the car, then measure off points either side of this, on the horizontal line, corresponding with the headlamp centres.

4 Switch on the main beam and check that the areas of maximum illumination coincide with the headlamp centre marks. If not, turn

the adjustment screws as necessary – the upper screw controls horizontal adjustment and the lower screw height adjustment (photo).

28 Lamp bulbs – renewal

Note: *Lamp bulbs should always be renewed with ones of identical type and rating, as listed in the Specifications.*

Front sidelights

1 Each sidelight is located in a holder which is mounted in the headlight reflector unit. Access is from the engine compartment side, so raise and support the bonnet.

2 Twist the bulbholder anti-clockwise and remove it from the headlamp (photo).

3 Push and twist the bulb to remove it.

4 Refit in reverse order and check operation.

Front direction indicators – bumper mounted

5 Remove the two screws and withdraw the lens.

6 Push and twist the bulb to remove it, (photo).

7 The bulb can only be fitted in one position because of the offset pins. Make sure that the gasket is serviceable.

8 Check operation of indicator on completion.

Front direction indicators – wing mounted

9 Raise and support the bonnet. Pull the rubber cap from the rear of the indicator unit, press the holder lug towards the bulb and withdraw the holder.

10 Press and untwist the bulb to remove it from the holder.

11 Refit in reverse ensuring that the lug fully engages as the holder is fitted. Check the operation of the indicator on completion.

Front foglamp – bumper mounted

12 Undo the lens retaining screws and withdraw the lens. Withdraw the bulb, but do not touch it with the fingers if it is to be re-used. Refit in reverse order of removal.

Front foglamp – wing mounted

13 Open the bonnet. Rotate the cap at the rear of the lamp in an anti-clockwise direction and remove the cap (photo).

14 Unhook the bulbholder retaining clip and pivot it back out of the way. Detach the lead connector, then extract the bulb and holder. Do not touch the bulb with the fingers if it is to be re-used.

15 Refit in the reverse order ensuring that the

10

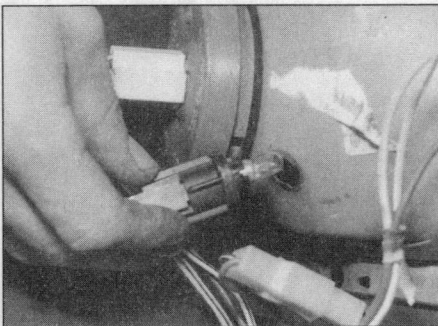

28.2 Sidelight bulb holder removal from the headlamp

28.6 Front indicator bulb removal – bumper mounted

28.13 Foglight cap removal (wing mounted)

28.17 Rear lamp cluster removed

28.18 Rear lamp cluster bulb locations

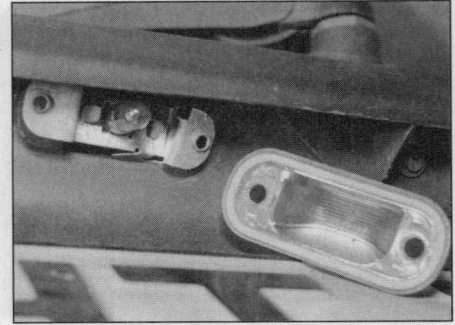

28.24 Number plate light lens removal – boot lid mounted

holder locates with the lug in the reflector. Fit the cap with the TOP marking uppermost.

16 Check the light for satisfactory operation on completion. Adjust the light if necessary by turning the screw indicated in photo 27.4.

Rear lamp cluster

17 Open the tailgate. Depress the plastic clips and withdraw the bulbholder (photo).
18 Push and twist the faulty bulb to remove it (photo).
19 Refit in reverse order and check operation of all lights in cluster.

Number plate light – bumper mounted

20 Prise the light out of the bumper.
21 Lever out the clips and remove the cover.

22 Push and twist the bulb to remove it.
23 Refit in reverse and check operation on completion.

Number plate light – boot lid mounted

24 Open the boot lid and remove the light retaining screws (photo).
25 Squeeze the plastic clips and remove the bulbholder (photo).
26 Push and twist the bulb to remove it.
27 Refit in the reverse order to removal and check operation on completion.

Interior/luggage compartment light

28 Prise out the light using a screwdriver. When removing the interior light, insert the

screwdriver at the opposite end to the switch.
29 Remove the festoon type bulb from the spring contacts (photo).
30 When fitting the bulb, make sure that the spring contacts are tensioned sufficiently to make good contact with the bulb.
31 Note that the luggage compartment light may be located on the rear panel or beneath the top panel, or in the Santana integral with the rear combination light on each side.

Glovebox lamp

32 Open the glovebox and prise out the switch/bulbholder.
33 Push and twist the bulb to remove it.

29 Horn – removal and refitting

1 The horns are located beneath the right-hand side of the engine compartment (photo).
2 Disconnect the battery negative lead.
3 Disconnect the wiring from the relevant horn then unbolt it from the bracket.
4 Refitting is a reversal of removal.
5 To check the horn press contact on the steering wheel prise free the press pad from the spoke. The wires can then be disconnected from the contact plates if required (photo).
6 Refit in reverse and on completion check operation of horn.

28.25 Number plate light bulb and holder assembly – boot lid mounted

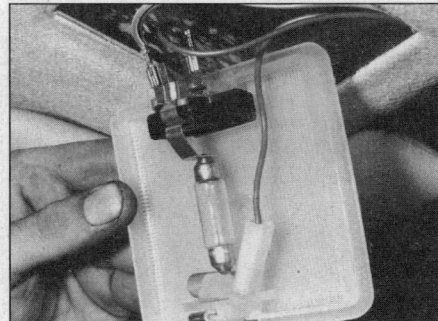

28.29A Interior light unit removed – front

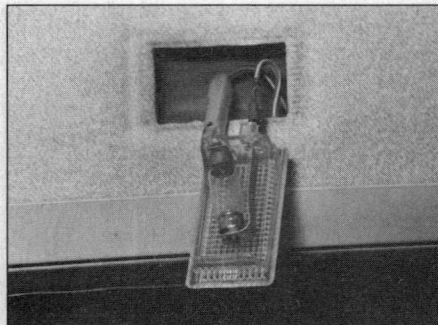

28.29B Interior light unit removed – rear

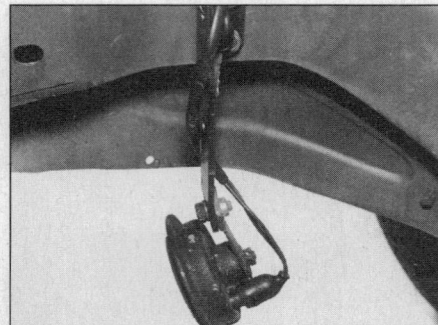

29.1 Horn location and mounting bracket

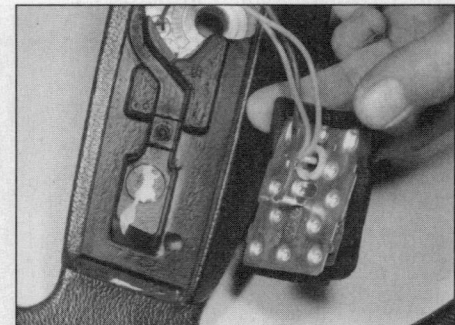

29.5 Horn press contact removed from steering wheel spoke

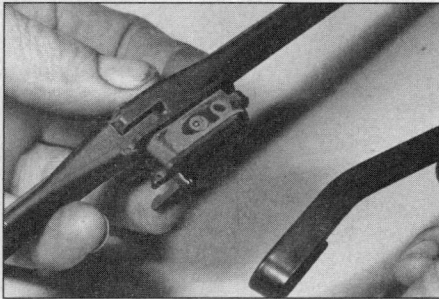

30.5 Removing a wiper blade from the wiper arm

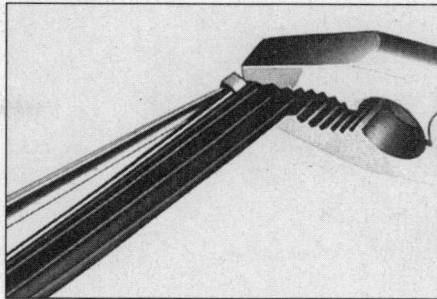

Fig. 10.46 Wiper blade rubber removal (Sec 30)

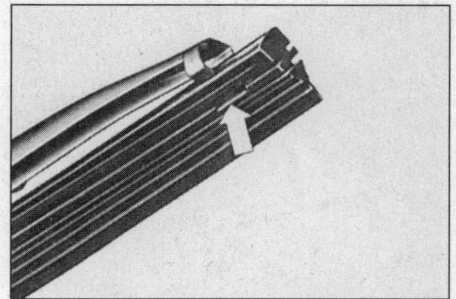

Fig. 10.47 Engage wiper blade rubber with lugs in retaining slots (arrowed) (Sec 30)

30 Wiper blades – renewal

1 The wiper blades should be renewed when they no longer clean the windscreen effectively.
2 If the rubbers only are worn they can be renewed by pressing together the steel strips on the rubber enclosed side (using pliers) then sliding the rubber from the blade complete with strips (Fig. 10.46).
3 Press the replacement rubber into the lower blade clips then locate both strips into the first rubber groove. The strip slits must face the rubber and locate in the rubber groove lugs. Press the steel strips and the rubber together and fit the upper clips. The lugs on each side must engage with the securing slots (arrowed) in the rubber (Fig. 10.47).

4 To renew the wiper blade and rubber complete, lift the wiper blade and arm from the windscreen.
5 Depress the plastic clip and withdraw the blade from the hooked end of the arm (photo).
6 Insert the new blade, making sure that the plastic clip is engaged.

31 Windscreen wiper motor – removal and refitting

1 Disconnect the battery negative lead.
2 Prise the pushrods off the wiper motor crank.
3 Mark the crank in relation to the motor spindle, then unscrew the nut and remove the crank from the spindle.
4 Unbolt the wiper motor from the frame and disconnect the wiring (photo).

5 Refitting is a reversal of removal. If a new wiper motor is being fitted, switch on the ignition and run the motor for a short period before switching it off. The motor is now in its parked position and the crank should be fitted to the spindle in the position shown in Fig. 10.48.

32 Windscreen wiper arms and linkage – removal and refitting

1 Prise up the covers and unscrew the nuts securing the wiper arms to the spindles (photo).
2 Carefully lever off the wiper arms (photo).
3 Unscrew the spindle bearing nuts and remove the rubber spacers (photo).
4 Disconnect the battery negative lead.
5 Unscrew the wiper frame mounting bolt and withdraw the linkage assembly, at the same

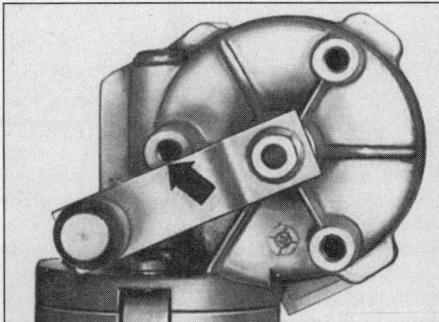

Fig. 10.48 Wiper motor crank alignment position when parked (Sec 31)

31.4 Windscreen wiper motor and lead connectors

32.1A Prise the wiper arm cover free . . .

32.1B . . . and unscrew the retaining nut

32.2 Withdraw the wiper arm

32.3 Wiper arm spindle, nut and rubber spacer

10

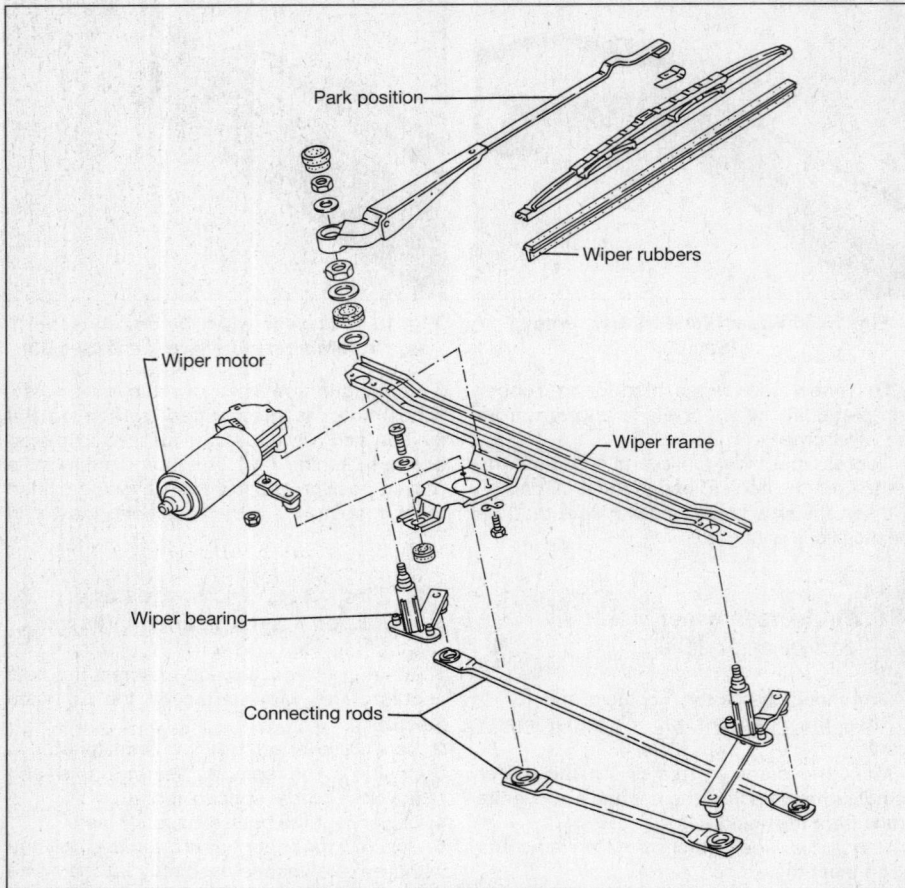

Fig. 10.49 Windscreen wiper components (Sec 32)

Park position

Wiper rubbers

Wiper motor

Wiper frame

Wiper bearing

Connecting rods

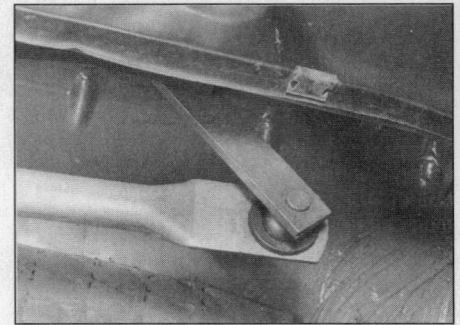

32.7 Pushrod and spindle arm joint

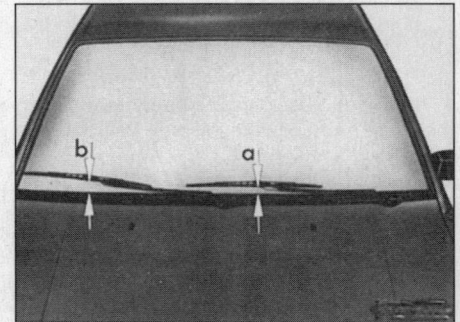

Fig. 10.50 Windscreen wiper arm and blade
PARK settings (Sec 32)

a = 36 mm (1.4 in)   b = 63 mm (2.5 in)

time disconnecting the wiring from the motor.
6 Remove the wiper motor from the frame, with reference to Section 31.
7 Prise the pushrods from the spindle arms (photo).
8 If the spindle bearings are to be removed, saw off the rivets. After fitting the new bearings clinch the new rivets onto the wiper frame.
9 Refitting is a reversal of removal, but smear a little molybdenum disulphide grease in the pushrod ball sockets. Before fitting the wiper arms in their parked position, switch on the

ignition and run the wiper motor for a short period before switching it off.
10 On refitting the wiper arms and blades check that their parked position is as shown in Fig. 10.50.

33 Rear wiper blade – renewal

The rear wiper blade is renewed in the same manner as that described for the windscreen in Section 30.

34 Rear wiper motor –
removal and refitting

1 Disconnect the battery earth lead.
2 Raise and support the tailgate, prise free the tailgate trim clips and detach the trim panel.
3 Use a suitable screwdriver and lever the connecting rod free from the motor arm pivot (photo).
4 Detach the wiper motor wiring connector and the earth lead (photo).
5 Undo the motor support frame mounting bolts and withdraw the motor and frame assembly (photo).

34.3 Prise free the connecting rod from the
rear wiper motor arm

34.4 Rear wiper motor wiring connection
and earth lead to tailgate

34.5 Rear wiper mounting bolts (arrowed)

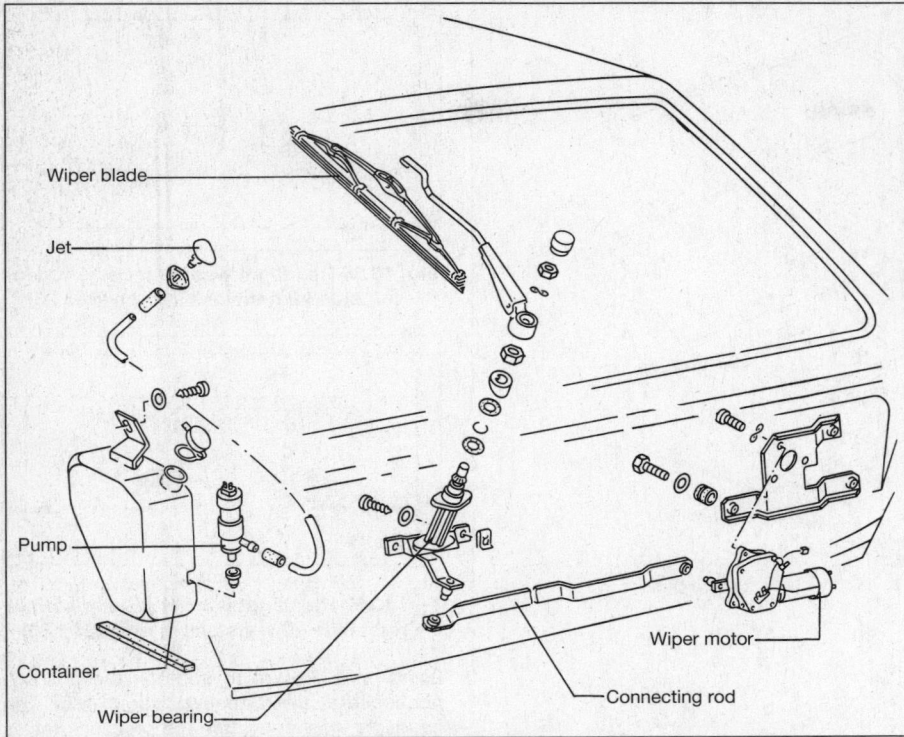

Fig. 10.51 Rear window wiper components (Secs 34 and 35)

Labels on figure:
Wiper blade
Jet
Pump
Container
Wiper bearing
Wiper motor
Connecting rod

35.4 Rear wiper arm bearing unit

the wiper operation and also the wiper arm blade and arm rest position when parked (see Fig. 10.52).

36 Windscreen, rear window and headlight washer systems

1 The component parts of each system are similar, consisting of a fluid reservoir, pump unit, hoses and connections and adjustable jets. Where headlight washers are fitted, the fluid reservoir used is of larger capacity and serves both the headlights and the windscreen washer systems although each system operates independently as required (photo).
2 The systems are shown in Figs. 10.51, 10.54 and 10.55.
3 It is important not to allow the reservoirs to run dry and when topping up use a screenwash additive that contains an anti-freeze content. Do not use cooling system anti-freeze as this can damage the paintwork.
4 Should the washers of a system fail to operate check that the hoses, connections and seals are secure and in good condition also that the pump is operational.
5 If the jets become blocked, detach and clean them by applying compressed air through them, do not pick and probe the jets

6 If the wiper motor is to be renewed unbolt and remove it from the support frame.
7 Refit in the reverse order to removal. If a new motor is being fitted, run the motor for a short period on completion then switch it off and check that the wiper blade rest position is as shown in Fig. 10.52. If adjustment is necessary check that the motor crank arm is positioned as shown in Fig. 10.53 with the earth connecting rivet just visible (arrowed). Adjust the crank arm or wiper arm if necessary.

35 Rear wiper arm and linkage – removal and refitting

1 To remove the wiper arm, proceed as described in Section 32. paragraphs 1 and 2.
2 Unclip and detach the rear trim panel from the tailgate.
3 Prise free and detach the connecting rod from the wiper bearing and wiper motor crank arm pivots.
4 Remove the retaining screws and withdraw the wiper bearing unit (photo) .
5 Refit in the reverse order of removal. Lubricate the connecting rod ball sockets before fitting the rod. On completion check

Fig. 10.52 Rear window wiper arm and blade alignment position when parked (Sec 34)

a = 27mm (1.06in)

Fig. 10.53 Rear window wiper motor crank arm position when parked (Sec 34)

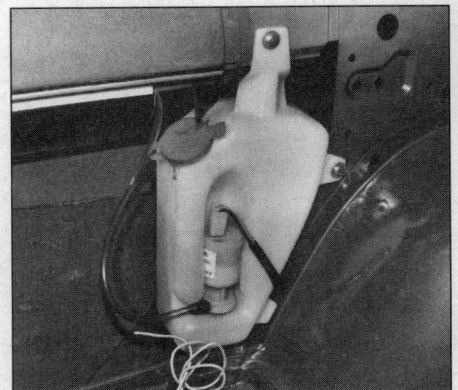

36.1 Rear window washer reservoir and pump unit – Variant (Estate)

10

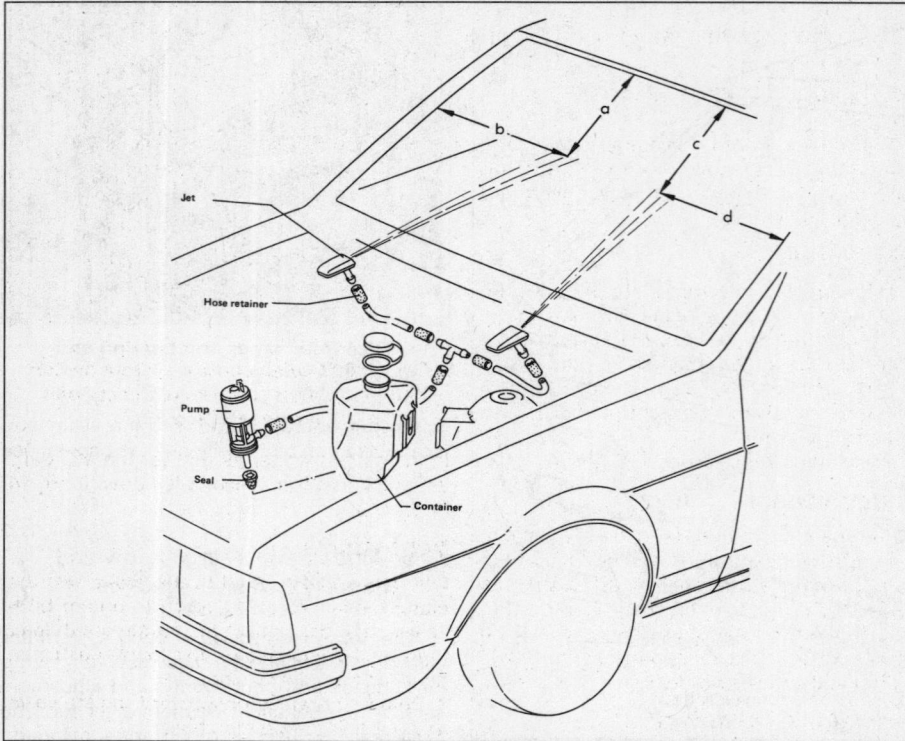

Fig. 10.54 Windscreen washer components and jet adjustment positions (Sec 36)

a = 434 mm (17.1 in)   b = 450 mm (17.7 in)   c = 435 mm (17.1 in)   d = 320 mm (12.5 in)

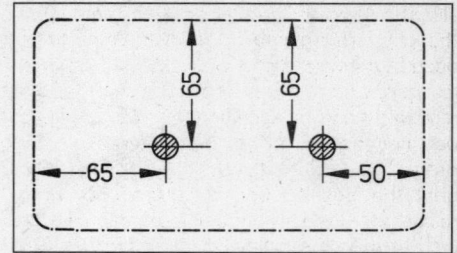

Fig. 10.55 Headlight washer system components (Sec 36)

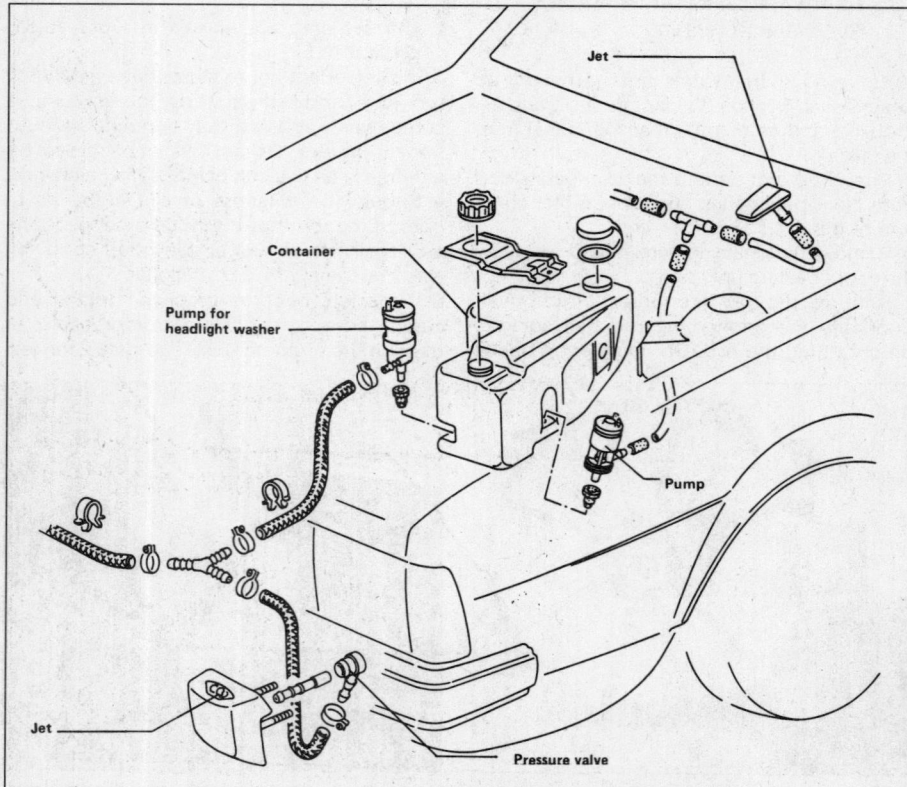

Fig. 10.56 Headlight washer jet adjustment – left side (dimensions in mm) (Sec 36)

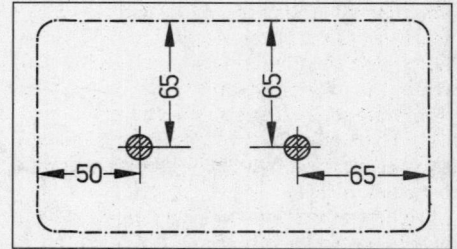

Fig. 10.57 Headlight washer jet adjustment – right side (dimensions in mm) (Sec 36)

using wire. Failure to unblock the jets will necessitate their renewal and with the headlight washer system the pressure valves should also be renewed at the same time. Check that the jets are adjusted in accordance with the illustrations (Figs. 10.56 and 10.57).

37 Cruise control system – description and operation

This system is available on models produced from August 1983 and is designed to assist the driver on long distance and motorway driving (in ideal conditions) by holding the speed of the vehicle at a selected level. The speed level can only be set at speeds above 30 kph (19 mph) and on automatic transmission models in the driving range D or 2.

The system is not designed for use in heavy traffic conditions or in other adverse driving conditions such as on snow or ice.

The system is actuated by means of control buttons on the indicator arm, the system being brought into operation by moving the

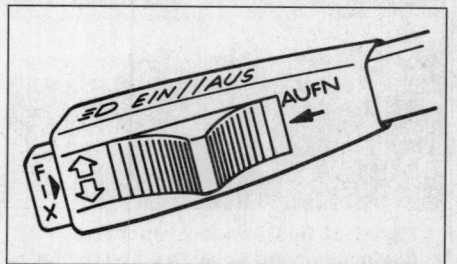

Fig. 10.58 Cruise control system actuating controls (Sec 37)

sliding button to the EIN position then when the desired cruising speed is reached the push button in the end of the stalk is pressed in briefly and released. The foot can then be removed from the accelerator pedal. Should it be necessary to further increase the programmed speed press the FIX button in until the desired speed is reached, then release it. Alternatively the speed can also be increased in the normal manner by pressing down on the accelerator pedal but using this method, when the accelerator pedal is released, the speed will again return to the programmed level originally set (Fig. 10.58).

The system will temporarily cancel when the brake or clutch pedals are depressed or when the speed drops well below the programmed level such as when climbing a hill in a high gear. The programmed speed can later be resumed by moving the slide button to the left (AUFN).

The system programme can be cancelled by moving the slide button to the right (AUS) or switching off the ignition.

Do not under any circumstances move the gear lever into neutral without operating the clutch pedal when the cruise control system is engaged or serious damage to the engine could well result!

If the system shows any malfunction at any time have it checked by your VW dealer.

38 Cruise control system components – removal and refitting

The cruise control system layout and components are shown in Fig. 10.59. The following components can be removed and refitted as described.

Switch unit

1 This unit is located behind the dash trim panel on the passenger side. Disconnect the battery earth lead, remove the trim panel for access to the switch, then detach the switch wiring connections. Remove the screws securing the switch unit and withdraw the switch (Fig. 10.60).
2 Refit in the reverse order of removal.

Speed sensor

3 The speed sensor unit is located on the rear face of the speedometer on the instrument panel. Remove the instrument panel as described in Section 19, then unscrew and remove the two retaining screws and pull the sensor unit from the speedometer (Fig. 10.24).
4 Refit in the reverse order to removal.

Indicator/CCS switch

5 The combined indicator and cruise control system (CCS) switch is removed and refitted

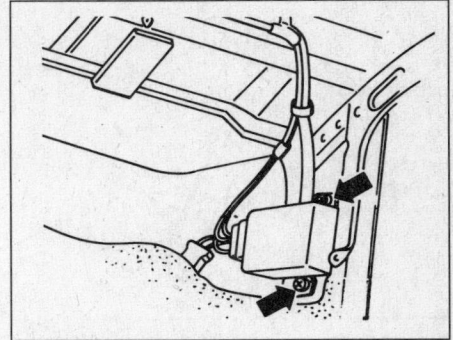

Fig. 10.60 Cruise control system switch unit location and retaining screws (arrowed) (Sec 38)

in the same manner as that for the standard indicator switch assembly described in Section 21.

Vent valves

6 A vent valve is fitted to the brake and the clutch pedal. To gain access to one or both valves, first disconnect the battery earth lead then unclip and detach the lower dash trim panel on the driver's side.
7 Detach the wiring connector from the valve and pull free the vacuum hose.
8 If available use the VW special tool, number 2041 and pull the valve unit free from its bush.

Fig. 10.59 Cruise control system components and layout (Sec 38)

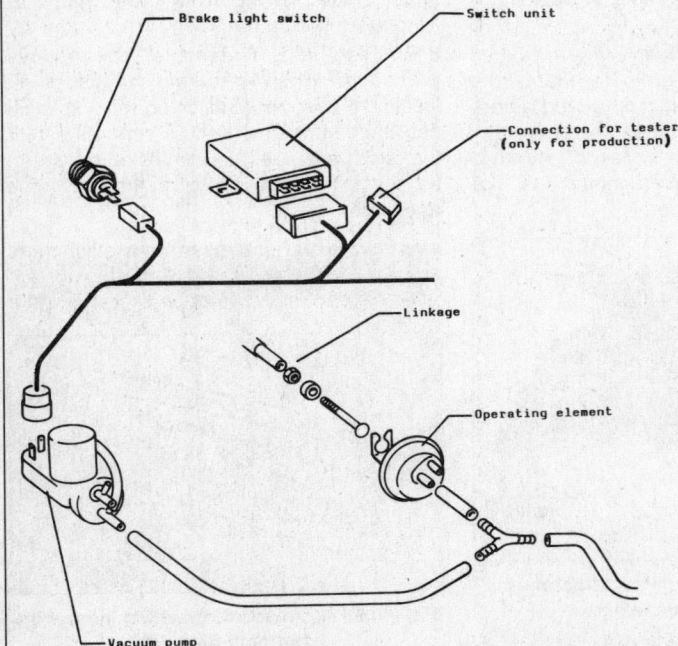

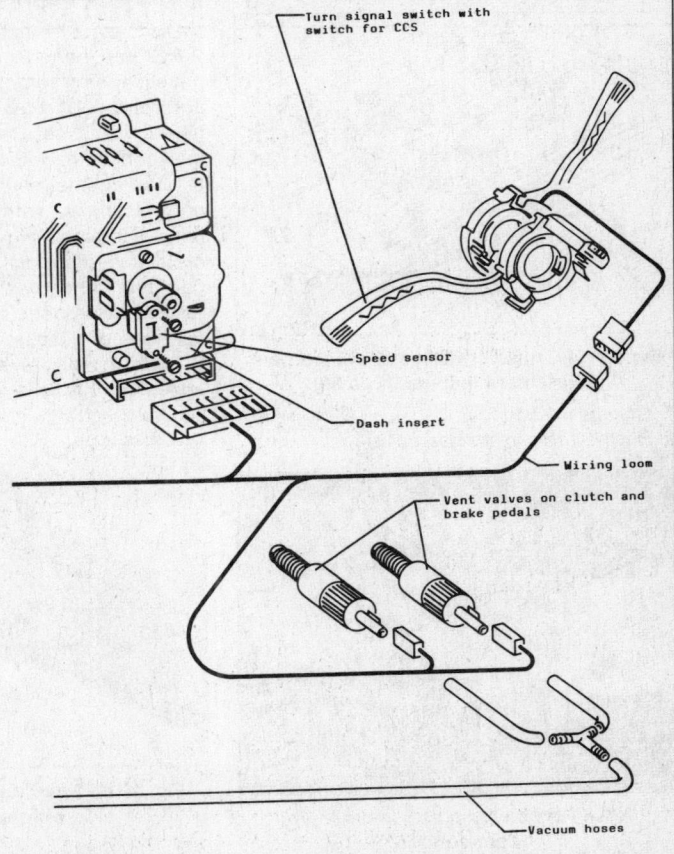

Brake light switch
Switch unit
Connection for tester (only for production)
Linkage
Operating element
Vacuum pump
Turn signal switch with switch for CCS
Speed sensor
Dash insert
Wiring loom
Vent valves on clutch and brake pedals
Vacuum hoses

10

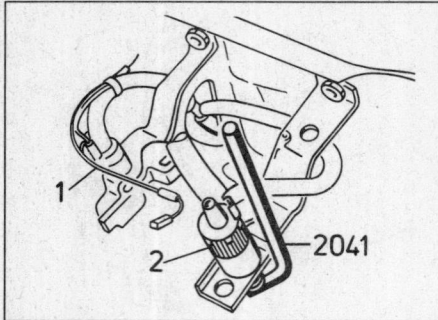

Fig. 10.61 Cruise control system vent valve removal using VW special tool 2041 (Sec 38)

1   Clutch pedal   2   Brake pedal

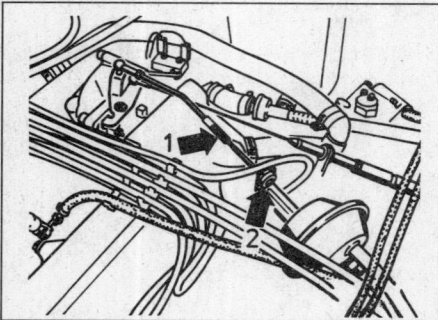

Fig. 10.62 Cruise control vent valve refitting using 10 mm socket (Sec 38)

1   Clutch pedal   2   Brake pedal

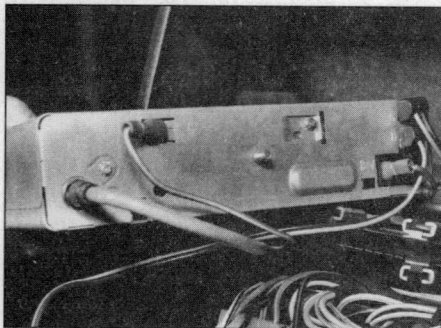

Fig. 10.63 Cruise control system operating element (Sec 38)

Once removed the valve must be renewed as the thread will have been damaged on removal (Fig. 10.61).

9 Refit the valve by initially screwing it into its location bracket by hand. When it has started, use a 10 mm socket as shown and push the valve through the bush to the stop (Fig. 10.62).

10 The valve is adjusted by pulling the appropriate pedal back fully to the stop.

Operating element

11 Unclip or unscrew the linkage, pull off the vacuum hose and remove the unit (arrowed in

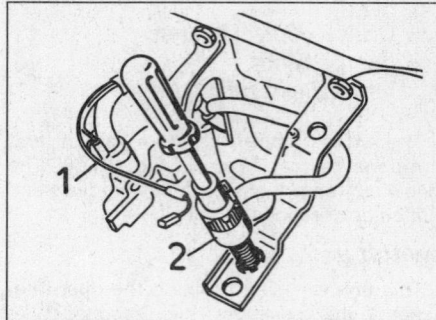

Fig. 10.64 Cruise control system element adjustment linkage (Sec 38)

1 Linkage adjuster nut
2 Bush-to-stop-plate clearance

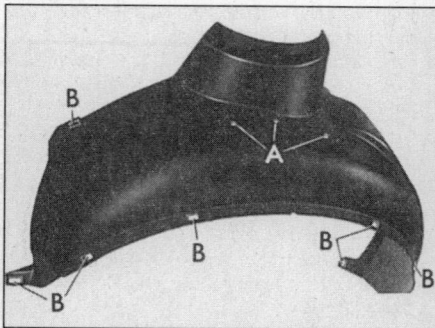

39.4 Radio wiring and aerial lead connections

Fig. 10.63) to allow removal of the element.

12 Refit in the reverse order of removal, but adjust the element linkage in the following manner.

13 Run the engine until its normal operating temperature is reached, then with the engine running at its normal idle speed turn the linkage adjuster nut to provide a bush to stop plate clearance of 0.1 to 0.3 mm (0.004 to 0.011 in).

39 Radio – removal and refitting

1 Disconnect the battery earth lead.

2 Pull free the radio control knobs. It may be necessary on some models to depress the control knob retaining pin on the shaft (through an access hole in the knob) to enable the knobs to be withdrawn.

3 Unscrew the retaining nuts from the radio control shafts, withdraw the washers and remove the radio trim panel.

4 Remove the radio attachment screws then partially withdraw the radio to enable the respective leads to be disconnected before fully removing the unit (photo).

5 Refit in the reverse order of removal ensuring that the lead connections are securely made.

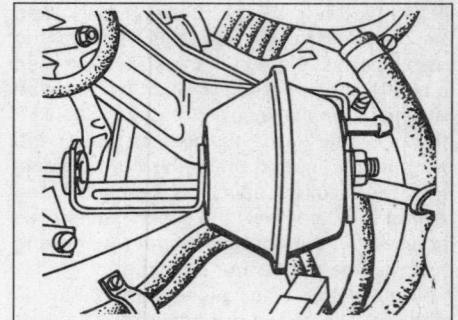

Fig. 10.65 Front wing lining attachment points (Sec 40)

A Screws   B Spreader clips

40 Radio aerial – removal and refitting

1 Disconnect the battery earth lead.

2 Remove the radio as described in the previous Section.

3 Detach and remove the instrument panel cluster as described in Section 19.

4 Pull the aerial lead through the bulkhead towards the plenum chamber.

5 Detach and remove the wheel arch lining from under the wing panel concerned. The lining is secured by five spreader clips and three Phillips screws. To remove the clips, drive the lock pin through the clip and then pull the clip free using suitable pliers (Fig. 10.65).

6 Reach up under the wing panel and pull the aerial lead through (Fig. 10.66).

7 Remove the aerial union nut and pass the aerial down through the wing panel to withdraw it.

8 Refitting is a reversal of the removal procedure. When re-routing the lead avoid locating it near the windscreen wiper linkages and the heater and fresh air controls behind the dash panel. Refit grommets to protect the aerial where it passes through the body panel apertures.

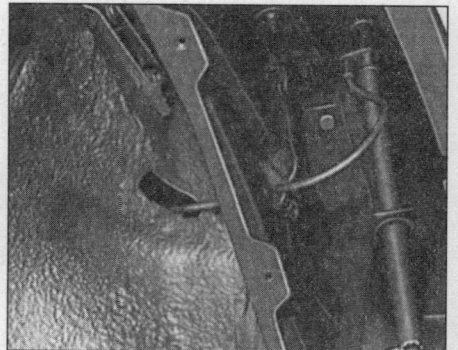

Fig. 10.66 Aerial location within inner wing housing (Sec 40)

42.3 Remove the heater motor cover

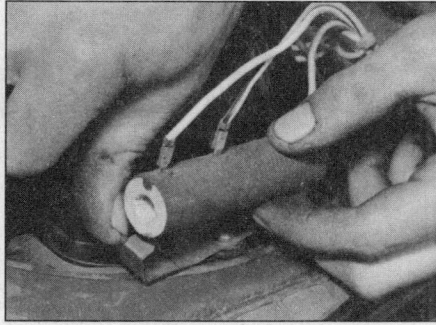

42.4 Unclip the heater motor resistor . . .

42.5. . . then unscrew and remove the resistor holder and securing screw

41 Mobile radio equipment – interference-free installation

Refer to the Haynes *In-car Entertainment Manual.*

42 Heater motor unit – removal and refitting

1 Disconnect the battery earth lead.
2 Carefully prise free the plastic screen cover between the windscreen and the bulkhead.
3 Unscrew the nuts and remove the motor cover (photo).
4 Unclip and remove the resistor leaving the wires connected (photo).
5 Remove the screw from the resistor holder (photo).
6 Withdraw the motor by tilting it on its side, then disconnect the wiring (photo).
7 Refitting is a reversal of removal. Check that the wiring connections are secure before refitting the motor cover (photo).
Note: *For details concerning the heater control cables and the heater unit refer to Chapter 12, Sections 36 and 37.*

43 Heater control switch – removal and refitting

1 Disconnect the battery earth lead.
2 Pull free the heater and fresh air control knobs.
3 Where fitted remove the radio, otherwise unclip and withdraw the glovebox above the heater controls.
4 Prise free and remove the rocker switch trim panel (below the heater unit).
5 Remove the retaining screws and withdraw the heater control panel .
6 The control switch wire multi-connector can now be detached, the switch retaining clips

42.6 Heater motor removal

compressed and the control switch withdrawn from the panel.
7 Refit in the reverse order to removal.

44 Heated rear screen – care and repair

1 Care should be taken to avoid damage to the element for the heated rear window.
2 Avoid scratching with rings on the fingers when cleaning, and do not allow luggage to rub against the glass.
3 Do not stick labels over the element on the inside of the glass.
4 A voltmeter or ohmmeter can be used to assist with location of defects. Look for an open-circuit or a sudden change in voltage, reading along the resistance wires.
5 If the element grids do become damaged, a special conductive paint is available from most motor factors to repair it.
6 To repair, degrease the affected area and wipe dry.
7 Apply tape at each side of the conductor, to mask the adjacent area.
8 Shake the repair paint thoroughly, and apply a thick coat with a fine paint brush. Allow to dry between coats. and do not apply more than three.

42.7 Heater motor installed with wiring reconnected

9 Allow to dry for at least one hour, before removing the tape.
10 Rough edges may be trimmed with a razor blade if necessary, after a drying time of some hours.
11 Do not leave the heated rear window switched on unnecessarily. as it draws a high current from the electrical system.
12 If the driver's exterior rear view mirror is electrically heated, this operates only during the period when the rear screen element is switched on.

45 Wiring diagrams – description

1 The wiring diagrams included at the end of this Manual are of the current flow type where each wire is shown in the simplest line form without crossing over other wires.
2 The fuse/relay panel is at the top of the diagram and the combined letter/figure numbers appearing on the panel terminals refer to the multiplug connector in letter form and the terminal in figure form.
3 Internal connections through electrical components are shown by a single line.
4 The encircled numbers along the bottom of the diagram indicate the earthing connecting points as given in the key.

10

Fault finding – electrical system

Starter fails to turn engine

☐ Battery discharged or defective
☐ Battery terminal and/or earth leads loose
☐ Starter motor connections loose
☐ Starter solenoid faulty
☐ Starter brushes worn or sticking
☐ Starter commutator dirty or worn
☐ Starter field coils earthed
☐ Starter armature faulty
☐ Automatic stop/start system fault (where applicable)

Starter turns engine very slowly

☐ Battery discharged
☐ Starter motor connections loose
☐ Starter brushes worn or sticking

Starter noisy

☐ Pinion or ring gear teeth badly worn
☐ Mounting bolts loose

Battery will not hold charge

☐ Plates defective
☐ Electrolyte level too low
☐ Alternator drivebelt slipping
☐ Alternator or regulator faulty
☐ Short in electrical circuit

Ignition light stays on

☐ Alternator faulty
☐ Alternator drivebelt broken

Ignition light fails to come on

☐ Warning bulb blown
☐ Warning light open circuit
☐ Alternator faulty

Instrument readings increase with engine speed

☐ Voltage stabilizer faulty

Fuel on temperature gauge gives no reading

☐ Wiring open circuit
☐ Sender unit faulty

Fuel on temperature gauge gives maximum reading all the time

☐ Wiring short circuit
☐ Sender unit or gauge faulty

Lights inoperative

☐ Bulb blown
☐ Fuse blown
☐ Switch faulty
☐ Wiring open circuit
☐ Connection corroded

Failure of component motor

☐ Fuse blown
☐ Wiring loose or broken
☐ Brushes sticking or worn
☐ Armature shaft bearings seized
☐ Field coils faulty
☐ Commutator dirty or burnt
☐ Armature faulty

Failure of an individual component

☐ Wiring loose or broken
☐ Fuse blown
☐ Switch faulty
☐ Component faulty

Chapter 11 Suspension and steering

For modifications, and information applicable to later models, see Supplement at end of manual

Contents

Degrees of difficulty

Easy, suitable for novice with little experience	**Fairly easy,** suitable for beginner with some experience	**Fairly difficult,** suitable for competent DIY mechanic	**Difficult,** suitable for experienced DIY mechanic	**Very difficult,** suitable for expert DIY or professional

Specifications

Front suspension
Type .... Independent with coil spring struts incorporating telescopic shock absorbers, lower wishbones and anti-roll bar

Rear suspension
Type .... Transverse torsion axle incorporating trailing arms, telescopic shock absorbers with externally mounted coil springs

Steering
Type .... Rack-and-pinion, tie-rods connected to coupling attached to rack, power assistance or steering damper fitted to some models

Turning circle .... 10.7 m (35.1 feet)

Power steering
Overall ratio .... 18.5 to 1
Steering wheel turns lock to lock 3.45
Pump fluid type .... ATF Dexron
Capacity .... 0.65 litre (1.1 Imp pint)

Front wheel alignment
Total toe-in .... +10' ± 10' (1.0 ± 1.0 mm/0.04 ± 0.04 in)
Toe angle difference, left and right, at 20° lock .... –50' ± 30'
Camber .... –40' ± 30'
Maximum camber difference between sides .... 30'
Castor .... +30' ± 30'
Maximum castor difference between sides .... 30'

Rear wheel alignment
Camber .... –1° 40' ± 20'
Maximum camber difference between sides .... 30'
Total toe-out .... 25 ± 15'
Maximum toe-out difference between sides .... 25'

11

Wheels

Type .... Pressed steel or light alloy
Size .... 5J x 13 or 5 1/2 J x 13

Tyres

Size .... 165 SR 13 82S or 185/70 SR 13 84 S

Pressures – bar (lbf/in²) cold tyres:	Front	Rear
Half load	1.8 (26)	1.8 (26)
Full load	1.9 (27)	2.3 (33)*
*Rear pressure, full load, Estate model	2.6 (37)	

Note: Increase pressures by 0.2 bar (3 lbf/in²) for winter grade tyres

Torque wrench settings

	Nm	lbf ft
Front suspension		
Strut to body	60	44
Strut bearing to piston rod	50	37
Shock absorber screw cap	150	110
Suspension balljoint to strut	50	37
Suspension balljoint to wishbone	65	48
Wishbone to subframe	60	44
Subframe to body	70	52
Anti-roll bar	25	18
Splash plate to wheel bearing housing	10	7
Rear suspension		
Trailing arm bracket to body	45	33
Trailing arm to bracket pivot bolt	70	52
Shock absorber to rear axle	70	52
Shock absorber to body nut	35	25
Brake backplate/stub axle	60	44
Steering		
Flange tube (manual) to pinion	25	18
Flange tube (power assisted) to pinion	30	22
Steering wheel	40	30
Column tube bolt (cheese head)	20	15
Steering lock bolt	10	7
Coupling bolts (power assisted)	25	18
Flange tube to coupling clamp (power assisted)	30	22
Bearing flange (power assisted)	10	7
Steering gear to body nut	35	26
Steering gear to side body nut	20	15
Tie-rod end to bracket nut	45	33
Tie-rod end balljoint nut	30	22
Tie-rod end balljoint lock nuts	40	30
Damper to steering gear bolt	35	26
Steering gear adjuster plate bolts	20	15
Power steering pressure line union	40	30
Power steering return line union	30	22
Power steering valve body bolts	20	15
Roadwheel bolts	110	81

1 General description

The front suspension is of independent type, incorporating coil spring struts and lower wishbones. The struts are fitted with telescopic shock absorbers and both front suspension units are mounted on a subframe. An anti-roll bar is fitted to the lower wishbones (Fig. 11.1).

The rear suspension consists of a transverse torsion axle with trailing arms rubber bushed to the body. The axle is attached to the lower ends of the shock absorbers, which act as struts, since they incorporate mountings for the coil springs (Fig. 11.2).

The steering is of rack-and-pinion type mounted on the bulkhead. The tie-rods are attached to a single coupling which is itself bolted to the steering rack. Power assistance is fitted to some models, and a steering damper is fitted to some models with manual steering.

2 Maintenance – suspension and steering

The following routine maintenance procedures should be undertaken at the specified intervals given at the start of this manual.

Tyres

1 Check and if necessary adjust the tyre pressures. Check the condition and general wear characteristics of the tyres and if

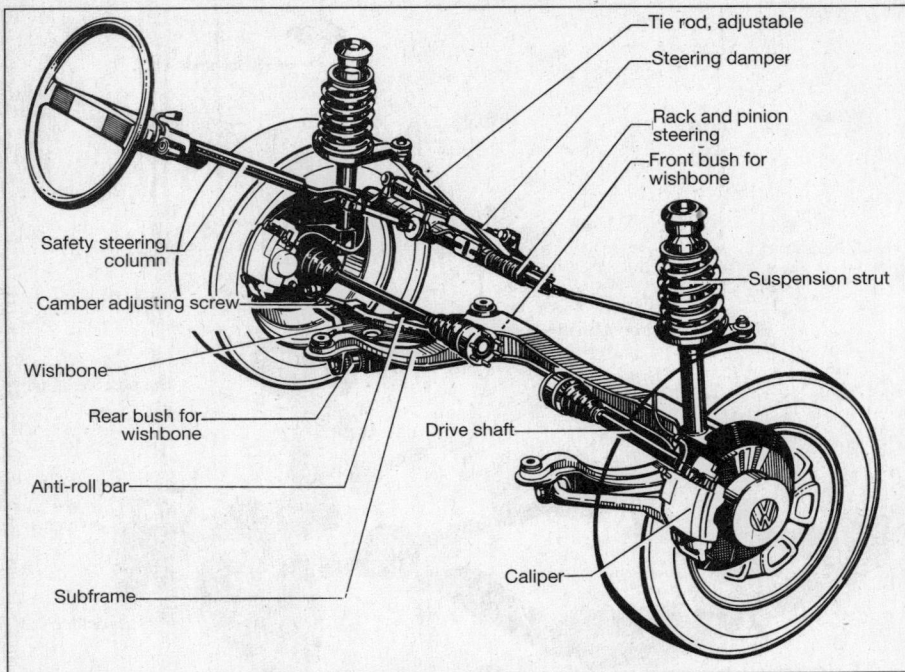

Fig. 11.1 Front suspension and steering layout (Sec 1)

Labels on Fig. 11.1:
- Tie rod, adjustable
- Steering damper
- Rack and pinion steering
- Front bush for wishbone
- Suspension strut
- Safety steering column
- Camber adjusting screw
- Wishbone
- Rear bush for wishbone
- Drive shaft
- Anti-roll bar
- Caliper
- Subframe

Fig. 11.3 Tyre tread wear indicator strips show when tyres need renewal (Sec 2)

Fig. 11.4 Power steering fluid reservoir (Sec 2)

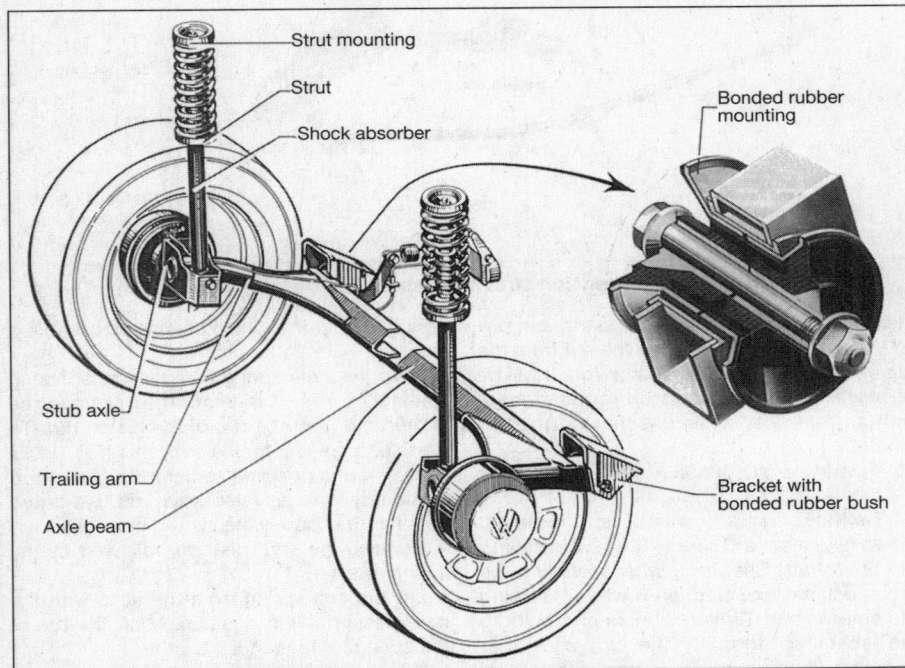

Fig. 11.2 Rear suspension layout (Sec 1)

Labels on Fig. 11.2:
- Strut mounting
- Strut
- Shock absorber
- Bonded rubber mounting
- Stub axle
- Trailing arm
- Axle beam
- Bracket with bonded rubber bush

2.2 Front suspension and steering check points (arrowed)

Steering

3 Inspect the steering tie-rod end balljoints for signs of excessive wear and the dust covers for splits or deterioration and leakage. Inspect the steering gear bellows for signs of leakage or splitting. Check that all steering joints and mountings are secure.

Power assisted steering

4 Check the power steering reservoir fluid level and top up using only the specified type of fluid if necessary. Check the power steering hoses and joints for condition and security. Check the condition and tension adjustment of the power steering pump drivebelt. Adjust or if necessary renew it as described in Section 22.

11

wearing unevenly have the steering and suspension alignment checked. Refer to Section 24 for further details on wheels and tyres.

Suspension

2 Raise and support each end of the vehicle in turn and inspect the suspension and steering components for signs of excessive wear or damage. Inspect the suspension balljoints for wear and the dust covers for any signs of splits or deterioration. Renew if necessary. Check the wishbone and anti-roll bar mounting/pivot bushes for signs of excessive wear and/or deterioration and again renew if necessary. Check the shock absorbers for signs of leakage and the suspension to body mountings for signs of corrosion (photo).

3.7 Suspension balljoint and clamp bolt (viewed from rear)

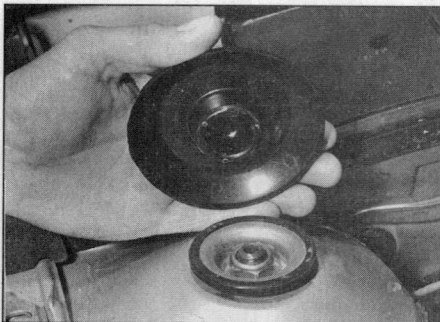

3.12 Removing the front suspension strut dust cap

3 Front suspension strut – removal and refitting

1 Prise the cap from the centre of the roadwheel, then unscrew the driveshaft nut. *This is tightened to a very high torque and no attempt to loosen it must be made unless the full weight of the car is on the road wheels.*

2 Loosen the four bolts securing the roadwheel.

3 Jack up the car and support the body securely on stands, or wood blocks.

4 Remove the roadwheel.

5 Remove the brake caliper with reference to Chapter 9, but do not disconnect the hydraulic hose. Support the caliper on a stand without straining the hose.

6 Detach the hose bracket from the strut.

7 Unscrew and remove the suspension balljoint clamp bolt, noting that its head faces forward (photo).

8 Using a balljoint removing tool, as described in Section 14, unscrew the nut and disconnect the tie-rod end from the strut/wheel bearing housing.

9 Unscrew the anti-roll bar mounting bolts and remove the bar, with reference to Section 6.

10 Remove the driveshaft nut, then push the wishbone down from the strut.

11 Using a hub puller, press the driveshaft out of the front wheel hub.

12 Support the suspension strut from below,

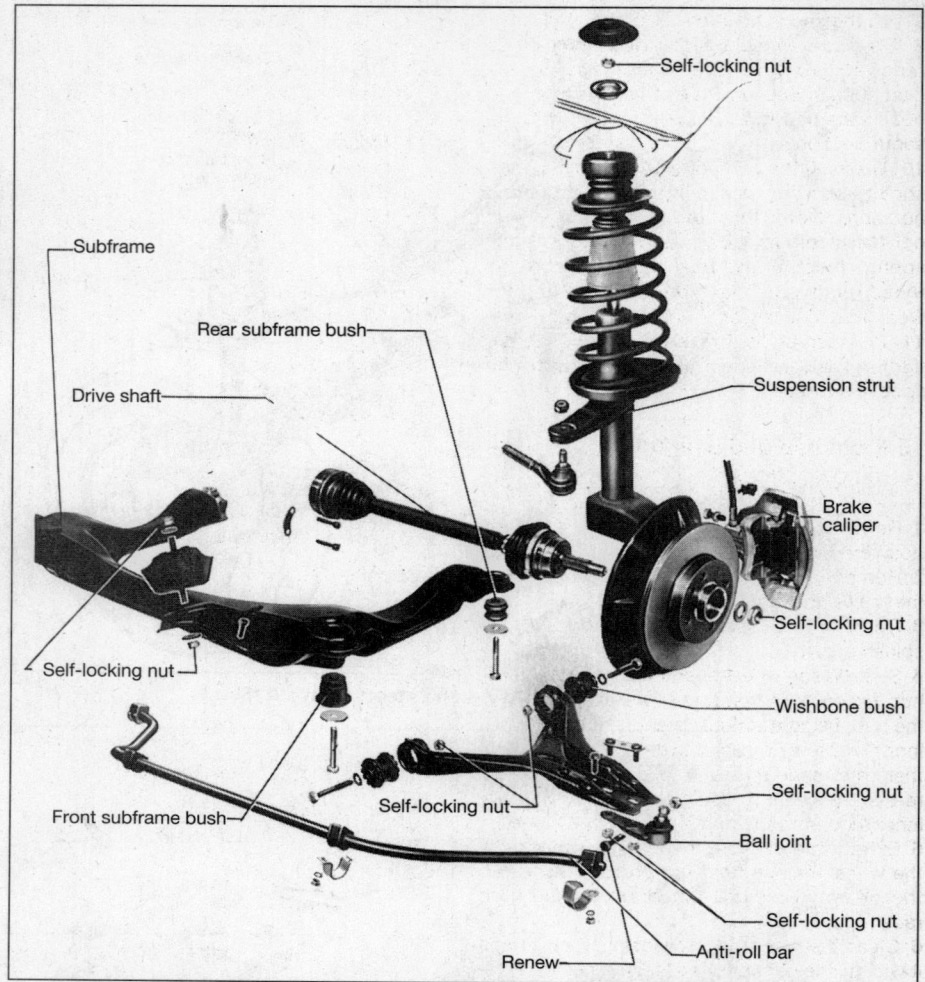

Fig. 11.5 Front suspension strut and associated components (Sec 3)

then, working in the engine compartment, pull off the dust cap and unscrew the nut from the top of the shock absorber piston rod. It will be necessary to hold the piston rod stationary with an Allen key while unscrewing the nut (photo).

13 Lower the suspension strut from under the car.

14 Refitting is a reversal of removal. However, refer to Chapter 8 when installing the driveshaft, Chapter 9 when installing the brake caliper, and Section 6 when installing the anti-roll bar. Tighten all nuts and bolts to the specified torque. The upper rubber damping ring should be dusted with talcum powder before fitting.

4 Front suspension strut – dismantling and reassembly

1 Do not attempt to dismantle the suspension strut unless a spring compressor has been fitted. If you have no special compressor, take the strut to a garage for dismantling.

2 With the spring compressor in place on the

spring, clamp the wheel bearing housing in a vice.

3 Compress the spring until the upper spring retainer is free of tension, then remove the slotted nut from the top of the piston rod. To do this, a special tool is available (Fig. 11.7). However it is possible to hold the piston rod stationary with an Allen key, and use a peg spanner to unscrew the slotted nut.

4 Remove the strut bearing, followed by the spring retainer.

5 Lift the coil spring from the strut with the compressor still in position. Mark the top of the spring for reference.

6 Withdraw the bump stop components from the piston rod, noting their order of removal.

7 Move the shock absorber piston rod up and down through its complete stroke and check that the resistance is even and smooth. If there are any signs of seizing or lack of resistance, or if fluid has been leaking excessively, the shock absorber should be renewed. To do this, unscrew the screw cap using a hexagon box spanner, or the special VW tool No 40-201 A. The cap is very tight, so the tool used must fit accurately.

8 Pull the shock absorber from the strut.

9 If the strut wheel bearing housing is to be renewed, remove the wheel bearing as described in Section 5 then fit the bearing and hub to the new housing, with reference to the same Section.

10 Clean all the components in particular the shock absorber recess in the wheel bearing housing. Note that the coil springs are normally colour coded, and therefore new springs must always bear the same code as those removed. Check the strut bearing for wear, and if necessary, renew it.

11 Reassembly is a reversal of removal, but tighten the screw cap and slotted nut to the specified torques.

5 Front hub and bearing –
removal and refitting

1 Remove the front suspension strut as described in Section 3.

2 Remove the crossheaded screw and tap the brake disc from the hub (Fig. 11.8).

3 Remove the screws and withdraw the splash guard.

4 Support the wheel bearing housing with the hub facing downward, and press or drive out the hub, using a suitable mandrel. The bearing inner race will remain on the hub, and therefore, once removed, it is not possible to re-use the bearing. Use a puller to remove the inner race from the hub.

5 Extract the circlips, then, while supporting the wheel bearing housing, press or drive out the bearing, using a mandrel on the outer race.

6 Clean the recess in the housing, then smear it with a little general purpose grease. Where a new wheel bearing kit has been obtained, the kit will contain a sachet of Molypaste. Smear some Molypaste onto the bearing seat (not the bearing).

7 Fit the outer circlip, then support the wheel bearing housing and press or drive in the new bearing, using a metal tube *on the outer race only*.

8 Fit the inner circlip making sure that it is correctly seated.

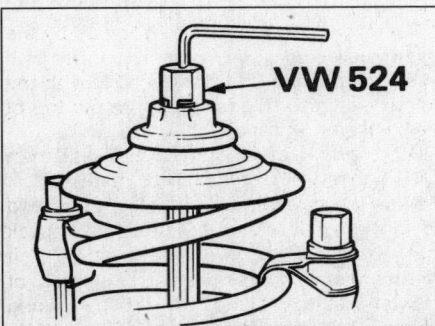

Fig. 11.7 Using the special tool to unscrew the slotted nut from the front suspension. A peg spanner and Allen key will suffice if necessary – see text (Sec 4)

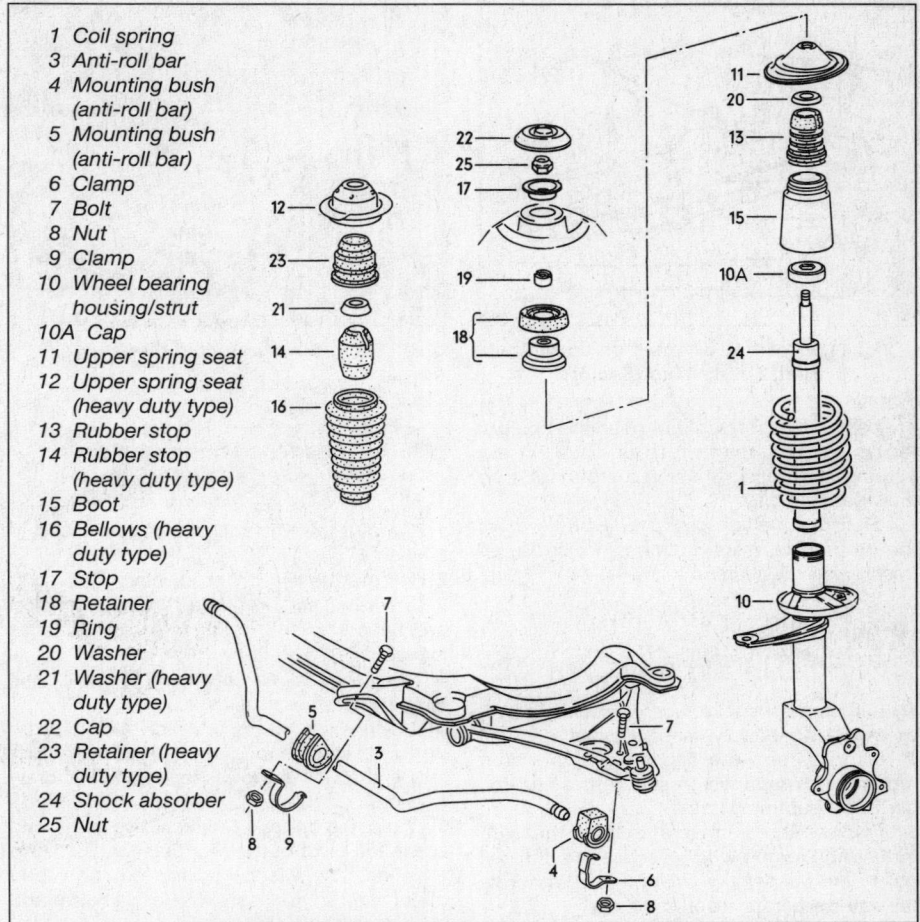

1 Coil spring
3 Anti-roll bar
4 Mounting bush (anti-roll bar)
5 Mounting bush (anti-roll bar)
6 Clamp
7 Bolt
8 Nut
9 Clamp
10 Wheel bearing housing/strut
10A Cap
11 Upper spring seat
12 Upper spring seat (heavy duty type)
13 Rubber stop
14 Rubber stop (heavy duty type)
15 Boot
16 Bellows (heavy duty type)
17 Stop
18 Retainer
19 Ring
20 Washer
21 Washer (heavy duty type)
22 Cap
23 Retainer (heavy duty type)
24 Shock absorber
25 Nut

Fig. 11.6 Exploded view of the front suspension strut (Sec 4)

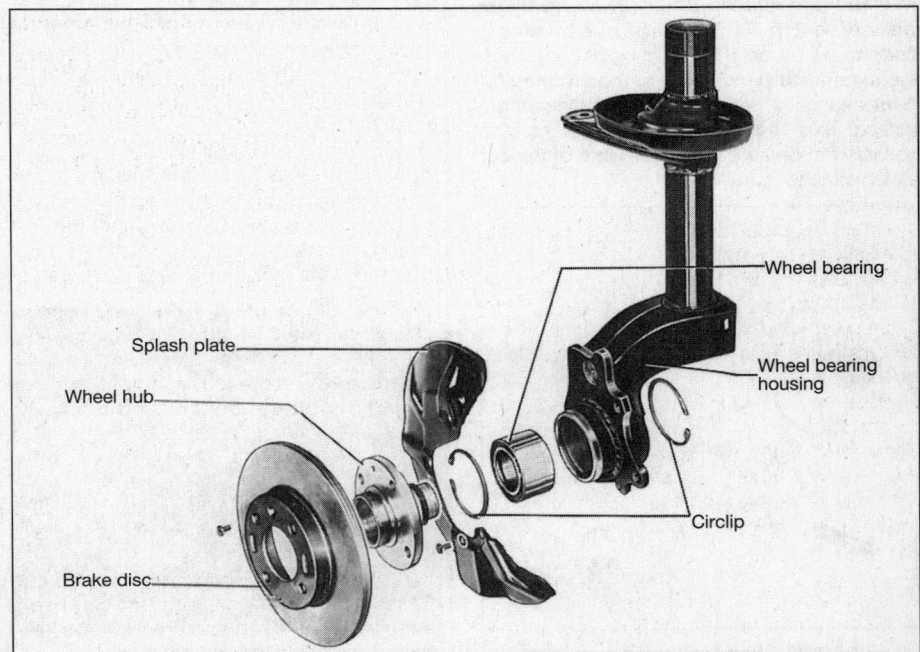

Fig. 11.8 Front hub and bearing components (Sec 5)

11

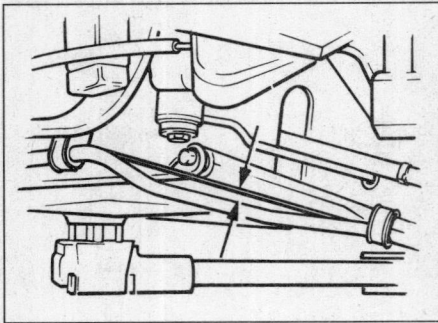

Fig. 11.9 Position of offset on correctly fitted anti-roll bar (Sec 6)

6.2 Anti-roll bar front mounting

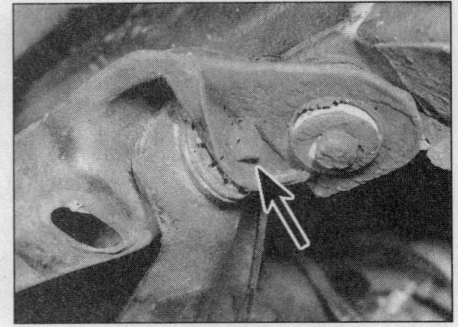

7.5 Front suspension wishbone pivot bolt (arrowed)

9 Position the hub with its bearing shoulder facing upward, then press or drive on the bearing and housing, using a metal tube *on the inner race only*.

10 Refit the splash guard and brake disc, then refit the front suspension strut, as described in Section 3.

6 Anti-roll bar – removal and refitting

1 Jack up the front of the car and support it on axle stands. Apply the handbrake.

2 Unscrew the nuts and extract the front mounting clamps which secure the anti-roll bar to the subframe (photo).

3 Unscrew the nuts and extract the side mounting clamps which secure the anti-roll bar to the suspension wishbones. Withdraw the anti-roll bar from under the car.

4 Refitting is a reversal of removal, but make sure that the anti-roll bar is fitted the correct way round, with the offset away from the wishbones (Fig. 11.9). Initially, the mounting bolts must be only partially tightened. Then the car should be lowered to the ground and 'bounced' a few times to settle the mountings, before fully tightening the bolts to the specified torque with the full weight of the car on its wheels.

7 Front suspension wishbone – removal, overhaul and refitting

1 Remove the anti-roll bar as described in Section 6.

2 Remove the relevant roadwheel.

3 Unscrew and remove the suspension balljoint clamp bolt, noting that its head faces forward.

4 Tap the wishbone downward to release the balljoint from the strut.

5 Unscrew and remove the pivot bolts from the subframe, noting the position of their heads, then lower the wishbone and withdraw it (photo).

6 Check the balljoint for excessive wear, and check the pivot bushes for deterioration. Also examine the wishbone for damage and distortion. If necessary, the balljoint and bushes should be renewed.

7 To renew the balljoint, first outline its exact position on the wishbone. This is important as the holes in the wishbone are elongated to allow camber adjustment. Unscrew the nuts and remove the balljoint and clamp plate. Fit the new balljoint in the exact outline, and tighten the nuts.

8 To renew the pivot bushes, use a long bolt together with a metal tube and washers to pull each bush from the wishbone eyes. Fit the new bushes using the same method, but first dip them in soapy water.

9 Refitting the wishbone is a reversal of

removal, but delay tightening the pivot bolts until the weight of the car is on the suspension. Refer to Section 6 when refitting the anti-roll bar. Have the front wheel camber angle checked and, if necessary, adjusted by a VW dealer.

8 Rear axle – removal and refitting

1 Jack up the rear of the car, and support it with axle stands positioned beneath the underbody. Chock the front wheels and remove the rear wheels.

2 Unhook the rubber mounting rings and lower the rear exhaust silencer as far as possible. Support the silencer on an axle stand.

3 Unscrew the nut from the handbrake lever rod and remove the compensator bar.

4 Prise the handbrake cable guide bushes from the underbody brackets, and remove the cables from the rear brackets and trailing arm pivot retainer each side (photo).

5 Remove the brake fluid reservoir filler cap and tighten it down onto a piece of polythene sheeting placed over the reservoir filler hose in order to prevent the loss of fluid. Alternatively fit brake hose clamps to both rear brake hoses.

6 Disconnect both rear brake hoses from the trailing arms by unscrewing the rigid brake line union nuts, while holding the hose end

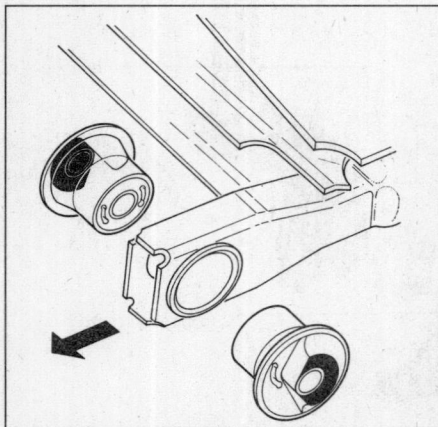

Fig. 11.10 Diagram showing correct positioning of rear axle pivot bushes (Sec 8)

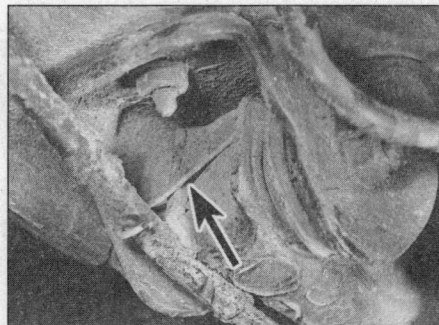

8.4 Handbrake cable locating retainer on trailing arm pivot (arrowed)

8.6 Detach the handbrake cable and brake line from trailing arm

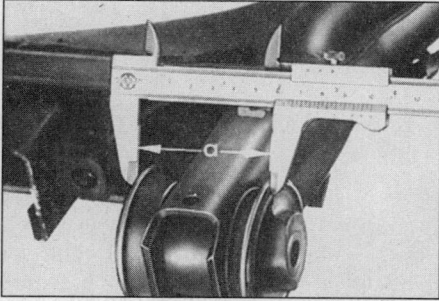

Fig. 11.11 Pivot bushes installed dimension (a) (Sec 8)

a = 61.6 to 62.0 mm (2.42 to 2.44 in)

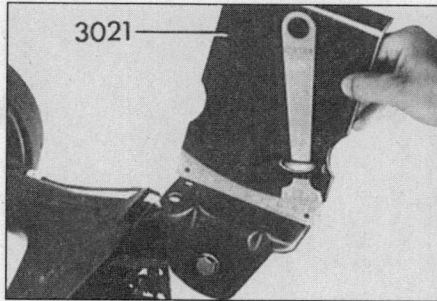

Fig. 11.12 Bracket to axle angle must be 15 to 19°. Adjust using VW tool 3021 if available (Sec 8)

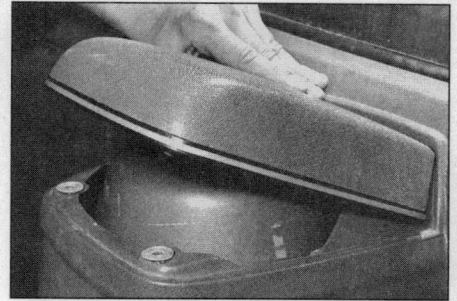

9.1 Detach the rear shock absorber upper mounting trim cover (Estate shown) . . .

unions stationary with a spanner. Remove the clamp plates (photo).

7 Loosen the nuts on the pivot bolts at the front of the trailing arms.

8 Unhook the brake pressure regulator spring from the rear axle.

9 Support the rear axle with a trolley jack, and place axle stands beneath the trailing arms.

10 Unscrew and remove the rear shock absorber lower mounting bolts.

11 Unscrew the nuts and remove the pivot bolts from the trailing arms.

12 With the help of an assistant if necessary, lower the rear axle to the ground while guiding the handbrake cable over the exhaust pipe.

13 If necessary, remove the stub axles, with reference to Section 11, also remove the brake lines and handbrake cables, if required. The pivot bushes may be renewed using a long bolt and nut, metal tube, and packing washers, but make sure that the bush gaps are in line with the trailing arms and press the bushes in flush (Fig. 11.10).

14 The trailing arm pivot brackets can be removed from the chassis member each side by unscrewing the three retaining nuts. If the brackets are to be refitted to the trailing arms prior to refitting the rear axle unit, they must be secured at an angle of between 15 to 19° to the axle as shown in Fig. 11.12.

15 Refitting is a reversal of removal, but delay tightening the rear axle and shock absorber mounting bolts until the weight of the car is on the suspension. Bleed the hydraulic system and adjust the handbrake cable, as described in Chapter 9.

9 Rear suspension strut/shock absorber – removal and refitting

1 Detach the trim cover from the top of the shock absorber within the luggage compartment (photo). On early Saloon models it will be necessary to remove the rear seat,

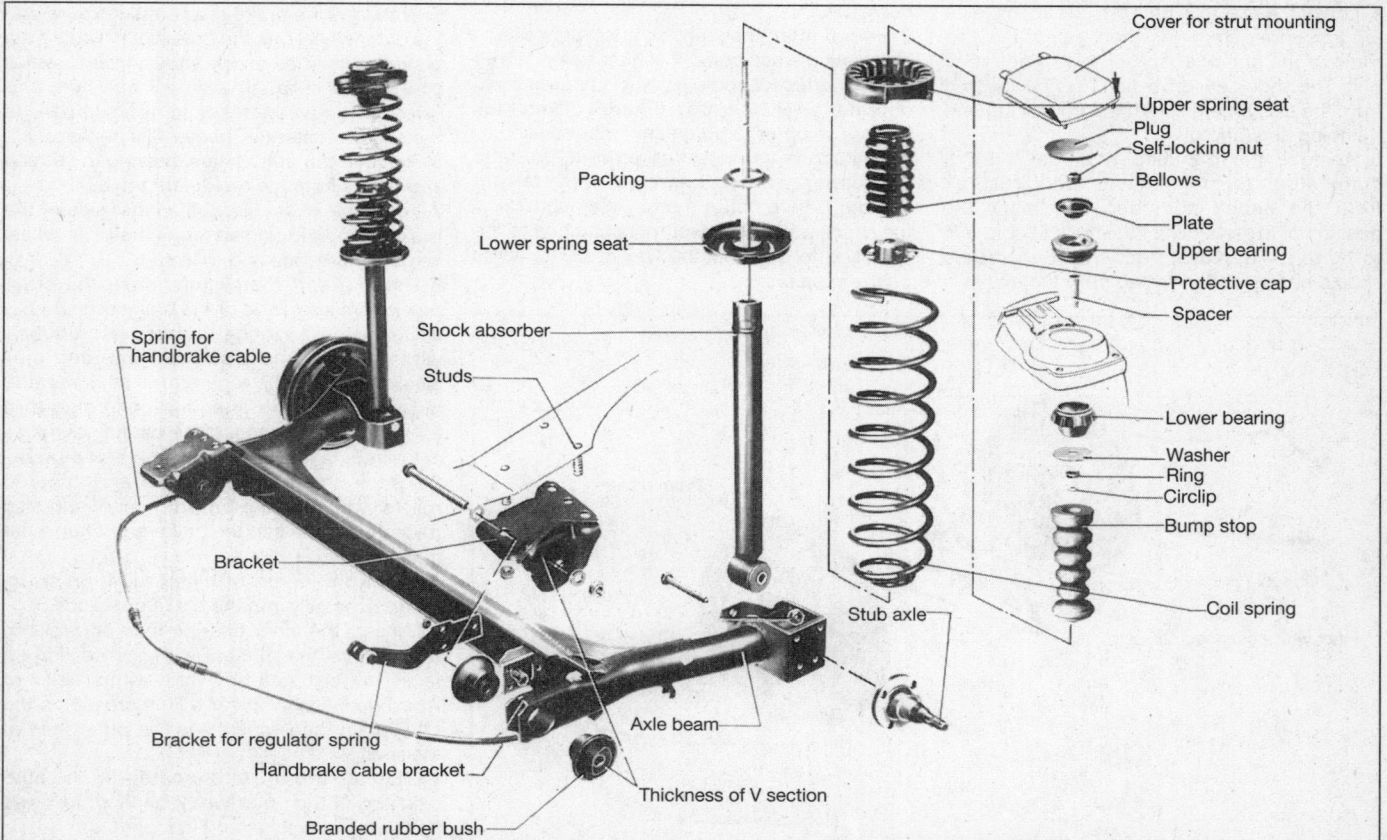

Spring for handbrake cable

Packing

Lower spring seat

Shock absorber

Studs

Bracket

Bracket for regulator spring

Handbrake cable bracket

Branded rubber bush

Thickness of V section

Axle beam

Stub axle

Cover for strut mounting

Upper spring seat

Plug

Self-locking nut

Bellows

Plate

Upper bearing

Protective cap

Spacer

Lower bearing

Washer

Ring

Circlip

Bump stop

Coil spring

Fig. 11.13 Rear suspension and strut assembly components (Sec 9)

11

9.2 . . . for access to the retaining nut

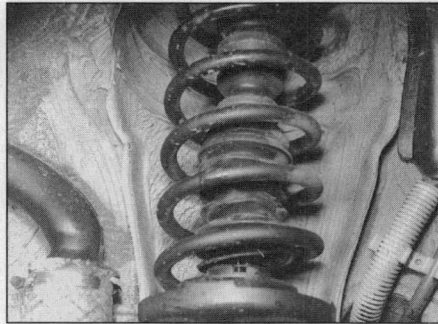

9.4 Rear suspension strut and coil spring

10.13A Lubricate the hub and stub axle . . .

backrest and parcel shelf to gain access to the upper end of the strut mounting. On later models produced from March of 1982 the rear parcel shelf was modified in that it has a diagonal 'break line' at each forward corner and it can be prised up at this point and hinged back to provide access to the strut upper mounting.

2 Unscrew the nut from the top of the shock absorber piston rod and remove the dished washer and rubber bearing (photo).

3 Jack up the rear of the car using a trolley jack beneath the rear axle, and support the car with axle stands positioned beneath the underbody. Chock the front wheels.

4 Lower the rear axle until the coil spring is no longer under tension then unscrew and remove the shock absorber lower mounting bolt. The shock absorber can now be lowered from the car, together with the coil spring and damping ring (photo).

5 Remove the damping ring, coil spring, bump stop components and spring retainer from the shock absorber, but mark the position of the retainer if the shock absorber is to be re-used. **Note:** *Both shock absorbers should not be disconnected from the rear axle*

at the same time, as the rear brake hydraulic hoses and the axle beam bushes may be damaged.

6 Clean all the components, and examine them for wear and damage. Note that the coil springs are normally colour coded, and therefore new springs must always bear the same code as those removed. With the shock absorber vertical, move the piston rod up and down through its complete stroke and check that the resistance is even and smooth. If there are any signs of seizing, or lack of resistance, or if fluid has been leaking excessively, the shock absorber should be renewed. Renew all components as necessary.

7 Before fitting new shock absorbers check their action when vertical – after being stored on their sides for some time, it is necessary to operate the piston rods through several full strokes, in order to purge any trapped air.

8 Refitting is a reversal of removal, but dust the upper rubber bearings with talcum powder before fitting them. Delay tightening the shock absorber lower mounting bolt to the specified torque until the weight of the car is on the suspension.

10 Rear wheel bearings – removal, refitting and adjustment

1 Chock the front wheels, then jack up the rear of the car and support it on axle stands. Remove the rear roadwheel, and release the handbrake.

2 Remove the cap from the centre of the brake drum by tapping it on alternate sides with a screwdriver or blunt chisel (Fig. 11.14).

3 Extract the split pin and remove the locking ring.

4 Unscrew the nut and remove the thrust washer.

5 Withdraw the brake drum, making sure that the outer wheel bearing does not fall out. If the brake drum binds on the shoes, insert a small screwdriver through a wheel bolt hole and lever up the wedged key in order to release the shoes – refer to Chapter 9, if necessary.

6 Remove the outer wheel bearing inner race and rollers from the brake drum/hub.

7 Lever the oil seal from the inner side of the brake drum/hub and withdraw the inner wheel bearing inner race and rollers.

8 Using a soft metal drift, drive the outer races from each side of the brake drum/hub.

9 Clean the bearings and brake drum/hub with paraffin, and also wipe the stub axle clean. Examine the tapered rollers, inner and outer races, brake drum/hub, and stub axle for wear and damage. If the bearing surfaces are pitted, renew them. Obtain a new oil seal.

10 Pack the bearing cages and tapered rollers with a lithium based grease, and also pack the grease into the brake drum/hub inner cavity.

11 Using a length of metal tube, drive the outer races fully into the brake drum/hub.

12 Insert the inner bearing race and rollers, then locate the oil seal with the sealing lip facing inward, and drive it in with a block of wood until flush. Smear a little grease on the oil seal lip then wipe clean the outer face of the seal.

13 Locate the brake drum/hub on the stub axle and fit the outer wheel bearing, followed by the thrust washer and nut (photos).

14 Tighten the nut firmly while turning the drum/hub, then back off the nut until the

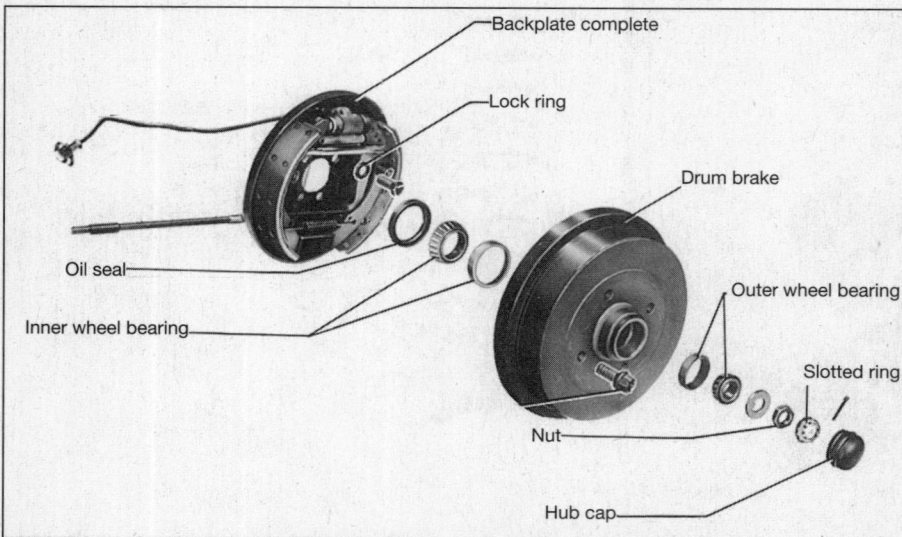

Fig. 11.14 Rear wheel hub and bearing components (Sec 10)

Backplate complete

Lock ring

Drum brake

Oil seal

Inner wheel bearing

Outer wheel bearing

Slotted ring

Nut

Hub cap

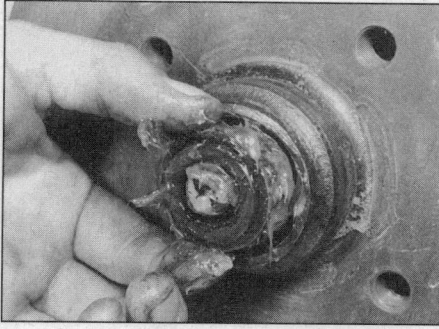
10.13B . . . locate the bearing . . .

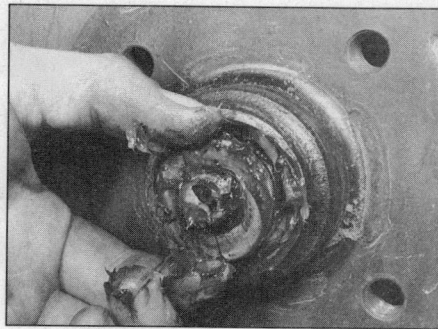
10.13C . . . the thrust washer and nut

10.14 Check that the thrust washer can still be rotated after fitting the nut

thrust washer can just be moved by pressing on it with a screwdriver. Do not lever or twist the screwdriver in an attempt to move the thrust washer (photo).

15 Fit the locking ring without moving the nut, and install a new split pin (photo).

16 Tap the grease cap onto the drum/hub (photo).

17 Refit the roadwheel and lower the vehicle to the ground.

11 Rear stub axle – removal and refitting

1 Proceed as described in the previous Section in paragraphs 1 to 5 inclusive.

2 Remove the brake shoes as described in Chapter 9, Section 7.

3 Remove the brake fluid reservoir filler cap, and tighten it down onto a piece of polythene sheeting over the reservoir filler hole in order to prevent the loss of fluid in the following procedure. Alternatively fit a brake hose clamp on the hydraulic hose between the rear axle and underbody.

4 Unscrew the brake pipe union on the rear of the wheel cylinder.

5 Unscrew the bolts and withdraw the brake backplate and stub axle from the rear axle mounting plate, at the same time disconnecting the handbrake cable.

6 Refitting the stub axle is a reversal of the removal procedure, but make sure that the mating faces are clean. Tighten the bolts to the specified torque. Refit the rear brake shoes referring to Chapter 9.

7 Refit the brake drum and adjust the wheel bearings as described in the previous Section.

8 Bleed the brake hydraulic system as described in Chapter 9 then depress the footbrake pedal firmly to set the rear brake shoes.

9 Refit the roadwheel and lower the vehicle to the ground.

12 Steering wheel – removal and refitting

1 Centralise the front wheels to the straight-ahead position.

2 Prise free the horn bar from the centre of the steering wheel.

3 The indicator (turn signal) switch should be in the neutral position.

4 Mark the steering wheel and inner column in relation to each other, then unscrew and remove the retaining nut and washer (photo).

5 Withdraw the steering wheel from the inner column splines. If it is tight, ease it off by rocking from side to side.

6 Refitting is a reversal of removal, but align the previously made marks, and tighten the retaining nut to the specified torque. Make sure that the turn signal lever is in neutral, otherwise the switch arm may be damaged.

13 Steering column – removal, overhaul and refitting

1 Disconnect the battery earth lead.

2 Remove the steering wheel as described in Section 12.

3 Detach and remove the lower trim panel beneath the dashboard.

4 Remove the lower shroud from the upper end of the steering column. The shroud is secured by recessed screws.

5 Disconnect the multi-plugs from the ignition switch and combination switch.

6 Remove the three screws, and withdraw the combination switch (Fig. 11.15).

7 On manual steering models unscrew and remove the clamp bolt securing the flange tube to the steering gear pinion, then, using a soft metal drift, drive the flange tube upward until it is disconnected from the inner steering column. If necessary the flange tube can be removed completely by removing the clamp and pulling it through the grommet (photo).

8 On power-assisted steering models mark the relative positions of the pinion and flange tube coupling, then remove the coupling flange clamp bolt and self-locking nut. Note that all steering column self-locking nuts must be renewed on reassembly. Unscrew the column mounting bolts – one of these requires an Allen key, and the remaining shear-bolt requires the use of an 8.5 mm drill

10.15 Fit the lock ring and split pin to secure . . .

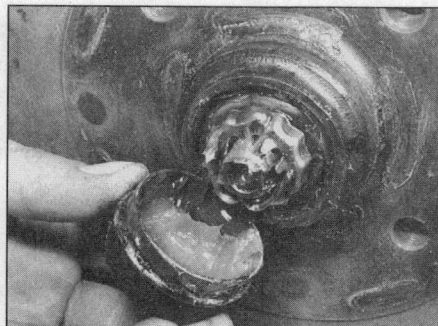
10.16 . . . then refit the grease cap

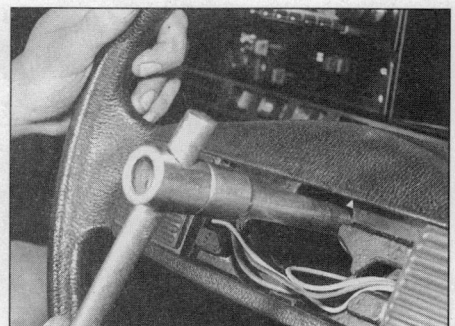
12.4 Unscrew and remove the steering wheel nut

11

Steering wheel

Horn bar

Steering column switch

Slip ring

Trim

Spring (6)

Lock washer (7)

Contact ring (5)

Steering lock housing (2)

Support ring (4)

Choke knob

Column tube (1)

Cheese head screw

Shear bolt

Flange tube

Bush

Trim

Steering column (3)

Clamp

Grommet

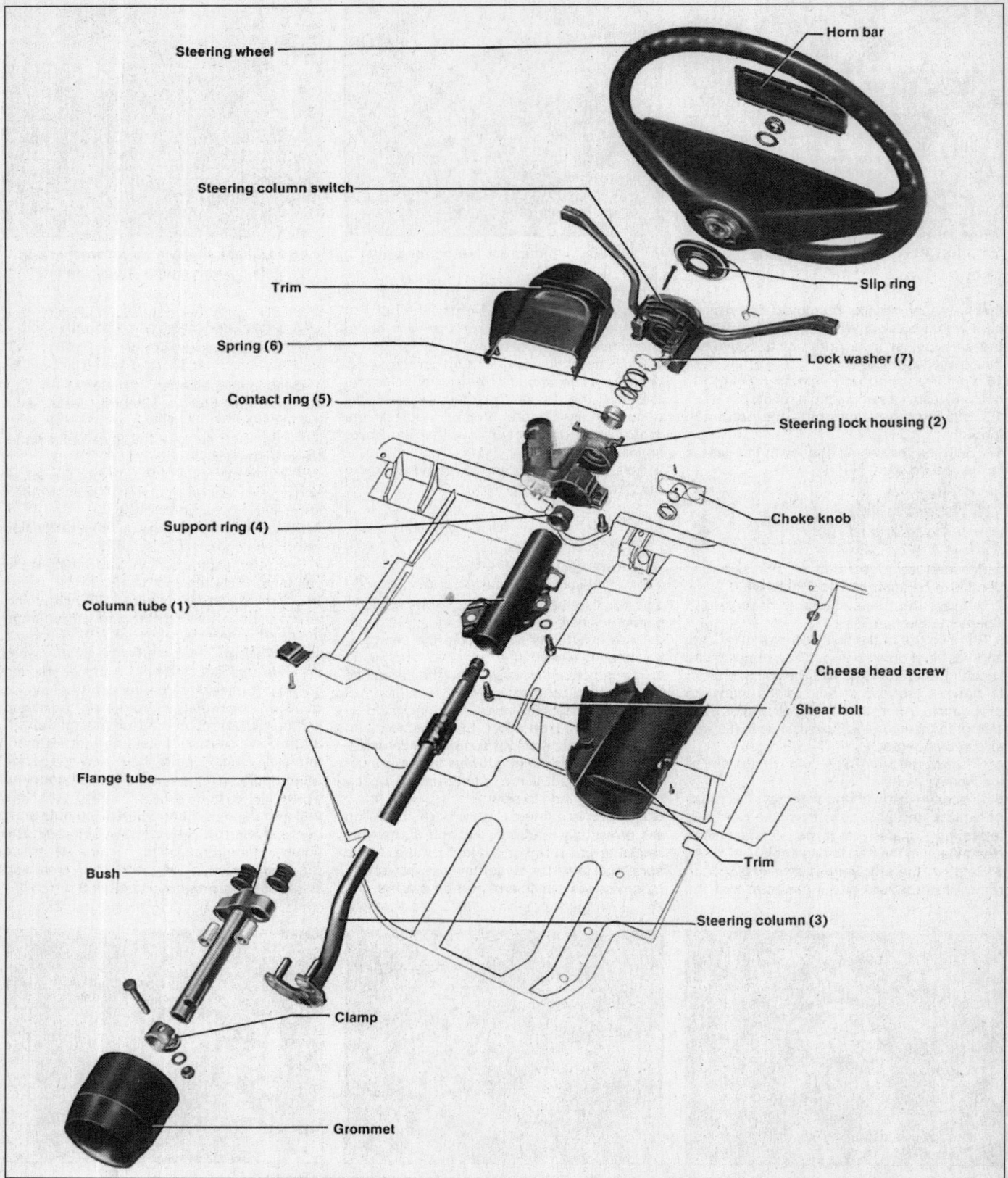

Fig 11.15 Steering column assembly components – manual steering (Sec 13)

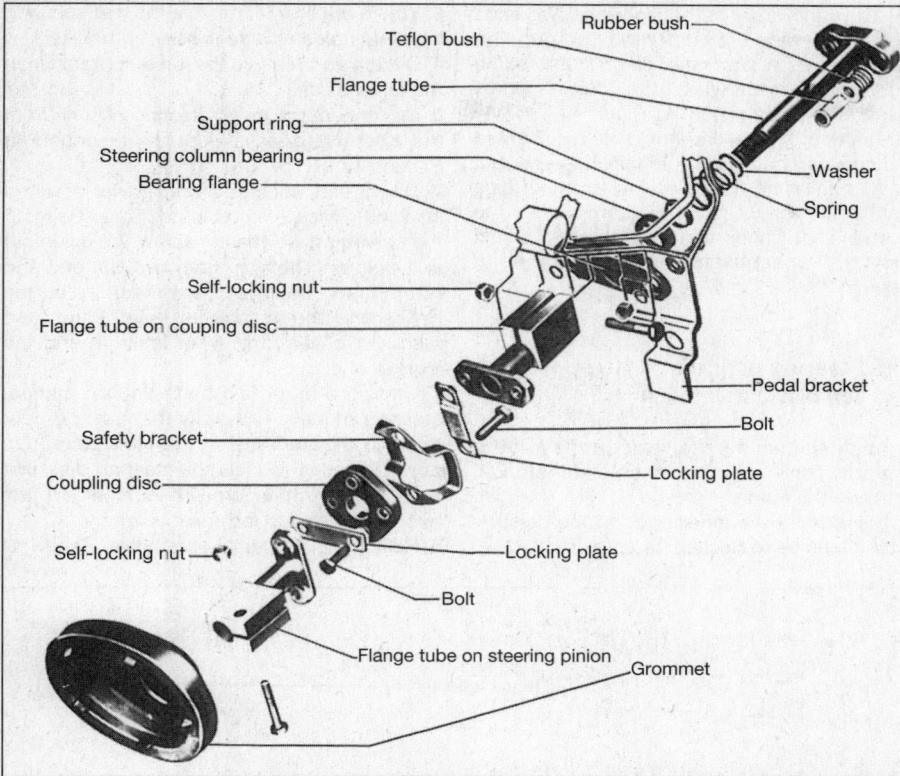

Fig. 11.16 Lower steering column components – power steering models (Sec 13)

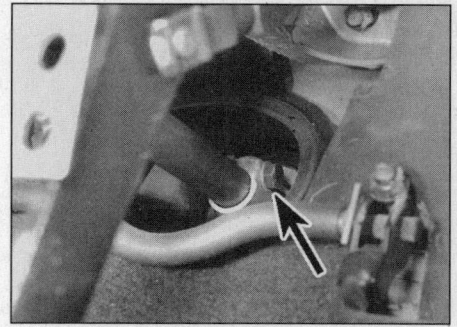

13.7 Manual steering flange tube clamp and bolt (arrowed)

them for wear and damage. Obtain new components as necessary, including a new shear-bolt and lockwasher.

16 Reassembly and refitting are a reversal of the dismantling and removal procedures, but the following points should be noted. Push the steering lock housing fully onto the outer column before tightening the bolt. Press the lockwasher onto the inner column so that it compresses the spring and contact ring against the stop. Before tightening the column mounting bolts make sure that the inner column is fully entered in the flange tube, and the flange tube is connected to the steering gear pinion. Except where Teflon sleeves are fitted, lubricate the flange tube bushes with a little molybdenum disulphide grease. If the flange tube is dismantled on power-assisted steering models, note that the bearing flange must be refitted with the lug towards the centre of the vehicle (Fig. 11.17).

14 Tie-rod – removal and refitting

1 The left-hand tie-rod is adjustable for length, but the right-hand tie-rod may not be adjustable on some models.
2 Apply the handbrake, jack up the front of the car, and support it on axle stands. Remove the roadwheel.
3 Unscrew the balljoint nut from the outer end of the tie-rod and use a separator tool to detach the tie-rod from the strut (photos).

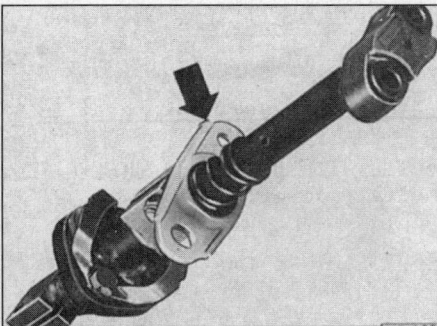

Fig. 11.17 Bearing flange lug must point to centre of vehicle (Sec 13)

to remove the head, after which the shank can be unscrewed (photo).
9 Withdraw the steering column assembly.
10 Commence dismantling the steering column by prising off the lockwasher and removing the spring.
11 Lever off the contact ring.
12 Withdraw the inner column from the outer column, and remove the rubber support ring.
13 Unscrew the clamp bolt from the steering lock housing and slide the housing from the outer column.
14 Note that the steering column on power-assisted steering models will have an additional bearing at its bottom end, but is otherwise the same as on manual models.
15 Clean all the components, and examine

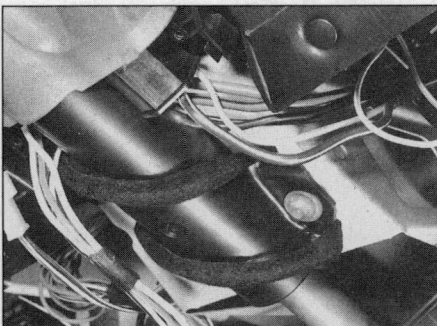

13.8 Column tube mounting showing shear-bolt

14.3A Using ball joint separator tool to remove tie-rod from strut

14.3B Steering tie-rod end showing rubber boot

11

14.4 Steering tie-rod inner connections and damper unit

4 Disconnect the inner end of the tie-rod by unscrewing the locking nuts and withdrawing the stud plate from both tie-rod inner ends. The self-locking nuts must be renewed on reassembly (photo).

5 If renewing the end of an adjustable tie-rod, measure the distance between the two ends before screwing the old tie-rod end out, and then screw in the new one to the same dimension, otherwise the front wheel alignment will be disturbed.

6 Refitting is a reversal of removal. Tighten the nuts and bolts to the specified torque, but delay tightening the tie-rod inner mountings until the weight of the car is on the suspension. Check and, if necessary, adjust the front wheel alignment as described in Section 23.

15 Steering gear unit – removal and refitting

1 Apply the handbrake, jack up the front of the car, and support it on axle stands. Remove the roadwheels.

2 Disconnect the inner ends of the tie-rods with reference to Section 14, paragraph 4.

3 Where applicable, remove the bolt securing the steering damper to the tie-rod bracket.

4 Detach and remove the lower trim panel on the driver's side.

5 Disconnect the lower steering column from the steering gear pinion as described in Section 13.

6 On models equipped with power assisted steering, place a suitable container beneath the steering gear, then unscrew the union nut and bolt, and detach the hydraulic feed and return lines. Recover the washers, plug the hoses and the ventilation hole in the fluid reservoir cover to prevent leakage and the ingress of dirt.

7 Undo the mounting bolts and on manual steering models, withdraw the steering gear through the aperture in the side panel. On power steering models the steering gear unit is removed from underneath together with the hose and reservoir (as applicable).

8 Refitting is a reversal of removal. Renew all

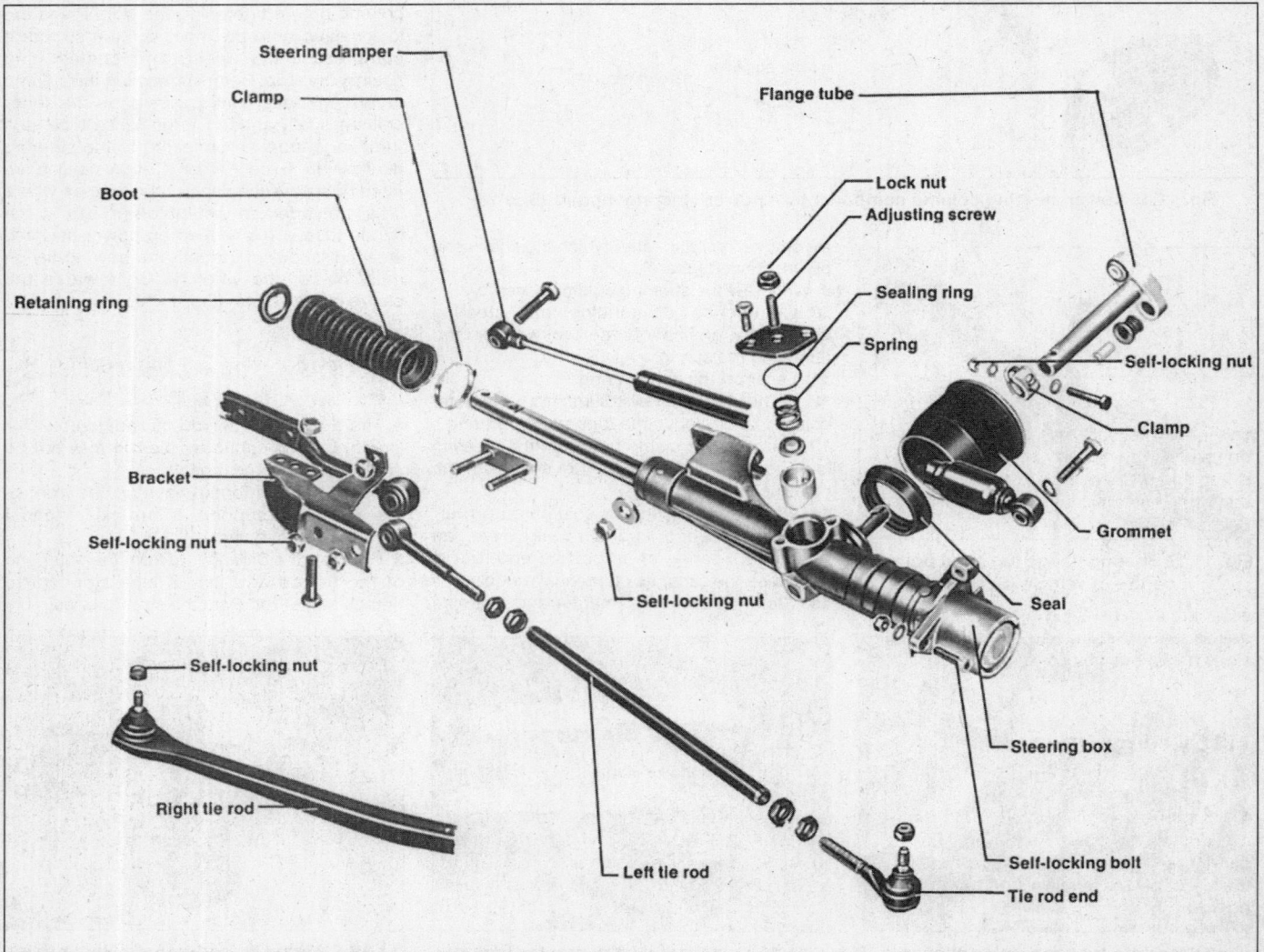

Fig. 11.18 Steering gear unit and associated components – manual steering (Sec 15)

Bellows

Hose clip

Notched ring

Seal retainer

Pressure line

Return flow line

Retaining ring

Valve body

Seal

Circlip

Ball bearing

Axial needle bearing

Thrust washer

O-ring

Intermediate cover

O-rings

O-ring

Tie rod bracket

Self-locking nuts

Steering rack seal

Cap

O-ring

Steering rack

Self-locking nut

Thrust piece

Cheese head bolt

Steering gear housing

Self-locking nut

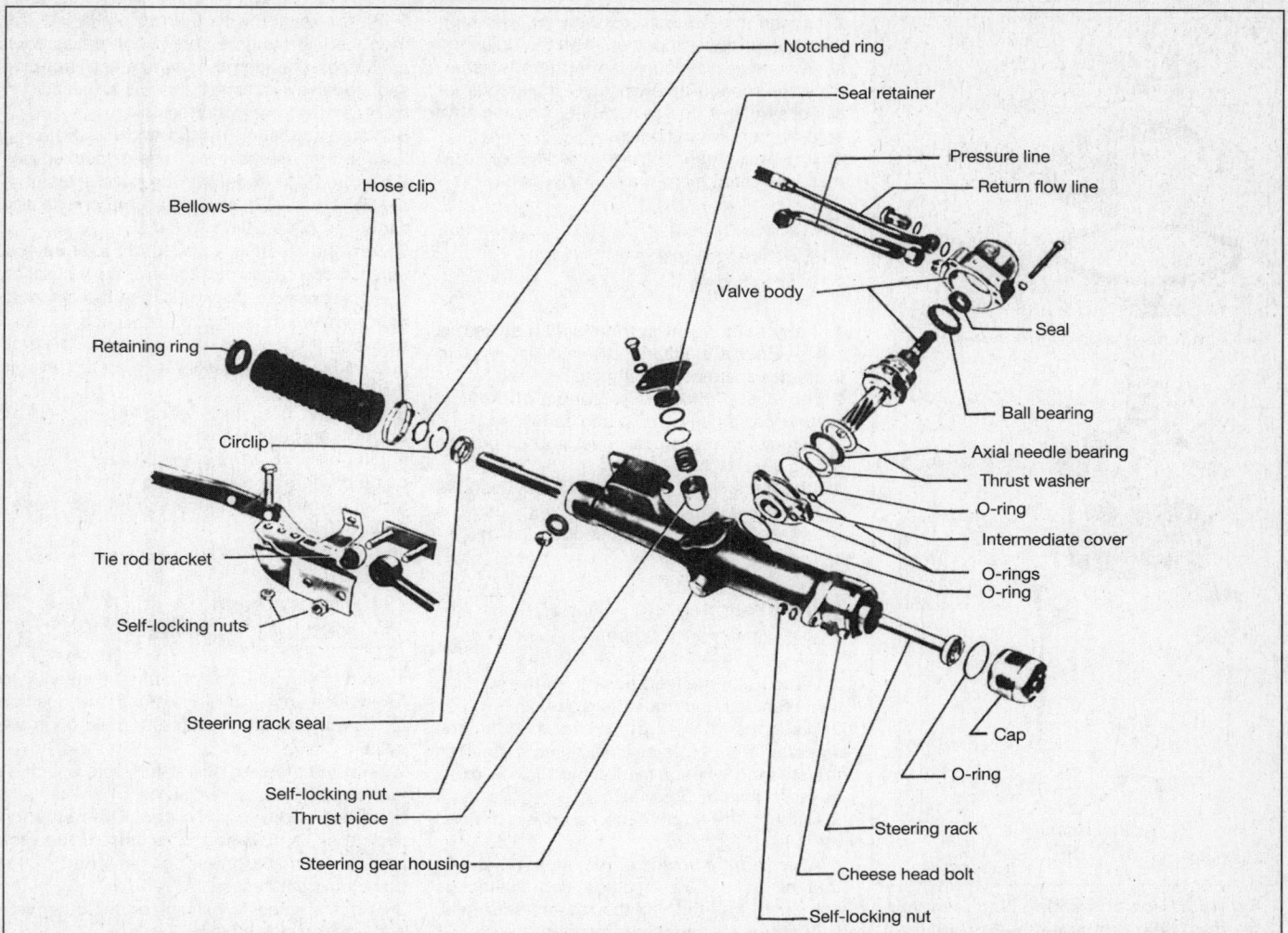

Fig. 11.19 Steering gear unit and associated components – power assisted steering (Sec 15)

self-locking nuts. Delay tightening all nuts and bolts until the weight of the car is on the suspension. Check and, if necessary, adjust the front wheel alignment as described in Section 23.

9 On power steering models unplug the hoses and reservoir cap ventilation hole, connect the hoses taking care not to let dirt enter the system. Top up the system fluid as described in Section 20 and check for any signs of leakage on completion.

16 Steering damper – testing, removal and refitting

1 A steering damper is fitted to models which do not have power steering, and it is connected between the tie-rod bracket and the side of the engine compartment (see photo 14.4).

2 Repair of the steering damper is not possible. If the steering is sensitive to road shocks, remove the damper and check its

operation. The damper is attached by a bolt at each end.

3 Test the damper by moving the piston in and out by hand, over the whole range of its travel. If the movement is not smooth and with a uniform resistance, fit a new damper.

17 Steering gear bellows – renewal

1 The steering gear bellows can be removed and refitted with the steering gear unit in situ or removed from the vehicle.

2 Unbolt the tie-rod bracket from the end of the steering rack.

3 Prise the bellows from the retaining ring, then remove the ring.

4 Release the clip and slide off the bellows.

5 Smear the rack with steering gear grease, then fit the new bellows, clip, and ring. If necessary, a worm drive clip may be fitted in place of the crimped type, with the screw toward the bulkhead.

6 Refit the tie-rod bracket, and tighten the bolts to the specified torque.

18 Steering gear (manual) – adjustment

1 If there is any undue slackness in the steering gear, resulting in noise or rattles, the steering gear should be adjusted as follows, with reference to photo 18.1.

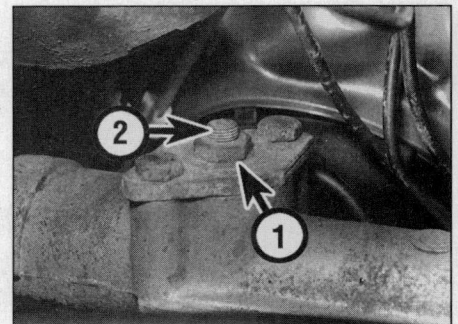

18.1 Manual steering gear adjustment locknut (1) and adjusting screw (2)

11

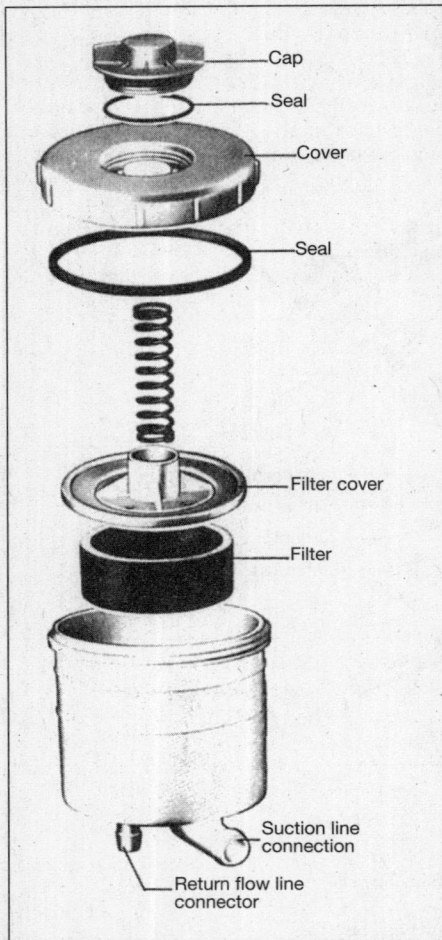

Fig. 11.20 Power steering fluid reservoir and filter components (Sec 19)

2 Loosen the locknut and, with the wheels in the straight-ahead position, turn the adjusting screw until it just touches the thrust washer. On some models an Allen key will be required.
3 Tighten the locknut while holding the adjusting screw stationary.
4 Turn the steering from lock to lock, and make sure that there are no tight spots.

19 Power steering – filter renewal

1 If any component in the steering system is renewed, or if the fluid in the steering system is changed, a new filter should be fitted.
2 The filter is fitted in the bottom of the fluid reservoir and can be renewed as follows.
3 Remove the reservoir cover and lift out the spring (Fig. 11.20).
4 Lift off the filter cover and take out the filter.
5 Fit a new filter and sealing ring using a reversal of the removal procedure.

20 Power steering – fluid draining and refilling

1 Disconnect the feed hose from the steering fluid reservoir and drain the reservoir.
2 Disconnect the return hose from the reservoir and put its open end into a jar. Turn the steering wheel from lock to lock to expel as much fluid as possible.
3 Discard the fluid which has been drained off.
4 After fitting a new filter (Section 19), ensure that all hoses are in place and their clips tightened, and then fill the system with fresh fluid of the approved specification.

5 Fill the reservoir to the top with fluid and then start the engine and switch off as soon as it fires, repeating the starting and stopping sequence several times; this will cause fluid to be drawn into the system quickly.
6 Watch the level of fluid and keep adding fluid so that the reservoir is never sucked dry. When the fluid ceases to drop as a result of the start/stop sequence, start the engine and allow it to run at idling speed.
7 Turn the steering from lock to lock several times, being careful not to leave the wheels on full lock because this will cause the pressure in the system to build up.
8 Watch the level of the fluid in the reservoir and add fluid if necessary to keep the level at the MAX mark.
9 When the level stops falling and no more air bubbles appear in the reservoir, switch the engine off and fit the reservoir cap. The level of fluid will rise slightly when the engine is switched off, but the rise should not exceed 10 mm (0.39 in).

21 Power steering – checking for fluid leaks

1 With the engine running, turn the steering to full lock on one side and hold it in this position to allow maximum pressure to build up in the system.
2 With the steering still at full lock, check all joints and unions for signs of leaks, and tighten if necessary. To check the steering rack seal, remove the inner end of the rack bellows from the steering gear, and pull it back to reveal the seal.
3 Turn the wheel to full lock on the other side and again check for leaks.
4 If a leak is found from the steering pinion, the valve housing seal and intermediate cover O-rings must be renewed.
5 If a steering rack seal in the housing is suspected, loosen off the bellows hose clip and slide the bellows to one side. If there is fluid in the bellows then the rack seals and O-rings must be renewed. The steering gear unit must be removed and dismantled to renew the seals.

22 Power steering pump – removal, refitting and adjustmen

1 The power steering pump and associated fittings will differ according to engine type, see Fig. 11.22 or 11.23 as applicable. If the drivebelt is in need of adjustment proceed as described in paragraph 8.
2 If the pump is suspected of malfunctioning it is advisable to check the pressure and flow limiting valve. Inspect the valve pistons and pump housing drillings for wear and any signs of dirt. The pistons must operate freely in the

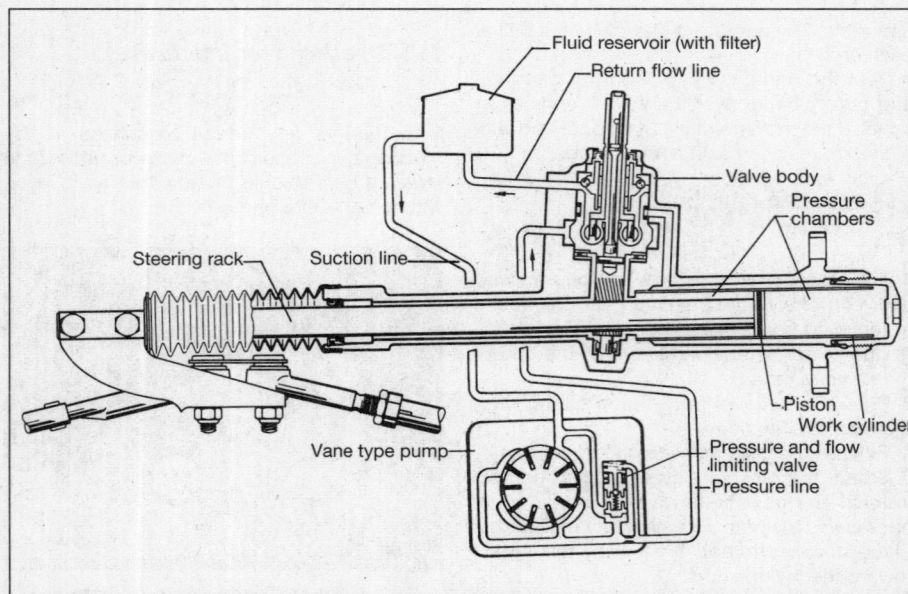

Fig. 11.21 Power steering layout and hydraulic circuits (Sec 21)

housing drillings. Do not dismantle the valve unit. Renew it if found to be defective and use a new washer (Fig. 11.24).

3 It is also possible to obtain a repair kit of seals which can be fitted after removing the pump backplate. More serious defects will require the renewal of the pump as follows.

4 First remove the alternator drivebelt (Chapter 10) and on five-cylinder models remove the retaining screws and detach the pump drivebelt cover.

5 Remove the pump drivebelt by loosening

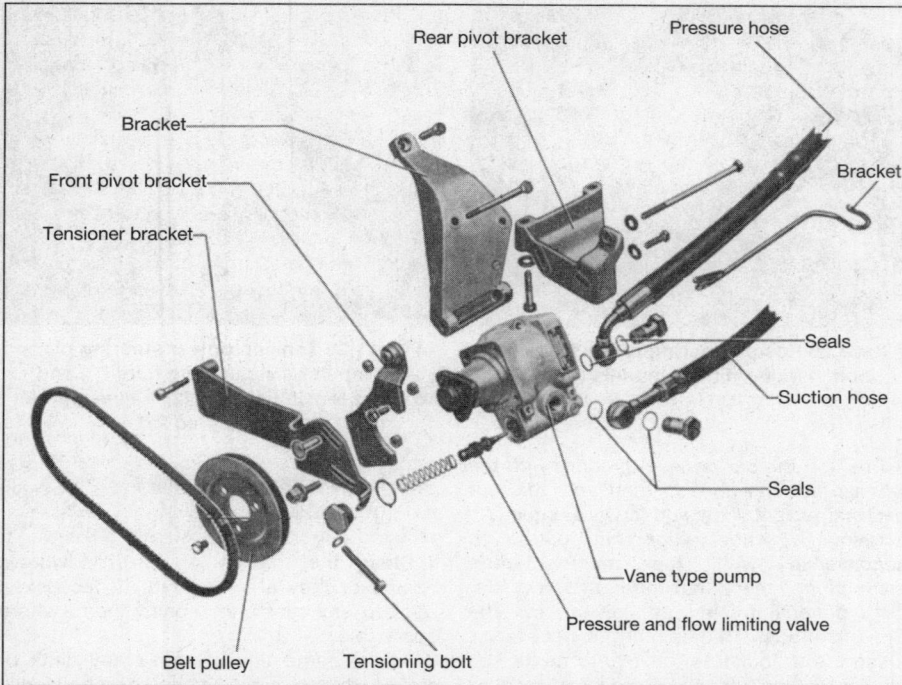

Fig. 11.22 Power steering pump and associated components – four-cylinder models (Sec 22)

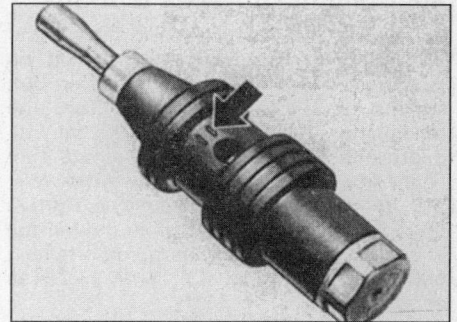

Fig. 11.24 Power steering valve piston tolerance group marking must be the same if renewed (Sec 22)

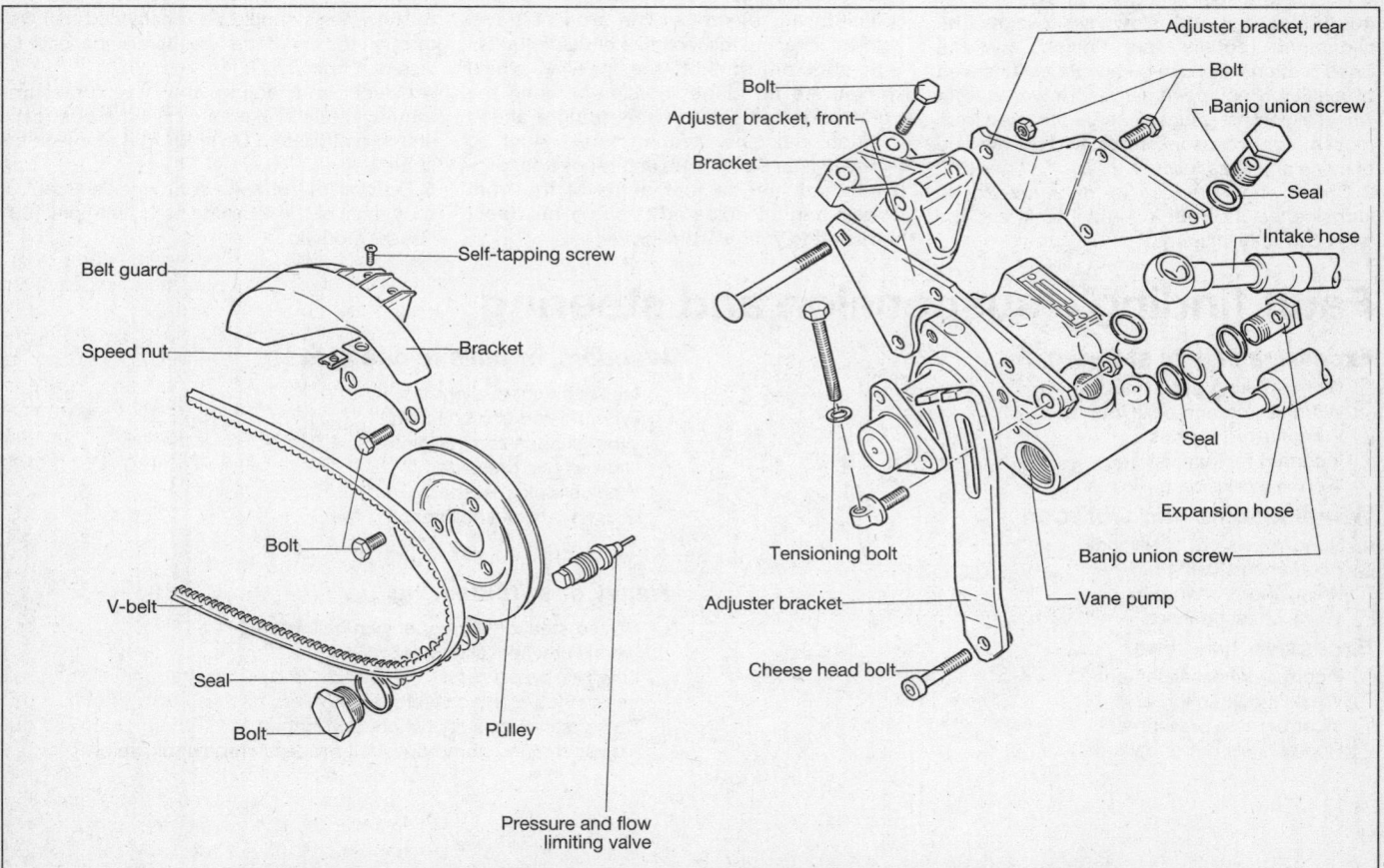

Fig. 11.23 Power steering pump and associated components – five-cylinder models (Sec 22)

11

off the unit retaining bolt and nuts and the tensioning bolt.

6 Place a container beneath the pump, then unscrew the union nut or bolt from the hydraulic lines. Do not twist the union nut line – if necessary unscrew the pump from the line when it is removed. Unscrew and remove the pivot bolt, and remove the pump.

7 Refitting is a reversal of removal, but tension the drivebelt as described in paragraph 8 and top up with new fluid and bleed the system as described in Section 20.

Adjusting the drivebelt tension

8 To adjust the drivebelt tension, the pump unit retaining nuts and bolts should be loosened. Also loosen the adjuster bolt locknut on the pump bracket. Turn the tensioning bolt until the belt can be depressed approximately 10.0 mm (0.4 in) under firm thumb pressure midway between the crankshaft and pump pulleys. Tighten the adjusting bolt locknut when the tension is correct. Also tighten the pump retaining nuts and bolts (Fig. 11.25 or 11.26 as applicable).

23 Wheel alignment – checking and adjusting

1 Accurate wheel alignment is essential for good steering and slow tyre wear. The alignment details are given in the Specifications and can be accurately checked by a suitably equipped garage. However, front wheel alignment gauges can be obtained from most motor accessory stores, and the method of using one is as follows.

2 Check that the car is only loaded to kerbside weight, with a full fuel tank, and the tyres correctly inflated.

Fig. 11.25 Power steering pump tensioning bolt – four-cylinder models (Sec 22)

3 Position the car on level ground, with the wheels straight-ahead, then roll the car backwards 12 ft (4 m), and forwards again.

4 Using a wheel alignment gauge in accordance with the manufacturer's instructions, check that the front wheel toe-out dimension is as given in the Specifications. If adjustment is necessary, loosen the locknuts or clamp bolts, as applicable, and turn the tie-rod (or tie-rods) as required, then tighten them. Where both tie-rods are adjustable, both must be turned by equal amounts. Make sure that the tie-rod end balljoints are central in the arcs of travel before tightening the locknuts or clamp bolts.

5 If, after adjustment, the steering wheel spokes are no longer horizontal when the front roadwheels are in the straight-ahead position, then the steering wheel must be removed (see Section 12) and repositioned.

6 Although the camber angle of the front wheels can be adjusted this is a task best entrusted to your VW dealer.

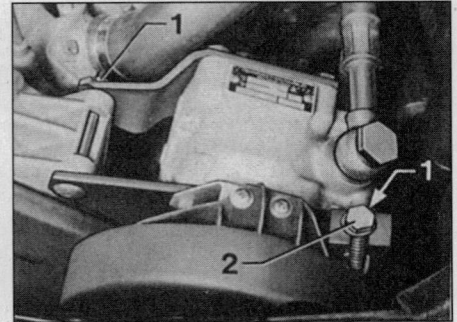

Fig. 11.26 Tension power steering pump drivebelt by slackening nuts (1) and turning tensioning bolt (2) – five-cylinder models (Sec 22)

24 Roadwheels and tyres – general

1 Clean the insides of the roadwheels whenever they are removed. If necessary, remove any rust and repaint them, where applicable.

2 At the same time, remove any flints or stones which may have become embedded in the tyres. Examine the tyres for damage and splits. Where the depth of tread is almost down to the legal minimum, renew them.

3 The wheels should be rebalanced half way through the life of the tyres to compensate for loss of rubber.

4 Check and adjust the tyre pressures regularly, and make sure that the dust caps are correctly fitted. Do not forget to check the spare tyre.

5 Do not interchange roadwheels (steel or alloy rims) or the retaining bolts, with pre 1981 Passat models.

Fault finding – suspension and steering

Excessive play in steering
☐ Worn steering gear
☐ Worn tie-rod end balljoints
☐ Worn tie-rod bushes
☐ Incorrect rack adjustment
☐ Worn suspension balljoints

Wheel wobble and vibration
☐ Roadwheels out of balance
☐ Roadwheels damaged
☐ Weak shock absorbers
☐ Worn wheel bearings

Excessive tyre wear
☐ Incorrect wheel alignment
☐ Weak shock absorbers
☐ Incorrect tyre pressures
☐ Roadwheels out of balance

Wanders, or pulls to one side
☐ Incorrect wheel alignment
☐ Worn tie-rod end balljoints
☐ Worn suspension balljoints
☐ Uneven tyre pressures
☐ Weak shock absorber
☐ Broken or weak coil spring

Heavy or stiff steering
☐ Seized steering or suspension balljoint
☐ Incorrect wheel alignment
☐ Low tyre pressures
☐ Leak of lubricant in steering gear
☐ Power steering faulty (where applicable)
☐ Power steering pump drivebelt broken (where applicable)

Chapter 12 Bodywork and fittings

For modifications, and information applicable to later models, see Supplement at end of manual

Contents

Degrees of difficulty

Easy, suitable for novice with little experience	**Fairly easy,** suitable for beginner with some experience	**Fairly difficult,** suitable for competent DIY mechanic	**Difficult,** suitable for experienced DIY mechanic	**Very difficult,** suitable for expert DIY or professional

1 General description

The body is of unitary all-steel construction, and incorporates computer calculated impact crumple zones at the front and rear, with a central safety cell passenger compartment. During manufacture the body is undersealed and treated with cavity wax injection. In addition all open box members are sealed.

There are two body styles available for the Passat, the four-door Hatchback and four-door Estate. The Santana model is only available as a four-door Saloon.

2 Maintenance - bodywork and underframe

The general condition of a vehicle's bodywork is the one thing that significantly affects its value. Maintenance is easy, but needs to be regular. Neglect, particularly after minor damage, can lead quickly to further deterioration and costly repair bills. It is important also to keep watch on those parts of the vehicle not immediately visible, for instance the underside, inside all the wheel arches, and the lower part of the engine compartment.

The basic maintenance routine for the bodywork is washing - preferably with a lot of water, from a hose. This will remove all the loose solids which may have stuck to the vehicle. It is important to flush these off in such a way as to prevent grit from scratching the finish. The wheel arches and underframe need washing in the same way, to remove any accumulated mud, which will retain moisture and tend to encourage rust. Paradoxically enough, the best time to clean the underframe and wheel arches is in wet weather, when the mud is thoroughly wet and soft. In very wet weather, the underframe is usually cleaned of large accumulations automatically, and this is a good time for inspection.

Periodically, except on vehicles with a wax-based underbody protective coating, it is a good idea to have the whole of the underframe of the vehicle steam-cleaned, engine compartment included, so that a thorough inspection can be carried out to see what minor repairs and renovations are necessary. Steam-cleaning is available at many garages, and is necessary for the removal of the accumulation of oily grime, which sometimes is allowed to become thick in certain areas. If steam-cleaning facilities are not available, there are some excellent grease solvents available which can be brush-applied; the dirt can then be simply hosed off. Note that these methods should not be used on vehicles with wax-based underbody protective coating, or the coating will be removed. Such vehicles should be inspected annually, preferably just prior to Winter, when the underbody should be washed down, and any damage to the wax coating repaired. Ideally, a completely fresh coat should be applied. It would also be worth considering the use of such wax-based protection for injection into door panels, sills, box sections, etc, as an additional safeguard against rust damage, where such protection is not provided by the vehicle manufacturer.

After washing paintwork, wipe off with a chamois leather to give an unspotted clear finish. A coat of clear protective wax polish will give added protection against chemical pollutants in the air. If the paintwork sheen has dulled or oxidised, use a cleaner/polisher combination to restore the brilliance of the shine. This requires a little effort, but such dulling is usually caused because regular washing has been neglected. Care needs to be taken with metallic paintwork, as special non-abrasive cleaner/polisher is required to avoid damage to the finish. Always check that the door and ventilator opening drain holes and pipes are completely clear, so that water can be drained out (photo). Brightwork should be treated in the same way as paintwork. Windscreens and windows can be kept clear of the smeary film which often appears, by the use of proprietary glass cleaner. Never use any form of wax or other body or chromium polish on glass.

12

2.4 Door drain holes must be kept clear

3 Maintenance -
upholstery and carpets

Mats and carpets should be brushed or vacuum-cleaned regularly, to keep them free of grit. If they are badly stained, remove them from the vehicle for scrubbing or sponging, and make quite sure they are dry before refitting. Seats and interior trim panels can be kept clean by wiping with a damp cloth. If they do become stained (which can be more apparent on light-coloured upholstery), use a little liquid detergent and a soft nail brush to scour the grime out of the grain of the material. Do not forget to keep the headlining clean in the same way as the upholstery. When using liquid cleaners inside the vehicle, do not over-wet the surfaces being cleaned. Excessive damp could get into the seams and padded interior, causing stains, offensive odours or even rot.

HAYNES HiNT *If the inside of the vehicle gets wet accidentally, it is worthwhile taking some trouble to dry it out properly, particularly where carpets are involved. Do not leave oil or electric heaters inside the vehicle for this purpose.*

4 Minor body damage - repair

Note: *For more detailed information about bodywork repair, Haynes Publishing produce a book by Lindsay Porter called "The Car Bodywork Repair Manual". This incorporates information on such aspects as rust treatment, painting and glass-fibre repairs, as well as details on more ambitious repairs involving welding and panel beating.*

Repairs of minor scratches in bodywork

If the scratch is very superficial, and does not penetrate to the metal of the bodywork, repair is very simple. Lightly rub the area of the scratch with a paintwork renovator, or a very fine cutting paste, to remove loose paint from the scratch, and to clear the surrounding bodywork of wax polish. Rinse the area with clean water.

Apply touch-up paint to the scratch using a fine paint brush; continue to apply fine layers of paint until the surface of the paint in the scratch is level with the surrounding paintwork. Allow the new paint at least two weeks to harden, then blend it into the surrounding paintwork by rubbing the scratch area with a paintwork renovator or a very fine cutting paste. Finally, apply wax polish.

Where the scratch has penetrated right through to the metal of the bodywork, causing the metal to rust, a different repair technique is required. Remove any loose rust from the bottom of the scratch with a penknife, then apply rust-inhibiting paint to prevent the formation of rust in the future. Using a rubber or nylon applicator, fill the scratch with bodystopper paste. If required, this paste can be mixed with cellulose thinners to provide a very thin paste which is ideal for filling narrow scratches. Before the stopper-paste in the scratch hardens, wrap a piece of smooth cotton rag around the top of a finger. Dip the finger in cellulose thinners, and quickly sweep it across the surface of the stopper-paste in the scratch; this will ensure that the surface of the stopper-paste is slightly hollowed. The scratch can now be painted over as described earlier in this Section.

Repairs of dents in bodywork

When deep denting of the vehicle's bodywork has taken place, the first task is to pull the dent out, until the affected bodywork almost attains its original shape. There is little point in trying to restore the original shape completely, as the metal in the damaged area will have stretched on impact, and cannot be reshaped fully to its original contour. It is better to bring the level of the dent up to a point which is about 3 mm below the level of the surrounding bodywork. In cases where the dent is very shallow anyway, it is not worth trying to pull it out at all. If the underside of the dent is accessible, it can be hammered out gently from behind, using a mallet with a wooden or plastic head. Whilst doing this, hold a suitable block of wood firmly against the outside of the panel, to absorb the impact from the hammer blows and thus prevent a large area of the bodywork from being "belled-out".

Should the dent be in a section of the bodywork which has a double skin, or some other factor making it inaccessible from behind, a different technique is called for. Drill several small holes through the metal inside the area - particularly in the deeper section. Then screw long self-tapping screws into the holes, just sufficiently for them to gain a good purchase in the metal. Now the dent can be pulled out by pulling on the protruding heads of the screws with a pair of pliers.

The next stage of the repair is the removal of the paint from the damaged area, and from an inch or so of the surrounding "sound" bodywork. This is accomplished most easily by using a wire brush or abrasive pad on a power drill, although it can be done just as effectively by hand, using sheets of abrasive paper. To complete the preparation for filling, score the surface of the bare metal with a screwdriver or the tang of a file, or alternatively, drill small holes in the affected area. This will provide a really good "key" for the filler paste.

To complete the repair, see the Section on filling and respraying.

Repairs of rust holes or gashes in bodywork

Remove all paint from the affected area, and from an inch or so of the surrounding "sound" bodywork, using an abrasive pad or a wire brush on a power drill. If these are not available, a few sheets of abrasive paper will do the job most effectively. With the paint removed, you will be able to judge the severity of the corrosion, and therefore decide whether to renew the whole panel (if this is possible) or to repair the affected area. New body panels are not as expensive as most people think, and it is often quicker and more satisfactory to fit a new panel than to attempt to repair large areas of corrosion.

Remove all fittings from the affected area, except those which will act as a guide to the original shape of the damaged bodywork (eg headlight shells etc). Then, using tin snips or a hacksaw blade, remove all loose metal and any other metal badly affected by corrosion. Hammer the edges of the hole inwards, in order to create a slight depression for the filler paste.

Wire-brush the affected area to remove the powdery rust from the surface of the remaining metal. Paint the affected area with rust-inhibiting paint, if the back of the rusted area is accessible, treat this also.

Before filling can take place, it will be necessary to block the hole in some way. This can be achieved by the use of aluminium or plastic mesh, or aluminium tape.

Aluminium or plastic mesh, or glass-fibre matting, is probably the best material to use for a large hole. Cut a piece to the approximate size and shape of the hole to be filled, then position it in the hole so that its edges are below the level of the surrounding bodywork. It can be retained in position by several blobs of filler paste around its periphery.

Aluminium tape should be used for small or very narrow holes. Pull a piece off the roll, trim it to the approximate size and shape required, then pull off the backing paper (if used) and stick the tape over the hole; it can be overlapped if the thickness of one piece is insufficient. Burnish down the edges of the tape with the handle of a screwdriver or similar, to ensure that the tape is securely attached to the metal underneath.

Bodywork repairs - filling and respraying

Before using this Section, see the Sections on dent, deep scratch, rust holes and gash repairs.

Many types of bodyfiller are available, but generally speaking, those proprietary kits which contain a tin of filler paste and a tube of resin hardener are best for this type of repair. A wide, flexible plastic or nylon applicator will be found invaluable for imparting a smooth and well-contoured finish to the surface of the filler.

Mix up a little filler on a clean piece of card or board - measure the hardener carefully (follow the maker's instructions on the pack), otherwise the filler will set too rapidly or too slowly. Using the applicator, apply the filler paste to the prepared area; draw the applicator across the surface of the filler to achieve the correct contour and to level the surface. As soon as a contour that approximates to the correct one is achieved, stop working the paste - if you carry on too long, the paste will become sticky and begin to "pick-up" on the applicator. Continue to add thin layers of filler paste at 20-minute intervals, until the level of the filler is just proud of the surrounding bodywork.

Once the filler has hardened, the excess can be removed using a metal plane or file. From then on, progressively-finer grades of abrasive paper should be used, starting with a 40-grade production paper, and finishing with a 400-grade wet-and-dry paper. Always wrap the abrasive paper around a flat rubber, cork, or wooden block - otherwise the surface of the filler will not be completely flat. During the smoothing of the filler surface, the wet-and-dry paper should be periodically rinsed in water. This will ensure that a very smooth finish is imparted to the filler at the final stage.

At this stage, the "dent" should be surrounded by a ring of bare metal, which in turn should be encircled by the finely "feathered" edge of the good paintwork. Rinse the repair area with clean water, until all of the dust produced by the rubbing-down operation has gone.

Spray the whole area with a light coat of primer - this will show up any imperfections in the surface of the filler. Repair these imperfections with fresh filler paste or bodystopper, and once more smooth the surface with abrasive paper. Repeat this spray-and-repair procedure until you are satisfied that the surface of the filler, and the feathered edge of the paintwork, are perfect. Clean the repair area with clean water, and allow to dry fully.

HAYNES HiNT *If bodystopper is used, it can be mixed with cellulose thinners, to form a really thin paste which is ideal for filling small holes.*

The repair area is now ready for final spraying. Paint spraying must be carried out in a warm, dry, windless and dust-free atmosphere. This condition can be created artificially if you have access to a large indoor working area, but if you are forced to work in the open, you will have to pick your day very carefully. If you are working indoors, dousing the floor in the work area with water will help to settle the dust which would otherwise be in the atmosphere. If the repair area is confined to one body panel, mask off the surrounding panels; this will help to minimise the effects of a slight mis-match in paint colours. Bodywork fittings (eg chrome strips, door handles etc) will also need to be masked off. Use genuine masking tape, and several thicknesses of newspaper, for the masking operations.

Before commencing to spray, agitate the aerosol can thoroughly, then spray a test area (an old tin, or similar) until the technique is mastered. Cover the repair area with a thick coat of primer; the thickness should be built up using several thin layers of paint, rather than one thick one. Using 400-grade wet-and-dry paper, rub down the surface of the primer until it is really smooth. While doing this, the work area should be thoroughly doused with water, and the wet-and-dry paper periodically rinsed in water. Allow to dry before spraying on more paint.

Spray on the top coat, again building up the thickness by using several thin layers of paint. Start spraying at one edge of the repair area, and then, using a side-to-side motion, work until the whole repair area and about 2 inches of the surrounding original paintwork is covered. Remove all masking material 10 to 15 minutes after spraying on the final coat of paint.

Allow the new paint at least two weeks to harden, then, using a paintwork renovator, or a very fine cutting paste, blend the edges of the paint into the existing paintwork. Finally, apply wax polish.

5 Major body damage – repair

Where serious damage has occurred or large areas need renewal due to neglect, it means that complete new panels will need welding in, and this is best left to professionals. If the damage is due to impact, it will also be necessary to completely check the alignment of the bodyshell, and this can only be carried out accurately by a VW dealer using special jigs. If the body is left misaligned, it is primarily dangerous as the car will not handle properly, and secondly, uneven stresses will be imposed on the steering, suspension and possibly transmission, causing abnormal wear, or complete failure, particularly to such items as the tyres.

6 Door rattles – tracing and rectification

1 The most common cause of door rattles is a striker plate which is worn, loose, or misaligned, but other causes may be:
(a) Loose door handles, window winder handles, door hinges, or door stays
(b) Loose, worn or misaligned door lock components
(c) Loose, or worn remote control mechanism
(d) Loose window glass
2 If the striker catch is worn, as a result of door rattles, fit a new striker before adjusting the door.
3 If door hinges are worn significantly, new hinges should be fitted.

7 Front door trim panel – removal and refitting

1 Unscrew and remove the locking knob.
2 If a manual door mirror adjuster is fitted, pull free the adjuster knob. On models with an electrically operated door mirror/heater, detach the battery earth lead, lever up and remove the switch from the door panel. Detach the lead connectors from the switch (Fig. 12.1).
3 Prise back the door pull handle retaining screw cover, undo the screw and remove the handle by pulling it downwards (photo).

Fig. 12.1 Electrically operated door mirror switch removal from trim panel (Sec 7)

7.3 Removing an armrest retaining screw

12

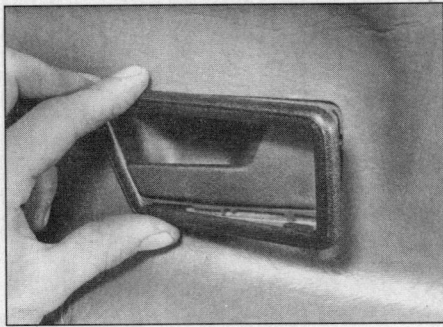

7.4 Removing the door release catch bezel

Fig. 12.2 Door trim removal tool dimensions (Sec 7)

a = 130 mm (5.1 in)   d = 2 mm (0.08 in)
b = 100 mm (3.9 in)   e = 0.5 mm (0.02 in)
c = 18 mm (0.70in)   f = 30 mm (1.18 in)

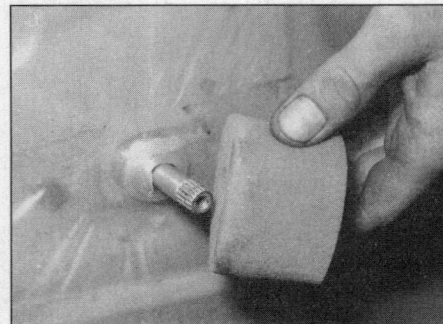

Fig. 12.3 Front door clip locations (Sec 7)

a = 30 mm (1.18 in)   c = 330 mm (12.9 in)
b = 300 mm (11.8 in)

4 Remove the door catch release bezel by pushing it rearwards (photo).
5 Remove the retaining screw and withdraw the window regulator handle and escutcheon (manual window regulator type).
6 On earlier models, withdraw the door pocket after removing its retaining screws. On later models the pocket is secured to the panel from inside, but where a radio speaker unit is fitted into the pocket, it can be prised free and the wires disconnected.
7 On earlier models, where the door speaker unit is fitted direct to the door trim panel, the speaker can be left in position and the speaker wires detached after partial removal of the panel.

8 Unscrew and remove the trim panel retaining screws on the leading and trailing edges (one screw to each edge). On later models the bottom edge is also secured by a retaining screw.
9 It should now be possible to pull the trim panel away from the door enabling it to be lifted clear. If the panel is reluctant to move, fabricate a trim removal tool to the dimensions shown in Fig. 12.2. Prise free the panel with the tool by inserting it between the trim panel and the door/window seal aperture at the appropriate points indicated. Avoid levering at those points indicated with an

asterisk (at fixed corner windows). These points will have to be carefully pulled free from the window slot seal (Figs. 12.3, 12.4 and 12.5).
10 With the panel removed, withdraw the spring insulator from the window regulator shaft (where applicable) and peel back the waterproof sheet for access to the inner door (photos).
11 Refitting is the reverse of removal. Make sure that the panel clips are aligned with the holes before tapping them in with the palm of the hand.

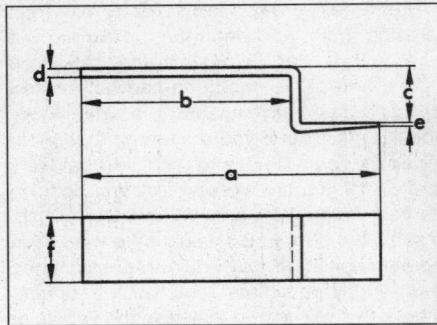

Fig. 12.4 Rear door clip locations (Sec 7)

a = 30 mm (1.18 in)   c = 330 mm (12.9 in)
b = 300 mm (11.8 in)

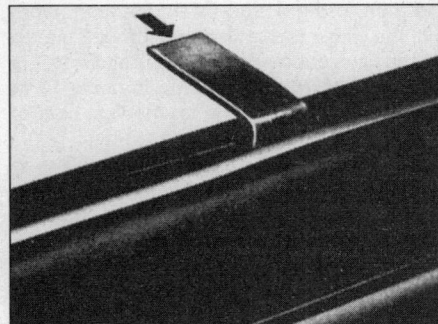

Fig. 12.5 Lever with clip to release door trim panel (Sec 7)

8 Front and rear door locks – removal and refitting

1 Using an Allen key unscrew and remove the two lock retaining screws (photo) and withdraw the lock approximately 10 to 12 mm (0.4 to 0.5 in).
2 Insert a screwdriver through the hole in the bottom of the lock to retain the lock lever in the 90° position.
3 Unhook the remote control rod from the lock lever.
4 Pull the lock out of the sleeve and disconnect the operating rod.
5 Refitting is a reversal of removal (Figs. 12.6 and 12.7).

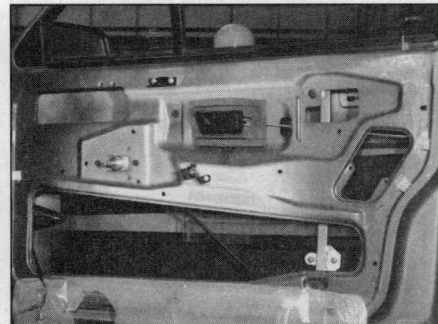

7.10A Remove the sponge insulator from the window regulator shaft . . .

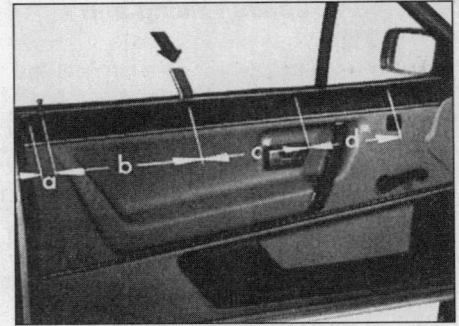

7.10B . . . and peel back the waterproof sheet

8.1 Front door lock location

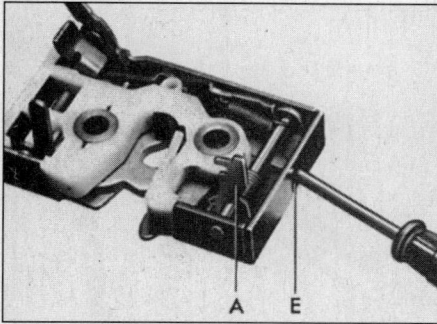

Fig. 12.6 Inner view of door lock showing screwdriver inserted through hole (E) to retain the lock lever (A) (Sec 8)

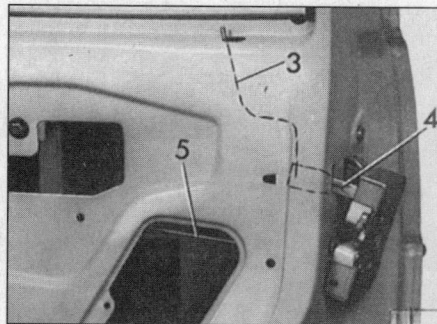

Fig. 12.7 Door lock installation showing rod (3), sleeve (4) and pullrod (50) (Sec 8)

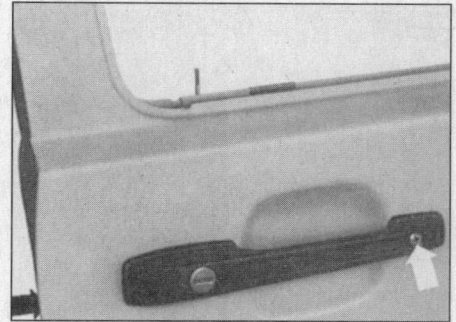

Fig. 12.8 Exterior door handle retaining screw locations (Sec 9)

9 Door exterior handles – removal and refitting

1 Prise the plastic insert from the handle, starting from the front.
2 Remove the handle retaining screws, one from the front of the handle and the other from the edge of the door. The handle can now be withdrawn from the door (Fig. 12.8).
3 Refitting is a reversal of removal.

10 Front door – dismantling and reassembly

1 Remove the door trim panel and the external door handle as described in Sections 7 and 9.
2 On models with power operated windows and door locks, remove the relevant components as described in Sections 34 and 35.

3 Remove the door lock inner control latch by pulling the retaining cleat from the location hole in the inner panel then sliding the control latch unit forwards. The latch seal can be pulled free (photo).
4 Remove the door lock as described in Section 8, also the lock rod and sleeve.
5 Lower the window so that the regulator lifter to window attachment, on its lower edge, is accessible through the inner door aperture. Undo the two screws to disengage the window from the lifter and lower it into the door cavity (photo).

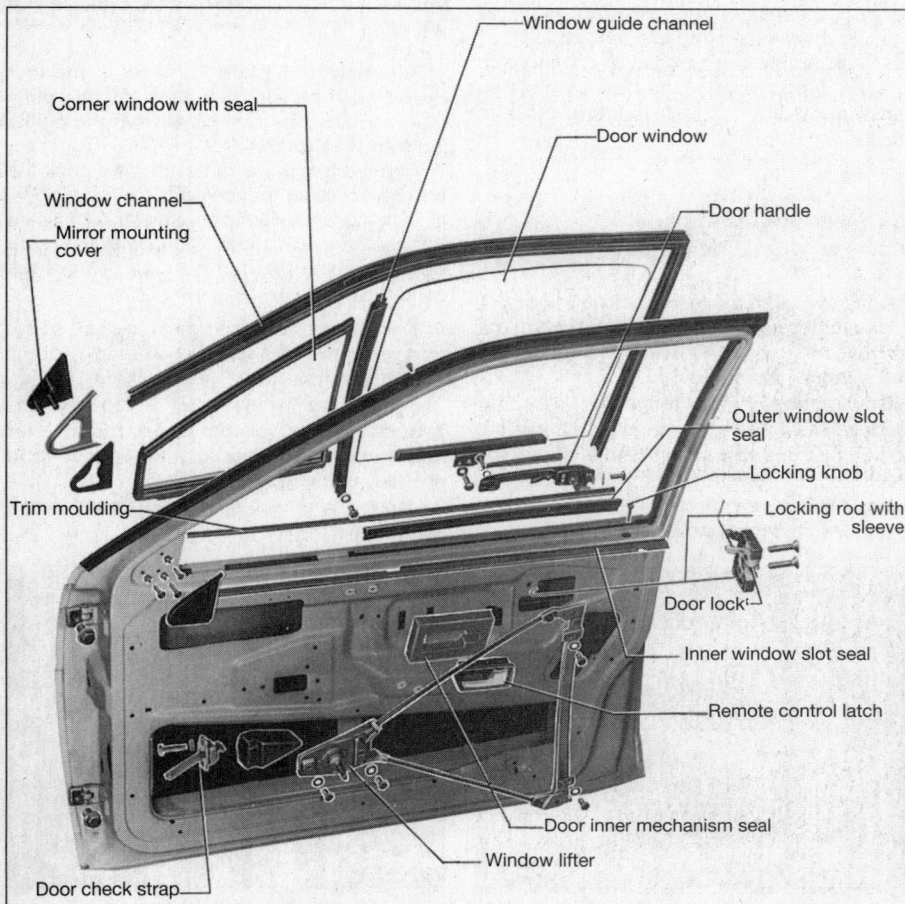

10.3 Door lock inner control latch showing retaining cleat (1) and latch unit (2)

Window guide channel
Corner window with seal
Window channel
Mirror mounting cover
Door window
Door handle
Outer window slot seal
Locking knob
Locking rod with sleeve
Trim moulding
Door lock
Inner window slot seal
Remote control latch
Door inner mechanism seal
Window lifter
Door check strap

Fig. 12.9 Exploded view of front door (Sec 10)

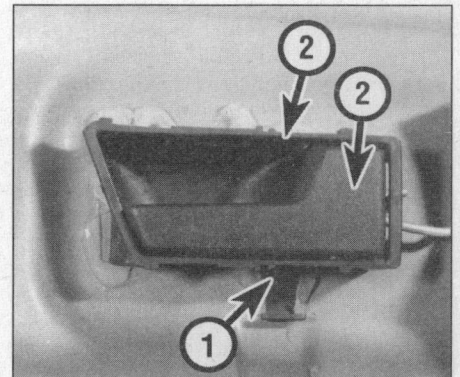

10.5 Window lifter to lower attachment screws

10.6A Window lifter (regulator) retaining bolts

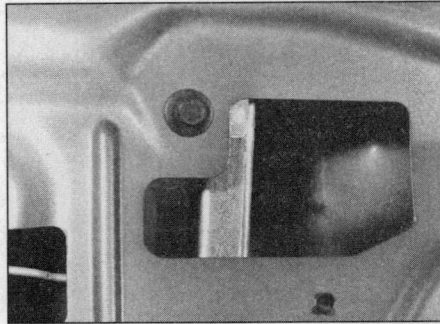

10.6B Window lifter upper retaining bolt

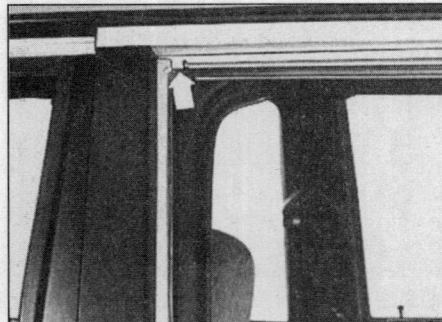

Fig. 12.10 Front door window guide channel fixing screw locations (arrowed) (Sec 10)

6 Unscrew the four bolts retaining the window lifter (regulator) unit and remove the lifter unit (photos).

7 Remove the door mirror as described in Section 25.

8 Remove the window inner slot seal.

9 Detach the window guide channel. This is secured by a screw and washer at the top and a bolt and washer at the bottom. Remove the channel on early models. On later models lower it into the door cavity (Fig. 12.10).

10 Unclip the corner window seal lip in the trim moulding area, then carefully remove the window and seal.

11 Pull free and remove the trim moulding together with the outer slot seal.

12 If still in position carefully lever free the door mirror mounting packing which is secured with double sided adhesive tape.

13 Bend back the tab securing the outer window surround trim moulding at the top rear corner and using a suitable plastic wedge tool, prise free the moulding which is secured with double sided adhesive tape. Remove the window channel (Figs. 12.11 and 12.12).

14 Tilt the window and remove it from the door.

15 Remove the door check strap by prising free the pivot pin retaining clip and drifting out the pin. Withdraw the rubber cover in the door cavity and remove the check strap retaining bolts. Withdraw the strap unit (photos).

16 Reassembly is a reversal of dismantling. Check the operation of the door catch, lock and window lifter prior to refitting the trim panel.

11 Rear door trim panel – removal and refitting

1 Unscrew and remove the locking knob.

2 Unclip the press fit trim strip from the armrest then unscrew the retaining screws and withdraw the armrest.

3 On models with manually operated windows, fully close the window and note the position of the regulator handle. Prise the plastic cover from the centre of the handle, remove the screw, and pull off the handle, together with the bezel.

4 Remove the door catch bezel by pushing it rearwards and disengaging it.

5 The trim panel is secured by two screws, one on the leading and one on the trailing edge. Remove the screw caps and undo each screw.

6 Referring to Section 7, make up the tool described in paragraph 9, then carefully prise free the rear door trim panel from the points indicated in Fig. 12.4.

7 With the trim panel removed, peel back the waterproof sheet for access to the inner door.

8 Refitting is a reversal of removal. Make sure that the panel clips are aligned with the holes before tapping them in with the palm of the hand.

12 Rear door – dismantling and reassembly

1 Remove the door trim panel, the exterior door handle and door lock as described in previous Sections.

2 Unclip and detach the remote control catch and remove the seal.

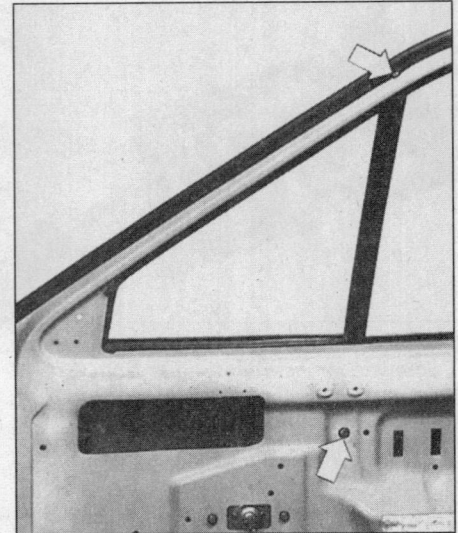

Fig. 12.11 Door window surround securing tab (arrowed) (Sec 10)

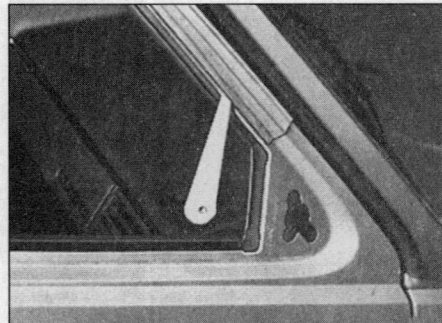

Fig. 12.12 Lower trim moulding with plastic wedge to remove (Sec 10)

10.15A Door check strap cover

10.15B Door check strap

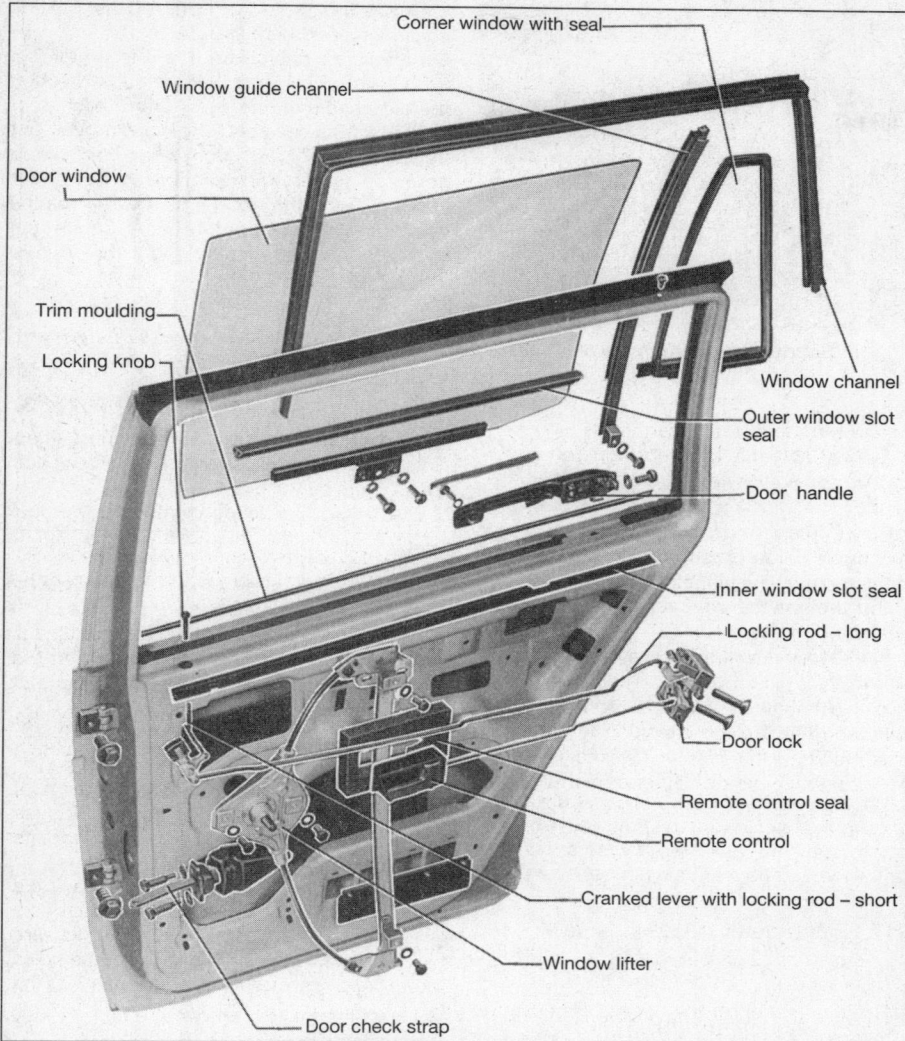

Fig. 12.13 Exploded view of rear door (Sec 12)

Labels for Fig. 12.13:
- Corner window with seal
- Window guide channel
- Door window
- Trim moulding
- Locking knob
- Window channel
- Outer window slot seal
- Door handle
- Inner window slot seal
- Locking rod – long
- Door lock
- Remote control seal
- Remote control
- Cranked lever with locking rod – short
- Window lifter
- Door check strap

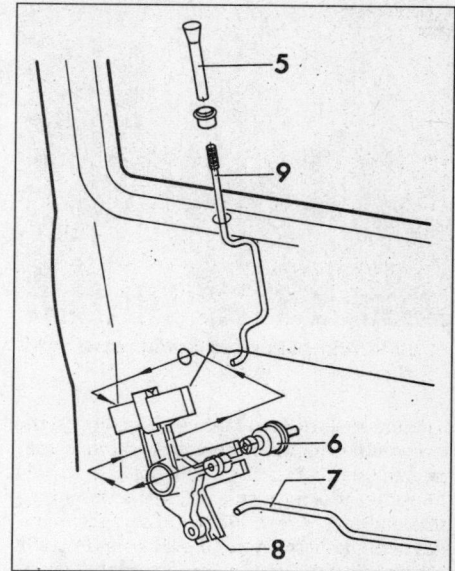

Fig. 12.14 Rear door cranked lever unit and associated components (Sec 12)

5 Locking knob   8 Cranked lever
6 Spreader rivet   9 Locking rod (short)
7 Locking rod (long)

locking system, disconnect the vacuum hose from the door switch element, detach the protective bellows and withdraw the hose from the door.

HAYNES HINT *If the same door is to be refitted, mark round the hinges on the door, using a pencil or a fine tipped ballpoint pen. This simplifies realignment of the door.*

3 On models with power windows and door locks remove the relevant components with reference to Sections 34 and 35.
4 Unclip the cranked lever and locking rod from the inner door cavity at the front end, detach the long locking rod, withdraw the lever and short locking rod. Withdraw the long rod separately (Fig. 12.14).
5 Lower the window and working through the inner door aperture detach the window from the lifter (regulator) by undoing the two retaining screws and spring washers. Lower the window into the door cavity.
6 Remove the four retaining bolts with spring washers and withdraw the window lifter (regulator) unit.
7 Prise free and remove the inner window slotted seal and the outer seal.
8 Undo the bolt securing the bottom end of the guide channel and the screw at the top, and remove the channel upwards.
9 Release the corner window seal lip from the

trim moulding and remove the corner window complete with seal.
10 Extract the window channel from the frame. To remove the window, tilt it and lift it out.
11 To remove the door check strap, prise free the pivot pin retaining clip and drive out the pin. Pull free the rubber cover within the door cavity then undo the two retaining bolts and spring washers. Withdraw the check strap unit.
12 Reassemble in the reverse order to dismantling. Check the operation of the door catch, lock and window regulator prior to refitting the trim panel.

13 Doors – removal and refitting

1 Remove the trim panel as described in Sections 7 or 11.
2 On models fitted with the centralised

3 Remove the retaining clip from the pin in the door check strap and tap out the pin. Fit the pin and clip back into the clevis to prevent their being lost.
4 With an assistant supporting the weight of the door, or with the bottom of the door propped, remove the screws from the hinges and take the door off (photo).

13.4 Door hinge unit

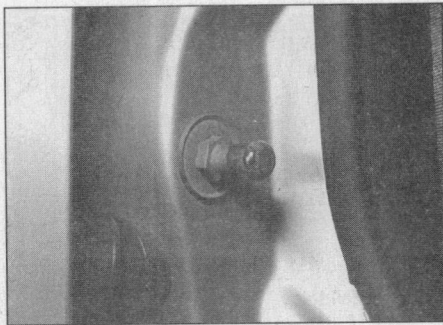

14.1 Front door striker location

16.3 Bonnet hinge and retaining nuts

4 Unclip the cable from the retainers in the engine compartment (Fig. 12.15).
5 Detach the cable from the release handle inside the vehicle and pull the cable through the bulkhead grommet and remove it (photo).
6 Refit in the reverse order to removal, but adjust the cable so that when the clamp screw is tightened the cable is not under tension. Bend the excess wire over. Pull the release lever to operate and check the satisfactory operation of the catch before closing the bonnet.

5 Refitting is a reversal of removal. If the same door is refitted, screw the screws in until they are just tight and line the hinges up with the marks made prior to removal. If using a different door, screw the hinge screws in until they are just tight, close the door and push it into the position which aligns its edge with the body contour. Move the door up and down or sideways as necessary to give an even gap between the edge of the door and the body pillars. When door alignment is correct, tighten the hinge screws fully, if necessary, adjust the striker as described in the following Section.

14 Door striker – adjustment

1 Mark round the door striker with a pencil, or a fine ballpoint pen (photo).
2 Fit a spanner to the hexagon on the striker and unscrew the striker about one turn so that the striker moves when tapped with a soft-headed hammer.
3 Tap the striker towards the inside of the car if the door rattles, or towards the outside of the car if the door fits too tightly, but be careful to keep the striker in the same horizontal line, unless it also requires vertical adjustment. Only move the striker a small amount at a time; the actual amount moved can be checked by reference to the pencil mark made before the striker was loosened.
4 When a position has been found in which the door closes firmly, but without difficulty, tighten the striker.

15 Front grille – removal and refitting

1 Undo and remove the single screw securing the grille lower side section (right or left will do), then pull free the side section from the grille.
2 Unclip the grille at the top edge and withdraw it from the opposing lower side grille section.
3 Refit in the reverse order to removal.

16 Bonnet – removal, refitting and adjustment

1 Support the bonnet in its open position, and place some cardboard or rags beneath the corners by the hinges.
2 Disconnect the windscreen washer pipe from the bonnet (photo).
3 Mark the location of the hinges with a pencil (photo).
4 With the help of an assistant, unscrew the nuts and withdraw the bonnet from the car.
5 Refitting is a reversal of removal, but adjust the hinges to their original positions so that the bonnet is level with the surrounding bodywork. If necessary the front of the bonnet height may be adjusted by screwing the striker located beneath the bonnet in or out.

17 Bonnet lock release cable – removal and refitting

1 Raise and support the bonnet. Remove the front grille (Section 15).
2 Reaching down through the aperture in the front cross panel unscrew and loosen off the cable to bonnet catch clamp screw.
3 Working from the front, unscrew and remove the cable retainer securing screw to the left of the catch.

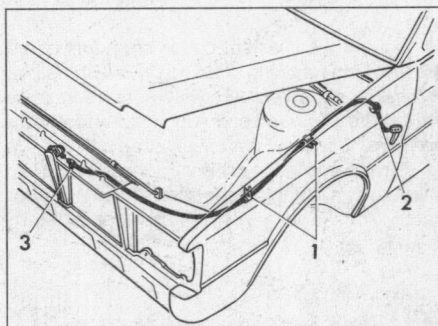

Fig 12.15 Bonnet lock release cable routing (Sec 17)

1 Cable retaining clips   3 Clip
2 Seal grommet

18 Boot lid – removal, refitting and adjustment

1 Support the boot lid in its open position, and place some cardboard or rags beneath the corners by the hinges.
2 Disconnect the wiring loom and lower lock hose, as necessary, then mark the location of the hinges with a pencil.
3 With the help of an assistant, unscrew the nuts and withdraw the boot lid from the car.
4 Refitting is a reversal of removal, but adjust the hinges to their original positions so that the boot lid is level with the surrounding bodywork.

19 Boot lid lock – removal, refitting and adjustment

1 Open the boot lid and, working through the inner panel aperture, drive out the eccentric retaining pin.
2 Detach and remove the lock housing assembly, the component parts of which are shown in Fig. 12.16.
3 Insert the key to remove the lock cylinder from its housing.
4 Refit in the reverse order to removal. On completion adjust the lid lock by loosening off the striker plate so that it is located, but not fully tightened. Close the boot lid to centralise the striker plate, then re-open the lid and fully tighten the striker plate securing screws. If necessary readjust the rubber stops.

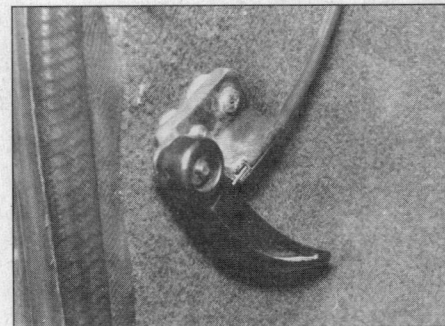

17.5 Bonnet release handle and cable

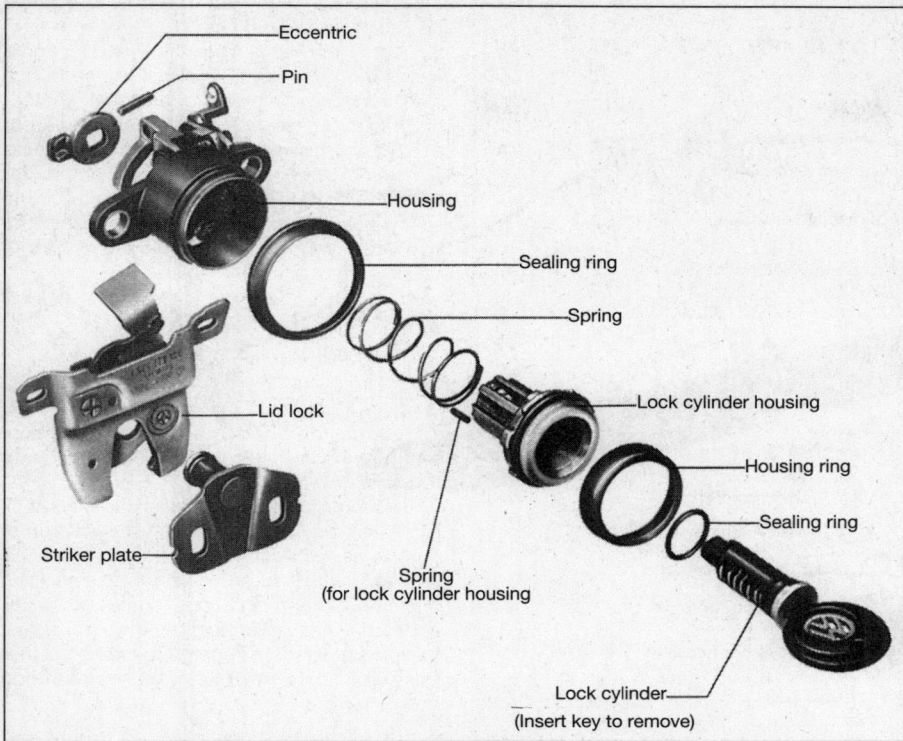

Fig. 12.16 Boot lid lock components (Sec 19)

Labels: Eccentric, Pin, Housing, Sealing ring, Spring, Lid lock, Lock cylinder housing, Housing ring, Sealing ring, Striker plate, Spring (for lock cylinder housing), Lock cylinder (Insert key to remove)

20 Tailgate (Hatchback) – removal, refitting and adjustment

1 Disconnect the battery negative lead.
2 Open the tailgate and support it.
3 Disconnect the cable for the rear light cluster.
4 Unclip and detach the luggage shelf stays from the tailgate.
5 Detach the rear window demister lead connector.
6 Get an assistant to support the tailgate, then disconnect the tailgate support struts (Fig. 12.17).
7 Prise free the tailgate hinge covers from the rear end of the roof headlining.
8 Mark the location of the hinges with a pencil, then, with the help of an assistant, unscrew the nuts and withdraw the tailgate from the car.
9 Refitting is a reversal of removal, but adjust the hinges to their original positions. Use the adjustment of the hinges on the tailgate to position the tailgate in relation to the rear panel, and the adjustment of the hinges on the roof panel to position the tailgate in relation to the side panels.

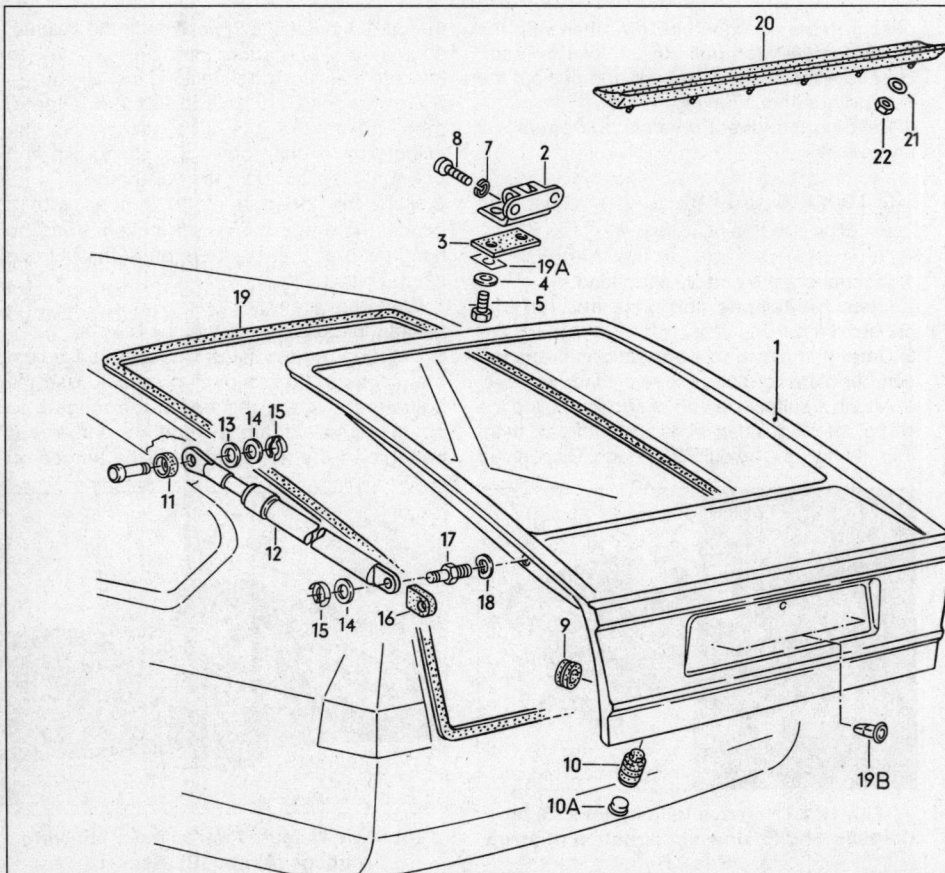

Fig. 12.17 Tailgate components – Hatchback (Sec 20)

1   Tailgate
2   Hinge
3   Gasket
4   Washer
5   Bolt
7   Lock ring washer
8   Screw
9   Grommet
10 Rubber stop
10A Spacer
11 Washer
12 Gas-filled strut
13 Washer
14 Washer (concave)
15 Lockwasher
16 Spacer
17 Pin
18 Spring washer
19 Gasket
19A Spacer shim
19B Clip
20 Deflectors (spoiler)
21 Washer
22 Nut

12

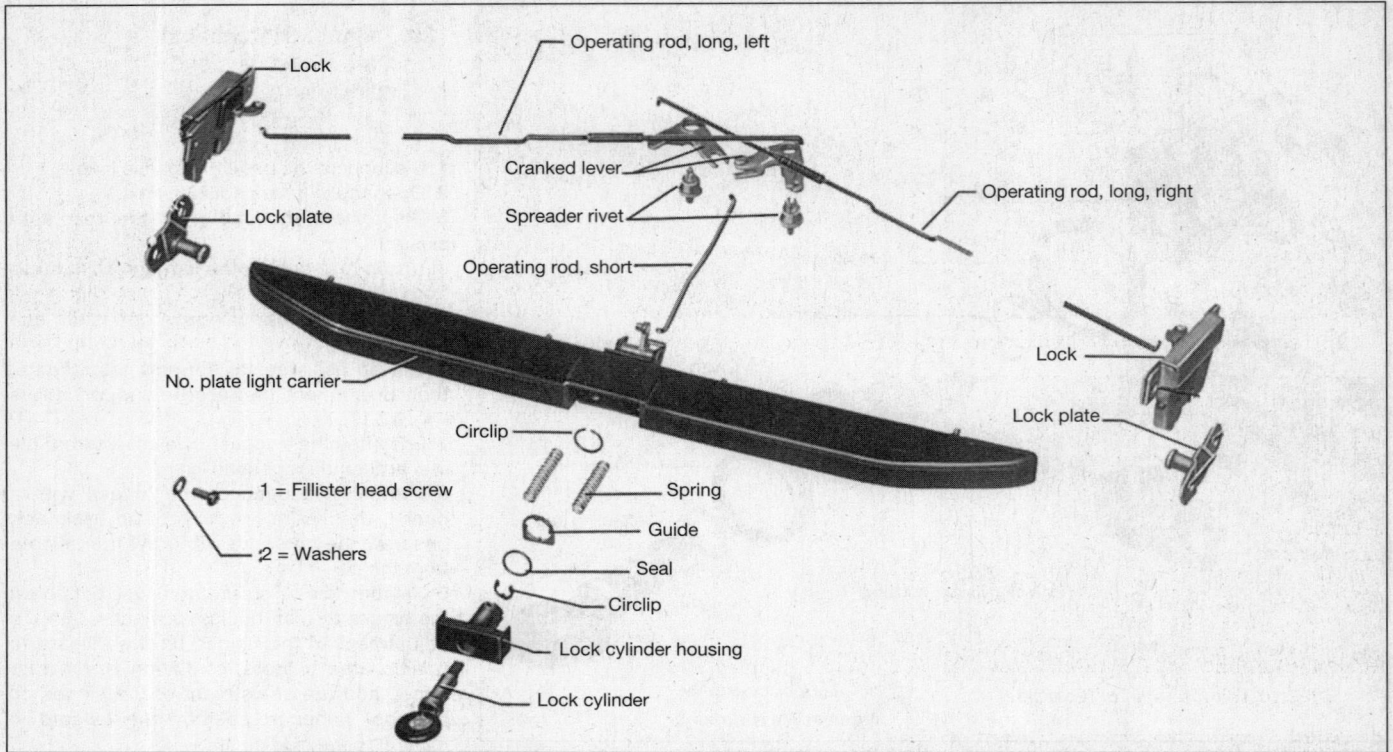

Fig. 12.18 Tailgate (rear flap) lock components – Hatchback (Sec 21)

21 Tailgate rear flap lock (Hatchback) – removal and refitting

1 Raise and support the tailgate and remove the trim panel from it.
2 Referring to Fig. 12.19, unclip and detach the lock operating rod.
3 Unclip and disengage the spreader rivets from the cranked lever, then detach the left and right-hand operating rods.
4 Remove the screws securing the lock to the rear flap, withdraw the lock and disengage the operating rod by pressing in direction indicated in Fig. 12.20.
5 To remove the lock cylinder and housing, prise free the securing circlip A, then with the key in the cylinder, pull out the lock cylinder. Prise free the second circlip B and pull out the lock cylinder housing (Fig. 12.21).
6 Refitting is a direct reversal of the removal procedure.

22 Tailgate (Estate) – removal and refitting

1 Disconnect the battery earth lead.
2 Open the tailgate and prise free the trim panel from it.
3 Detach the wiring connections from the number plate light and the rear window wiper.
4 Attach a suitable length of strong cord to the end of the dismantled wires and withdraw them from the tailgate by pulling through. Disconnect the cord and leave it in position in the tailgate, so that the procedure can be reversed to relocate the wires when refitting the tailgate.
5 Get an assistant to support the tailgate, then detach the gas-filled struts from the tailgate by levering down the spring clip and pulling the socket from the ball (photo).
6 Mark the location of the hinges with a pencil. Prise free the plastic covers from the headlining at the rear to expose the tailgate hinge bolts (photos).
7 With the assistant firmly supporting the tailgate, unbolt and remove the tailgate.
8 Refitting is reversal of removal, but adjust the hinges to their original positions. Use the adjustment of the hinges on the tailgate to position the tailgate in relation to the rear panel, and the adjustment of the hinges on

Fig. 12.19 Tailgate lock components (Sec 21)

1 Operating rod   3 Operating rod
2 Cranked lever   4 Operating rod

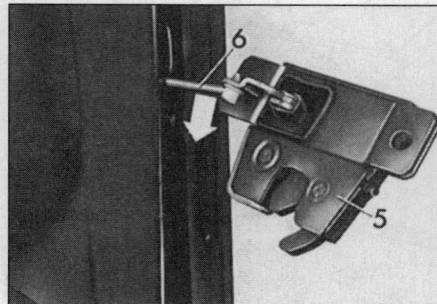

Fig. 12.20 Remove tailgate lock (5) by pressing rod (6) down in direction of arrow (Sec 21)

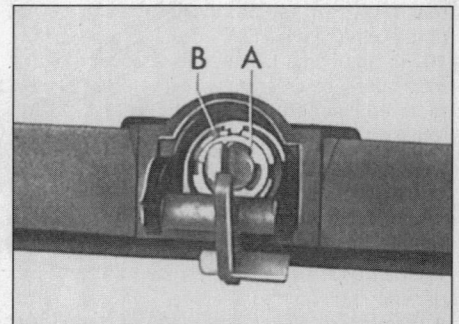

Fig. 12.21 Tailgate lock cylinder showing circlips (A) and (B) (Sec 21)

22.5 Tailgate strut-to-body attachment

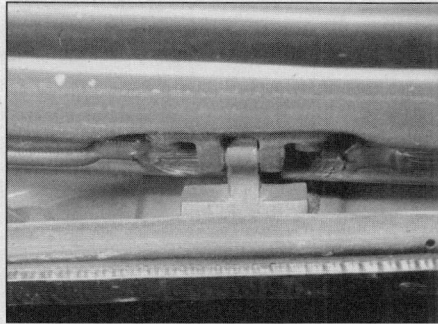

22.6A Mark the hinge positions (tailgate) then . . .

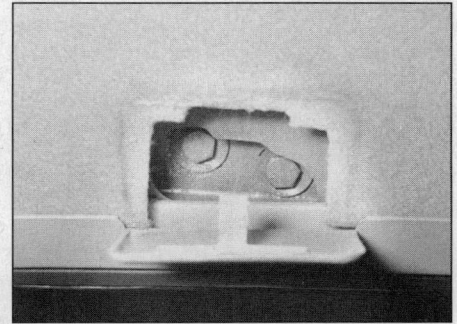

22.6B . . . remove the hinge securing bolts

22.10 Tailgate striker (Estate)

23.2 Tailgate lock cylinder viewed through inner panel aperture

the roof panel to position the tailgate in relation to the side panels.

9 Reattach the wires to the cord and pull them carefully into position through the tailgate, detach the cord and reconnect the wires. Check the number plate lights and the rear wiper for satisfactory operation before refitting the trim panel.

10 To adjust the tailgate striker loosen off the retaining screws and move the striker in the required direction then retighten the screws (photo).

23 Tailgate lock (Estate) – removal and refitting

1 Open the tailgate and remove the rear trim panel by prising free the plastic clips.

2 Compress the retaining lugs of the grip on the inside (photo) and withdraw the lock cylinder unit and grip.

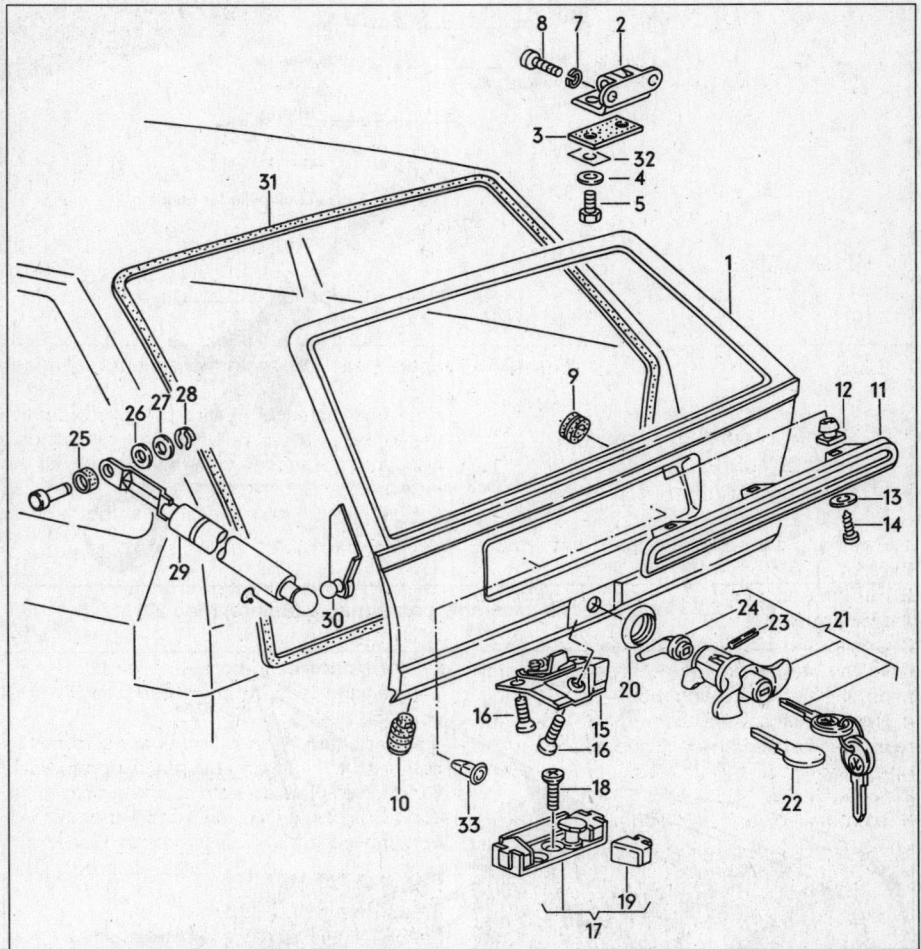

Fig. 12.22 Tailgate and associated components (Estate) (Sec 22)

1 Tailgate	12 Speed nut	22 Key
2 Hinge	13 Washer	23 Roll pin
3 Gasket	14 Screw	24 Eccentric
4 Washer	15 Lock unit	25 Spacer
5 Bolt	16 Screw	26 Washer
7 Spring washer	17 Striker plate	27 Concave washer
8 Screw	18 Screw	28 Lock washer
9 Grommet	19 Rubber stop	29 Strut
10 Stop	20 Seal washer	30 Ball mounting
11 Number plate cover	21 Lock unit	31 Gasket

12

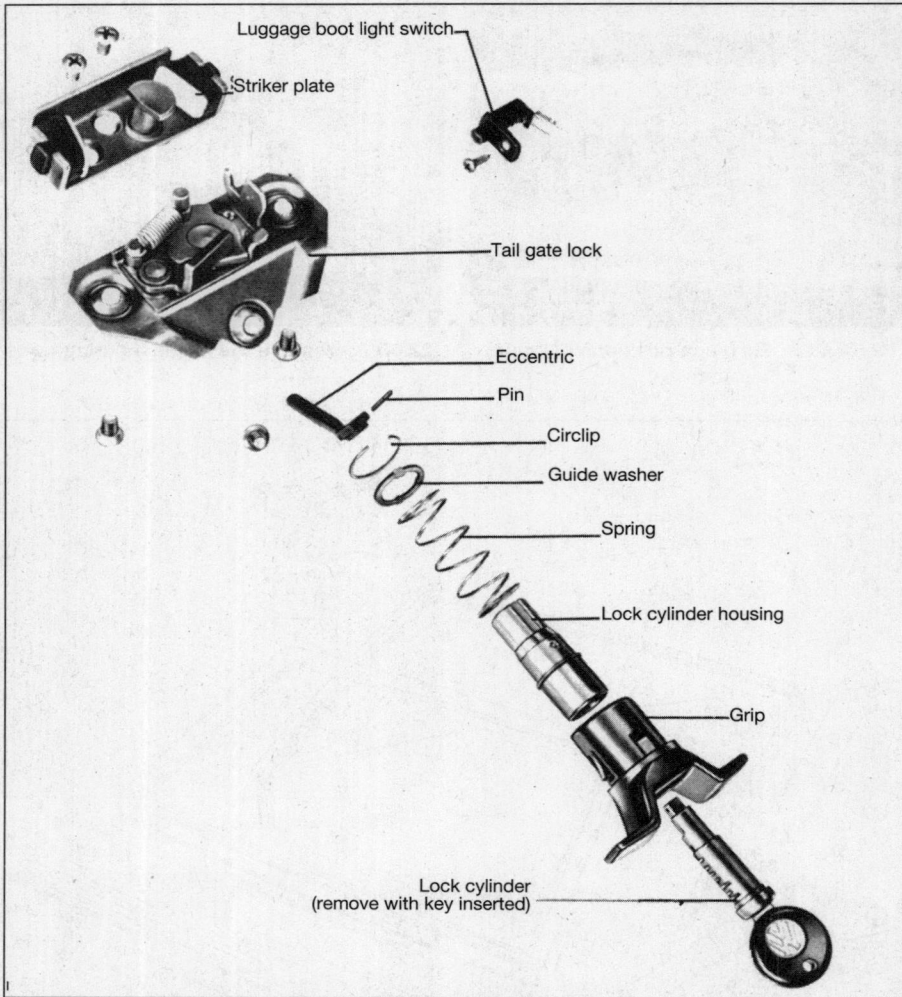

Fig. 12.23 Tailgate lock assembly components (Estate) (Sec 23)

- Luggage boot light switch
- Striker plate
- Tail gate lock
- Eccentric
- Pin
- Circlip
- Guide washer
- Spring
- Lock cylinder housing
- Grip
- Lock cylinder (remove with key inserted)

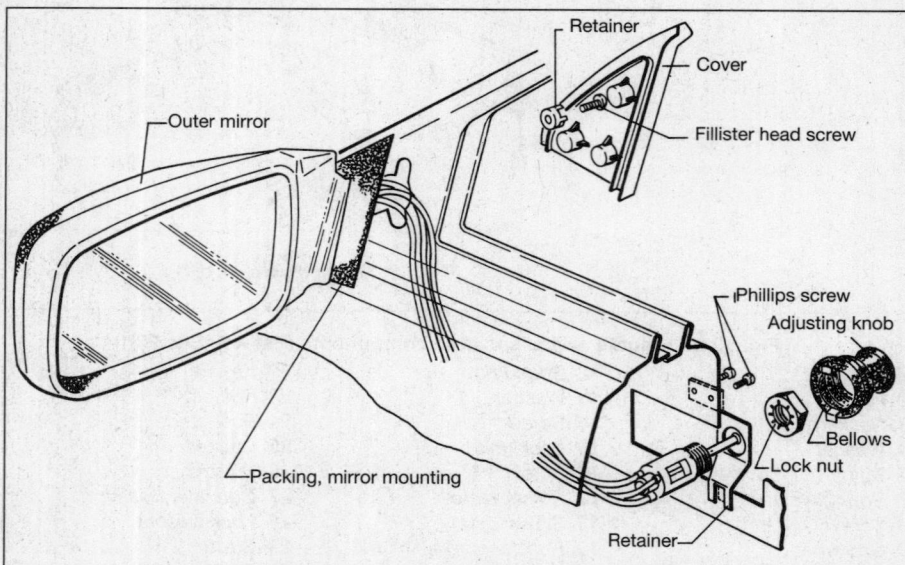

Fig. 12.24 External mirror and adjuster components – mechanical (Sec 25)

- Retainer
- Cover
- Fillister head screw
- Outer mirror
- Phillips screw
- Adjusting knob
- Bellows
- Lock nut
- Packing, mirror mounting
- Retainer

23.5 Tailgate lock unit and three securing screws

3 To remove the lock from the grip drive out the roll pin securing the eccentric to the lock, withdraw the eccentric.
4 Extract the circlip, the guide washer and spring and separate the lock cylinder housing from the grip (Fig. 12.23).
5 Undo the three retaining screws to remove the lock unit (photo).
6 Refit in the reverse order to removal. Apply a ring of sealant to the lock seal.

24 Tailgate/rear hatch support struts – removal and refitting

1 Raise and support the tailgate/rear hatch. An assistant or suitable support will be necessary if both struts are to be removed.
2 Prise free the circlips, remove the single washer at the top of the strut, two washers at the bottom end and remove the strut.
3 The strut is gas filled and no attempt must be made to repair or dismantle the strut. Renew if defective.
4 Refit in the reverse order of removal.

25 Door mirror – removal and refitting

Standard type

1 Prise free the trim cover.
2 Remove the retaining screws, withdraw the mirror and packing piece.
3 Refit in reverse order of removal. Adjust by pressing on glass as required.

Manual adjustment (internal) type

4 Pull free the adjusting knob and bellows.
5 Remove the door trim panel as described in Section 7.
6 Unscrew and remove the adjuster locknut then pass the adjuster through the retainer (Fig. 12.24).
7 Prise free the mirror internal cover, remove the three retaining screws together with the trim retainers whilst supporting the mirror.
8 Withdraw the mirror and feed the adjusting knob assembly up through the door cavity.
9 Refit in reverse order of removal and adjust mirror.

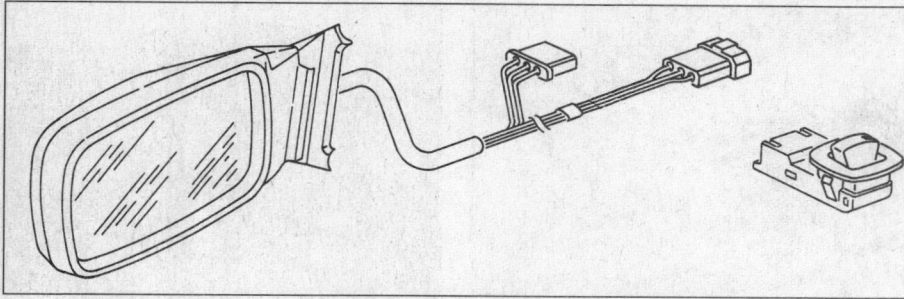

Fig. 12.25 Electrically adjusted external door mirror components (Sec 25)

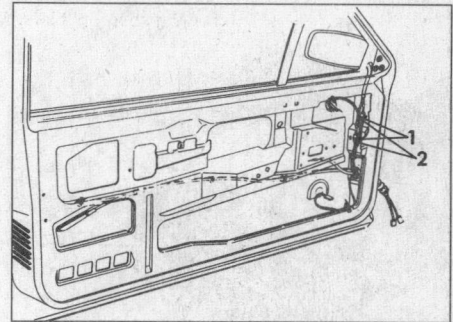

Fig. 12.26 Electrically adjusted/heated door mirror wiring clip (1) and connector (2) locations (Sec 25)

Electrically operated type

10 Disconnect the battery earth lead.

11 Remove the door trim panel as described in Section 7.

12 Disconnect the heated rear window/exterior rear view mirror and mirror adjuster switch lead connectors (Fig. 12.25).

13 Detach the mirror as described in paragraph 7 and withdraw the mirror feeding the cables up through the door cavity (Fig. 12.26).

14 Refit in reverse order of removal and adjust the mirror.

26 Bumpers – removal and refitting

1 Disconnect the direction indicator/rear number plate wiring and headlight washer equipment from the bumper, as applicable.

2 Unscrew and remove the centre mounting bolts located on the front underframe or in the rear luggage compartment (photo).

3 Slide the bumper from the side guides, keeping the bumper parallel to the front of the car.

4 If necessary, remove the screws and withdraw the corner sections, then unclip the cover from the bumper.

5 Refitting is a reversal of removal.

27 Windscreen and rear window glass – removal and refitting

The windscreen and rear window are directly bonded to the metalwork, and the removal and fitting of this glass should be left to your VW dealer or a specialist glass replacement company.

28 Facia – removal and refitting

1 Disconnect the battery earth lead.

2 Remove the steering wheel as described in Chapter 11.

3 Undo the two screws securing the central trim panel in position, withdraw the panel and detach the cigarette lighter lead.

4 Referring to Fig. 12.27, undo and remove the five passenger side trim panel retaining screws. Pull the trim back to release it from the lower retaining clip A and remove it.

5 Where a manual choke system is fitted, drive out the roll pin securing the choke control knob, remove the knob and the ring nut (Fig. 12.28).

6 Referring to Fig. 12.31, undo the three securing screws and pull the lower trim panel free from its lower edge retaining clip.

7 Remove the radio (if fitted) referring to Chapter 10 for details.

8 Pull free the heater and ventilation control knobs then carefully press out the trim and detach the controls.

9 Undo the screws from the locations arrowed in Fig. 12.30 then partially withdraw the instrument panel, detach the switch and

26.2 Front bumper retaining bolts

Fig. 12.27 Passenger side trim retaining screw locations (arrowed), and clip (A) – left-hand drive shown (Sec 28)

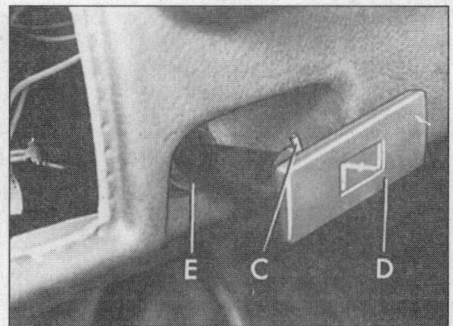

Fig. 12.28 Manual choke control grip (D), ring nut (E) and sleeve (C) (Sec 28)

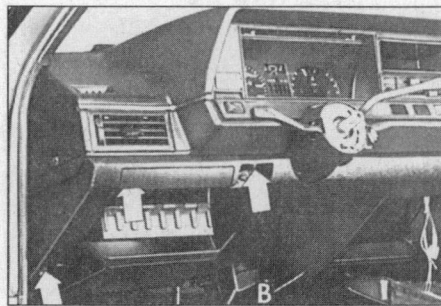

Fig. 12.29 Driver's side trim retaining screw locations (arrowed) and clip (B) – left-hand drive shown (Sec 28)

Fig. 12.30 Instrument panel retaining screw positions (left-hand drive shown) (Sec 28)

12

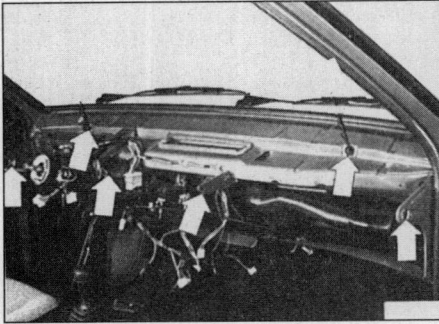

Fig. 12.31 Facia panel retaining screw positions (Sec 28)

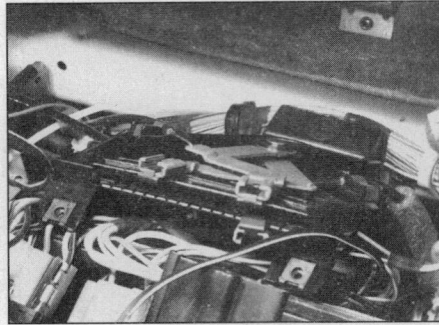

28.9 Disconnect the switch and warning lamp leads and connections from earthing strip

Fig. 12.32 Plenum chamber nut location (arrowed) (Sec 28)

warning lamp leads and remove the panel trim (photo).

10 Detach the main wiring loom connections and the speedometer cable by reaching behind the instrument panel and compressing the cable connector securing lugs.

11 Undo the upper securing screw and remove the instrument panel and detach the vacuum gauge from the rear face of the fuel consumption gauge.

12 Press the fresh air vent out (with care).

13 Unscrew and remove the screws from the positions shown in Fig. 12.31.

14 Peel back and remove the door A pillar trim.

15 Raise the bonnet and working at the bulkhead end undo and remove the two nuts from the plenum chamber (Fig. 12.32).

16 Withdraw the facia panel unit, detaching from the body attachment points and remove it through the door aperture.

17 Refitting is a reversal of the removal procedure.

29 Central console – removal and refitting

1 Disconnect the battery earth lead.

2 Remove the single retaining screw each side of the console (under the facia), and the single securing screw from the console rear end directly to the rear of the gear lever.

3 Prise free and release the gear lever gaiter from the console.

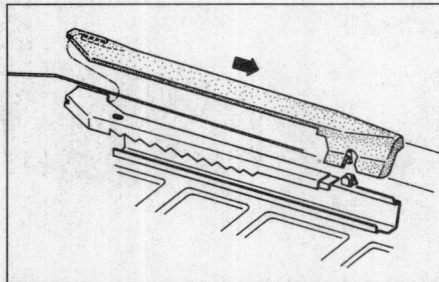

Fig. 12.33 Front seat removal – upper runner cover (Sec 30)

4 Lift the console partly clear, detach the wiring to the console switches as necessary and lift the console clear over the gear lever.

5 Refit in the reverse order to removal.

30 Seats – removal and refitting

Front seat – all models

1 Prise free the lower runner cover and clip, towards the rear end of the seat.

2 Pull the cover from the runner and then pull the seat forwards (Fig. 12.33).

3 Referring to Fig. 12.34, unscrew the cap nut, remove the washer and cheese head screw. Then releasing the securing rod, remove the seat rearwards.

4 Refitting is a reversal of the removal procedure, but the cap nut must be tightened to the manufacturer's recommended torque setting of 1.5 Nm (1.10 lbf ft).

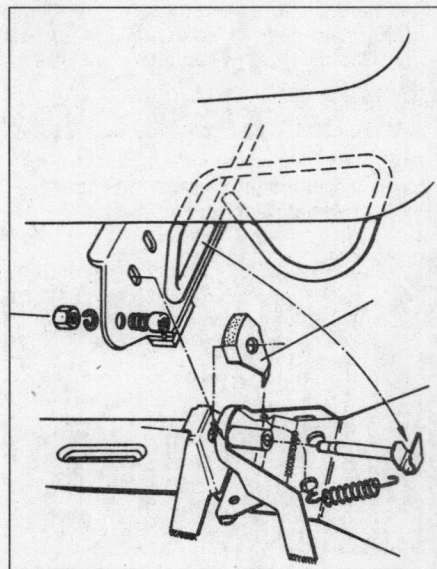

Fig. 12.34 Front seal securing rod and associated components (Sec 30)

Rear seat – Saloon

5 Remove the seat cushion by pressing on the pressure points each side at the front lower edge of the cushion, and lift the cushion out (Fig. 12.35).

6 Rotate the headrest guide sleeves through 90° and remove the headrests by pulling them upwards from the backrest.

7 At the bottom end of the seat backrest, bend the retaining tabs up (one each side), pull the backrest retainers out and upwards and remove the seat backrest.

8 Refitting is a reversal of the removal procedure, but ensure that the backrest retaining hooks fully engage.

Rear seat – Hatchback and Estate

9 Refer to Supplement.

31 Sunroof – general

1 The sunroof requires little in the way of periodic maintenance apart from the following:

(a) Clean out the drain channels occasionally by passing a suitable length of flexible cable (inner speedo cable is ideal) through the channels shown in Fig. 12.36. The drain hoses at the front are housed in the A pillars and the cleaning cable should be passed downwards through them from

Fig. 12.35 Pressure points (arrowed) for removal of rear seat cushion (Sec 30)

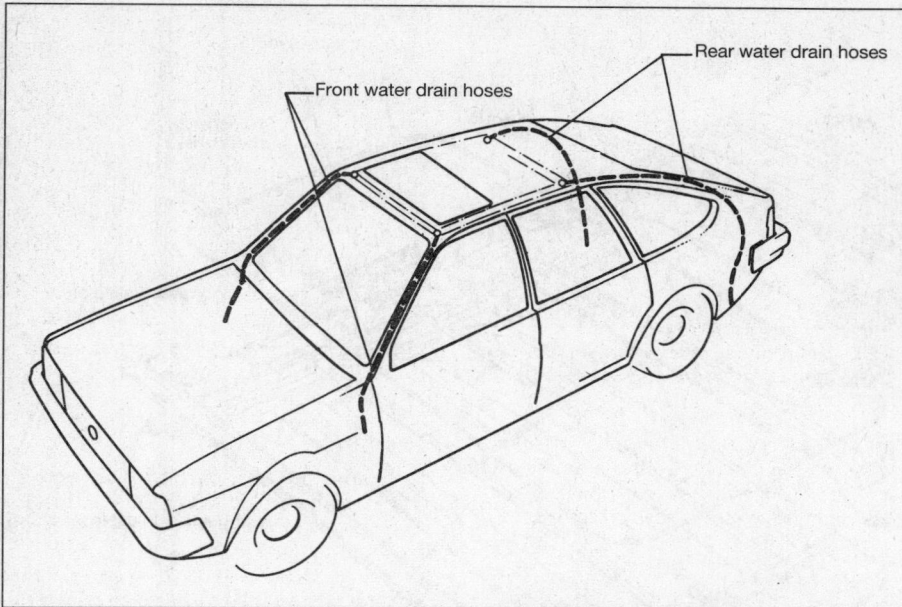

Fig. 12.36 Sunroof drain channel locations (Sec 31)

Front water drain hoses

Rear water drain hoses

Fig. 12.37 Electric sunroof showing emergency manual operation (Sec 31)

1 Unhook pin and swing it to rear
2 Take crank off rear panel
3 Insert crank in gear and turn

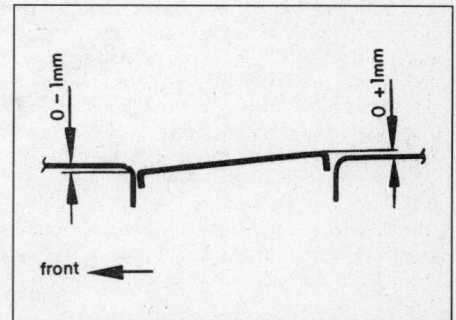

Fig. 12.38 Sunroof height adjustment (Sec 32)

the roof panel opening. The rear hoses are located within the body side panel apertures and should be cleaned by passing the cleaning cable upwards through them.

(b) The guide plates should be lubricated liberally with grease periodically to ensure ease of operation when opening and closing the roof panel.

(c) Check the sunroof panel for adjustment and adjust if necessary as described in Section 32.

2 If for any reason the sunroof panel is to be removed refer to Section 33. Removal and refitting of the associated fixed trim components around the aperture should be entrusted to your VW dealer as special gauges are required during assembly.

3 Where an electrically operated sunroof is fitted, the following procedures should be followed to open or close the roof panel should it malfunction at any time.

(a) Prise free the interior light/roof panel switch cover frame from the inner roof by inserting a suitable lever (ignition key will do) into the slot in the right-hand side of the light cover and levering it free.

(b) Remove the interior light by pressing the spring in towards the centre of the light (at the opposite end to the switch).

(c) Press out the roof panel switch from the inside.

(d) Work through the large aperture and push the spring clips at the trim panel front end rearwards and remove the panel.

(e) Extract the manual crank from its location within the panel.

(f) Unhook the pin (2 in Fig. 12.37) and rotate it rearwards, then engage the manual crank and wind it accordingly to open or close the roof panel.

32 Sunroof – adjustment only

1 The sunroof panel is correctly adjusted when it is positioned as shown in Fig. 12.38. If adjustment is necessary first remove the trim panel from the sliding/tilting roof as described in Section 33.

2 Prise the lock washer from the water guide plate retaining pin then detach the water deflector plate operating lever from the pin. This is necessary to ensure accurate adjustment.

3 Referring to Fig. 12.39, slacken off the mounting screws A and turn the central adjuster screw B to adjust the panel. With adjustment made recheck the height

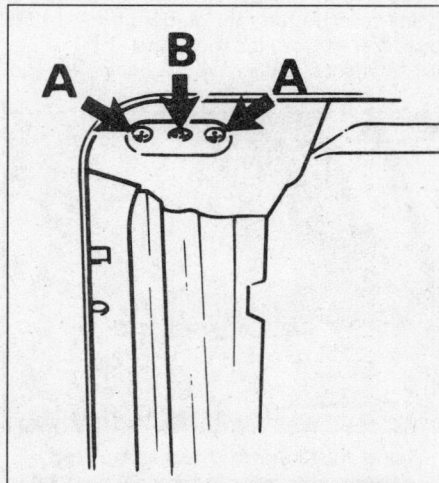

Fig. 12.39 Sunroof mounting screws (A) and adjusting screw (B) (Sec 32)

adjustment pushing up the roof panel slightly to allow for the trim panel. Retighten the mounting screws an equal amount and check that the front guide rail pin is flat against the adjuster screw.

4 Loosen off the guide plate mounting screws and push the rear edge of the roof panel to the correct height, then retighten the screws (Fig. 12.40).

5 If the roof fails to open and close in a parallel manner, the cable drive and rear guides are probably in need of adjustment. Special gauges are required to make this adjustment and such adjustments must be entrusted to your VW dealer.

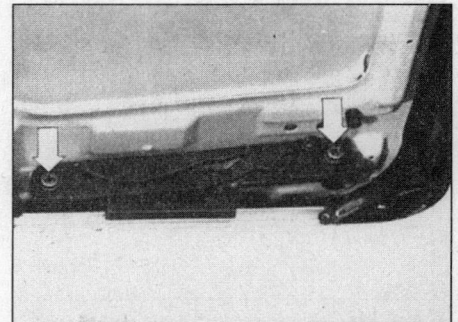

Fig. 12.40 Sunroof guide plate mounting screws (arrowed) (Sec 32)

12

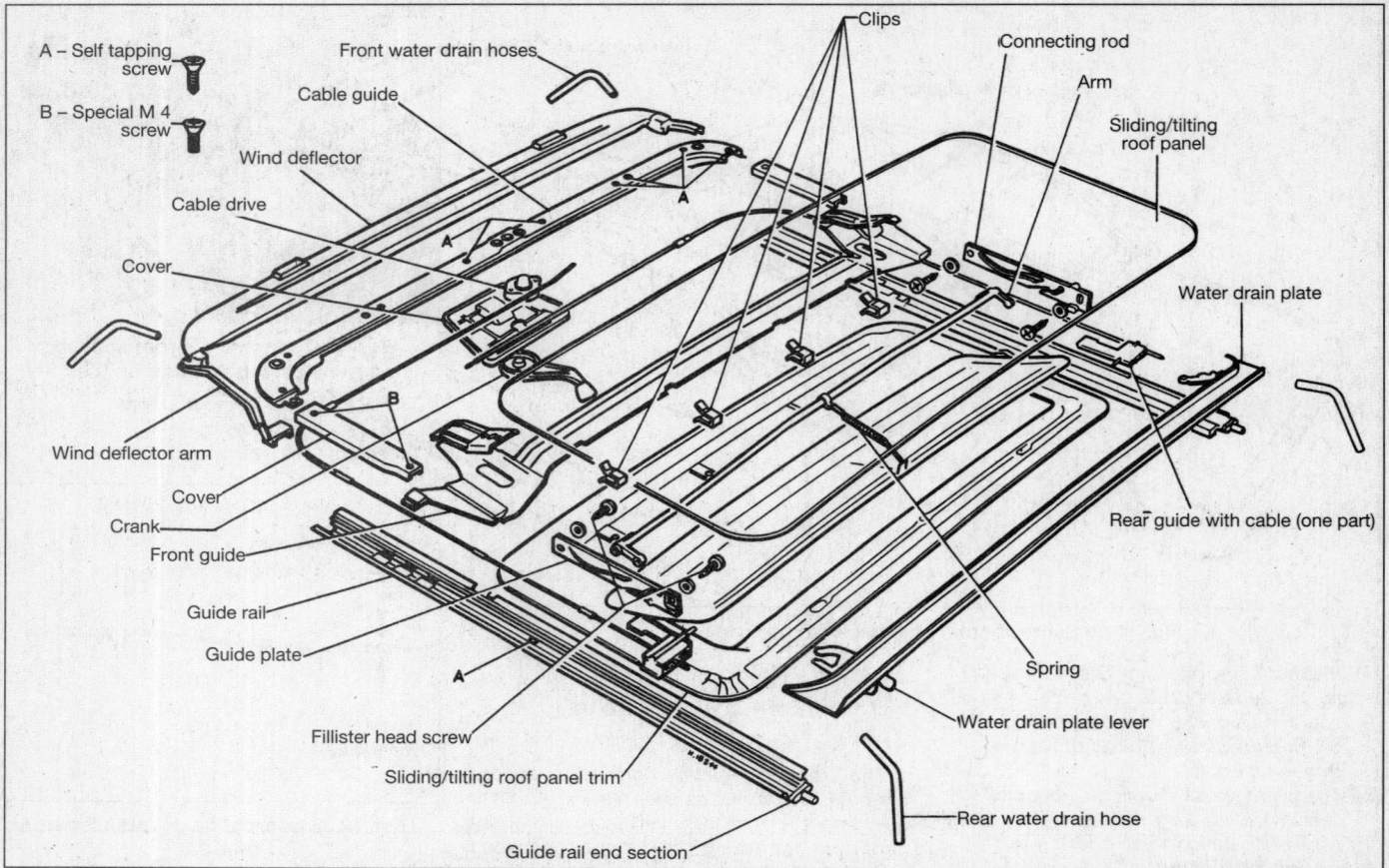

A – Self tapping screw
B – Special M 4 screw

Front water drain hoses
Cable guide
Wind deflector
Cable drive
Cover
Wind deflector arm
Cover
Crank
Front guide
Guide rail
Guide plate
Fillister head screw
Sliding/tilting roof panel trim
Guide rail end section

Clips
Connecting rod
Arm
Sliding/tilting roof panel
Water drain plate
Rear guide with cable (one part)
Spring
Water drain plate lever
Rear water drain hose

Fig. 12.41 Exploded view of the sunroof and associated components (Sec 33)

33 Sunroof – removal, refitting and adjustment

1 Tilt the sunroof using the crank.
2 Using a length of welding rod or similar, make up a hook to the dimensions shown in Fig. 12.42.
3 Carefully prise down the roof panel trim and insert the hook rod to unhook the coil spring from the trim, but take care not to damage the paintwork.
4 Shut the roof panel, unclip the trim panel from the front guide. Push the panel rearwards.
5 Unhook the tilt mechanism on the right-hand side and disconnect the connecting rods to remove them, then pull the left side mechanism free (Fig. 12.43).
6 Prise free the locking washer from the water guide plate retaining pin and press the water guide plate operating lever from the pin.
7 Undo and remove the roof panel to front guide retaining screws.
8 Move the roof panel forwards using the crank to slide the rear guide pin out of the plate then remove the roof panel.
9 Commence refitting by positioning the roof panel into place, then crank the rear guide back to engage the guide pin with the guide plate.
10 Locate and hand tighten only the roof panel to front guide securing screws.
11 The roof panel must now be adjusted for position to provide an equal clearance at the front and rear as shown in Fig. 12.38.
12 Slacken off the guide plate retaining screw and push the water deflector plate into position so that it engages with the retainer pin then fit the locking bolt to secure. Tighten the guide plate retaining screw.
13 Press the left side tilt mechanism into the guide plate so that it engages on the rear guide pin then push the tilt mechanism connecting rod into engagement (on the central pins).
14 Crank the roof panel to the tilt position then refit the coil spring using the previously made hook tool.
15 Refit the roof panel trim by moving the panel forwards to provide a gap at the front edge of 10 to 15 mm (0.4 to 0.6 in) between the panel and the roof aperture edge. Push the trim panel up a fraction at the front and simultaneously slide it forwards to engage the four retaining clips with the front guide.
16 Using the special hooked tool reconnect the coil spring with the panel trim.

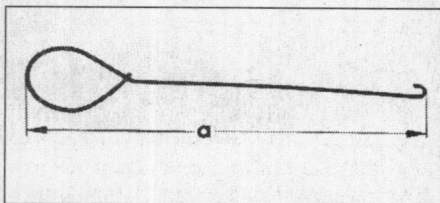

Fig. 12.42 Sunroof coil spring removal hook (Sec 33)

a = 300 mm (11.8 in)

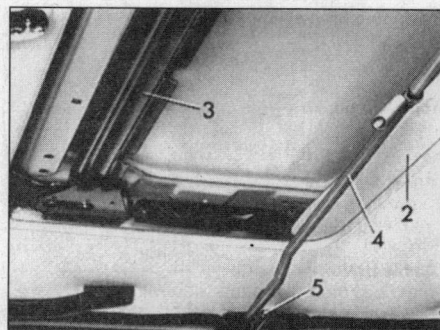

Fig. 12.43 Sunroof and associated components prior to removal (Sec 33)

2 Panel trim
3 Front guide
4 Connecting rod
5 Tilt mechanism

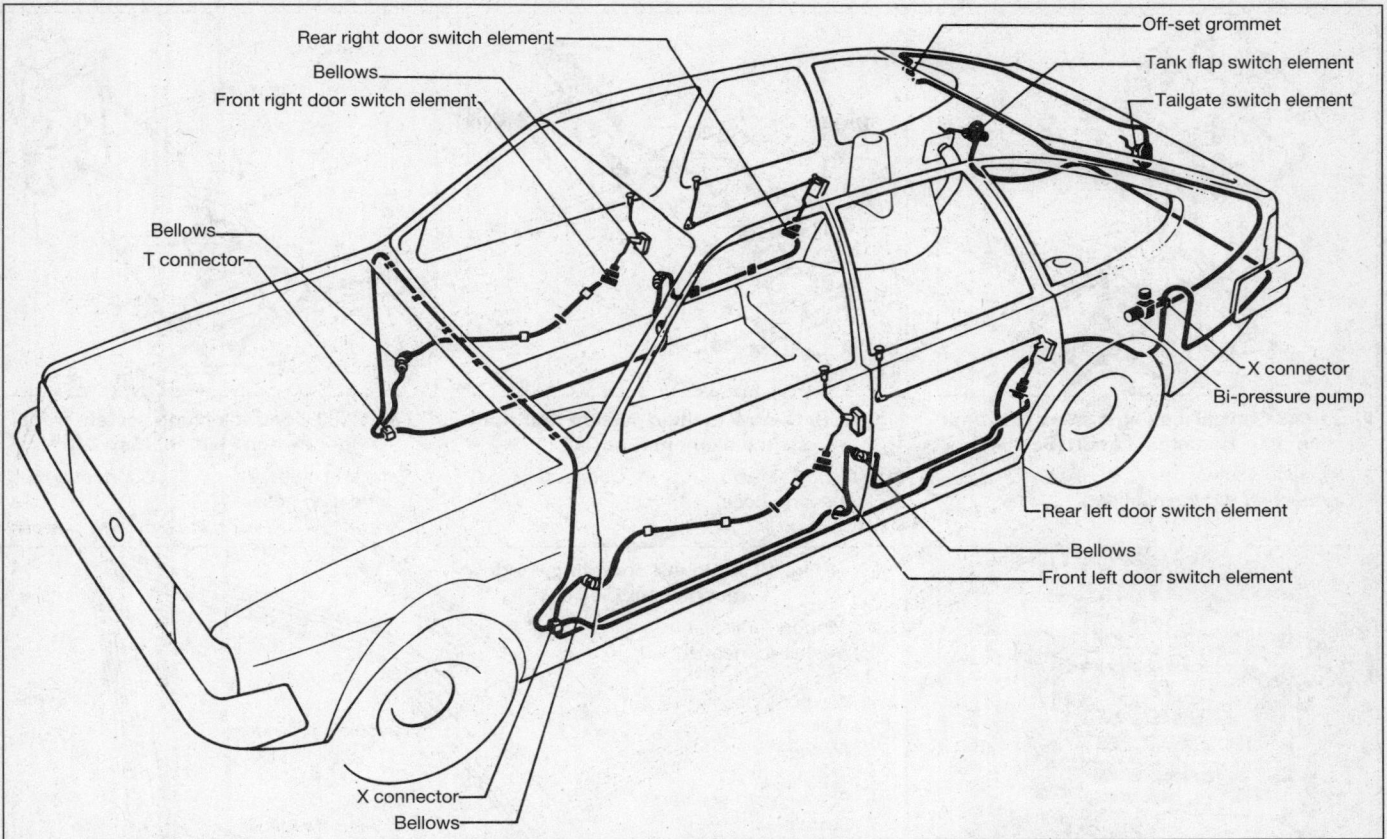

Rear right door switch element
Bellows
Front right door switch element

Bellows
T connector

Off-set grommet
Tank flap switch element
Tailgate switch element

X connector
Bi-pressure pump

Rear left door switch element
Bellows
Front left door switch element

X connector
Bellows

Fig. 12.44 Central locking system layout (Sec 34)

34 Central door locking system – general

1 Certain models are equipped with a central door locking system which automatically locks all doors and the rear tailgate/boot lid in unison with the manual locking of either front door.

2 The system is vacuum operated by means of a bi-pressure pump which is electrically

controlled. The system layout is shown in Fig. 12.44.

3 Any fault in the system is most likely to be caused by a leak in one of the hoses, in which case the hose and connections should be checked and repaired as necessary. Access to the appropriate actuator is generally fairly straightforward after removal of the relevant trim panel.

4 The bi-pressure pump unit is located behind the left side trim panel in the luggage compartment and is housed in a left and right-hand bracket (Fig. 12.45).

5 In the event of a system fault, first check that the fuse S7 is in good condition and securely connected before removing any part of the system.

6 If the fuse is in good condition but the pump fails to operate, the electrical part of the system can be checked using a 12V test light. Remove the side trim panel for access to the pump then detach the wiring multi-connector from the pump unit.

7 With the driver's door lock knob in the open

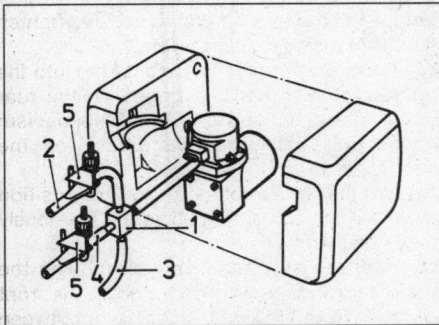

Fig. 12.45 Central locking system bi-pressure pump and connections (Sec 34)

1 Four-way connector
2 Tank filler flap hose
3 Door switches hose
4 Tailgate hose
5 Retaining clips

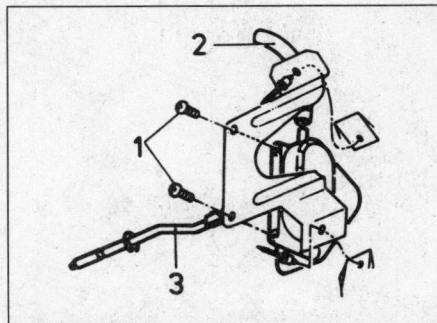

Fig. 12.46 Tailgate switch element (central locking system) showing retaining screws (1), hose connection (2) and connecting rod (3) (Sec 34)

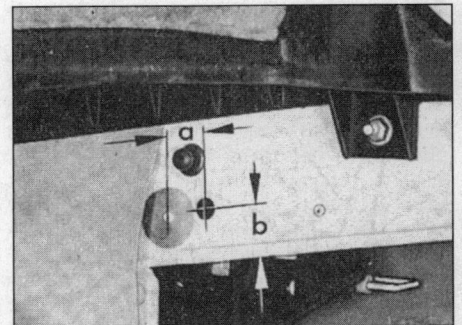

Fig. 12.47 Central locking system fuel tank filler flap switch retaining rivet locations in luggage compartment. Use 10 mm dia. drill at points shown (Sec 34)

a = 20mm (0.78in)   b = 25mm (0.98in)

12

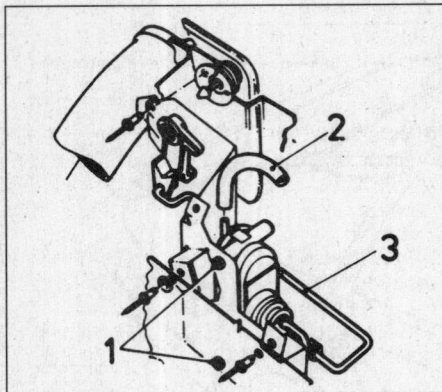

Fig. 12.48 Central locking system fuel tank filler flap element location (Sec 34)

1 Retaining screws   2 Hose
3 Connecting rod (on element)

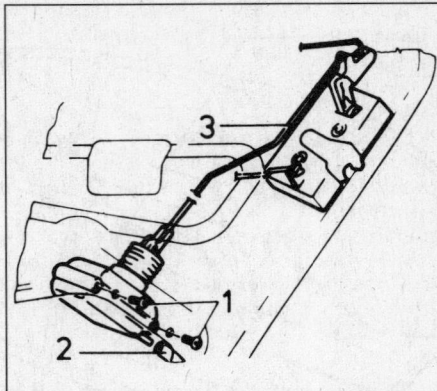

Fig. 12.49 Central locking system rear door switch element (Sec 34)

1 Retaining screws   3 Connecting rod
2 Connecting hose

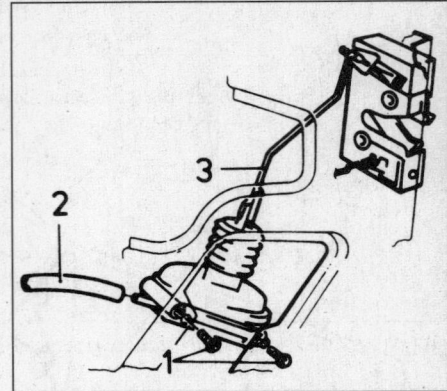

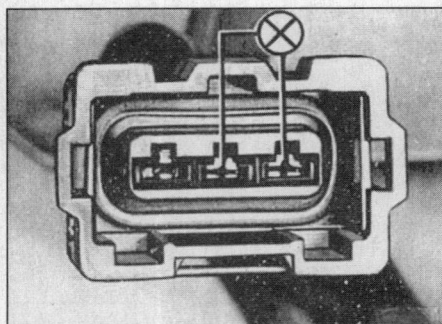

Fig. 12.50 Central locking system front door switch element (Sec 34)

1 Retaining screws   3 Connecting rod
2 Connecting hose

Fig, 12.51 Central locking system bi-pressure pump wiring connector terminals – test light check (Sec 34)

Fig. 12.52 Power operating window mechanism (Sec 35)

1 Window lifter unit
2 Washer (concave)
3 Bolt
4 Washer (concave)
5 Nut
6 Washer
7 Screw
8 Shake-proof washer
9 Switch
11 Relay plate
12 Relay
13 Fuse box
14 Fuse
15 Bellows
16 Bracket
17 Connector housing

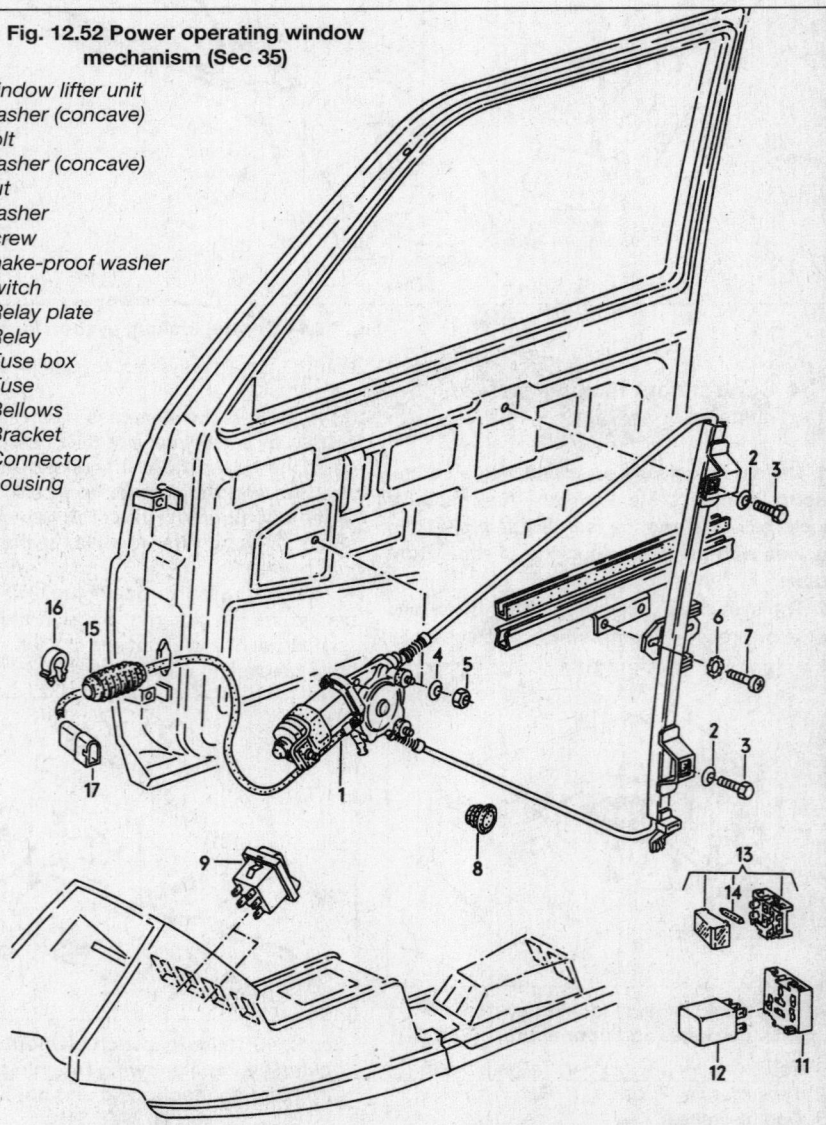

position, connect up the test light leads to the right-hand and central contacts of the connector (Fig. 12.51). If the test lamp fails to light then the driver's door switch element is defective or there is a wiring fault. If the test lamp lights up the driver's door switch unit or the bi-pressure pump is defective.

8 To find out which, connect the test lamp up between the left-hand and central connector terminals with the door lock knob in the closed position. If the test lamp fails to light then the door switch element or wiring are at fault, but if the lamp does light up then the pump unit is defective.

9 It should be noted that if the vehicle has not been used for a long period or where a new bi-pressure pump unit has been fitted, the system may have to be activated several times before becoming operational. Under normal circumstances the system locks should operate within two seconds of actuation. As a general guide if the pump operates for a longer period than five seconds a leak may be present in the system. Do not operate the pump for periods in excess of thirty five seconds (continuously).

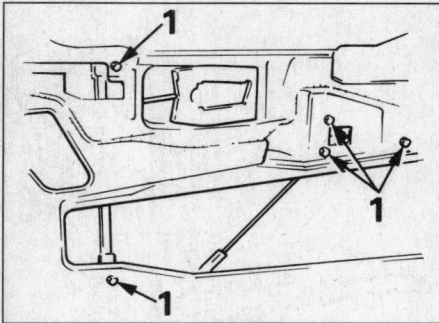

Fig. 12.53 Power window securing screws (1) (Sec 35)

35 Power operated windows – general

1 Certain models are equipped with power (electric) operated windows which can be raised or lowered when the ignition is switched on.
2 If a fault develops, first check that the system fuse has not blown.
3 Access to the electric motors and regulator mechanism is gained by removing the door trim panels as described in Sections 7 or 11.

4 The window lifter (regulator) unit which consists of the motor, cables and guide rail can be removed by first undoing the window securing bolts and lowering the window down into the door cavity (Fig. 12.52).
5 Detach the wiring connector then undo and remove the window lifter retaining screws shown in Fig. 12.53. Lower and remove the window lifter unit from the lower door aperture.
6 Refitting is a reversal of the removal procedure, but route the upper cable under the guide rail securing bracket.
7 Check window operation before refitting the door trim panel.

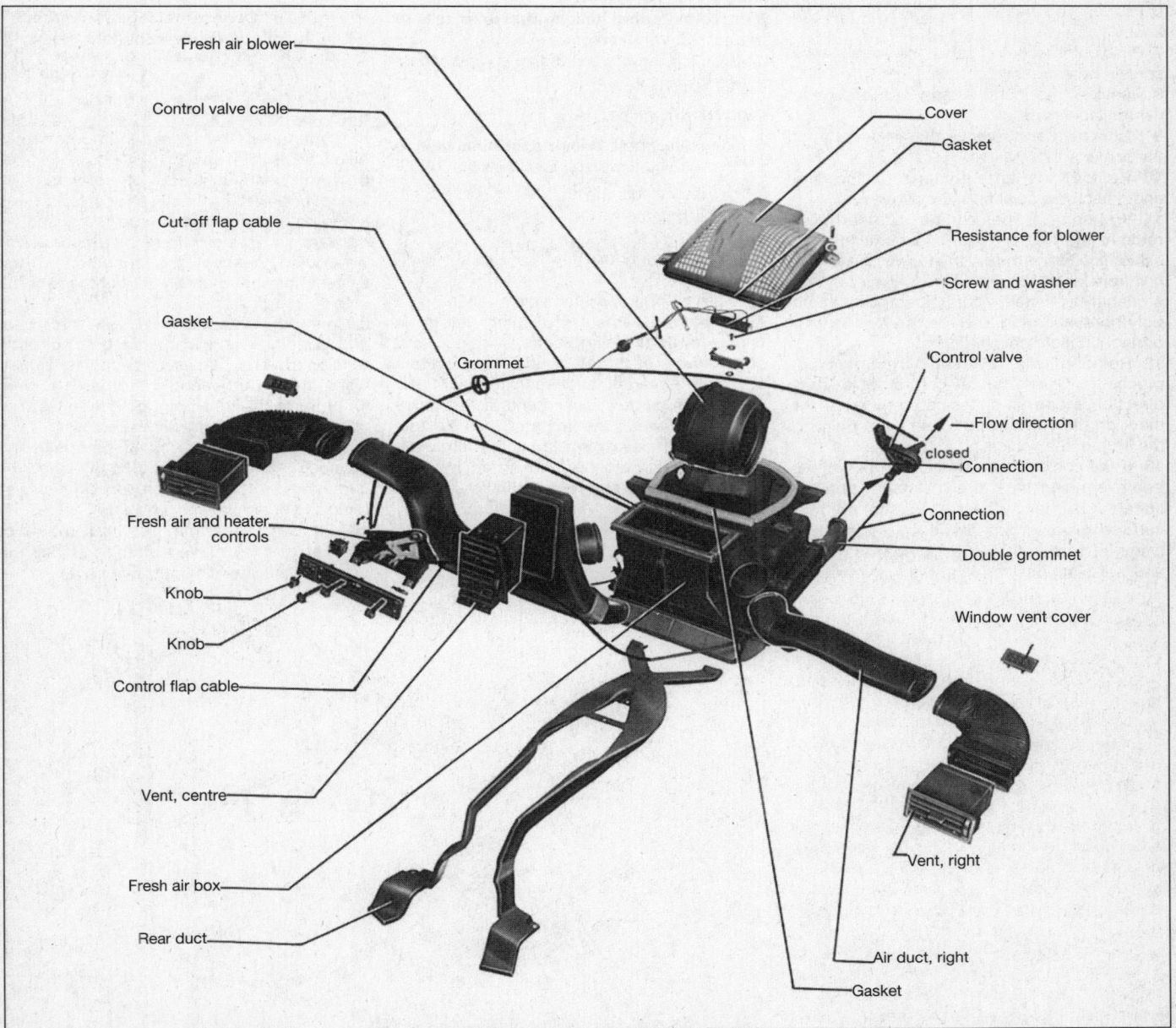

Fig. 12.54 Heater assembly components (Sec 36)

12

36 Heater unit – removal and refitting

Note: *If the heater/fresh air blower motor unit only is to be removed, refer to Chapter 10, Section 42.*

1 Disconnect the battery negative lead.
2 Drain the cooling system as described in Chapter 2.
3 Disconnect the heater hoses at the bulkhead.
4 Detach the temperature control cable from the heater valve.
5 Remove the central console and the left and right trim from below the facia.
6 Pull off the heater and fresh air control knobs.
7 Prise out the control panel, then remove the screws securing the controls.
8 Remove the screws and withdraw the central facia panel.
9 Disconnect and remove the air ducts from the heater.
10 Remove the large heater retaining clip, and detach the cowl from the air plenum.
11 Working in the engine compartment remove the heater mounting bolts located either side of the motor, then lower the heater unit from inside the car. Since there will still be a quantity of water in the heater, place some polythene sheeting, or rags, on the floor to protect the floor covering.
12 Refitting is a reversal of the removal procedure. Check that the control valve fitting direction is correct as indicated by an arrow mark on the valve body which must point to the front of the vehicle.
13 If the coolant hose location grommet has been removed from the bulkhead, apply a suitable rubber adhesive to the contact surface groove. Aid fitting by inserting a length of cord in the grommet sealing groove and pull it into position from within the vehicle.

Lubricate the hoses with washing-up liquid to ease their fitting through the grommet.
14 Before refitting the central console and the trim panels, check the adjustment of the heater control cables as described in the following Section.

37 Heater control cables – adjustment

Control valve cable

1 Move the lower control lever fully to the left.
2 Press the control valve lever to the closed position, then secure the control cable.

Cut-off flap cable

3 Move the upper right control lever fully to the left.
4 Press the cut-off control flap up and secure the control cable.

Control flap cable

5 Move the upper left lever fully to the right.
6 Press the control flap in the direction of the engine and secure the control cable (photo).

38 Air conditioning system

1 This is an option on certain models.
2 Whenever overhaul of a major nature is being undertaken to the engine and components of the air conditioning system obstruct the work and such items of the system cannot be unbolted and moved aside sufficiently within the limits of their flexible connecting pipes, to avoid such obstruction the system should be discharged by your dealer or a competent refrigeration engineer.
3 As the system must be completely evacuated before recharging, the necessary vacuum equipment to do this is only likely to be available at a specialist.

37.6 Heater control flap cable connection

4 The refrigerant fluid is Freon 12 and although harmless under normal conditions, contact with the eyes or skin must be avoided. If Freon comes into contact with a naked flame, then a poisonous gas will be created which is injurious to health.
5 The air conditioner is designed so that it will not operate if the outside temperature is below 5°C (41°F).
6 In hot weather, a pool of water may be found under the vehicle after it has been standing. This is normal and is caused by condensation dripping from the evaporator.
7 Regularly inspect the condition of the system hoses and check the tightness of the unions.
8 Check the compressor drivebelt for wear and condition and keep the belt correctly tensioned. The belt is correctly tensioned when it can be deflected through 10.0 mm (0.39 in) at the mid-point of its longest run under moderate thumb pressure.
9 On four-cylinder engined models the drivebelt tension is adjusted by unbolting the two halves of the compressor pulley and removing or adding spacer shims.
10 On five-cylinder models, the compressor bracket can be moved after releasing the pivot and adjuster slot bolts.

Chapter 13 Supplement:
Revisions and information on later models

Contents

Degrees of difficulty

Easy, suitable for novice with little experience	**Fairly easy,** suitable for beginner with some experience	**Fairly difficult,** suitable for competent DIY mechanic	**Difficult,** suitable for experienced DIY mechanic	**Very difficult,** suitable for expert DIY or professional

1 Introduction

This Supplement contains information which has become available since the manual was first produced, in particular for the period from late 1984 to 1986. However some information also applies to pre-1984 models.

In order to use the Supplement to the best advantage it is suggested that it is referred to before the main Chapters of the manual: this will ensure that any relevant information can be noted and incorporated within the procedures given in Chapters 1 to 12. Time and cost will therefore be saved and the particular job will be completed correctly.

13

2 Specifications

These are revisions of, or supplementary to, the Specifications at the beginning of each of the preceding Chapters.

Four-cylinder engine

1.6 and 1.8 litre carburettor engines (1986-on model year)

Valve timing at 1 mm lift/zero clearance:	**1.6**	**1.8**
Inlet opens BTDC	3°	3°
Inlet closes ABDC	19°	33°
Exhaust opens BBDC	27°	41°
Exhaust closes ATDC	5°	5°

Fuel and exhaust systems

2.0 litre fuel injection engine (from March 1986)

System pressure	5.6 to 6.0 bar (81.2 to 87.0 lbf/in²)
Idle speed	850 ± 50 rpm
Fuel octane requirement (minimum)	98 RON
CO content	1.0 ± 0.5%

Ignition system

Spark plugs

Type (1986 models on)	
1.6 litre engine	Bosch W8DTC, Champion N9BYC
1.8 and 2.0 litre engine	Bosch W6DTC, Champion N7BYC
Spark plug electrode gap	0.8 mm (0.032 in)

Clutch

Cable free play (at clutch pedal)

1986-on models	Approximately 20.0 mm (0.787 in)

Type

(January 1985-on five-cylinder models with 093 gearbox)	Hydraulic actuation
Clutch fluid type (hydraulic clutch)	Hydraulic fluid to FMVSS 116 DOT 3 or DOT 4

Manual gearbox and final drive

Ratios – gearbox 093 code 7K

Final drive	4.70 : 1
1st	2.85 : 1
2nd	1.52 : 1
3rd	0.97 : 1
4th	0.70 : 1
5th	0.56 : 1
Reverse	3.17 : 1

Automatic transmission and final drive

Code numbers/letters – application

089/KAB (October 1982 to August 1983)	Fitted to 1.6 litre engine models

Ratios (089/KAB)

Final drive	3.73 : 1
1st	2.71 : 1
2nd	1.50 : 1
3rd	1.00 : 1
Reverse	2.43 : 1

Electrical system

Alternator drivebelt tension

Toothed rack nut torque	9.0 Nm (6.6 lbf ft)

Suspension and steering

Front wheel alignment

Toe angle difference, left and right, at 20° lock – power steering models from January 1984, chassis No. 33ZEE 125984	– 1° ± 30'
Castor – power steering models from January 1984 chassis No. 33ZEE 125984	+ 1° ± 20'

Torque wrench settings

	Nm	**lbf ft**
Front anti-roll bar link	20	15
Front suspension subframe:		
Stage 1	35	26
Stage 2	Further 90° turn	Further 90° turn

3 Engine

1.6 and 1.8 litre carburettor engines (1986-on) modifications

1 As from the 1986 model year, 1.6 and 1.8 litre carburettor engines are fitted with hydraulic tappets which automatically take up the clearance between the valve stems and camshaft. Engines so equipped have a sticker as shown in Fig. 13.1 on the cylinder head cover.

2 Instead of an oil deflector inside the cylinder head cover, a full length plate is located over the camshaft bearing cap nuts. The camshaft runs in four bearings instead of the previous five, and the former No. 4 camshaft bearing location serves as an oil return for the hydraulic tappets.

3 A different camshaft is fitted to hydraulic tappet engines due to the modified bearing arrangement and the valve timing (see 'Specifications'). For engines fitted with hydraulic tappets, the camshaft is identified by a letter 'B located between the No. 1 cams on the 1.6 litre engine, and a letter 'H' in the same position for 1.8 litre engines.

4 Engines manufactured before January 1985 have an oil pump with 26.0 mm (1.024 in) wide gears. However, all hydraulic tappet engines must be fitted with the later increased capacity oil pump with 30.0 mm (1.181 in) wide gears.

Hydraulic tappets – checking and renewal

5 When starting the engine from cold there may be a little noise from the tappets until the normal oil pressure is re-established. However, if the noise persists, a faulty tappet is indicated. To locate the tappet, run the engine to normal operating temperature indicated by the electric cooling fan switching on, increase the speed to 2500 rpm for 2 minutes, then stop the engine.

6 Remove the cylinder head cover, then turn the engine until the peak of cam No. 1 is pointing upwards.

7 Using a tapered wooden or plastic rod,

Fig. 13.1 Diagram of cylinder head cover sticker depicting engines fitted with hydraulic tappets (Sec 3)

press on the tappet (Fig. 13.3). If the amount of free travel exceeds 0.1 mm (0.004 in), the tappet is faulty and should be renewed.

8 Repeat the procedure on the remaining tappets noting the results as necessary.

Note: *After fitting new hydraulic tappets the engine must not be started for approximately 30 minutes, otherwise there is a danger of the valve heads touching the pistons.*

Valves – servicing

9 Due to the working tolerances of the hydraulic tappets it is important to observe the minimum dimensions between the end of the valve stems and the cylinder head upper surface, as shown in Fig. 13.4, when re-grinding the valve heads and seats. If the dimension is less than the minimum, the hydraulic tappets will not function correctly and the end of the valve stem must be ground squarely to correct it.

4 Cooling system

Water pump sealing ring – modifications

1 When fitting a VW reconditioned water pump, it is important to fit the correct size

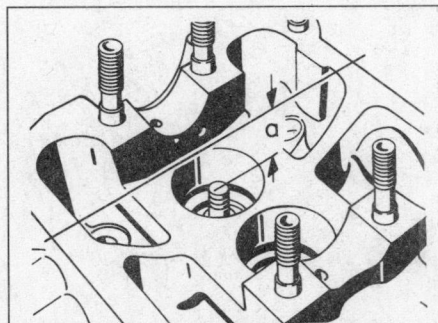

Fig. 13.2 Oil deflector plate for engines fitted with hydraulic tappets (Sec 3)

1 Engine oil filler cap   3 Oil deflector plate
2 Cylinder head cover

sealing ring. On some exchange pumps the sealing ring groove is enlarged for fitting of a 5 mm diameter ring instead of the standard 4 mm diameter ring.

2 Water pumps with this modification have the number '5' stamped on the housing next to the mounting flange.

5 Fuel and exhaust systems

Pressure surge switch (fuel injection models) – checking

1 Disconnect the wiring plug from the idle switch (see next sub-section) and connect an ohmmeter across the plug terminals (Fig. 13.5).

2 Open the throttle and check that zero resistance is obtained. If not, adjust or if necessary renew the idle switch as described in the next sub-section .

3 Disconnect the wiring plug from the pressure surge switch and connect a voltmeter between the plug No. 1 terminal and earth (Fig. 13.6).

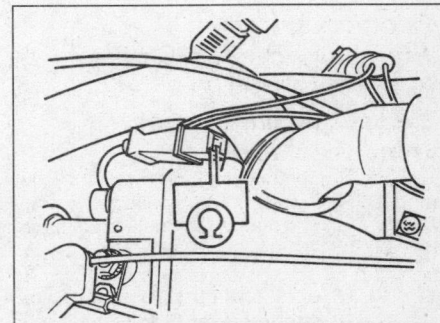

Fig. 13.3 Checking the free travel of a hydraulic tappet (Sec 3)

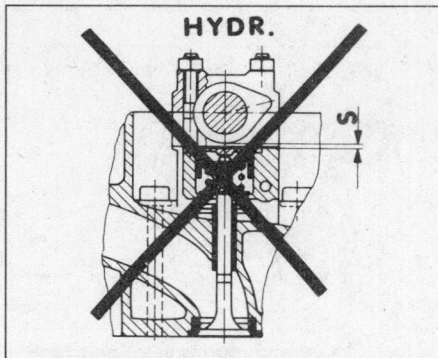

Fig. 13.4 Minimum dimension 'a' between valve stem and cylinder head surface (Sec 3)

Inlet valve 33.8 mm (1.331 in)
Exhaust valve 34.1 mm (1.343 in)

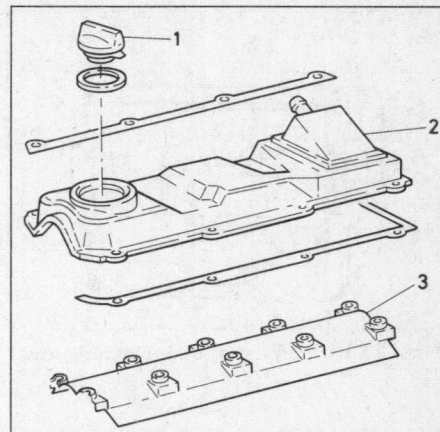

Fig. 13.5 Ohmmeter connection to the idle switch (Sec 5)

13

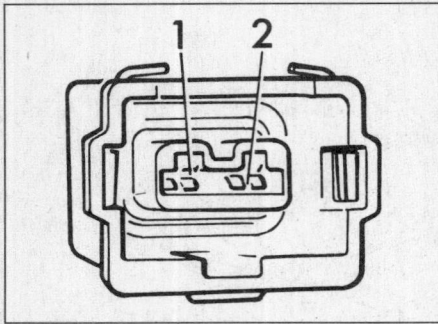

Fig. 13.6 Wiring plug terminal identification (Sec 5)

1 and 2 See text

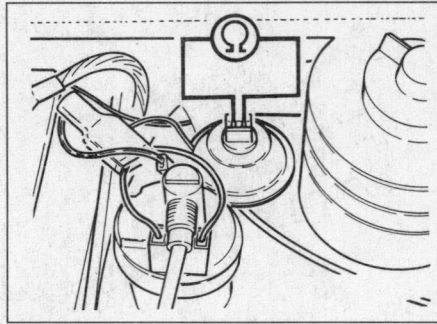

Fig. 13.7 Ohmmeter connection to the pressure surge switch (Sec 5)

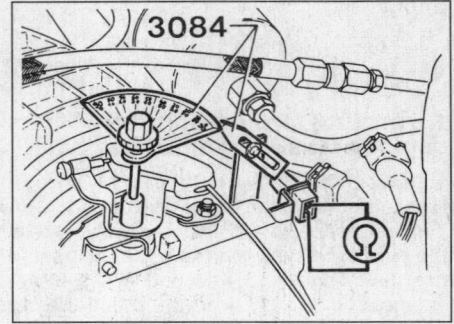

Fig. 13.8 Adjusting the idle switch (Sec 5)

4 With the ignition switched on, open the throttle and check that battery voltage is indicated on the voltmeter. If not, there is an open circuit in the supply wire.

5 Switch off the ignition and connect an ohmmeter across the pressure surge switch terminals. With the throttle closed, infinite resistance should be registered indicating open contacts.

6 Run the engine, then open the throttle quickly. A resistance of 1 ohm should be registered briefly.

7 Disconnect the wiring plug from the cold start valve and connect a voltmeter between the cold start valve plug No. 1 terminal and earth (Fig. 13.6).

8 Run the engine then open the throttle quickly. The voltage should rise and fall between 1 and 10 volts. If not, there is an open circuit in the supply wire.

9 Connect the voltmeter across the cold start valve plug terminals (1 and 2), then operate the starter motor. With the engine cold, battery voltage should be registered. However, with the engine coolant temperature above 40°C (1 04°F), zero voltage should be registered. If not, there may be an open circuit in the supply wire, or the thermo-time switch may need checking as described in Chapter 3, Section 31.

Idle switch (fuel injection models) – checking

10 The idle switch is located on the throttle lever, and is part of the digital idle stabilizer (DIS) system. To check it, first secure a suitable protractor and pointer as shown in Fig. 13.8 – the protractor must be on the first stage throttle shaft.

11 Disconnect the wiring from the idle switch, and connect an ohmmeter across the switch terminals.

12 Open the throttle approximately 20° then close it slowly. Between 2.5° and 1.0° from the closed position the ohmmeter should register infinite resistance, indicating that the contacts have opened. If not, either adjust the position of the switch or renew it. To adjust the switch, it is necessary to remove the throttle housing (see Chapter 3).

Carburettor (Type 2E2) – choke valve plate gap adjustment

13 The choke valve plate gap may be altered if necessary by bending the operating lever with a suitable slotted tool (Fig. 13.9).

14 Remove the choke cover and using a rubber band, hold the choke valve plate closed.

15 Hold the primary throttle valve plate open at 45°. A nut of suitable size (M6) fitted between the diaphragm unit rod and the fast

idle speed screw will facilitate setting the valve plate (Fig. 13.10).

16 Check the valve plate opening using a twist drill of suitable diameter. The gap should be between 6.0 and 6.6 mm (0.23 to 0.26 in).

17 If adjustment is needed, grip one end (A) of the operating lever while bending at (B) – see Fig. 13.11.

18 Finally, check the choke pull-down system as described in Chapter 3.

K-Jetronic fuel injection system – modifications

19 As from March 1986, the following modifications have been made to the fuel injection system.

20 When ordering spare parts, make sure that the production date of the car is specified.

(a) Higher injector pressures
(b) Conical seat connections instead of banjo type unions
(c) Metal fuel lines to injectors
(d) Modified pressure accumulator, thermo-time switch and cold start valve

6 Ignition system

Distributor rotor arm – modifications

1 As from the 1986 model year, 1.6 and 1.8 litre engines with hydraulic tappets are

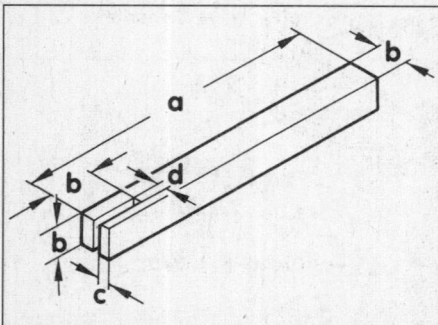

Fig. 13.9 Special tool for adjusting choke operating lever (Sec 5)

a 80.0 mm (3.12 in)   c 2.5 mm (0.10 in)
b 8.0mm (0.31in)   d 2.0 mm (0.08 in)

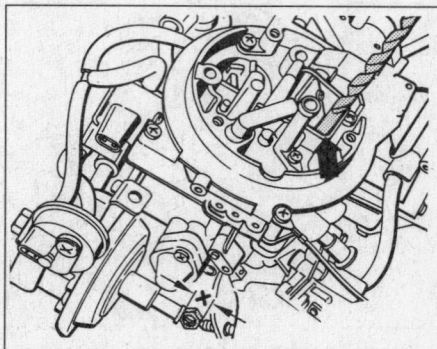

Fig. 13.10 Using nut to hold primary throttle valve plate open (Sec 5)

x = 10.0 mm (0.39 in)

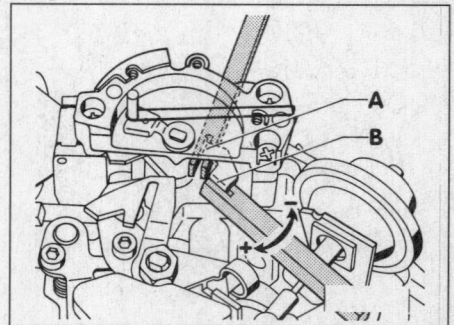

Fig. 13.11 Choke operating lever bending points A and B (Sec 5)

Clutch
Release bearing
Release lever
Push-rod
Fluid reservoir
Pedal bracket
Over-centre spring
Push-rod
Bleeder valve
Slave cylinder
Master cylinder
Clutch pedal
H9512

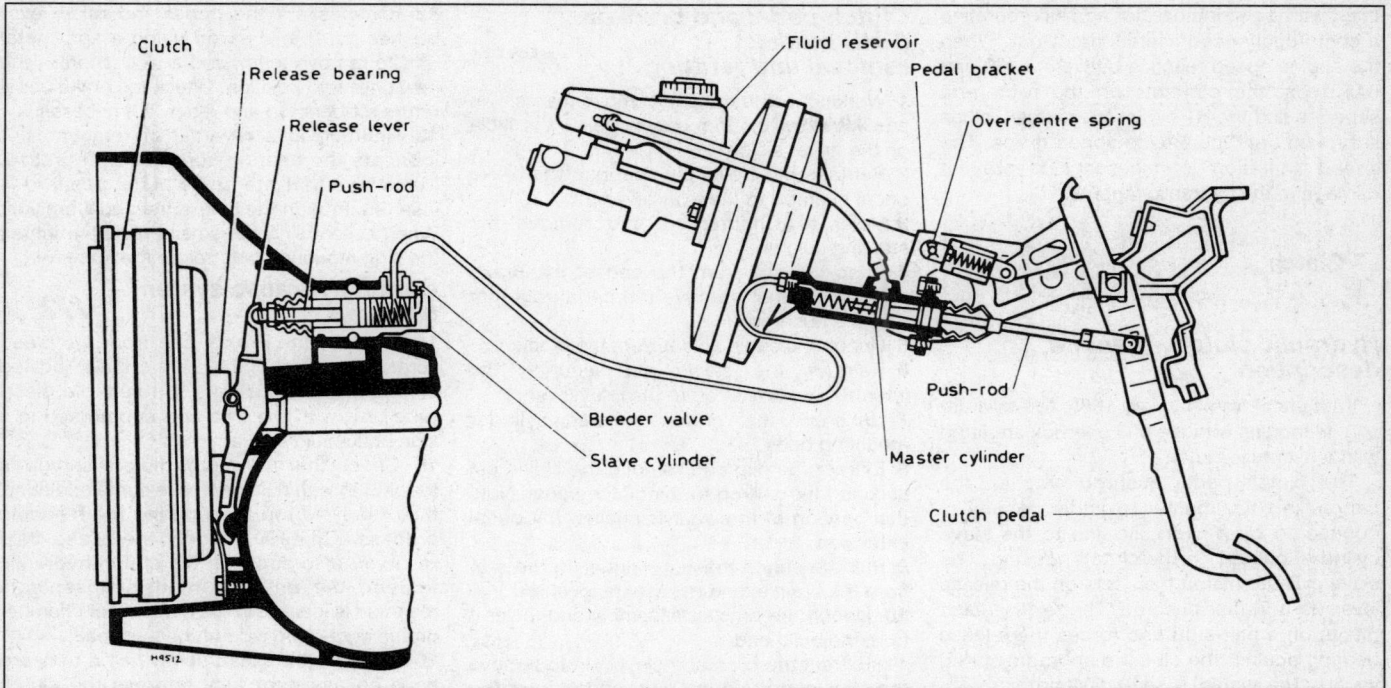

Fig. 13.12 Diagram of hydraulic clutch components (LHD model shown) (Sec 7)

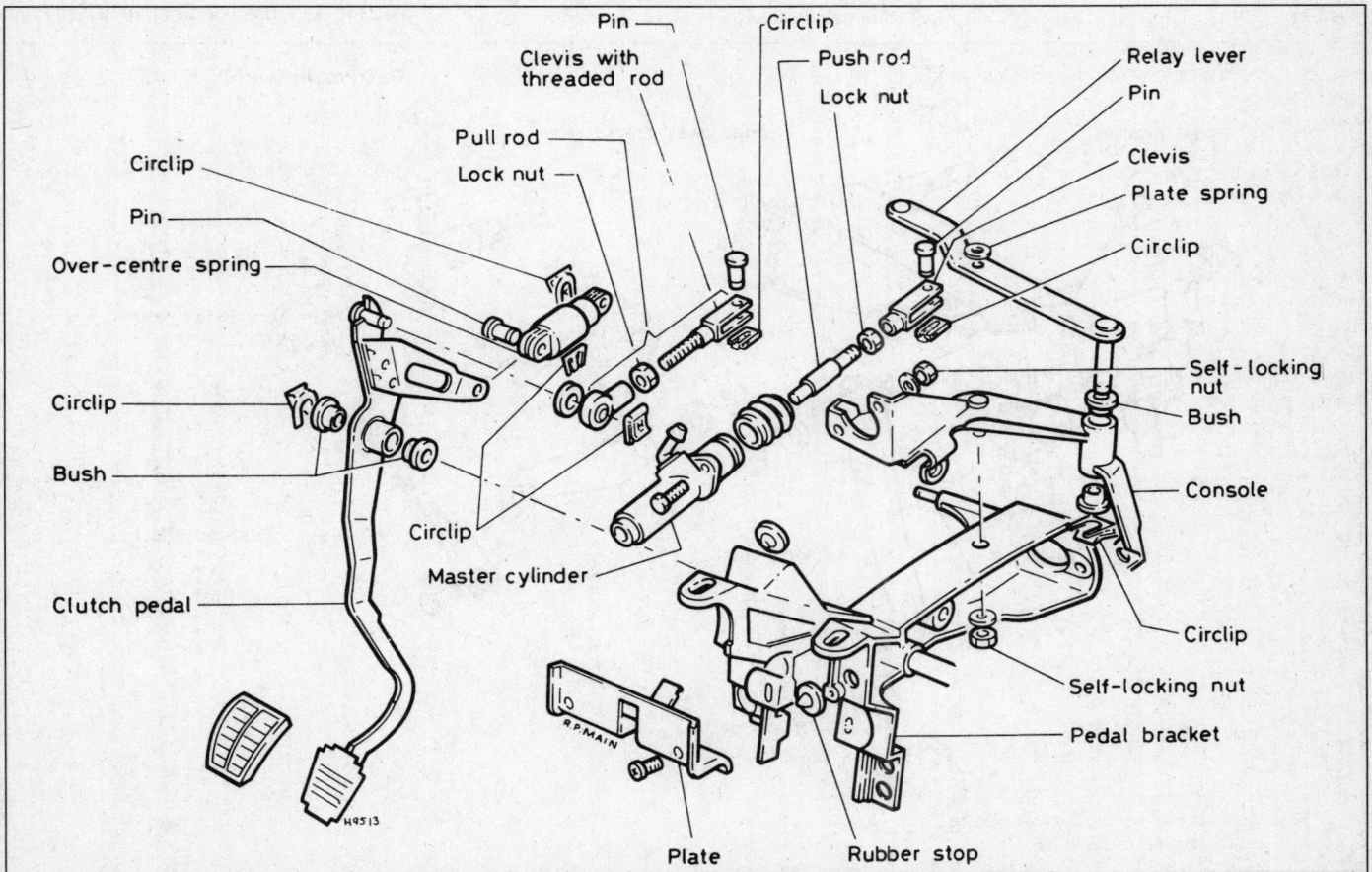

Pin
Clevis with threaded rod
Pull rod
Lock nut
Circlip
Pin
Over-centre spring
Circlip
Bush
Circlip
Clutch pedal
Circlip
Master cylinder
Circlip
Push rod
Lock nut
Relay lever
Pin
Clevis
Plate spring
Circlip
Self-locking nut
Bush
Console
Circlip
Self-locking nut
Pedal bracket
R.P.MAIN
H9513
Plate
Rubber stop

Fig. 13.13 Clutch pedal and bracket components – RHD models (Sec 7)

13

fitted with a distributor rotor arm incorporating a centrifugal speed limitation cut-out. When the engine speed reaches between 6150 and 6450 rpm the contacts on the rotor arm separate and the HT circuit to the spark plugs is broken until the engine speed drops. The speed limitation is necessary to prevent damage to the hydraulic tappets.

7 Clutch

Hydraulic clutch – general description

With effect from January 1985, five-cylinder engine models with the 093 gearbox are fitted with a hydraulic clutch.

The clutch pedal pushrod acts on the piston in the master cylinder which is connected by a hydraulic line to the slave cylinder located on the clutch housing. The slave cylinder piston then acts on the release lever via a further pushrod. The release lever pivots on a ball-stud and forces the release bearing against the clutch diaphragm spring fingers. The system is self-adjusting.

On LHD models the brake fluid reservoir doubles as a reservoir for the clutch hydraulic system.

Clutch pedal and bracket (RHD models) – removal and refitting

1 Working inside the car, remove the screws and withdraw the trim panels below the facia on the driver's side.
2 Remove the clevis pin securing the brake pedal pushrod to the relay lever.
3 Refer to Chapter 11 and remove the steering column.
4 Prise the clip from the end of the brake pedal pivot then remove the pedal together with the pushrod.
5 Unbolt the cable holder from the bracket.
6 Remove the clevis pin securing the intermediate brake rod to the relay lever.
7 Unscrew the clutch master cylinder mounting bolts.
8 Extract the clip and remove the clevis pin securing the pullrod to the clutch pedal. Note the location of the washer against the pedal extension.
9 Pull the clutch master cylinder to the rear from the bracket and release the pushrod.
10 Unbolt the bracket assembly and lower it from the bulkhead.
11 Extract the clips and clevis pin to remove the over-centre spring unit, and extract the clip to remove the clutch pedal. If necessary the two small brackets may be unbolted from the main bracket.

12 If necessary the pedal and relay lever bushes can be renewed using a soft metal drift to remove them, and a vice to press the new ones into position. Check the brake pedal return rubber stop and renew it if necessary.
13 Refitting is a reversal of removal, but lubricate the bearing surfaces with grease. Make sure that the brake pedal pushrod is inserted through the hole in the pedal bracket. When assembling the small brackets, tighten the side mounting bolt before the upper nut.

Clutch hydraulic system – bleeding

14 Remove the rubber cap from the bleed screw on the top of the slave cylinder located on the clutch housing. Fit a suitable bleed tube to it with the free end submersed in a little brake fluid in a jar.
15 Check that the clutch fluid reservoir is topped up with fluid and have some additional fluid ready to top it up during the bleeding process. On LHD models the brake fluid reservoir also supplies the clutch hydraulic system, but on RHD models a separate reservoir is located in the heater inlet chamber on the right-hand side of the bulkhead.
16 Unscrew the bleed screw half a turn and have an assistant fully depress the clutch pedal three or four times. Tighten the bleed screw with the clutch pedal depressed, then top up the fluid reservoir using new fluid only.

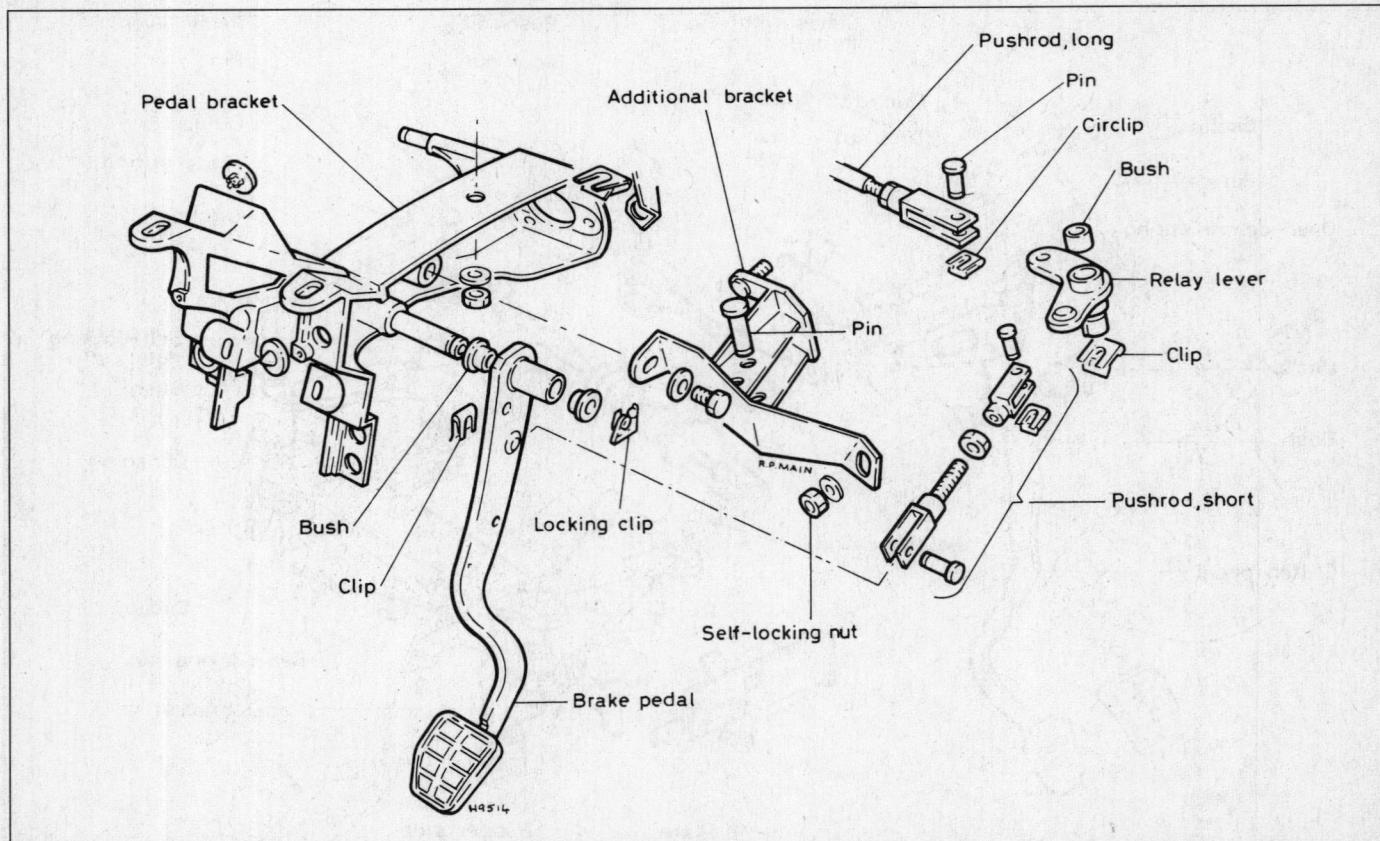

Fig. 13.14 Brake pedal and bracket components – RHD models (Sec 7)

17 Repeat the procedure in paragraph 15 two or three times then check that the clutch pedal movement is correct.

18 Disconnect the bleed tube and refit the rubber cap.

Clutch slave cylinder – removal, overhaul and refitting

19 The slave cylinder is retained in the clutch housing by a roll pin which engages the annular groove of the cylinder. First remove the clip, then use a suitable drift to drive out the roll pin.

20 Slide out the slave cylinder, together with the pushrod.

21 Loosen the hydraulic pipe union and unscrew the cylinder from the pipe. Plug the pipe to prevent loss of fluid.

22 Clean the cylinder, then prise off the rubber boot and remove the pushrod.

23 Withdraw the piston and spring from the mouth of the cylinder.

24 Clean the components and examine them for wear and deterioration. If the piston and bore are worn excessively, renew the complete cylinder, but if they are in good condition remove the seal from the piston and obtain a repair kit.

25 Dip the new seal in brake fluid and fit it on the piston using the fingers only to manipulate it into position. Make sure that the seal lip faces the spring end of the piston.

26 Insert the spring in the cylinder then dip the piston in brake fluid and carefully insert it.

27 Locate the small end of the rubber boot in the groove in the pushrod and fit the retaining ring.

28 Apply molybdenum disulphide grease to the ends of the pushrod, then fit it together with the boot to the slave cylinder.

29 Refitting is a reversal of removal, but finally bleed the hydraulic system as described in paragraphs 14 to 18.

Clutch master cylinder – removal, overhaul and refitting

30 Working inside the car, remove the screws and withdraw the trim panels below the facia on the driver's side.

31 Disconnect the pushrod clevis from the pedal (LHD models) or relay lever (RHD models).

32 Disconnect and plug the hydraulic fluid supply hose, then unscrew the outlet union and disconnect the pressure line.

33 Unbolt and remove the master cylinder.

34 Clean the cylinder then prise off the rubber boot and remove the pushrod.

35 Extract the circlip from the mouth of the cylinder and withdraw the washer, piston and spring, noting that the smaller end of the spring contacts the piston. Prise the inlet grommet from the cylinder.

36 Clean the components and examine them for wear and deterioration. If the piston and bore are worn excessively, renew the complete cylinder, but if they are in good condition remove the seals from the piston and obtain a repair kit.

37 Dip the new seals in brake fluid and fit them on the piston using the fingers only to manipulate them into position. Make sure that the seal lips face the spring end of the piston.

38 Smear the inlet grommet with brake fluid and locate it in the cylinder aperture.

39 Insert the spring in the cylinder large end first, then dip the piston in brake fluid, locate it on the spring, and carefully insert it.

40 Locate the washer, then fit the circlip in the groove.

41 Fit the pushrod and rubber boot, making sure that the boot is correctly located in the grooves. Apply a little grease to the pushrod end.

42 Refitting is a reversal of removal, but finally bleed the hydraulic system as described in paragraphs 14 to 18. Lubricate the clevis pin with a little grease. On LHD

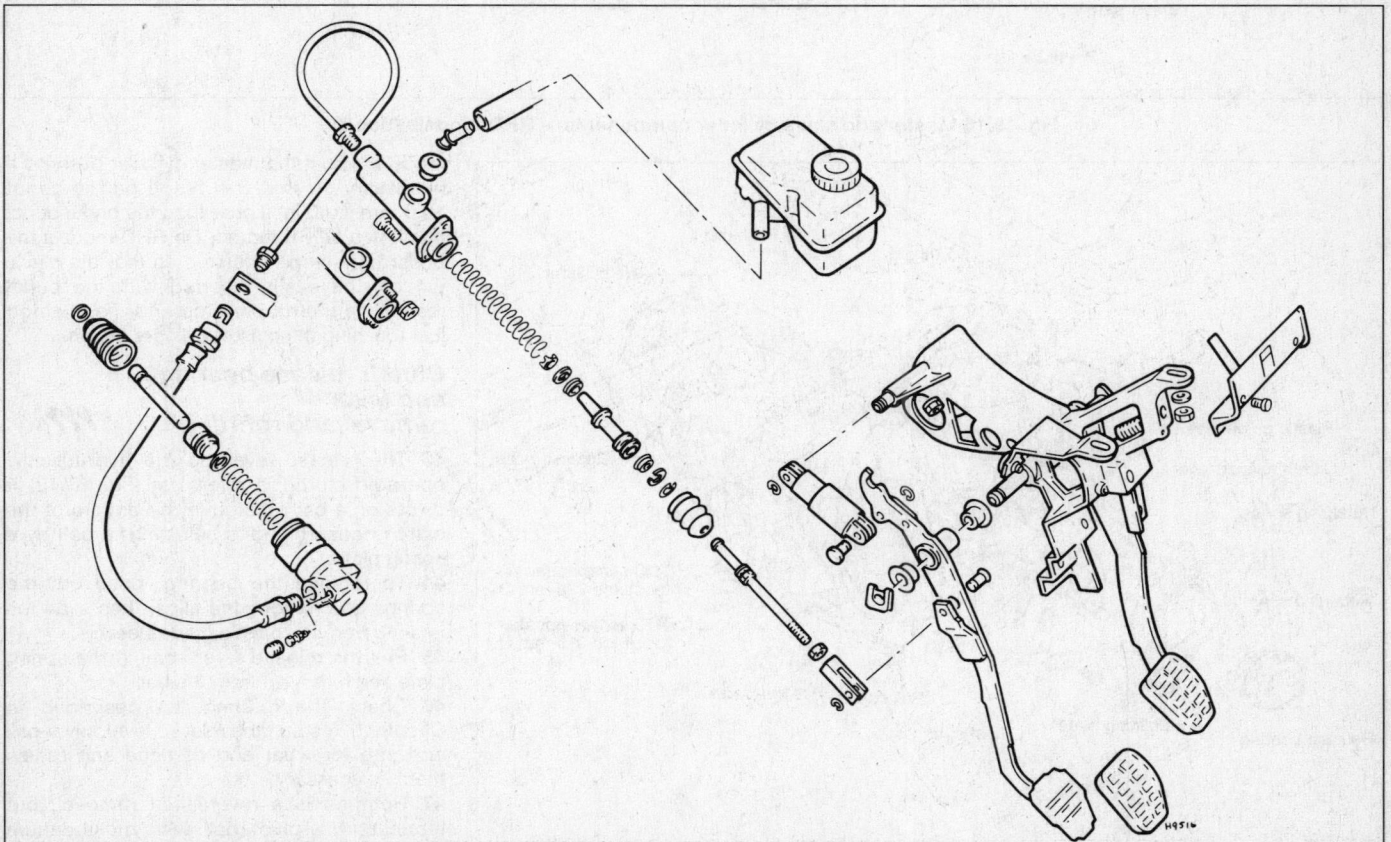

Fig. 13.15 Pedal and hydraulic components – LHD models (Sec 7)

13

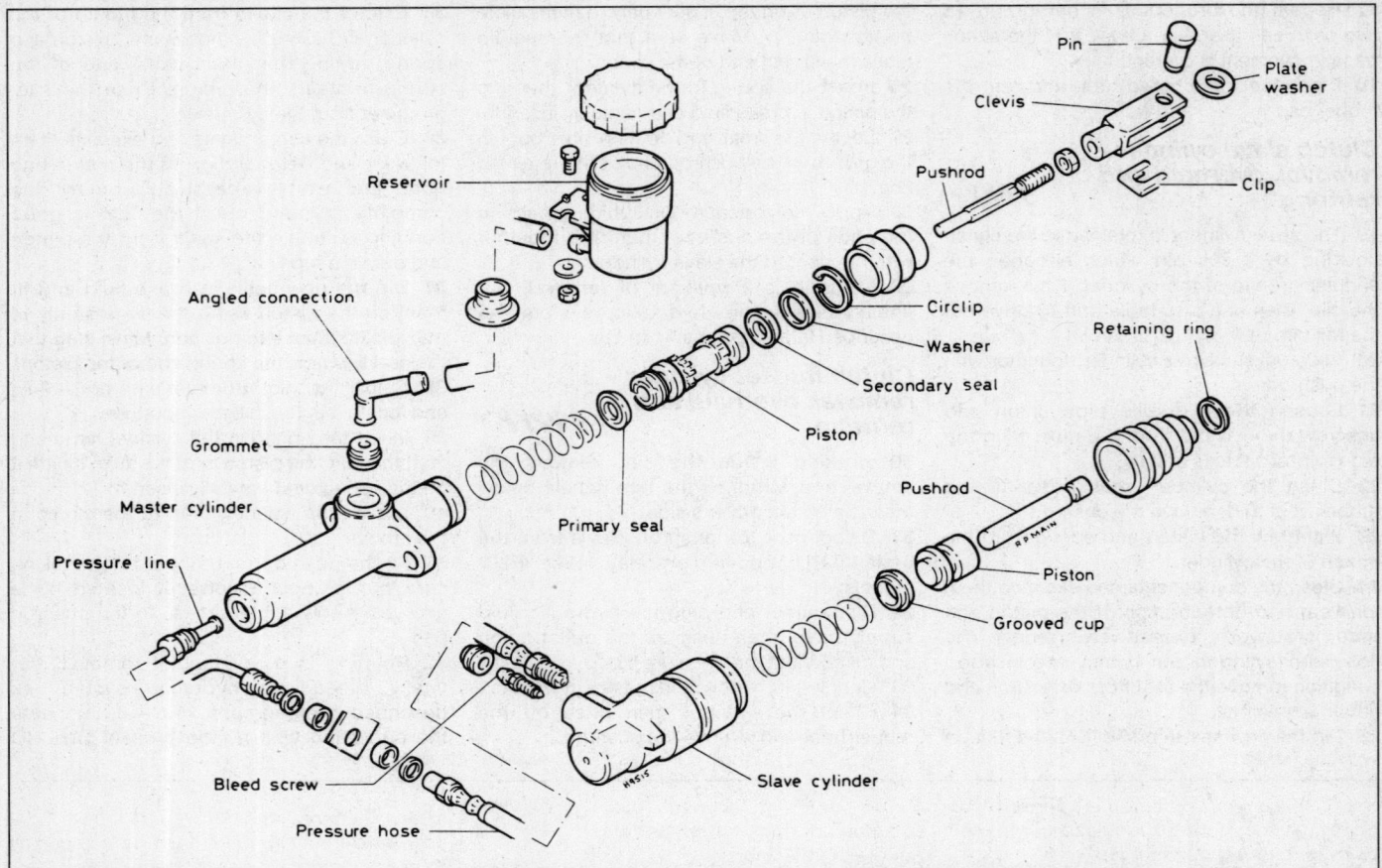

Fig. 13.16 Master and slave cylinder components – RHD models (Sec 7)

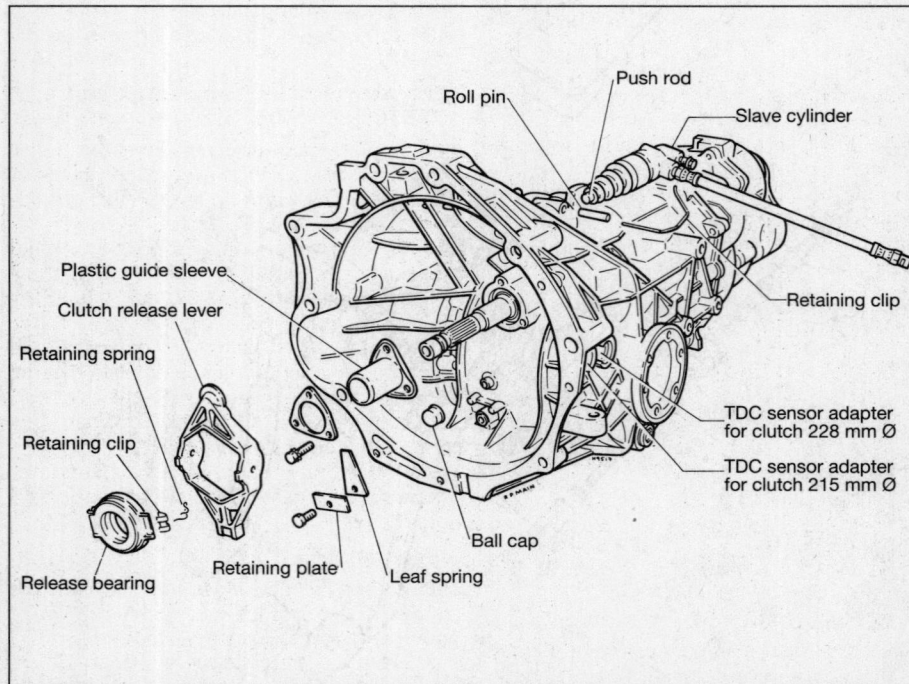

Fig. 13.17 Clutch release bearing and lever components for hydraulic clutch (Sec 7)

models, adjust the master cylinder pushrod if necessary so that the clutch pad is about 10.0 mm (0.40 in) higher than the brake pedal pad when fully returned. On RHD models the pushrod must be adjusted so that the clevis pin can be easily inserted with the pedal resting on the rubber stop and the pushrod just touching the master cylinder piston.

Clutch release bearing and lever – removal and refitting

43 The release lever on the hydraulically operated clutch is shown in Fig. 13.12; it pivots on a ball located in the bottom of the clutch housing and is held on the ball by a spring plate.

44 To remove the bearing, prise out the springs and remove the clips, then slide the bearing from the plastic guide sleeve.

45 Pull the release lever against the spring plate and remove it from the ball.

46 Check the bearing, as described in Chapter 5. Inspect the release lever, pivot ball and cap for wear and damage and renew them if necessary.

47 Refitting is a reversal of removal, but lubricate the pivot ball with molybdenum disulphide grease. **Do not,** however, lubricate the plastic cap or plastic guide sleeve.

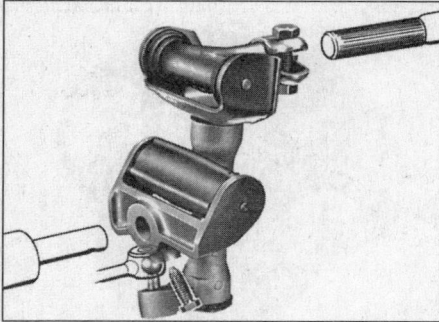

Fig. 13.18 Modified shift rod coupling (Sec 8)

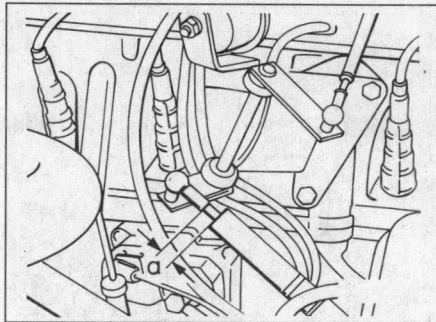

Fig. 13.19 Throttle pushrod dimension for kickdown position on 087 transmission (Sec 9)

a = 10.5 mm (0.413 in)

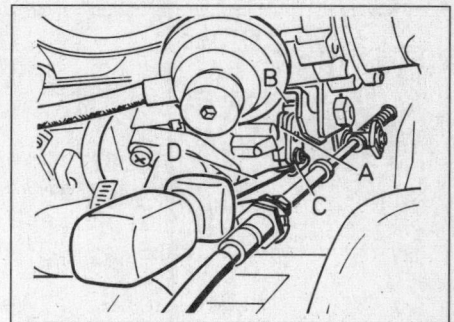

Fig. 13.20 Accelerator pedal and linkage adjustment on the 2E2 carburettor (Sec 9)

A Warm-up lever   C Lever
B Throttle valve operating pin D Screwdriver

8 Manual gearbox and final drive

Shift rod coupling – modification

1 As from late 1984, five-cylinder engine models with the 093 gearbox are equipped with a modified shift rod coupling (Fig. 13.18) incorporating additional rubber bushes to reduce the transmission noise inside the car.
2 Modified replacement couplings are also available for the 014 and 013 gearboxes, although new cars are currently being fitted with the earlier type.
3 Unlike the earlier type, the modified shift rod coupling cannot be dismantled for repair, and must therefore be renewed complete if worn excessively or damaged.

Gearshift linkage – lubrication

4 As from late 1984 the gearshift linkage is lubricated with molybdenum disulphide based grease instead of white lubricating paste. When reassembling earlier linkages all traces of paste should be removed before applying the new grease.

Gearbox – removal and refitting

5 When removing the gearbox on models fitted with a hydraulically operated clutch (see Section 7) it will be necessary to remove the clutch slave cylinder from the clutch housing

Fig. 13.21 Vacuum pump connected to outlet A with outlet B plugged (Sec 9)

and tie it to one side, leaving the hose still attached. References in Chapter 6 to the clutch cable are not, of course, applicable.

9 Automatic transmission and final drive

Accelerator pedal and linkage (087 transmission) – adjustment

1 Move the selector lever to P.
2 Apply the handbrake then run the engine to normal operating temperature.
3 With the engine switched off, loosen the knurled adjusting nut at the transmission end of the throttle cable.
4 Check that the choke is fully open on carburettor engines and that the throttle is closed.
5 Disconnect both rods from the throttle relay lever.
6 Turn the relay lever fully anti-clockwise, then adjust the length of the pullrod (engine side of relay lever) so that the socket is aligned with the ball. Tighten the locknuts, and refit the rod to the relay lever.
7 Refit the pushrod then loosen its clamp bolt, check that both the transmission and throttle levers are in their idle position, and tighten the clamp bolt again.

Fig. 13.22 Kickdown dimension on non-2E2 carburettor engines (Sec 9)

A = 8.0 mm (0.315 in)

8 Have an assistant fully depress the accelerator pedal to the kickdown position then turn the knurled nut until the transmission lever contacts the stop, and tighten the locknut.
9 Check that, with the accelerator pedal at the full throttle position, the throttle is fully open without compressing the spring in the pushrod. With the accelerator pedal at the kickdown position, the spring must be compressed (Fig. 13.19).

Accelerator pedal and linkage (089 transmission) – adjustment

10 Move the selector lever to P.
11 Apply the handbrake, then run the engine to normal operating temperature.
12 With the engine switched off, loosen the knurled adjusting nut at the transmission end of the accelerator pedal cable.
13 On carburettor engines remove the air cleaner.

Engines with 2E2 carburettor and overrun cut-off

14 Turn the warm-up lever so that the throttle valve operating pin is not making contact, then retain the warm-up lever in this position with a screwdriver (Fig. 13.20).
15 Pull the vacuum hoses from the three-point vacuum unit then connect a vacuum pump to outlet A (Fig. 13.21) and plug outlet B.
16 Operate the vacuum pump until the diaphragm pushrod stays in the overrun position and is not touching the cold idle adjusting screw. Engines without 2E2 carburettor and overrun cut-off
17 Check that the throttle is closed and on carburettor engines that the choke is fully open.

All engines

18 Loosen the locknuts at the bracket on the engine and adjust the outer cable position to eliminate all play. Tighten the locknuts.
19 Have an assistant fully depress the accelerator pedal to the kickdown position, then turn the knurled adjusting nut until the

13

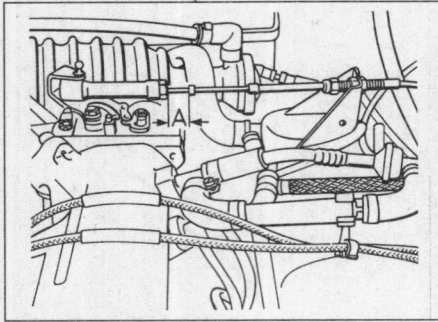

Fig. 13.23 Kickdown dimension on fuel injection engines (Sec 9)

A = 8.0 mm (0.315 in)

Fig. 13.24 Kickdown dimension on 2E2 carburettor engines (Sec 9)

a = 8.0 mm (0.315 in)

Fig. 13.25 Plastic skin to pierce when removing the battery cell plugs (Sec 10)

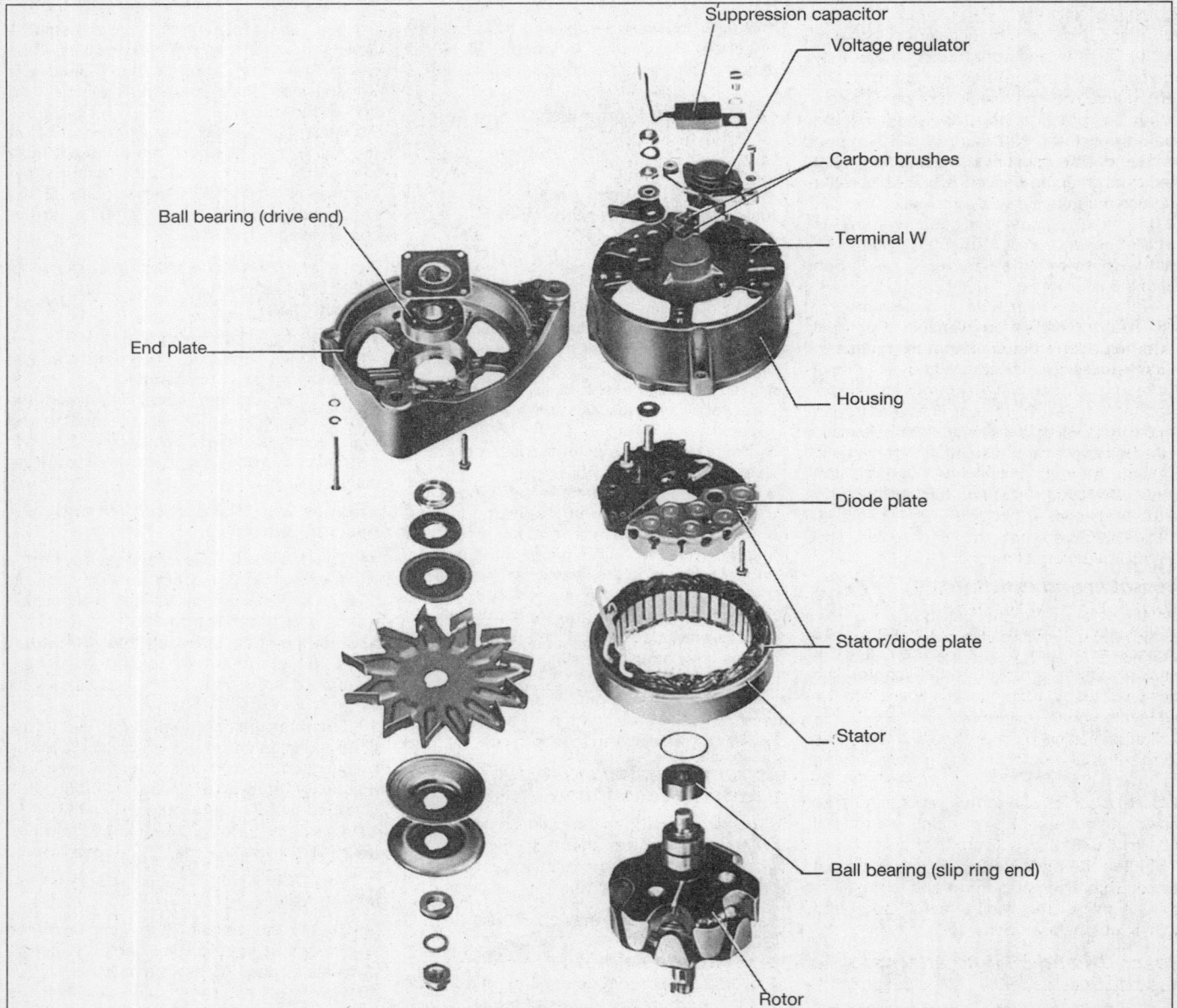

Fig. 13.26 Exploded view of later type Bosch 75 and 90 amp alternator (Sec 10)

10.4 Alternator adjusting link with toothed rack

transmission lever contacts the stop, and tighten the locknut.

20 Check that, with the accelerator pedal at the full throttle position, the throttle is fully open without compressing the spring at the carburettor. With the accelerator pedal at the kickdown position the spring must be compressed (Figs. 13.22 and 13.24).

21 Remove the vacuum pump and reconnect the hoses on engines with the 2E2 carburettor and overrun cut-off.

10 Electrical system

Battery – maintenance

1 On some later batteries the cell plugs are initially locked in position and a special procedure is necessary to remove them for checking the electrolyte level. Before removing the plugs however, check any conditions given in the battery warranty restricting their removal.

2 Using a small screwdriver, pierce the plastic skin (Fig. 13.25) and turn the top part of the plug as far as possible.

3 The plug can now be unscrewed and refitted in the normal manner.

Alternator drivebelt (from April 1985) – adjustment

4 With effect from April 1985, 1.6 and 1.8 litre engines are fitted with a toothed rack for tensioning the alternator drivebelt (photo). The tension is set by using a torque wrench on the nut located on the clamp bolt.

5 First loosen all pivot and adjustment bolts so that the alternator is free to move in and out from the engine.

6 Using a socket and torque wrench, turn the tensioning nut anti-clockwise until a torque of 9.0 Nm (6.6 lbf ft) is indicated.

7 Tighten the pivot bolt while checking that the alternator remains in the same position, then tighten the adjustment bolt and adjustment rail pivot bolt.

Bosch 75 and 90 amp alternator – description

8 The Bosch 75 and 90 amp alternator fitted from January 1985 is shown in Fig. 13.26 and

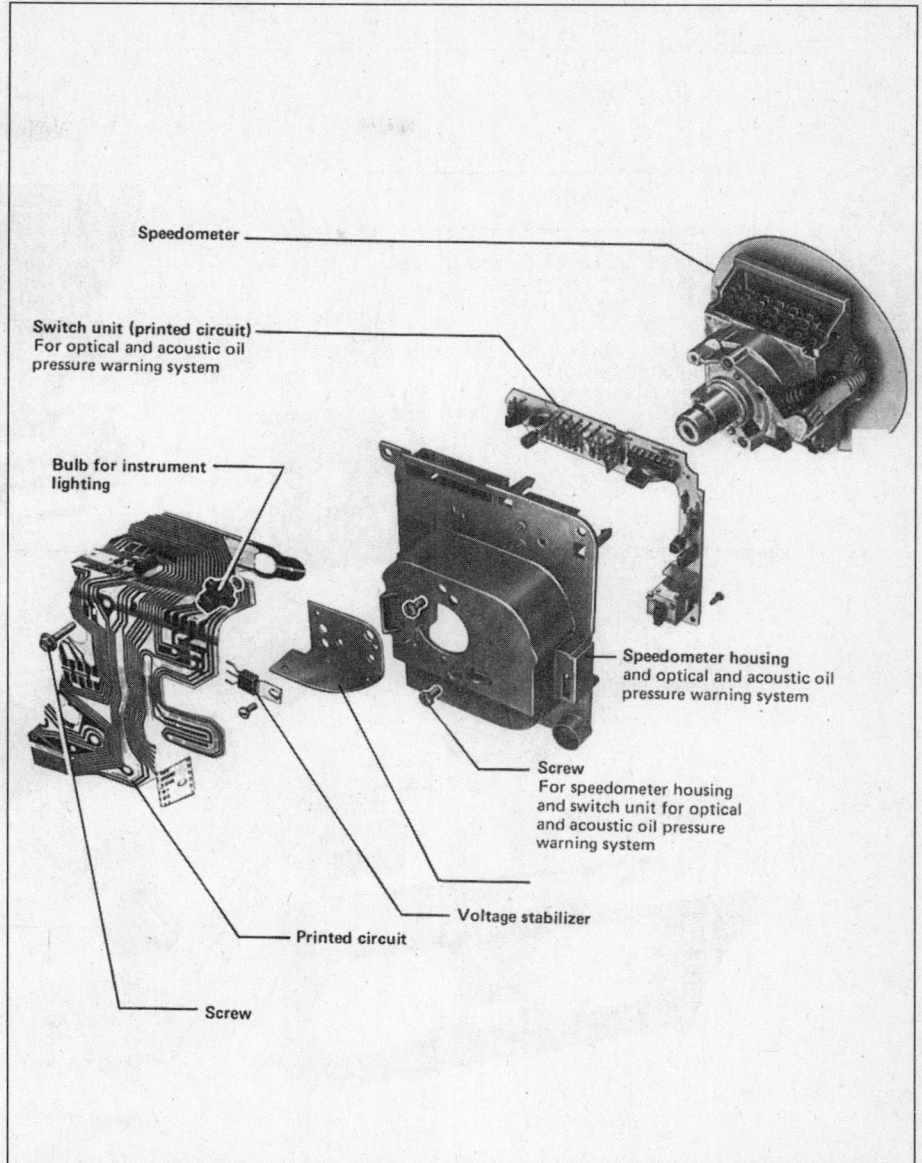

Fig. 13.27 Instrument panel components fitted to certain models after July 1983 (Sec 10)

apart from a few minor components is identical to the previous type. The servicing procedures are as given in Chapter 10.

Instrument panel cluster (later models) – description

9 The instrument panel cluster components for later models are shown in Figs. 13.27 and 13.28. Removal and refitting procedures are as given in Chapter 10.

Headlamp and foglamp units (later models) – removal and refitting

10 Remove the screw and withdraw the trim strip from under the headlamp and foglamp units (photo).

10.10 Removing headlamp trim strip

13

Printed circuit

Bulb for instrument lighting

Voltage stabiliser

Heat sink

Housing for speedometer and
switch unit for oil pressure
warning system

Switch unit (printed board) for
oil pressure warning system

Coolant temperature gauge

Warning lamp housing

Speedometer

Rev counter

LEDs

Retaining plate

Fuel gauge

Bulb

Cap (blue)

Diode holder

Dash insert

Digital clock

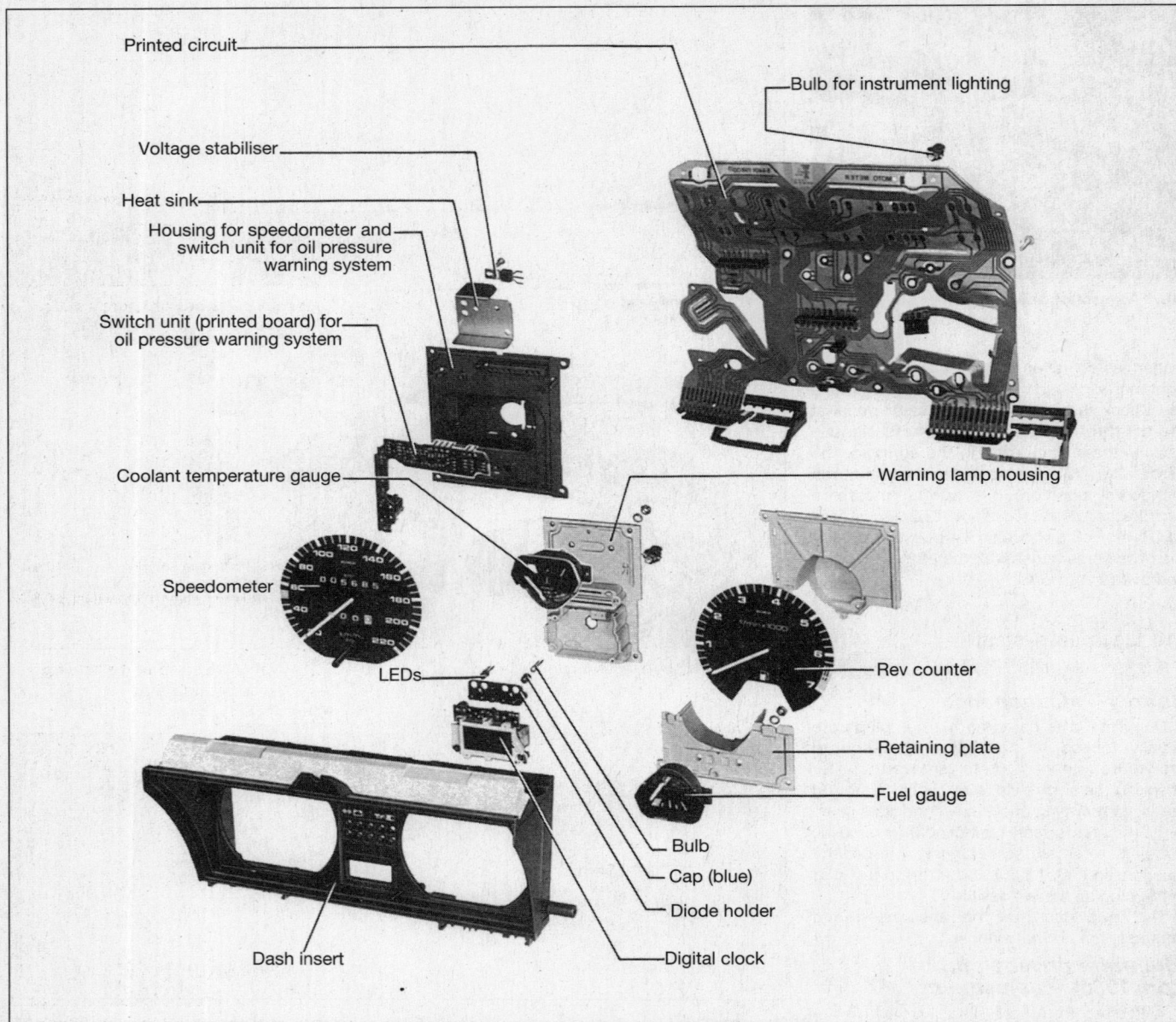

Fig. 13.28 Instrument panel components fitted to certain models after January 1985 (Sec 10)

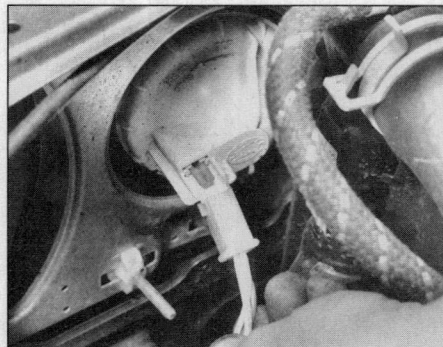

10.11A Disconnecting foglamp wiring

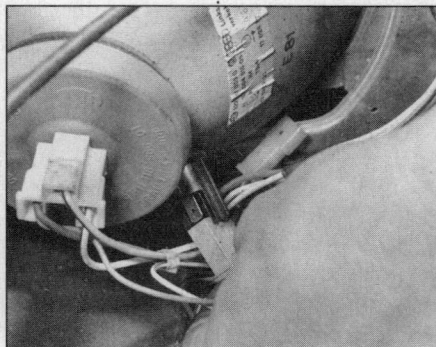

10.11B Disconnecting sidelamp wiring

11 With the bonnet open, disconnect the wiring from the foglamp, sidelamp and headlamp bulbs (photos).

12 Unscrew the top and bottom mounting screws and withdraw the lamp unit complete (photos).

13 To remove the headlamp unit from the frame, unscrew the beam adjustment screw (photo) and turn the retaining clips through 90°. To remove the foglamp unit turn the three clips through 90° (photos).

14 Refitting is a reversal of removal, but finally adjust the headlamp beam with reference to Chapter 10. Make sure that the 'TOP' mark on the foglamp cap is uppermost.

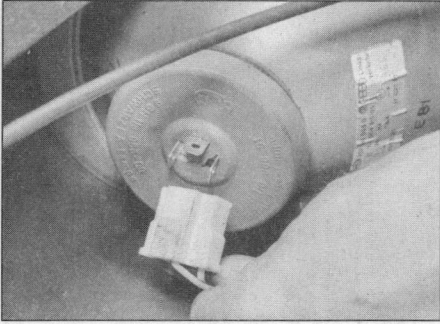

10.11C Disconnecting headlamp wiring

10.12A Unscrew the top mounting screws . . .

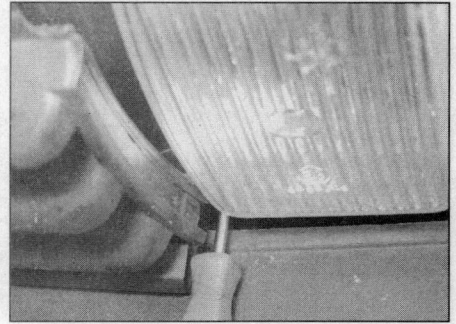

10.12B . . . inner bottom screws . . .

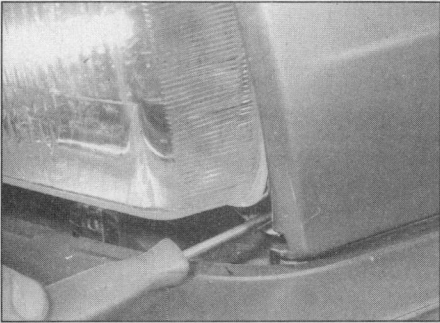

10.12C . . . and outer bottom screws . . .

10.12D . . . and withdraw the lamp unit

10.12E Complete lamp unit showing mounting frame

10.13A Headlamp beam adjustment screw location

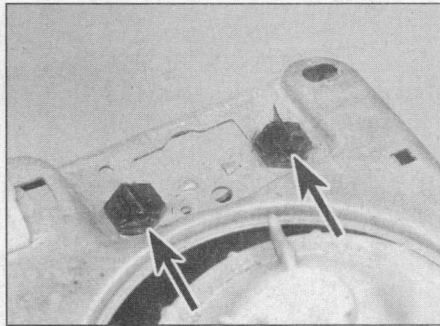

10.13B Foglamp unit top mounting clips . . .

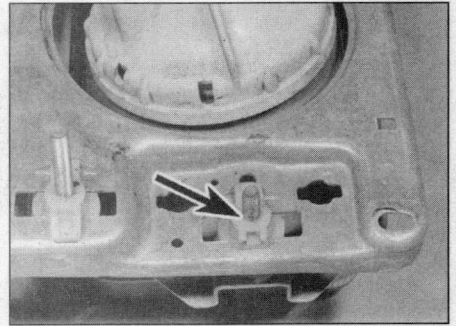

10.13C . . . and bottom mounting clip

Fig. 13.29 The foglamp cap 'Top' mask must be uppermost (Sec 10)

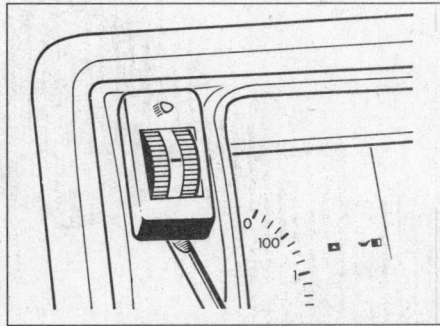

Fig. 13.30 Prising out the headlamp range control (See 10)

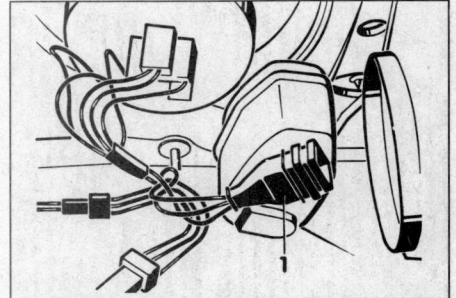

Fig. 13.31 Disconnecting a wiring plug (1) from a headlamp range control adjuster motor (See 10)

13

10.22 Depressing the retaining tab on the side marker lamp

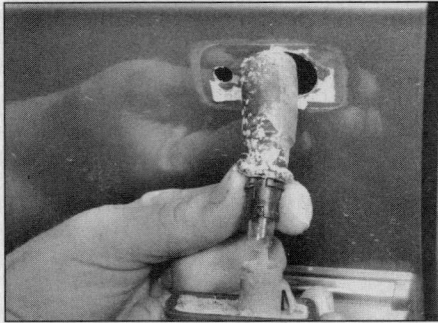

10.24A Pull out the bulbholder . . .

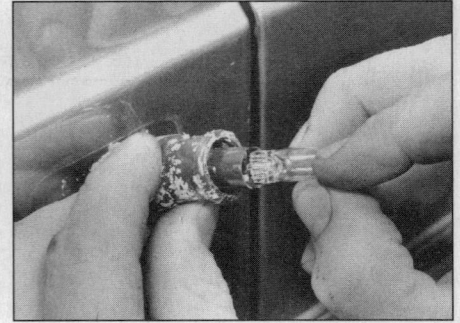

10.24B . . . and extract the wedge type bulb

Headlamp range control – description, removal and refitting

15 The headlamp range control is an optional extra on certain models as from January 1986. It consists of a facia mounted range control and two adjuster motors mounted on the rear of the headlamps. The control only functions on dipped beam and it is only possible to lower the beam from the normal setting (ie when carrying a heavy load).

16 Before adjusting the headlamp beam it is important to check that the range control is at its basic setting.

17 To remove the control from the facia, first disconnect the battery negative lead, then prise out the control and disconnect the wiring plug.

18 To remove the motors, working in the engine compartment, disconnect the wiring plugs from both adjuster motors.

19 Disengage the motors from the headlamp frames by turning the right-hand motor anti-clockwise and the left-hand motor clockwise.

20 Remove the headlamp units as previously described, then unscrew the adjustment screws and withdraw the motors.

21 Refitting is a reversal of removal.

Lamp bulbs – renewal

Side marker lamp

22 Using a thin screwdriver or a piece of plastic sheet, depress the plastic tab from the rear end of the lamp (photo). Use a piece of

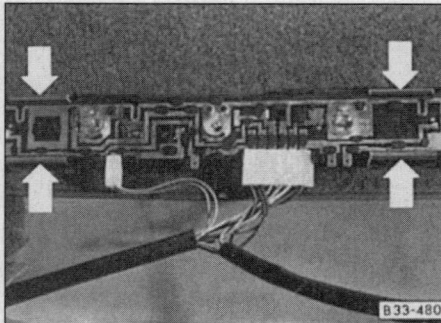

Fig. 13.32 Rear lamp cluster retaining tab locations on Saloon models (Sec 10)

card or rag to prevent damage to the paintwork.

23 Ease out the rear of the lamp and unhook the front tab from the hole.

24 Pull out the bulbholder, then extract the wedge type bulb from the holder (photos).

25 Refitting is a reversal of removal.

Rear lamp cluster – saloon

26 Open the boot lid, then squeeze together the plastic tabs as shown in Fig. 13.32 and withdraw the bulbholder.

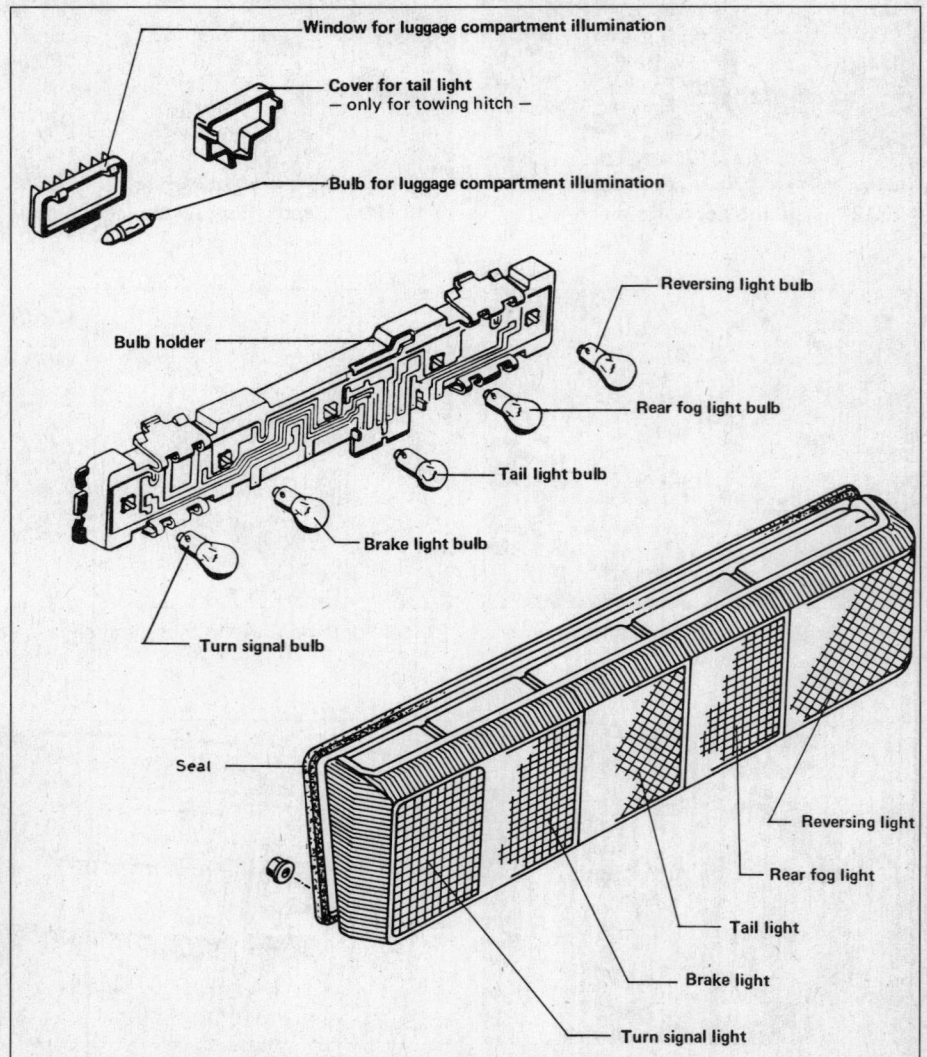

Fig. 13.33 Rear lamp cluster components on Saloon models (Sec 10)

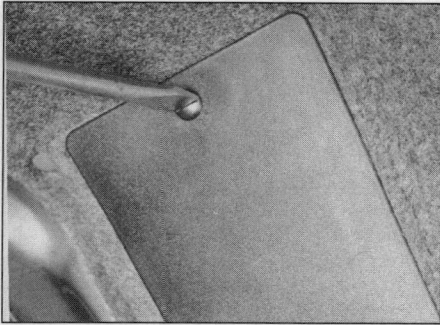
10.29 Removing inner trim panel cover for access to the tailgate rear lamp cluster

10.30A Squeeze together the plastic tabs . . .

10.30B . . . and remove the tailgate bulbholder

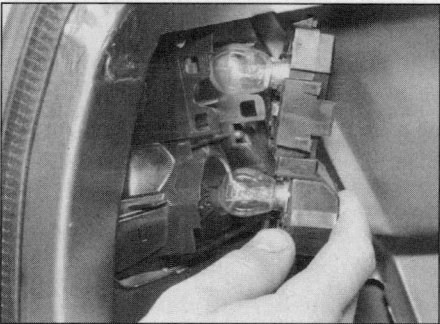
10.30C . . . or body corner bulbholder

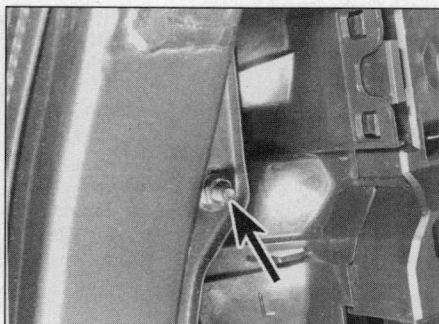
10.32A Unscrew the inner nuts . . .

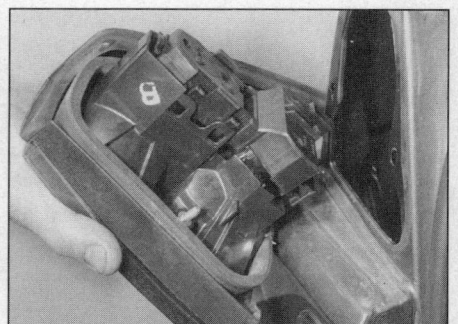
10.32B . . . and remove the lamp body

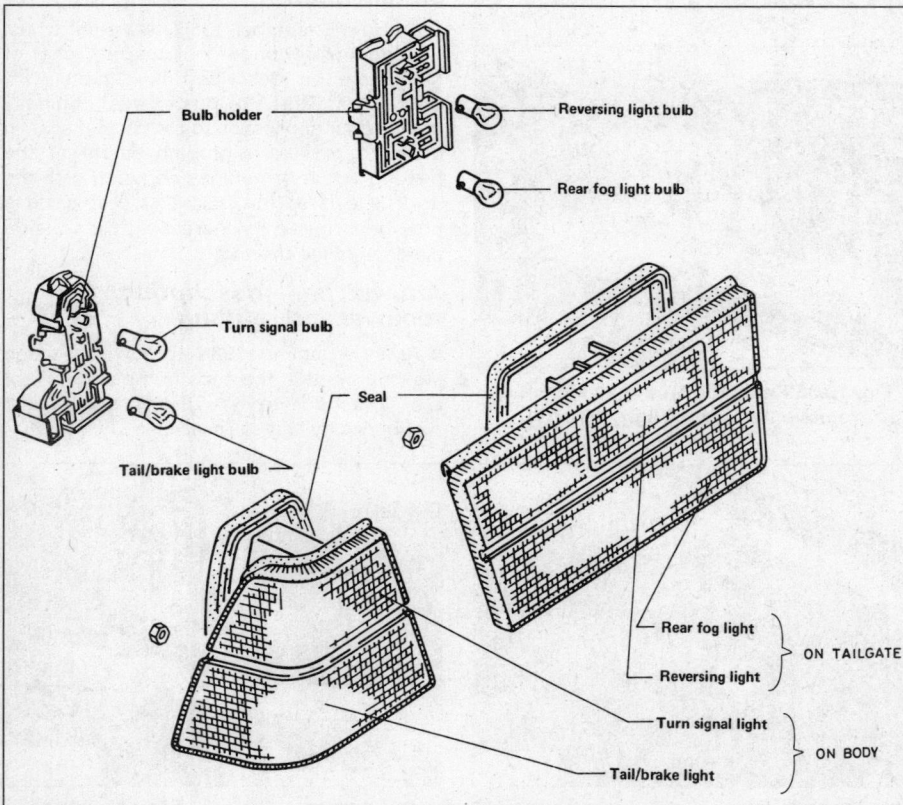
Fig. 13.34 Rear lamp cluster components on hatchback models from January 1985 (Sec 10)

27 Push and twist the bulb to remove it.
28 Refitting is a reversal of removal.

Rear lamp cluster – hatchback models (January 1985-on)

29 Open the tailgate and remove the inner trim panel from the body corner or tailgate inner panel. Where necessary turn the fastener 90° and pull off the trim cover (photo).
30 Squeeze together the plastic tabs and remove the bulbholder (photos).
31 Depress and twist the bulb to remove it.
32 To remove the lamp body, unscrew the inner nuts and withdraw the lamp from the rear of the car (photos).
33 Refitting is a reversal of removal.

Number plate light – Hatchback models (January 1985-on)

34 Remove the cross-head screws and withdraw the number plate light from the tailgate (photos).
35 Separate the bulbholder, then depress and twist the bulb to remove it.

Interior light – later models

36 Press the spring (Fig. 13.35) towards the switch and remove the light.
37 Renew the bulb as described in Chapter 10.

Rear reading lamp

38 Release the lamp by inserting a screwdriver in the direction shown in Fig. 13.36 against the retaining spring.

13

10.34A Number plate light mounting screws

10.34B Separating the bulbholder from the lens

10.44 Prise off the cover . . .

39 Renew the bulb as described for the interior light.

40 When refitting the lamp, insert the switch end first.

Luggage compartment light

41 On hatchback models, prise the light from the trim. On saloon models, depress the plastic tab and remove the lens.

42 Remove the festoon type bulb from the spring contacts.

43 Refitting is a reversal of removal.

Door mounted speakers – removal and refitting

44 Carefully prise off the speaker cover with a screwdriver (photo).

45 Remove the cross-head mounting

10.45 . . . for access to the door mounted speaker

Fig. 13.36 Insert a screwdriver as shown to remove the rear reading lamp (Sec 10)

screws, withdraw the speaker from the door trim panel, and disconnect the wiring plug, noting which way round the wires are connected (photo).

46 Refitting is a reversal of removal.

Headlamp dim-dip system

47 This system is fitted to later models (1986 on) to meet current UK regulations. The system is designed to prevent the vehicle being driven with only parking lamps on.

Fuses and relays – general

48 From July 1982 a revised fusebox and relay plate is fitted to all models. The new unit comprises a main relay plate and auxiliary plate clipped together to form one unit. On the

Fig. 13.35 Press the spring as shown to remove the interior light (Sec 10)

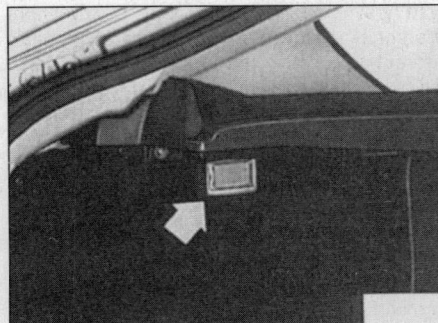

Fig. 13.37 Luggage compartment light on hatchback models (Sec 10)

main relay plate there is provision for 22 fuses and 11 relays. Additional fuses and relays are carried on the auxiliary plate, according to model, year and options fitted. Details of the fuse locations and circuits protected are given in the Specifications at the beginning of Chapter 10. The relay locations and circuits controlled are given with the appropriate wiring diagrams at the end of this Supplement.

11 Suspension and steering

Front hub and bearing – modification

1 As from December 1985, new front wheel bearing kits no longer included a sachet of Molypaste as described in Chapter 11, Section 5. The Molypaste was originally provided for application to the bearing seat, in order to prevent slight movement of the bearing which sometimes occurred with the front wheels on full lock. This problem has now been cured by increasing the outside diameter of the bearing.

Anti-roll bar – description, removal and refitting

2 As from January 1984 on power-assisted steering models, the ends of the anti-roll bar are attached to the front suspension wishbones by angled links.

Fig. 13.38 Luggage compartment light removal on Saloon models (Sec 10)

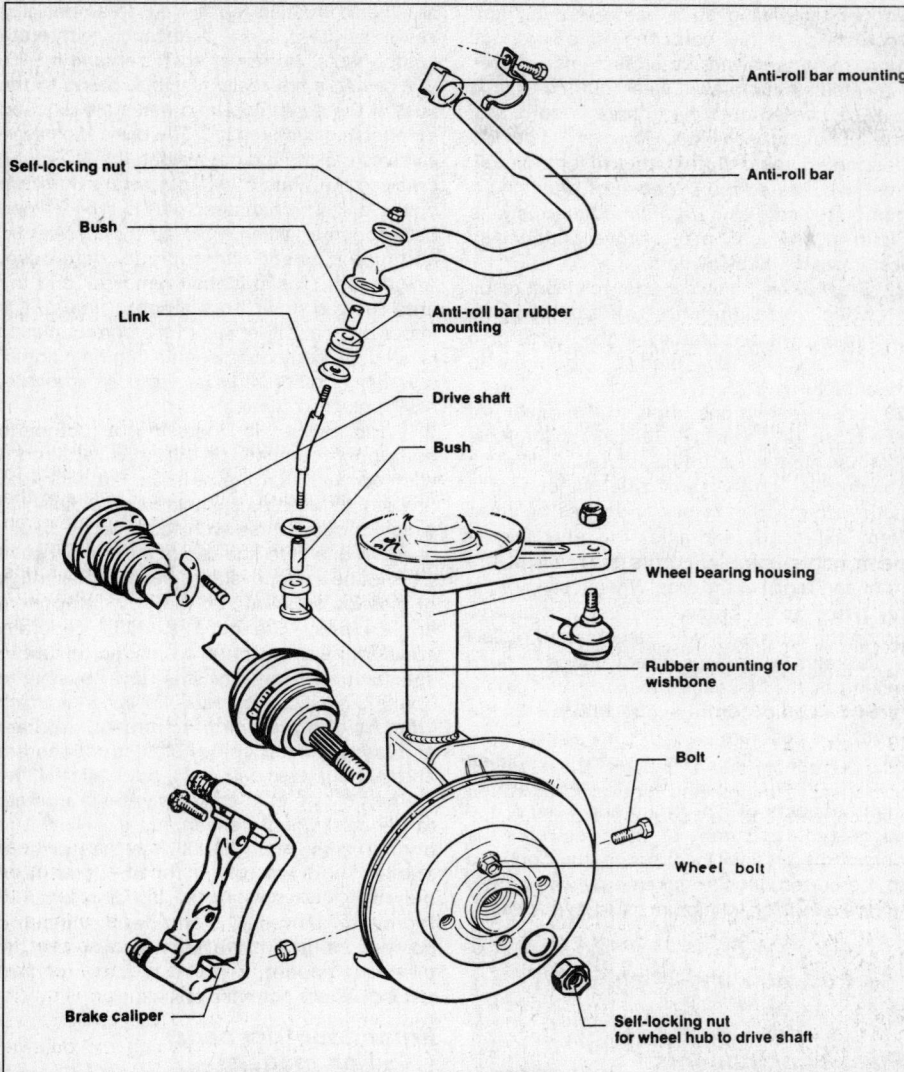

Fig. 13.39 Front suspension components showing anti-roll bar with angled link (Sec 11)

11.12A Prise off the centre pad . . .

11.1 2B . . . for access to the steering wheel retaining nut

3 When fitting the links, make sure that the shoulders are at right-angles to the surfaces of the wishbone and anti-roll bar.

4 The bushes in the anti-roll bar and wishbone can be renewed using the procedure described in Chapter 11, Section 7, but first cut off the collars with a hacksaw.

Front suspension subframe – general

5 The front suspension subframe mounting bolts should be renewed when removing and refitting the subframe.

6 The correct sequence for tightening the bolts is as follows:

Rear left
Rear right
Front left
Front right

7 Originally a torque of 70 Nm (44 lbf ft) was specified for the subframe bolts (see Chapter 11), but this has now been changed to 35 Nm (26 lbf ft) plus angle tightening of 90°. This applies to all model years.

Front shock absorber – renewal

8 New vehicles are installed at the factory with 'wet' type front suspension struts where the shock absorber consists of a piston and rod acting directly in the strut which is filled with hydraulic fluid.

9 The original piston, rod and seals cannot be renewed, but they can be replaced by a shock absorber cartridge. To do this, remove the piston and rod, as described in Chapter 11, then drain the fluid from the strut wheel bearing housing and clean the recess thoroughly.

10 Insert the new cartridge and tighten the screw cap.

11 With the cartridge upright fully operate the shock absorber piston rod through several strokes to purge it of any trapped air.

Steering wheel – removal and refitting

12 The accompanying photos show the steering wheel on a later model and the removal and refitting procedure is as given in Chapter 11.

Wheels and tyres – general care and maintenance

13 Wheels and tyres should give no real problems in use provided that a close eye is kept on them with regard to excessive wear or damage. To this end, the following points should be noted.

14 Ensure that tyre pressures are checked regularly and maintained correctly. Checking should be carried out with the tyres cold and not immediately after the vehicle has been in use. If the pressures are checked with the tyres hot, an apparently high reading will be obtained owing to heat expansion. Under no circumstances should an attempt be made to reduce the pressures to the quoted cold reading in this instance, or effective underinflation will result.

15 Underinflation will cause overheating of the tyre owing to excessive flexing of the casing and the tread will not sit correctly on the road surface. This will cause a consequent loss of adhesion and excessive wear, not to mention the danger of sudden tyre failure due to heat build-up.

16 Overinflation will cause rapid wear of the centre part of the tyre tread coupled with

13

reduced adhesion, harsher ride, and the danger of shock damage occuring in the tyre casing.

17 Regularly check the tyres for damage in the form of cuts or bulges, especially in the sidewalls. Remove any nails or stones embedded in the tread before they penetrate the tyre to cause deflation. If removal of a nail does reveal that the tyre has been punctured, refit the nail so that its point of penetration is marked. Then immediately change the wheel and have the tyre repaired by a tyre dealer. Do not drive on a tyre in such a condition. In many cases a puncture can be simply repaired by the use of an inner tube of the correct size and type. If in any doubt as to the possible consequences of any damage found, consult your local tyre dealer for advice.

18 Periodically remove the wheels and clean any dirt or mud from the inside and outside surfaces. Examine the wheel rims for signs of rusting, corrosion or other damage. Light alloy wheels are easily damaged by 'kerbing' whilst parking, and similarly steel wheels may become dented or buckled. Renewal of the wheel is very often the only course of remedial action possible.

19 The balance of each wheel and tyre assembly should be maintained to avoid excessive wear, not only to the tyres but also to the steering and suspension components. Wheel imbalance is normally signified by vibration through the vehicle's bodyshell, although in many cases it is particularly noticeable through the steering wheel. Conversely, it should be noted that wear or damage in suspension or steering components may cause excessive tyre wear. Out-of-round or out-of-true tyres, damaged wheels and wheel bearing wear/maladjustment also fall into this category. Balancing will not usually cure vibration caused by such wear.

20 Wheel balancing may be carried out with the wheel either on or off the vehicle. If balanced on the vehicle, ensure that the wheel-to-hub relationship is marked in some way prior to subsequent wheel removal so that it may be refitted in its original position.

21 General tyre wear is influenced to a large degree by driving style harsh braking and acceleration or fast cornering will all produce more rapid tyre wear. Interchanging of tyres may result in more even wear, but this should only be carried out where there is no mix of tyre types on the vehicle. However, it is worth bearing in mind that if this is completely effective, the added expense of replacing a complete set of tyres simultaneously is incurred, which may prove financially restrictive for many owners.

22 Front tyres may wear unevenly as a result of wheel misalignment. The front wheels should always be correctly aligned according to the settings specified by the vehicle manufacturer.

23 Legal restrictions apply to the mixing of tyre types on a vehicle. Basically this means that a vehicle must not have tyres of differing construction on the same axle. Although it is not recommended to mix tyre types between front axle and rear axle, the only legally permissible combination is crossply at the front and radial at the rear. When mixing radial ply tyres. textile braced radials must always go on the front axle, with steel braced radials at the rear. An obvious disadvantage of such mixing is the necessity to carry two spare tyres to avoid contravening the law in the event of a puncture.

24 In the UK, the Motor Vehicles Construction and Use Regulations apply to many aspects of tyre fitting and usage. It is suggested that a copy of these regulations is obtained from your local police if in doubt as to the current legal requirements with regard to tyre condition, minimum tread depth, etc.

12 Bodywork and fittings

Plastic components – repair

1 With the use of more and more plastic body components by the vehicle manufacturers (eg bumpers. spoilers, and in some cases major body panels), rectification of more serious damage to such items has become a matter of either entrusting repair work to a specialist in this field, or renewing complete components. Repair of such damage by the DIY owner is not really feasible, owing to the cost of the equipment and materials required for effecting such repairs. The basic technique involves making a groove along the line of the crack in the plastic, using a rotary burr in a power drill. The damaged part is then welded back together, using a hot-air gun to heat up and fuse a plastic filler rod into the groove. Any excess plastic is then removed, and the area rubbed down to a smooth finish. It is important that a filler rod of the correct plastic is used, as body components can be made of a variety of different types (eg polycarbonate, ABS, polypropylene).

2 If the owner is renewing a complete component himself, or if he has repaired it with epoxy filler, he will be left with the problem of finding a suitable paint for finishing which is compatible with the type of plastic used. At one time, the use of a universal paint was not possible, owing to the complex range of plastics encountered in body component applications. Standard paints, generally speaking, will not bond to plastic or rubber satisfactorily. However, it is now possible to obtain a plastic body parts finishing kit which consists of a pre-primer treatment, a primer and coloured top coat. Full instructions are normally supplied with a kit, but basically, the method of use is to first apply the pre-primer to the component concerned, and allow it to dry for up to 30 minutes. Then the primer is applied, and left to dry for about an hour before finally applying the special-coloured top coat. The result is a correctly-coloured component, where the paint will flex with the plastic or rubber, a property that standard paint does not normally posses.

Front door trim panel (1984-on models) – removal and refitting

3 The accompanying photos show minor differences for the removal and refitting of the front door trim panel on later models. The basic procedure remains as given in Chapter 12, Section 7.

12.3A Pull off the knob . . .

12.3B . . . and cover from the manual door mirror adjuster

12.3C Unscrew the locking knob . . .

12.3D .. and remove the plastic cover

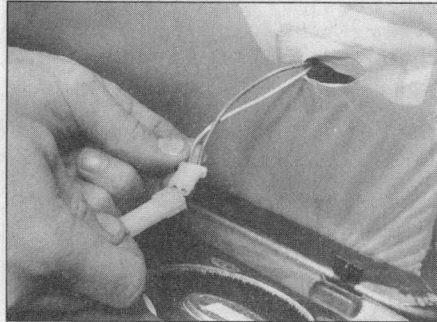

12.3E Disconnect the wiring from the door mounted speaker

12.4 Removing a trim strip from the front grille

Front grille (later models) – removal and refitting

4 Remove the two screws and slide the twin trim strips outwards from under the headlamp units (photo).
5 Release the top of the grille from the crossmember using a screwdriver to depress the plastic tabs (photos).
6 Unhook the bottom of the grille from the valance (photo).
7 Refitting is a reversal of removal, but press in the top of the grille until the clips snap into place.

Tailgate (Hatchback) – removal, refitting and adjustment

8 The procedure is as given in Chapter 12, Section 20. However, the following additional points should be noted.
9 Pull off the luggage shelf stays to disconnect them (photo).
10 Disconnect the tailgate support struts by prising out the retaining clips (photo).
11 When shut, the tailgate should be firmly in contact with the weatherstrip and rubber stop. To achieve this adjust the side strikers (or lock plates) as necessary, and also ensure that the locks engage the strikers centrally (photos).

Seat belts

12 The front and rear seat belts are of inertia reel type

12.5A Depress the plastic tabs with a screwdriver . . .

12.5B . . . to release the top of the grille from the crossmember . . .

12.6 . . . then unhook the bottom from the valance

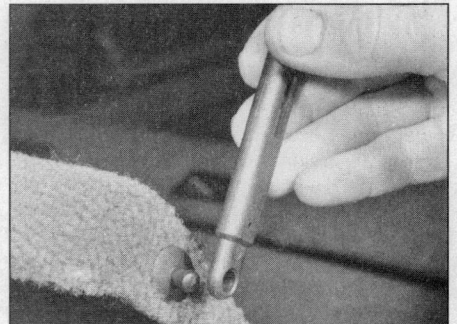

12.9 Disconnecting a luggage shelf stay

12.10 Disconnecting a tailgate support strut

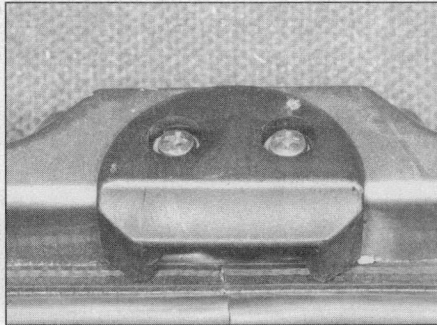

12.11A Tailgate rubber stop

12.11B Tailgate lock side striker

13

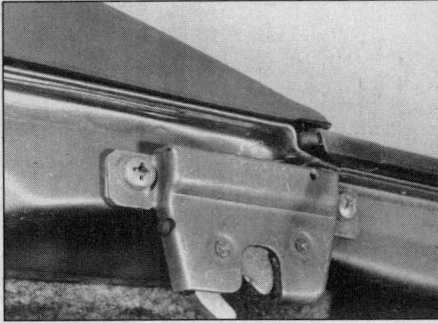

12.11C Tailgate lock

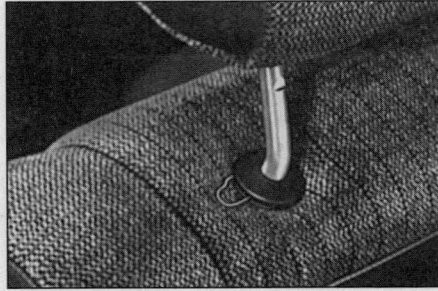

Fig. 13.40 Front head restraint pillar, bezel and spring clip (Sec 12)

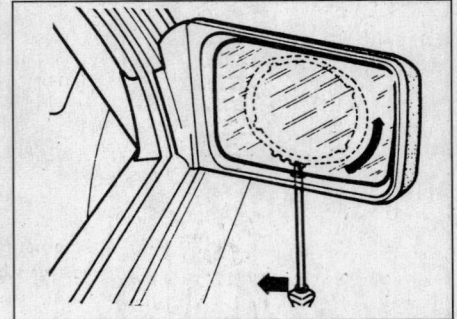

Fig. 13.41 Using a screwdriver to release the door mirror glass locking ring (Sec 12)

13 Examine the belts regularly for signs of fraying or damage. If evident renew the complete belt assembly. The belt should also be renewed if the belts have been strained due to the vehicle having been involved in a front end collision.

14 Clean the belts using only a damp cloth and liquid detergent. Never use solvents of any kind.

15 Whenever a seat belt is removed and refitted, the anchor bolt components (wave washer, spacer, anchor plate) must be fitted in their originally installed sequence.

Head restraints

16 To adjust the height of the head restraints, simply grip the sides and pull up or push down.

17 To remove a front head restraint, push the spring clips out of the grooved bezels at the base of the head restraint pillars. Pull the head restraint up and out.

18 When refitting, press the spring clips into the bezel grooves so that the straight parts of the clips are towards the rear of the car. Insert the head restraint pillars into their holes.

19 To remove a rear head restraint from Hatchback versions, release the backrest and tilt it forwards, then remove the head restraint by giving it a sharp jerk upwards.

20 To remove a rear head restraint from the Saloon, twist both bezels at the base of the pillars until the slots in the bezels point to the side. Pull the head restraint upwards and out.

21 After refitting the Saloon rear head restraints, twist the bezels until their slots point towards the front.

Rear seat (Hatchback and Estate) – removal and refitting

22 To remove the rear seat pull the cushion loop upwards and fold the seat cushion forwards.

23 Release the seat back and fold it forwards. The side pivot bracket screws may now be removed and the seat cushion and back withdrawn from the vehicle.

24 On certain later Hatchback and Estate models, the rear seat is of asymmetrical (60/40) split type but the removal procedure is similar to that described earlier.

Grab handles

25 The interior grab handles which are located above the door apertures can be removed if the end caps are prised out and the fixing screws extracted.

Exterior door mirror glass (electrically heated and adjustable type) – renewal

26 The mirror glass can be renewed without removing the complete mirror assembly from the door, using the following procedure .

27 Using a small screwdriver inserted through the slot in the base of the mirror body, engage the locking ring teeth and turn the locking ring anti-clockwise (Fig 13. 41).

28 With the locking ring released, withdraw the glass, disconnect the wiring and remove the glass from the mirror body.

29 Refitting is the reverse sequence to removal.

Battery

Alternator with voltage regulator

Starter

Ignition coil

Wiper motor — 2-speed

Electric motor

Switch (hand operated, on/off)

Switch (pressure-operated).

Switch (mechanically operated)

Switch (mechanically operated)

Switch (thermally operated)

Fuse

Solenoid valve

Sender unit for coolant temperature gauge

Sender unit for fuel gauge

Instrument

Clock

Bulb

Relay (electronically controlled)

Crossed wires (not connected)

Horn

Relay (with diode)

Cigarette lighter

Heated rear window

Interior light

Spark gap

Flat connector

Connector (multi-point)

Wiring connection, detachable

Wiring connection, fixed

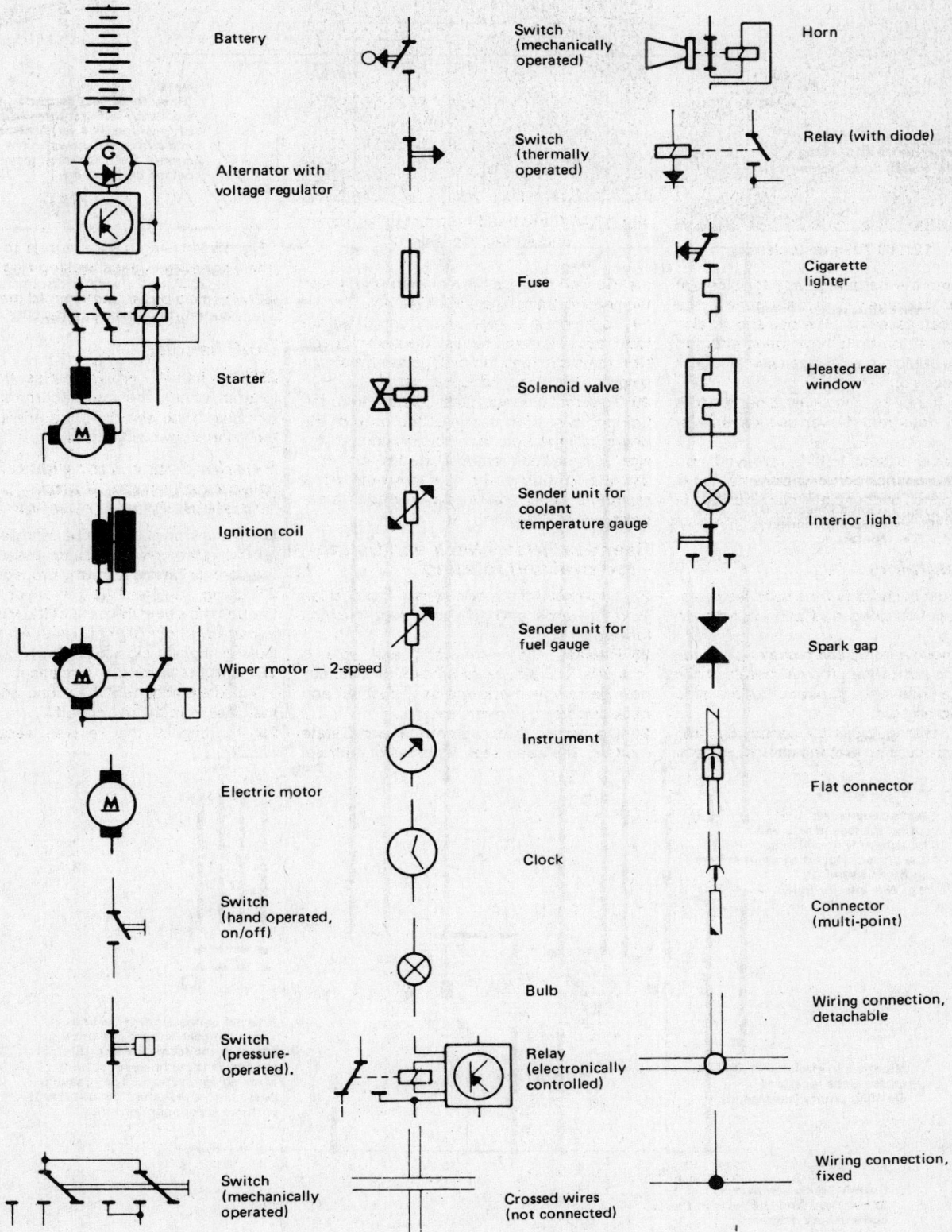

Symbols used in current flow diagrams

14

This area represents the relay plate with fuse holder

Note:
All **switches** and **contacts** are illustrated in the **mechanical off position**. The various contacts in a switch are shown in the current track to which they belong by function.

Numbers in squares indicate that a wire is discontinued in the diagram and refer to the current track where it is continued.

Wire cross section in mm²

Wiring colour codes
Primary colour
Tracer colour

Numbers/number combinations
These indicate the individual contacts in a multi-point connector, e.g. T 10/4
T 10 = ten-point connector
/4 = contact 4

Terminals with the numbers which are on the actual parts.

Wiring codes
black – sw
blue – bl
brown – br
green – gn
grey – gr
lilac – li
red – ro
white – we
yellow – ge

Symbols
(in this case: bulb)

Parts designation
Using the legend you will be able to find which part in the current flow diagram is referred to by this symbol, e.g. W = interior light

Internal connections (thin lines)
These connections are **not** to be found in the form of wires. Internal connections are however current-carrying connections. They make it possible to check the flow of current within a component or unit.

Numbers in circles
indicate the locations of earthing points (see legend)

Current track numbers
to help you find the parts in the current flow diagram (see legend)

Instructions for using current flow diagrams

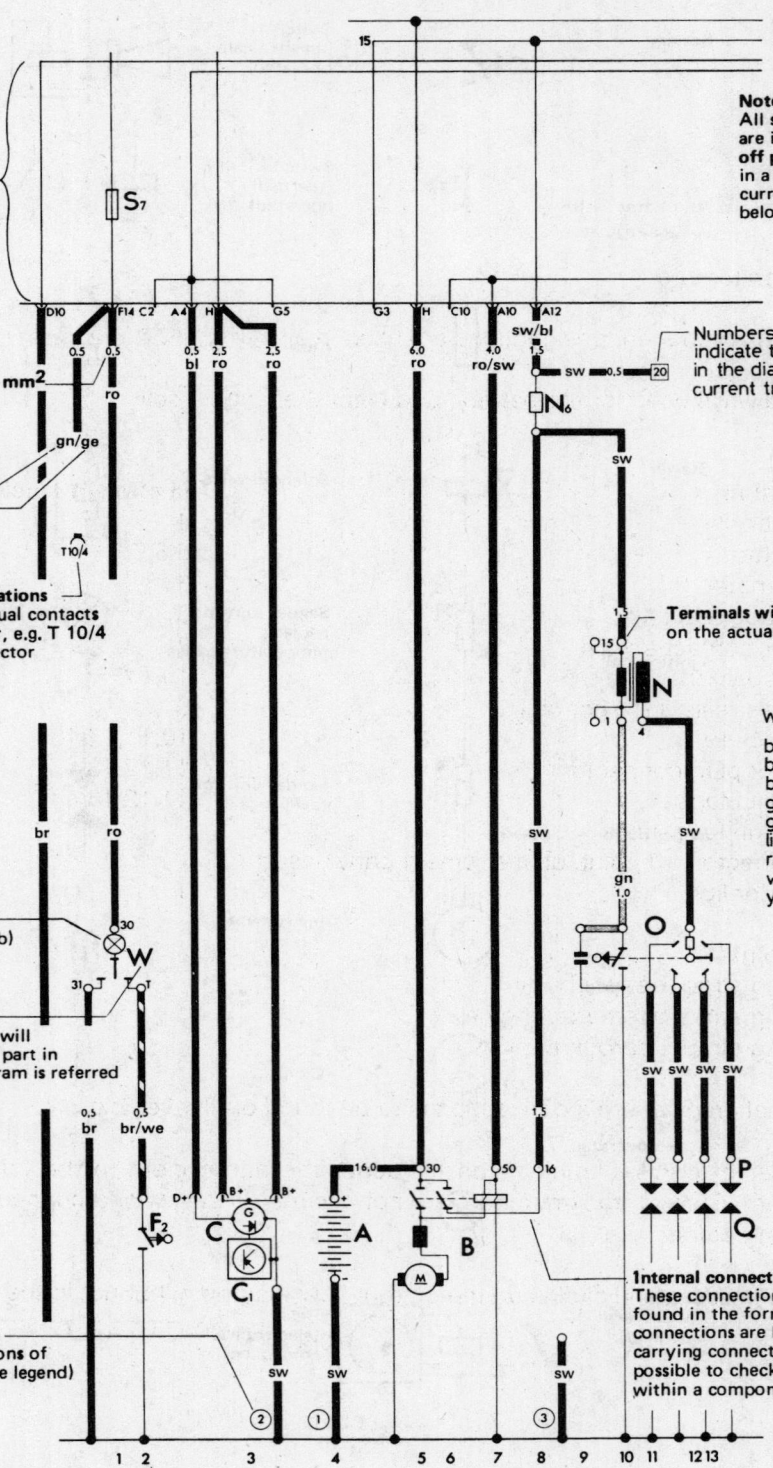

Specimen legend

The same component designations are used in all current flow diagrams

eg: A is always used for the battery, and N for the ignition coil

Designation	in current track*
A Battery	4
B Starter	5,6,7,8
C Alternator	3
C1 Voltage regulator	3
F2 Door contact switch	2
N Ignition coil	10,11
N6 Series resistance (for coil)	8
O Distributor	10,11,12,13
P Spark plug connector	11,12,13
Q Spark plugs	11,12,13
S7 Fuse in fuse box	
T10 Connector, 10 point, on instrument panel insert**	
W Interior light	1

Earth points*
1 Earthing strap, battery/body
2 Earthing strap, alternator/engine
3 Earthing strap, gearbox/chassis

* Explanation of where a connection is to be found on the vehicle

**Number of current track helps to locate the component in the current flow diagram. Current track numbers are not given for fuses, wiring connections, and earthing points

*** Earth points are in circles on the current flow diagrams, but not in the keys

Instructions for using current flow diagrams (continued)

14

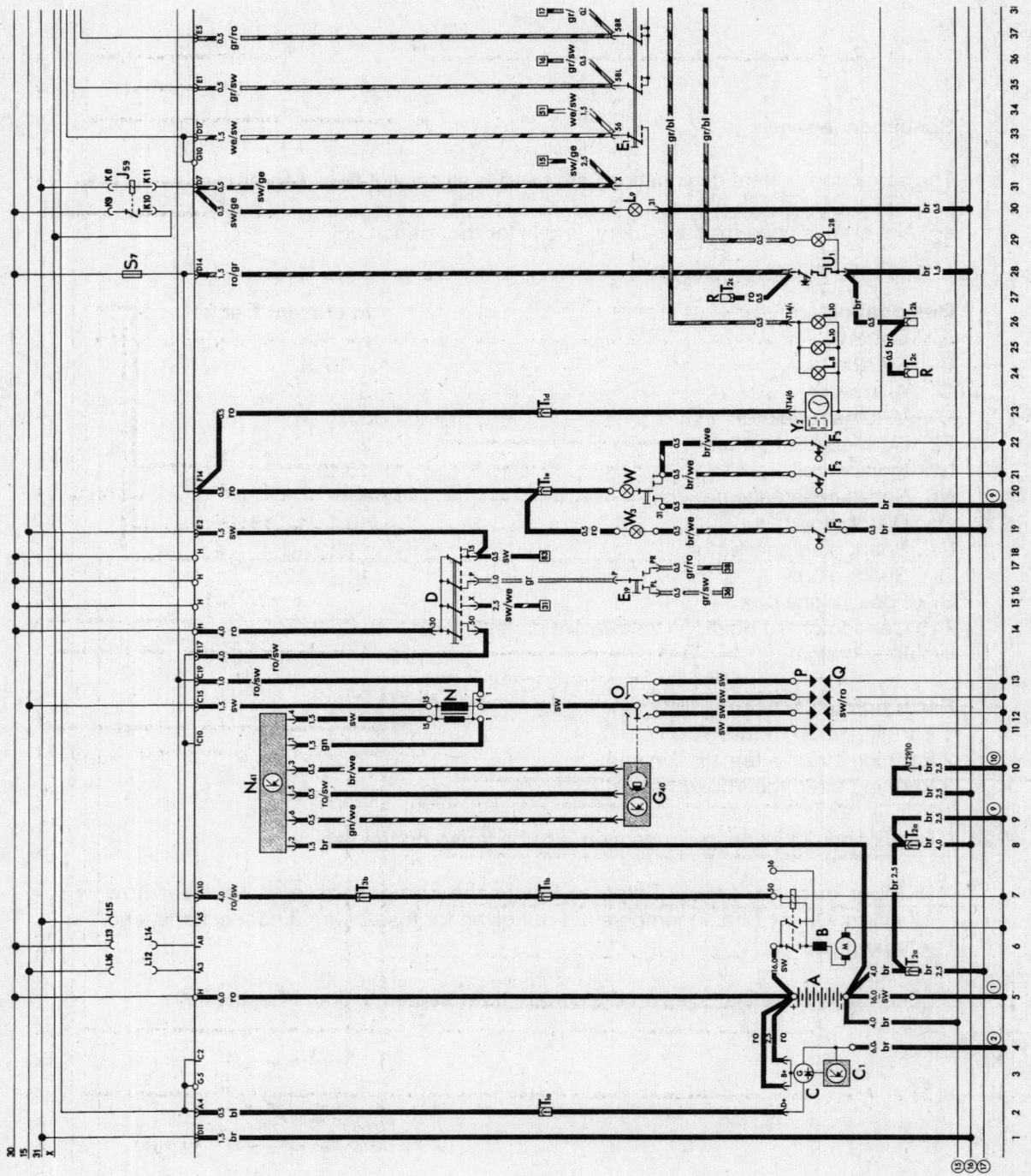

Wiring diagram for all 1981/82 models

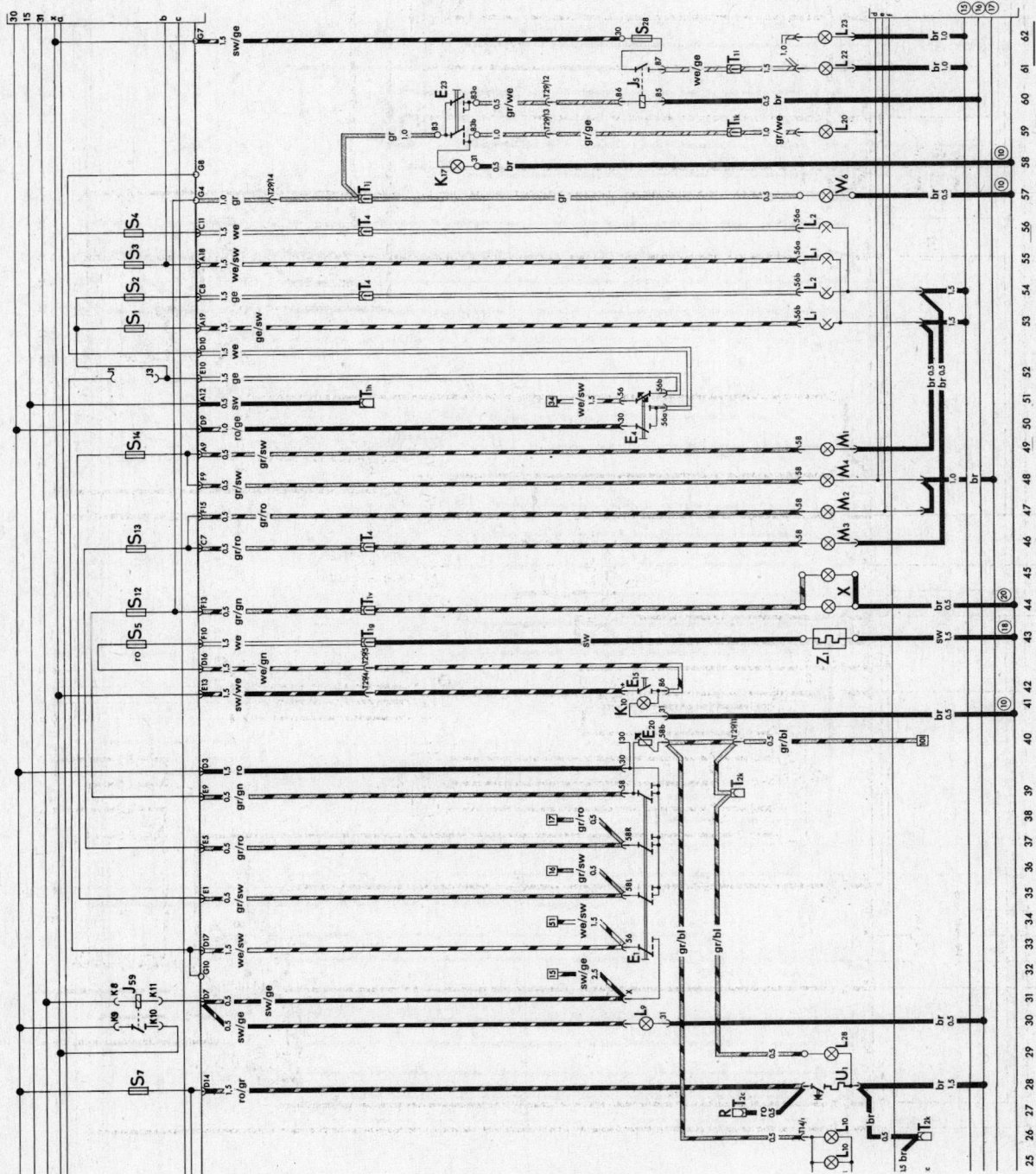

Wiring diagram for all 1981/82 models (continued)

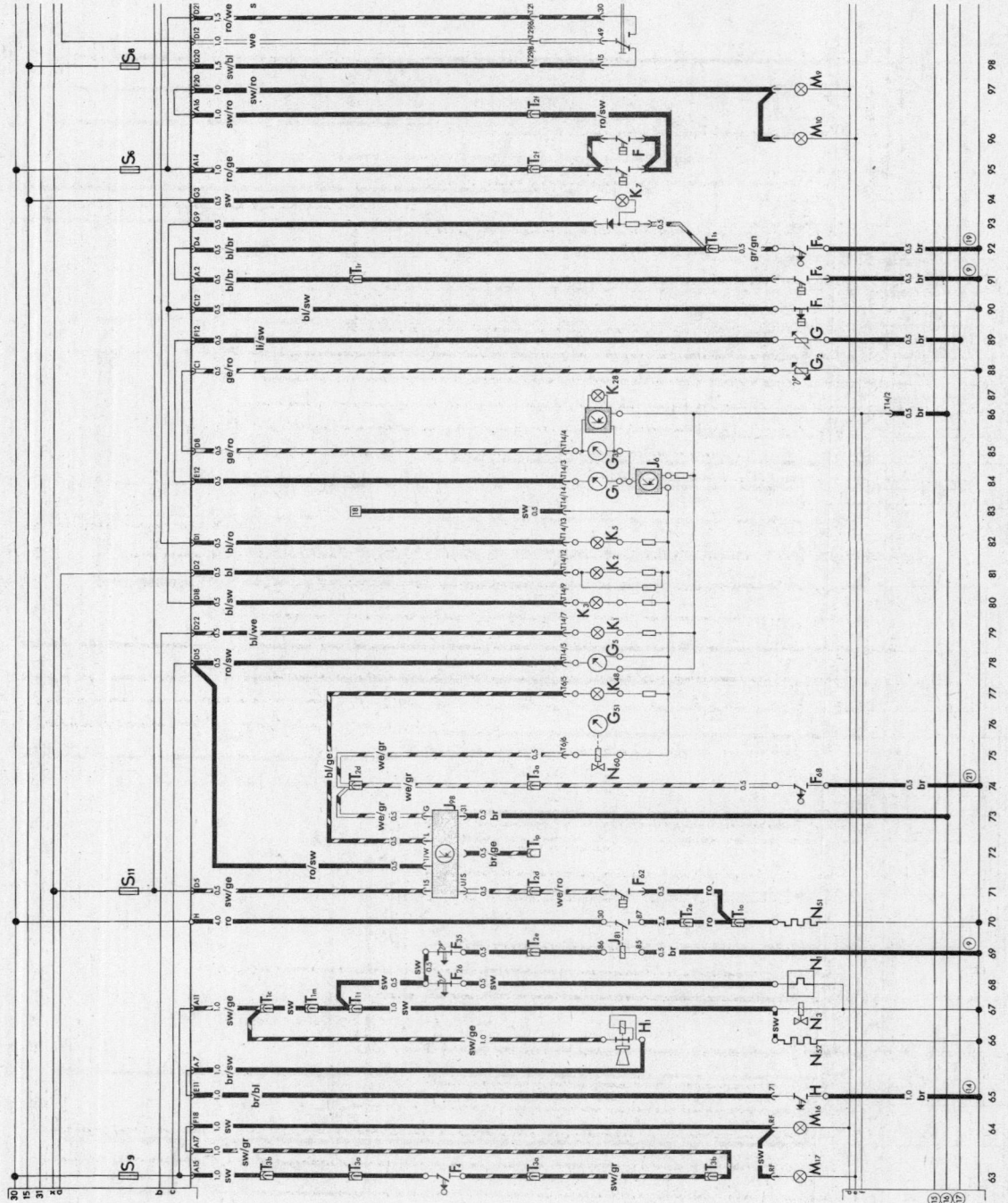

Wiring diagram for all 1981/82 models (continued)

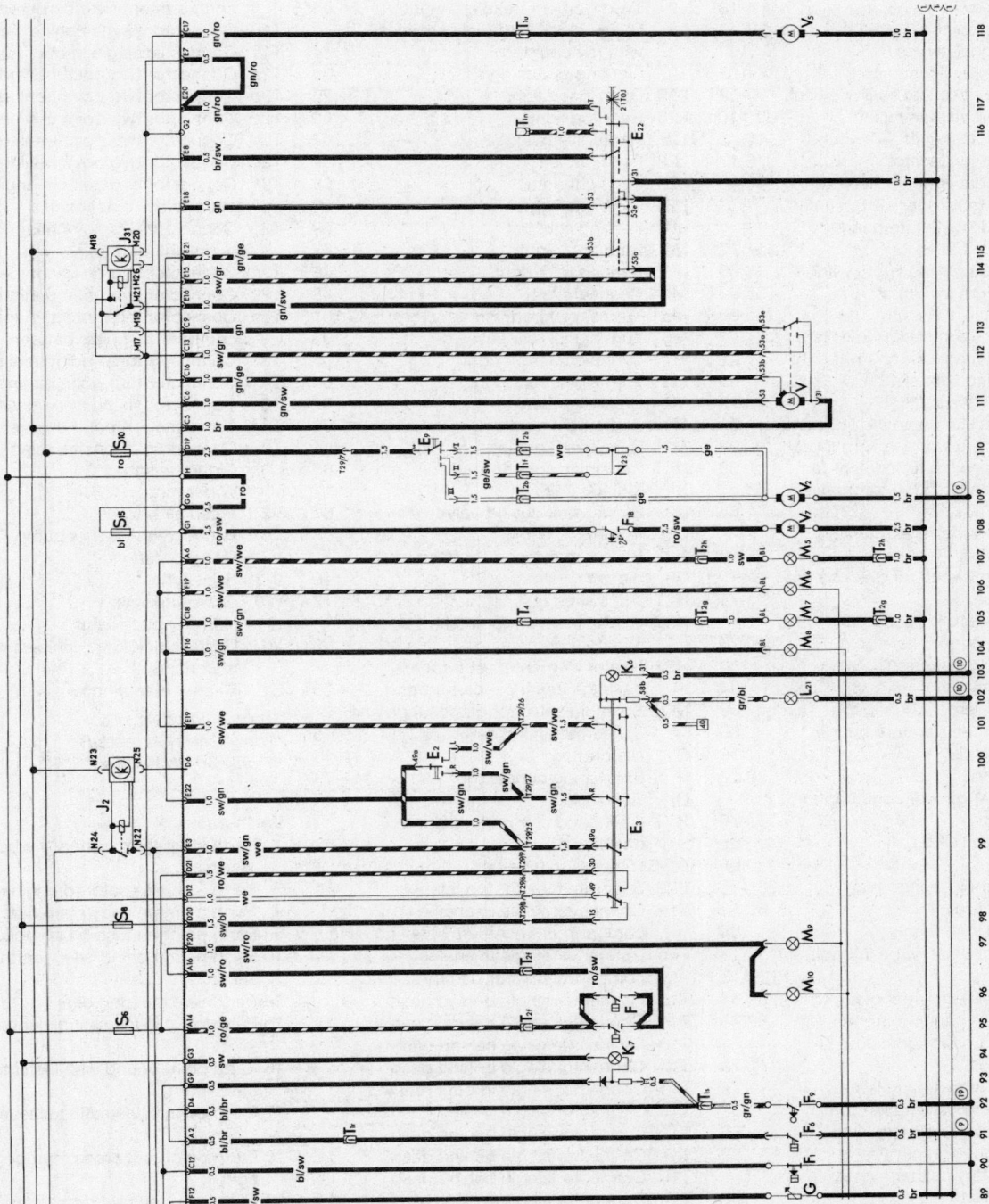

Wiring diagram for all 1981/82 models (continued)

14

Key to wiring diagram for all 1981/82 models

Designation		In current track
A	Battery	5
B	Starter	6,7
C	Alternator	2-4
C1	Voltage regulator	2-4
D	Ignition switch	14-18
E1	Lighting switch	30-40
E2	Turn signal switch	100
E3	Emergency light switch	97-102
E4	Dip and headlight flasher switch	50-52
E9	Fresh air blower switch	109,110
E15	Heated rear window switch	41-42
E19	Parking light switch	16,17
E20	Instrument lighting rheostat**	40
E22	Intermittent wiper switch (only GL model with intermittent operation)	114-117
E23	Fog and rear fog light switch*	58,60
F	Brake light switch	95 96
F1	Oil pressure switch	90
F2	Door contact switch front left	21
F3	Door contact switch front right	22
F4	Reversing light switch	63
F5	Switch for boot light*	19
F6	Brake system warning light switch*	91
F9	Handbrake warning light switch*	92
F18	Thermoswitch for coolant fan	108
F26	Thermo switch for automatic choke ***	68
F35	Thermo switch for manifold preheating ***	69
F62	Vacuum switch for gearchange indicator	71
F68	Gear switch for gearchange and consumption indicator	74
G	Fuel gauge sender	89
G1	Fuel gauge	84
G2	Coolant temperature gauge sender	88
G3	Coolant temperature gauge	85
G5	Rev counter	78
G40	Hall sender	9,10
G51	Gearchange and consumption indicator	76
H	Horn controls	65
H1	Horn	66
J2	Emergency flasher relay	99,100
J5	Fog light relay*	60,62
J6	Voltage stabilizer	84
J31	Relay for wash/wipe intermittent facility*	112-115
J59	Relief relay (for X contact)	30,31
J81	Relay for manifold preheating***	69,70
J98	Control unit for gearchange indicator	71-73
K1	High beam warning lamp	79
K2	Generator warning lamp	81
K3	Oil pressure warning lamp	80
K5	Turn signal warning lamp	82
K6	Emergency flasher system warning lamp	103
K7	Dual circuit brake and handbrake warning lamp*	93,94
K10	Heated rear window warning lamp	41
K17	Fog light warning lamp	58

Designation		In current track
K28	Coolant temperature warning lamp (too hot, red)	87
K48	Gearchange indicator lamp	77
L1	Twin filament headlight left	53,55
L2	Twin filament headlight right	54,56
L8	Digital clock light (only vehicles with rev counter)	24
L9	Lighting switch light	30
L10	Dash insert light	25,26
L20	Rear fog light*	59
L21	Heater lever light**	102
L22	Fog light left	61
L23	Fog light right*	62
L28	Cigarette lighter light*	29
M1	Parking light left	49
M2	Rear light right	47
M3	Parking light right	46
M4	Rear light left	48
M5	Turn signal front left	107
M6	Turn signal rear left	106
M7	Turn signal front right	105
M8	Turn signal rear right	104
M9	Brake light left	97
M10	Brake light right	96
M16	Reversing light left	64
M17	Reversing light right	63
N	Ignition coil	12
N3	By-pass air cut-off valve	67
N1	Automatic choke***	68
N23	Ballast resistance for fresh air blower	110
N41	TCI switch unit	8 - 12
N51	Heater element for intake manifold preheating***	66
N52	Heater element (part throttle passage heating – carburettor)	66
N60	Solenoid valve for gearchange and consumption indicator	75
O	Distributor	11-13
P	Spark plug connector	11-13
Q	Spark plug	11-13
R	Connection for radio (behind dash insert)	24,27
S1-S15	Fuses in fuse box	
S28	Separate fuse for fog lights	62
T1a	Connector single in engine compartment near left wheel housing	
T1b	Connector single in engine compartment near battery	
T1c	Connector single behind dash	
T1d	Connector single behind dash	
T1e	Connector sinyle behind dash	
T1f	Connector single behind dash	
T1g	Connector single in boot rear left	
T1h	Connector single behind dash	
T1j	Connector single behind dash	
T1k	Connector single behind dash	
T1l	Connector single behind dash	
T1m	Connector single in engine compartment near carburettor	
T1n	Connector single behind dash	
T1o	Connector single in engine compartment near carburettor	

Designation		In current track
T1p	Connector single behind dash	
T1r	Connector single behind dash	
T1s	Connector single behind dash	
T1t	Connector single in engine compartment near carburettor	
T1u	Connector single behind dash	
T1v	Connector single in boot rear left	
T2a	Connector two point behind dash	
T2b	Connector two point behind dash	
T2c	Connector two point behind dash	
T2d	Connector two point behind dash	
T2e	Connector two point behind dash	
T2f	Connector two point in engine compartment near servo	
T2g	Connector two point near right headlight	
T2h	Connector two point near left headlight	
T2k	Connector two point behind dash	
T3a	Connector three point in engine compartment near battery	
T3b	Connector three point behind dash	
T4	Connector four point behind dash	
T6/	Connector six point on dash insert	
T14/	Connector 14 point on dash insert	
T29/	Connector 29 point behind dash	
U1	Cigarette lighter**	28
V	Wiper motor	111-113
V2	Fresh air blower	109
V5	Windscreen washer pump	118
V7	Radiator fan	108
W3	Boot light*	19
W6	Glove box light*	57
X	Number plate light	44,45
Y2	Digital clock (only vehicles with rev counter)	23
Z1	Heated rear window	43

* only GL models
** only L and GL models
*** only 55 kW engine

Earth points

1 Earth strap from battery via body to gearbox
2 Earth wire alternator to engine block
9 Earth point on relay plate bracket
9 Earth point on fuse box bracket
10 Earth point behind dash (earth plate secured to dash)
14 Earth wire steering gear
15 Earth point in insulating sleeve of front loom
16 Earth point bound with insulating tape in dash loom
17 Earth point in insulating sleeve of rear loom
18 Earth point roof cross member upper right
19 Earth point on rear cross member (driver's side)
20 Earth point roof cross member upper left
21 Earth point on earth strap from battery via body to gearbox

Key to wiring diagram for 1982/83 Passat C, CL & GL5 models and Santana LX, GX & GX5 models

Designation		In current track
A	Battery	7
B	Starter	10–12
C	Alternator	4, 6
C1	Voltage regulator	4, 6
D	Ignition/starter switch	25–29
E1	Lighting switch	59-69
E2	Turn signal switch	100,101
E3	Emergency light switch	95-103
E4	Headlight dimmer/flasher switch	76,77
E9	Fresh air blower switch	115,116
E15	Heated rear window switch	93
E19	Parking light switch	27,28
E20	Instrument/instrument panel lighting control	69
E22	Intermittent wiper switch	121,122
E23	Fog light and rear fog light switch (only GL models)	87,88
F	Brake light switch	108,109
F1	Oil pressure switch (1.8 bar)	46
F2	Front left door contact switch	53
F3	Front right door contact switch	54
F4	Reversing light switch	112
F5	Luggage boot light switch (only Variant)	50
F9	Handbrake warning system switch (only GL models)	127
F10	Rear left door contact switch	52
F11	Rear right door contact switch	55
F18	Radiator fan thermoswitch	114
F22	Oil pressure switch (0.3 bar)	47
F26	Thermoswitch for choke	24
F34	Brake fluid level warning contact (only GL models)	126
F35	Thermoswitch for intake manifold preheating	23
F62	Gearshift indicator vacuum switch	48
F66	Coolant shortage indicator switch (from 63 kW engine)	135
F68	Gear switch for gearshift and consumption indicator	49
G	Fuel gauge sender	45
G1	Fuel gauge	40
G2	Coolant temperature gauge sender	44
G3	Coolant temperature gauge	41
G5	Rev counter	35
G40	Hall sender	14,15
G51	Consumption indicator	39,40
H	Horn control	130
H1	Dual horns (only GL models)	132,133
J2	Emergency light relay (relay 12)	98-100
J4	Dual tone horn relay (only GL models) (relay 6)	130, 131
J5	Fog light relay (only GL models) (relay 7)	86,88
J6	Voltage stabilizer	40
J31	Intermittent wash/wiper relay (relay 10)	121-124
J39	Headlight washer system relay (only GL models) (relay 15)	124,125
J59	Relief relay (for x contact) (relay 8)	60-63
J87	DIS (idling stabilisation) switch unit (relay 4)	14,15

Designation		In current track
J98	Gearshift indicator control unit	48,49
J114	Oil pressure monitor control unit	33-34
J120	Switch unit for coolant shortage indicator (from 63 kW engine) (relay 18)	134,135
K1	Main beam warning lamp	37
K2	Alternator warning lamp	30
K3	Oil pressure warning lamp	32
K5	Turn signal warning lamp	36
K6	Emergency light system warning lamp	103
K7	Dual circuit brake and handbrake warning lamp (only GL models)	128
K10	Heated rear window warning lamp	92
K17	Fog light warning lamp (only GL models)	89
K28	Coolant temp. warning lamp (overheating, red)	43
K48	Gearshift indicator warning lamp	38
L1	Twin filament headlight bulb left	78,80
L2	Twin filament headlight bulb right	79,81
L8	Clock light	64
L9	Lighting switch light	129
L10	Instrument panel insert light	65,66
L20	Rear fog light (only GL models)	84
L21	Heater controls light	102
L22	Fog light left (only GL models)	82
L23	Fog light right (only GL models)	83
L28	Cigarette lighter light	59
L39	Heated rear window switch	91
L40	Front and rear fog light switch (only GL models)	90
M1	Side light left	75
M2	Tail light right	73
M3	Side light right	72
M4	Tail light left	74
M5	Front left turn signal	105
M6	Rear left turn signal	104
M7	Front right turn signal	107
M8	Rear right turn signal	106
M9	Brake light left	111
M10	Brake light right	110
M16	Reversing light left	113
M17	Reversing light right	112
N	Ignition coil	18, 19
N1	Automatic choke	24
N3	By-pass air cut-off valve	23
N23	Series resistance for fresh air blower	116
N41	TCI control unit	13-16
N51	Heater element for intake manifold preheating	21
N52	Heater element (part throttle channel heating/carburettor)	22
N60	Solenoid valve for consumption indicator	39
P	Spark plug connector	16-19
Q	Spark plugs	16-19
R	Radio connection	
S1-S22	Fuses in fuse box	
S27	Separate fuse for rear fog light	
T1	Connector, single near battery	

Designation		In current track
T1a	Connector, single engine compartment front left	
T1b	Connector, single engine compartment centre	
T1c	Connector, single near battery	
T1d	Connector, single behind dash on right	
T1e	Connector, single near relay plate	
T1f	Connector, single behind dash	
T1g	Connector, single in boot rear left	
T1l	Connector, single behind relay plate	
T1m	Connector, single behind dash	
T1s	Connector, single behind relay plate	
T1v	Connector, single in tailgate	
T2g	Connector, two point engine compartment front right	
T2h	Connector, two point engine compartment front left	
T2k	Connector, two point behind dash	
T2k	Connector, two point engine compartment front	
T21	Connector, two point behind relay plate	
T29	Connector 29 pin behind dash	
T2b	Connector, two point behind dash	
T3a	Connector, three point engine compartment	
T3b	Connector, three point engine compartment centre/near battery	
T4	Connector, four point behind dash	
T14	Connector, 14 point on dash insert	
U1	Cigarette lighter	58
V	Windscreen wiper motor	117-120
V2	Fresh air blower	115
V5	Windscreen washer pump	123
V7	Radiator fan	114
V11	Headlight washer pump (only GL models)	125
W	Interior light, front	51-53
W3	Luggage boot light	50
X	Number plate light	70,71
Y2	Digital clock	56
Z1	Heated rear window	94

Earth points

1 Earth strap from battery to body
9 Earth point on fuse box bracket
10 Earth point behind dash (earth plate secured to dash)
14 Earth wire, steering box
15 Earth point in insulating sleeve of front loom
16 Earth point bound with insulating tape in dash loom
17 Earth point in insulating sleeve of rear loom

Fuse colour/capacities

S1– green – 30A
S14 – yellow – 20A
S4 – S6, S11 – S13, S16 – blue – 15A
S2, S3, S7– S10 , S17 – S22 – red – 10A

14

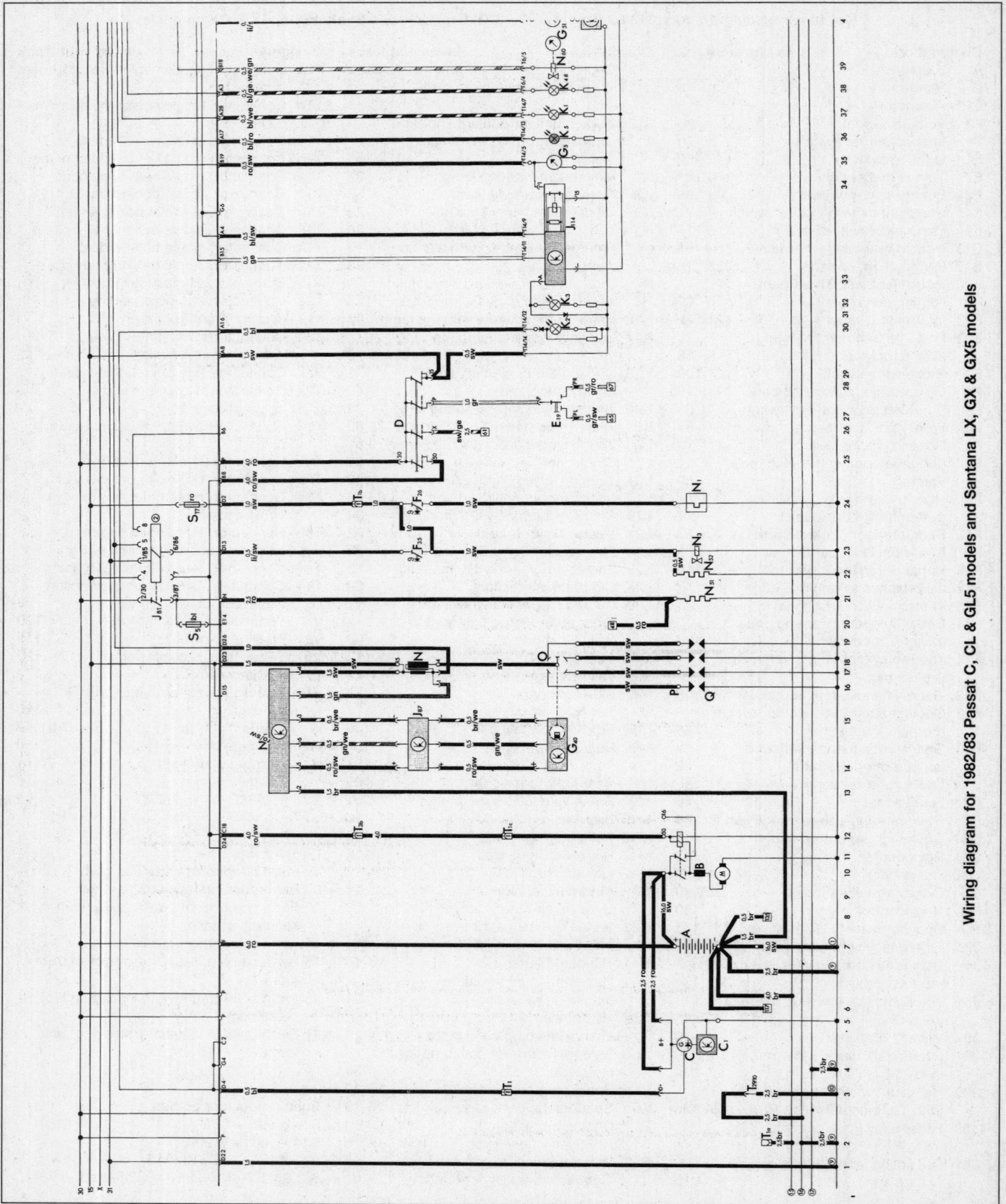

Wiring diagram for 1982/83 Passat C, CL & GL5 models and Santana LX, GX & GX5 models

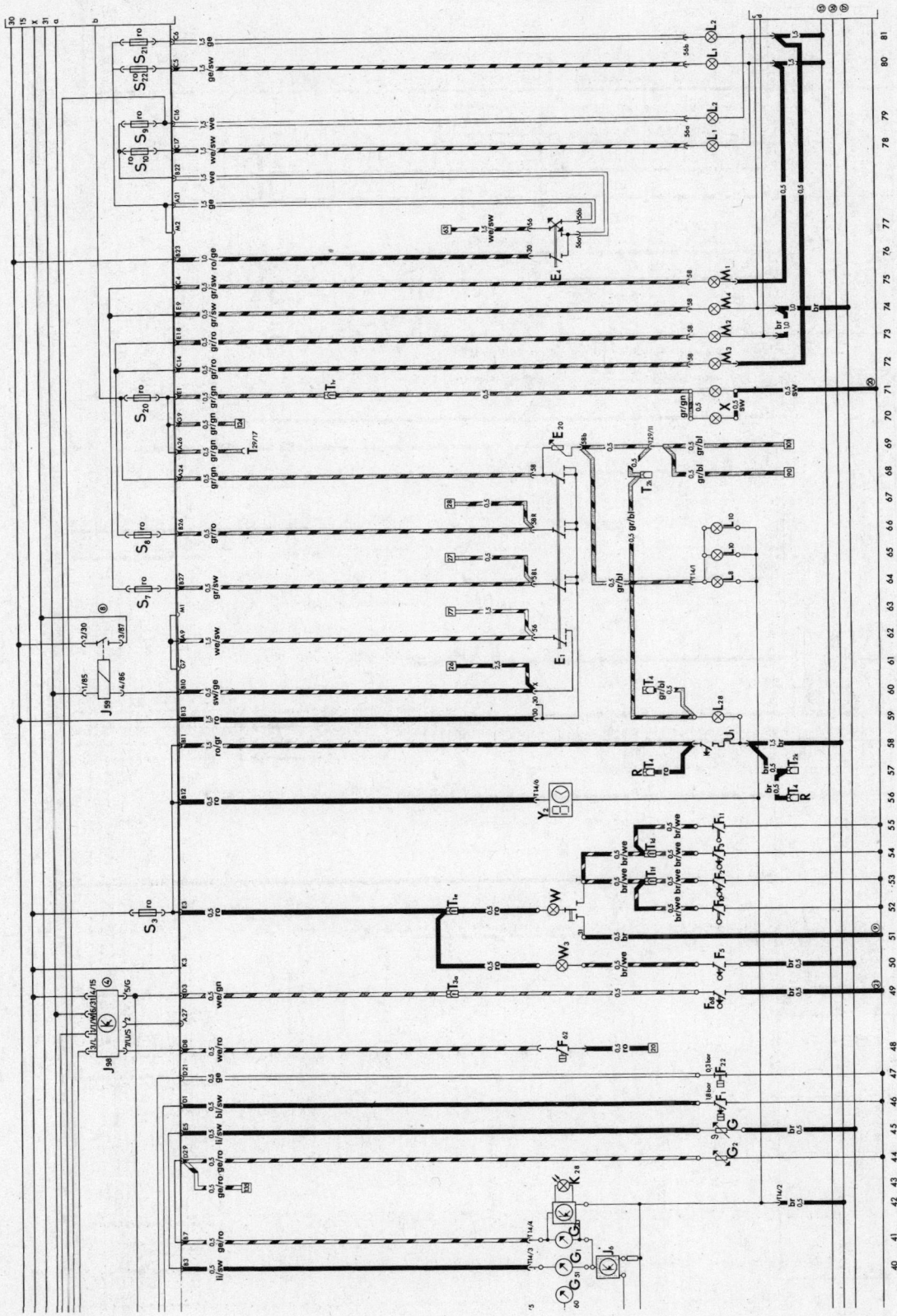

Wiring diagram for 1982/83 Passat C, CL & GL5 models and Santana LX, GX & GX5 models (continued)

14

Wiring diagram for 1982/83 Passat C, CL & GL5 models and Santana LX, GX & GX5 models (continued)

Wiring diagram for 1982/83 Passat C, CL & GL5 models and Santana LX, GX & GX5 models (continued)

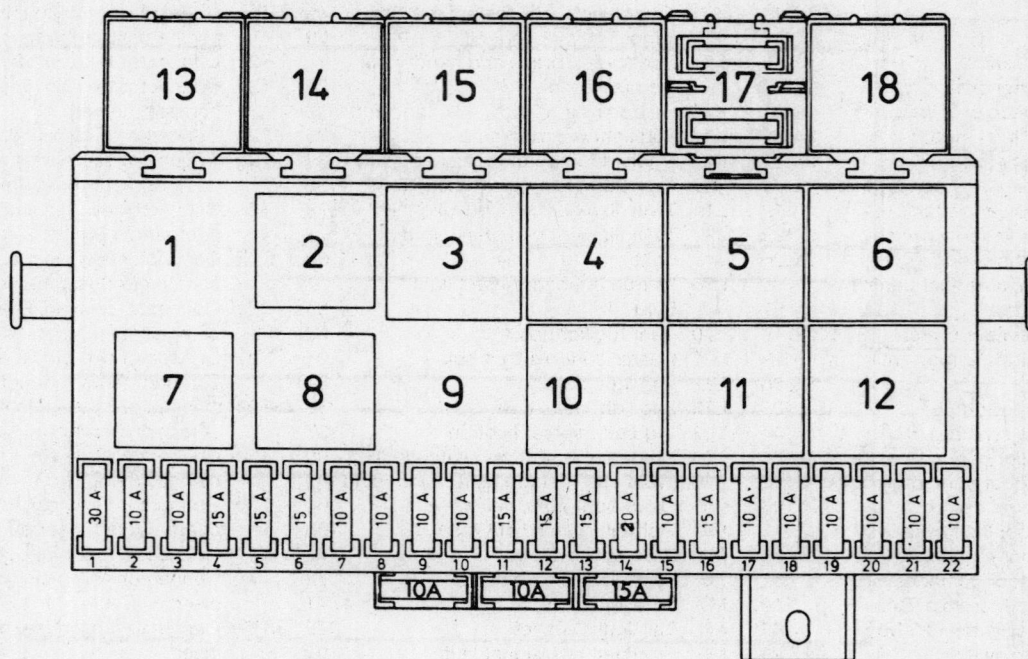

Relay locations for 1982/83 Passat C, CL & GL5 models and Santana LX, GX & GX5 models

1	Vacant
2	Intake pre-heating relay
3	Vacant
4	Switch unit for gearshift indicator
5	Vacant
6	Duel horn relay
7	Fog light relay
8	Relief relay for X contact
9	Vacant
10	Relay for wash-wipe intermittent facility
11	Vacant
12	Emergency light relay
13	Vacant
14	Vacant
15	Headlight washer relay
16	Vacant
17	Vacant
18	Switch unit for coolant shortage indicator

Key to wiring diagram for Passat CL5 & GL5 and Santana LX5 & GX5 from 1983

Designation		In current track
A	Battery	7
B	Starter	9-11
C	Alternator	5,6
C1	Voltage regulator	5,6
D	Ignition/starter switch	23-27
E1	Lighting switch	59-67
E3	Emergency light switch	94-99
E3	Turn signal switch	96-97
E4	Headlight dimmer/flasher switch	75,76
E9	Fresh air blower switch	112
E15	Heated rear window switch	92
E19	Parking light switch	25,26
E20	Instrument/instrument panel lighting control	67
E22	Intermittent wiper switch	115-117
E23	Fog light and rear fog light switch	86-88
F	Brake light switch	104,105
F1	Oil pressure switch (1.8 bar)	44
F2	Front left door contact switch	51
F3	Front right door contact switch	52
F4	Reversing light switch	108
F5	Luggage boot light switch	47
F10	Rear left door contact switch	50
F11	Rear right door contact switch	53
F18	Radiator fan thermo switch	110
F22	Oil pressure switch (0.3 bar)	43
F26	Thermo time switch	17,18
F34	Brake fluid level warning contact	127
F62	Gearshift indicator vacuum switch	45
F66	Coolant shortage indicator switch	123
F68	Gear switch for gearshift and consumption indicator	46
F89	Switch for acceleration enrichment	18
F9	Handbrake warning system switch	126
F92	Pressure switch	18
G	Fuel gauge sender	20
G1	Fuel gauge	37
G2	Coolant temperature gauge sender	41
G3	Coolant temperature gauge	38
G5	Rev. counter	32
G6	Electric fuel pump	19
G40	Hall sender	12,13
G51	Consumption indicator	36
H	Horn button	129
H1	Horn	131,132
J2	Emergency light relay	95-96
J4	Dual tone horn relay	129,130
J5	Fog light relay	85-87
J6	Voltage stabilizer	37
J17	Fuel pump relay	19-22
J31	Intermittent wash/wiper relay	115-118
J39	Headlight washer system relay	119-120
J59	Relief relay for x contact	58-60
J98	Gearshift indicator control unit	45,46
J114	Oil pressure monitor control unit	30,31
J120	Switch unit for coolant shortage indicator	122,123
K1	Main beam warning lamp	34
K2	Generator warning lamp	28
K3	Oil pressure warning lamp	29

Designation		In current track
K5	Turn signal warning lamp	33
K6	Emergency light system warning lamp	99
K7	Dual circuit brake and handbrake warning lamp	124
K10	Heated rear window warning lamp	92
K17	Fog light warning lamp	88
K28	Coolant temperature warning lamp	39
K48	Gearshift indicator warning lamp	35
L1	Twin filament bulb, headlight left	77,79
L2	Twin filament bulb, headlight right	78 80
L10	Instrument panel insert light bulb	62,63
L20	Rear fog light bulb	83
L21	Heater controls light bulb	98
L22	Fog light bulb, left	81
L23	Fog light bulb, right	82
L28	Cigarette lighter light bulb	57
L39	Heated rear window switch bulb	90
L40	Front and rear fog light switch bulb	89
L8	Clock light bulb, headlight	61
L9	Lighting switch light bulb	128
M1	Side light bulb, left	74
M2	Tail light bulb, right	72
M3	Side light bulb, right	71
M4	Tail light bulb, left	73
M5	Front left turn signal bulb	101
M6	Rear left turn signal bulb	100
M7	Front right turn signal bulb	103
M8	Rear right turn signal bulb	102
M9	Brake light bulb, left	107
M10	Brake light bulb, right	106
M16	Reversing light bulb left	109
M17	Reversing light bulb right	108
N	Ignition coil	15
N17	Cold start	18
N21	Auxiliary air valve	22
N23	Series resistance for fresh air blower	112
N41	TCI control unit	12-15
N60	Solenoid valve for consumption indicator	36
N9	Warm-up valve	21
O	Ignition distributor	14,15
P	Spark plug connector	14-16
Q	Spark plugs	14-16
R	Radio connection	
S1-S22	Fuses in fuse box	
S27	Separate fuse for rear fog light	84
T1	Connector single, engine compartment, left	
T1d	Connector single, behind dash panel, right	
T1e	Connector single, behind relay plate	
T1f	Connector single, behind dash panel	
T1g	Connector single, in luggage boot, left	
T1l	Connector single, in engine compartment, left	
T1m	Connector single, behind dash panel	
T1s	Connector single, behind relay plate	
T1u	Connector single, behind relay plate	

Designation		In current track
T1v	Connector single, in luggage boot, left	
T1x	Connector single, behind relay plate/ in engine compartment, centre	
T2	Connector two pin, behind relay plate	
T2a	Connector two pin, in engine compartment, front	
T2b	Connector two pin, behind dash panel	
T2g	Connector two pin, in engine compartment, front right	
T2h	Connector two pin, in engine compartment, front left	
T2k	Connector two pin, behind dash panel/near dual tone horns	
T2u	Connector two pin, in engine compartment, right	
T2v	Connector two pin, in engine compartment, front right	
T2x	Connector two pin, in engine compartment, front	
T3a	Connector three pin, in engine compartment, left	
T3b	Connector three pin, in engine compartment, centre	
T4	Connector four pin, behind dash panel	
T7/	Connector seven pin, on dash panel insert	
T7a/	Connector seven pin, on dash panel insert	
T7b/	Connector seven pin, on dash panel insert	
T7c/	Connector seven pin, on dash panel insert	
T29/	Connector 29 pin, behind dash panel	
U1	Cigarette lighter	56
V	Windscreen wiper motor	113,114
V2	Fresh air blower	111
V5	Windscreen washer pump	118
V7	Radiator fan	110
V11	Headlight washer pump	121
W	Interior light, front	49-51
W3	Luggage boot light	47
W6	Glovebox light	68
W11	Reading lamp, rear left	54
W12	Reading lamp, rear right	48
X	Number plate light	69,70
Y2	Digital clock	55
Z1	Heated rear window	93

Earth points

1 Earthing strap from battery
9 Earthing point near relay plate
15 Earthing point in insulating hose of front wiring loom
16 Earthing point in insulating hose of dash panel wiring loom
17 Earthing point in insulating hose of rear wiring loom
18 Earthing point in insulating hose of wiring loom for instruments
19 Earthing point near handbrake
20 Earthing point in a tailgate/boot lid
21 Earthing point on gearbox

14

Wiring diagram for Passat CL5 & GL5 and Santana LX5 & GX5 from 1983

Wiring diagram for Passat CL5 & GL5 and Santana LX5 & GX5 from 1983 (continued)

14

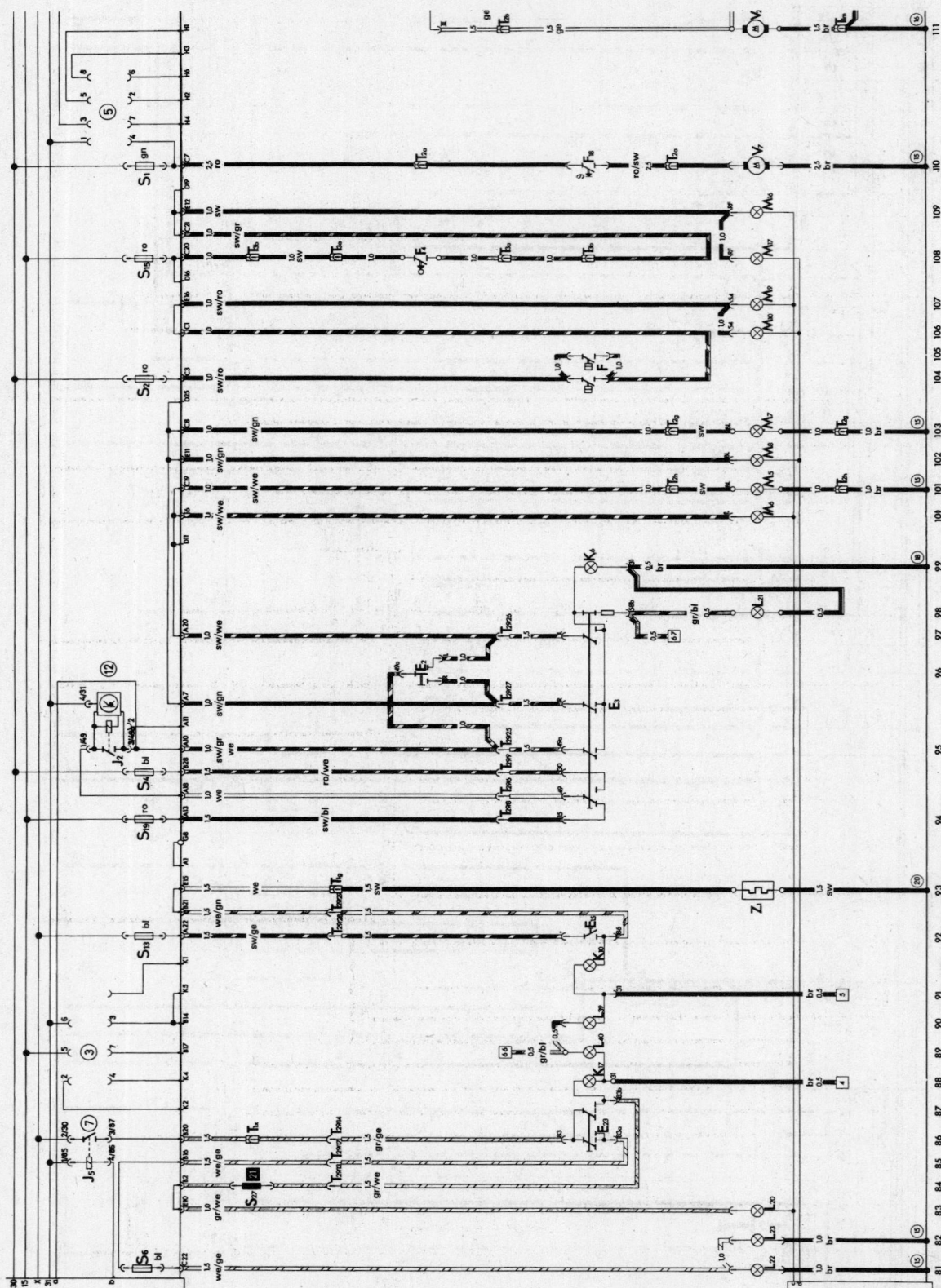

Wiring diagram for Passat CL5 & GL5 and Santana LX5 & GX5 from 1983 (continued)

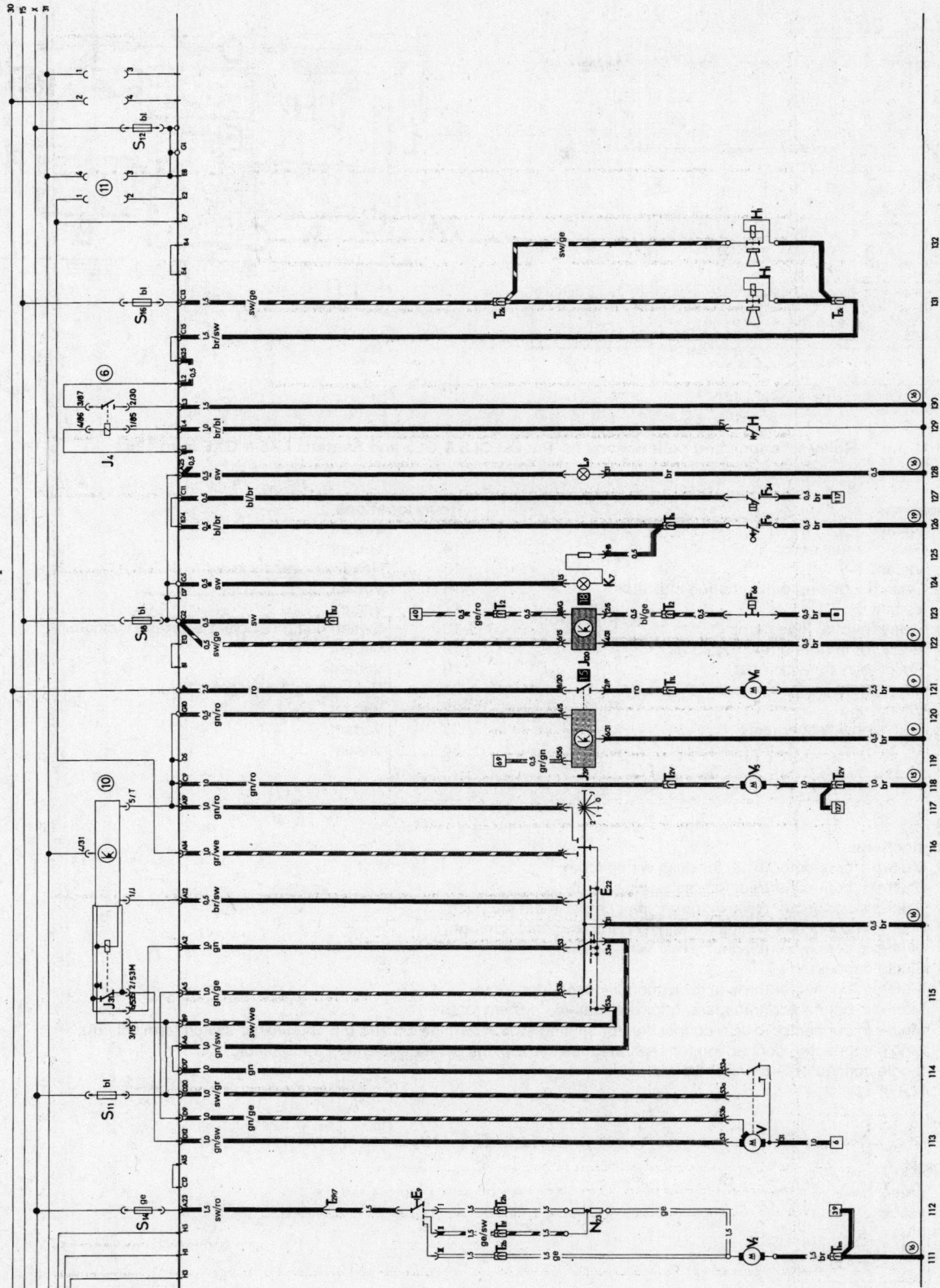

Wiring diagram for Passat CL5 & GL5 and Santana LX5 & GX5 from 1983 (continued)

14

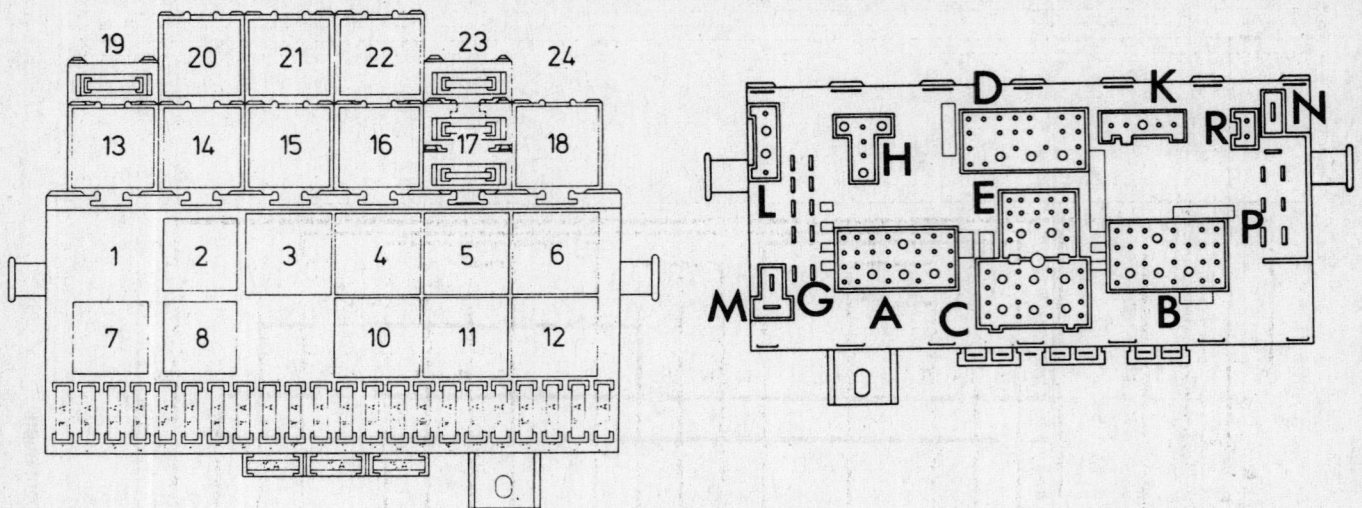

Relay locations and connections for Passat CL5 & GL5 and Santana LX5 & GX5 from 1983

Relay locations

1	Vacant
2	Relay for fuel pump
3	Vacant
4	Gearchange and consumption indicator
5	Vacant
6	Relay for dual tone horns
7	Relay for front and rear fog lights
8	Relief relay for X contact
10	Intermittent wash/wipe relay
11	Vacant
12	Turn signal flasher relay

Relay locations

13	Vacant
14	Vacant
15	Headlight washer relay
16	Vacant
17	Vacant
18	Switch unit for coolant shortage indicator
19	Vacant
20	Vacant
21	Rear fog light fuse (S17/10A)
22	Vacant
23	Vacant
24	Vacant

Relay connections

A	Multi-pin connector (blue) for dash wiring loom
B	Multi-pin connector lred) for dash wiring loom
C	Multi-pin connector (yellow) for engine compartment loom left
D	Multi-pin connector (white) for engine compartment loom right
E	Multi-pin connector (black) for rear wiring loom
G	Single connector
H	Multi-pin connector (brown) for air conditioner wiring loom
K	Multi-pin connector (transparent) for belt warning system loom
L	Multi-pin connector (black) connection for lighting switch terminal 56 and dip and flasher switch terminal 56b
N	Single connector, connection for separate fuse (glow plugs or heater element for manifold)
P	Single connector – terminal 30
R	Not in use

Fuse colours

30A	green
20A	yellow
15A	blue
10A	red

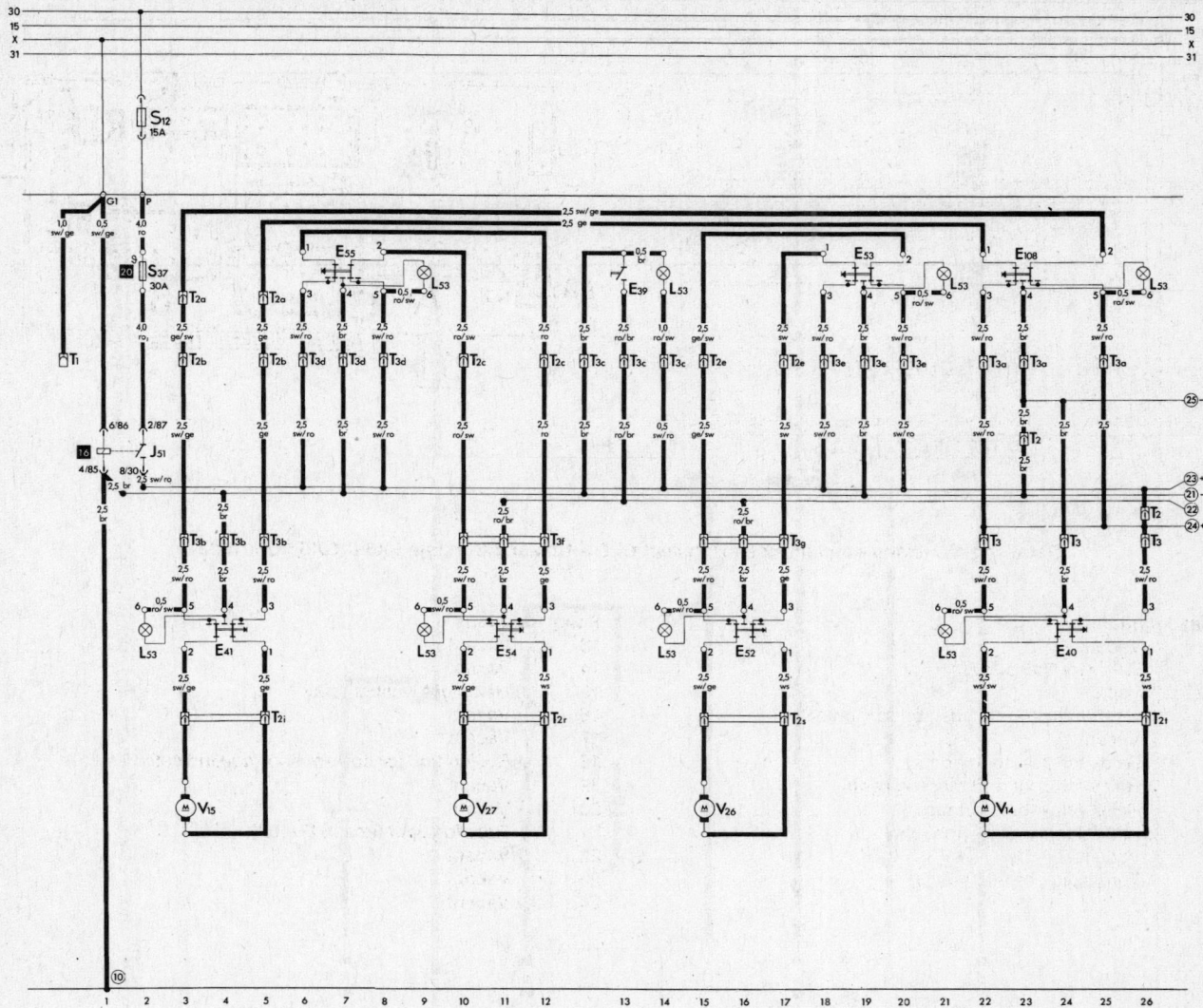

Wiring diagram for electric window lifters

Designation	In current track
E39 Window lifter switch	13
E40 Window lifter switch, left	22-26
E41 Window lifter switch, right	3-5
E52 Rear left window lifter switch (in door)	15-17
E53 Rear left window lifter switch (in console)	18-20
E54 Rear right window lifter switch (in door)	10-12
E55 Rear right window lifter switch (in console)	6-8
E108 Window lifter switch, right (in driver's door)	22-24
J51 Window lifter relay	1, 2
L53 Bulb – for window lifter switch light	2
S12 Fuse in fuse box	
S37 Separate fuse for window lifter (Automatic fuse)	2

Designation	In current track
T1 Connector single, behind relay plate	
T2 Connector 2 pin, behind relay plate	
T2a Connector 2 pin, in driver's door	
T2b Connector 2 pin, behind relay plate	
T2c Connector 2 pin, behind console	
T2e Connector 2 pin, behind console	
T2i Connector 2 pin, in front right door	
T2r Connector 2 pin, in rear right door	
T2s Connector 2 pin, in rear left door	
T2t Connector 2 pin, in front left door	
T3 Connector 3 pin, behind dash	
T3a Connector 3 pin, behind dash	
T3b Connector 3 pin, behind dash	
T3c Connector 3 pin, behind console	
T3d Connector 3 pin, behind console	
T3e Connector 3 pin, behind console	
T3f Connector 3 pin, on B pillar	
T3g Connector 3 pin, on B pillar	

Designation	In current track
V14 Window lifter motor, left	22
V15 Window lifter motor, right	3-5
V26 Window lifter motor, rear left	15
V27 Window lifter motor, rear right	

Earth crimped connections

10 Earth point, near relay plate
21 Earth point, in crimped connector of window lifter loom
22 Crimped connector in window lifter loom
23 Crimped connector in window lifter loom
24 Crimped connector in window lifter loom

Relay locations

16 Window lifter relay
20 Automatic fuse for window lifter

14

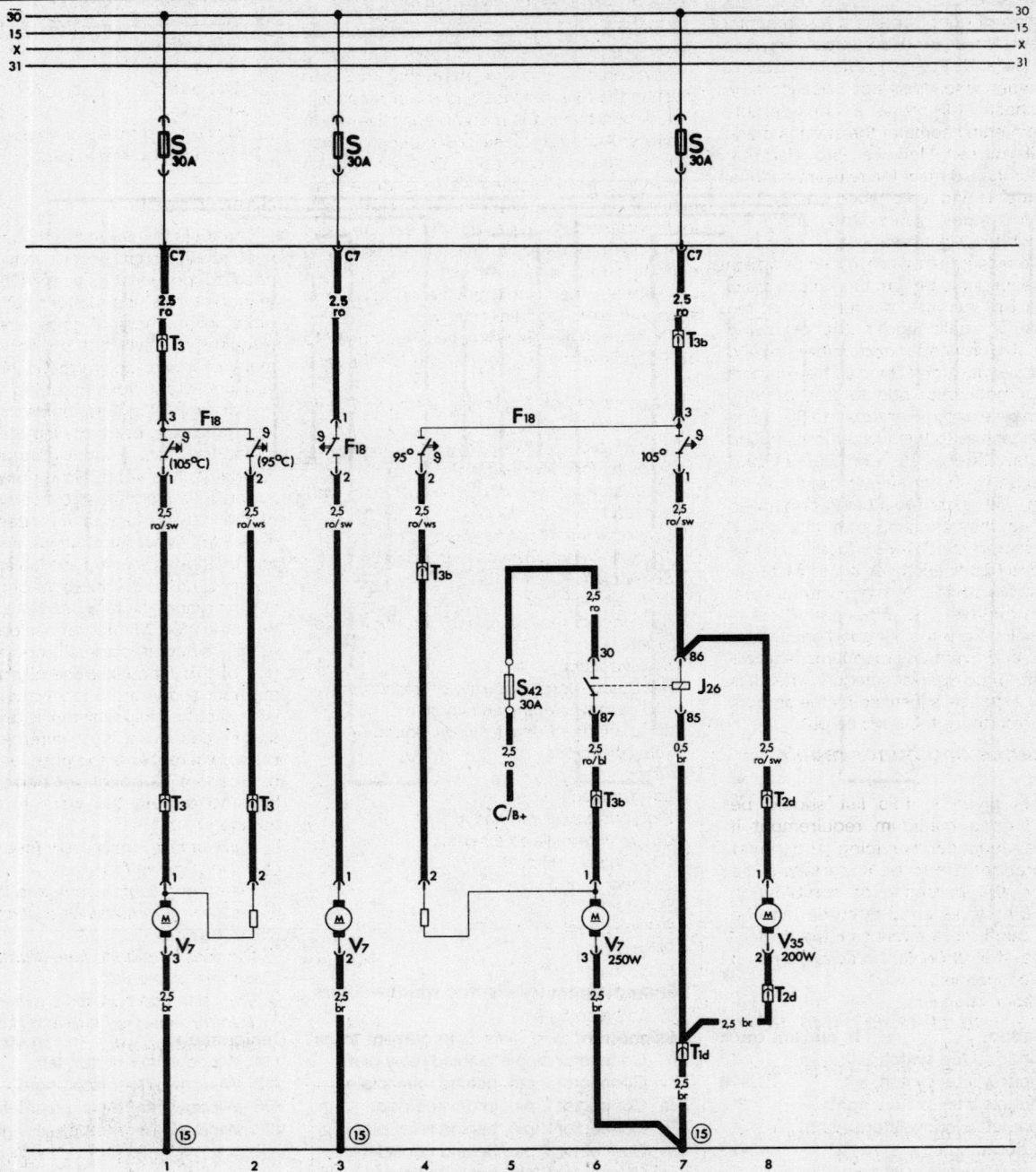

Wiring diagram for radiator fan – from January 1985

Designation	In current track	Designation	In current track	Designation	In current track
250 W and 200 W radiator fan wiring		T3 Connector 3 pin		**Single speed radiator fan wiring**	
C Alternator	5	V7 Radiator fan	4-6	F18 Radiator fan thermo switch	3
F18 Radiator fan thermo switch	4-7	V35 Radiator fan II	8	S1 Fuse in fuse box	3
J26 Radiator fan relay	6, 7	**2-speed radiator fan wiring**		V7 Radiator fan	3
S1 Fuse in fuse box	7	F18 Radiator fan thermo switch	1, 2		
S42 Separate fuse for radiator fan	5	S1 Fuse in fuse box	1	**Earth point**	
T1 Connector single		T3 Connector 3 pin		15 Earthing point in insulating sleeve of front	
T2 Connector 2 pin		V7 Radiator fan	1, 2	wiring loom	

Introduction

A selection of good tools is a fundamental requirement for anyone contemplating the maintenance and repair of a motor vehicle. For the owner who does not possess any, their purchase will prove a considerable expense, offsetting some of the savings made by doing-it-yourself. However, provided that the tools purchased meet the relevant national safety standards and are of good quality, they will last for many years and prove an extremely worthwhile investment.

To help the average owner to decide which tools are needed to carry out the various tasks detailed in this manual, we have compiled three lists of tools under the following headings: *Maintenance and minor repair*, *Repair and overhaul*, and *Special*. Newcomers to practical mechanics should start off with the *Maintenance and minor repair* tool kit, and confine themselves to the simpler jobs around the vehicle. Then, as confidence and experience grow, more difficult tasks can be undertaken, with extra tools being purchased as, and when, they are needed. In this way, a *Maintenance and minor repair* tool kit can be built up into a *Repair and overhaul* tool kit over a considerable period of time, without any major cash outlays. The experienced do-it-yourselfer will have a tool kit good enough for most repair and overhaul procedures, and will add tools from the *Special* category when it is felt that the expense is justified by the amount of use to which these tools will be put.

Maintenance and minor repair tool kit

The tools given in this list should be considered as a minimum requirement if routine maintenance, servicing and minor repair operations are to be undertaken. We recommend the purchase of combination spanners (ring one end, open-ended the other); although more expensive than open-ended ones, they do give the advantages of both types of spanner.

- ☐ Combination spanners:
  Metric - 8, 9, 10, 11, 12, 13, 14, 15, 16, 17, 19, 21, 22, 24 & 26 mm
- ☐ Adjustable spanner - 35 mm jaw (approx)
- ☐ Transmission drain plug key
- ☐ Set of feeler gauges
- ☐ Spark plug spanner (with rubber insert)
- ☐ Spark plug gap adjustment tool
- ☐ Brake bleed nipple spanner
- ☐ Screwdrivers:
  Flat blade - approx 100 mm long x 6 mm dia
  Cross blade - approx 100 mm long x 6 mm dia
- ☐ Combination pliers
- ☐ Hacksaw (junior)
- ☐ Tyre pump
- ☐ Tyre pressure gauge
- ☐ Oil can
- ☐ Oil filter removal tool
- ☐ Fine emery cloth
- ☐ Wire brush (small)
- ☐ Funnel (medium size)

Repair and overhaul tool kit

These tools are virtually essential for anyone undertaking any major repairs to a motor vehicle, and are additional to those given in the *Maintenance and minor repair* list. Included in this list is a comprehensive set of sockets. Although these are expensive, they will be found invaluable as they are so versatile - particularly if various drives are included in the set. We recommend the half-inch square-drive type, as this can be used with most proprietary torque wrenches. If you cannot afford a socket set, even bought piecemeal, then inexpensive tubular box spanners are a useful alternative.

The tools in this list will occasionally need to be supplemented by tools from the *Special* list:

- ☐ Sockets (or box spanners) to cover range in previous list
- ☐ Reversible ratchet drive (for use with sockets) *(see illustration)*
- ☐ Extension piece, 250 mm (for use with sockets)
- ☐ Universal joint (for use with sockets)
- ☐ Torque wrench (for use with sockets)
- ☐ Self-locking grips
- ☐ Ball pein hammer
- ☐ Soft-faced mallet (plastic/aluminium or rubber)
- ☐ Screwdrivers:
  Flat blade - long & sturdy, short (chubby), and narrow (electrician's) types
  Cross blade - Long & sturdy, and short (chubby) types
- ☐ Pliers:
  Long-nosed
  Side cutters (electrician's)
  Circlip (internal and external)
- ☐ Cold chisel - 25 mm
- ☐ Scriber
- ☐ Scraper
- ☐ Centre-punch
- ☐ Pin punch
- ☐ Hacksaw
- ☐ Brake hose clamp
- ☐ Brake bleeding kit
- ☐ Selection of twist drills
- ☐ Steel rule/straight-edge
- ☐ Allen keys (inc. splined/Torx type) *(see illustrations)*
- ☐ Selection of files
- ☐ Wire brush
- ☐ Axle stands
- ☐ Jack (strong trolley or hydraulic type)
- ☐ Light with extension lead

Special tools

The tools in this list are those which are not used regularly, are expensive to buy, or which need to be used in accordance with their manufacturers' instructions. Unless relatively difficult mechanical jobs are undertaken frequently, it will not be economic to buy many of these tools. Where this is the case, you could consider clubbing together with friends (or joining a motorists' club) to make a joint purchase, or borrowing the tools against a deposit from a local garage or tool hire specialist. It is worth noting that many of the larger DIY superstores now carry a large range of special tools for hire at modest rates.

The following list contains only those tools and instruments freely available to the public, and not those special tools produced by the vehicle manufacturer specifically for its dealer network. You will find occasional references to these manufacturers' special tools in the text of this manual. Generally, an alternative method of doing the job without the vehicle manufacturers' special tool is given. However, sometimes there is no alternative to using them. Where this is the case and the relevant tool cannot be bought or borrowed, you will have to entrust the work to a franchised garage.

- ☐ Valve spring compressor *(see illustration)*
- ☐ Valve grinding tool
- ☐ Piston ring compressor *(see illustration)*
- ☐ Piston ring removal/installation tool *(see illustration)*
- ☐ Cylinder bore hone *(see illustration)*
- ☐ Balljoint separator
- ☐ Coil spring compressors (where applicable)
- ☐ Two/three-legged hub and bearing puller *(see illustration)*

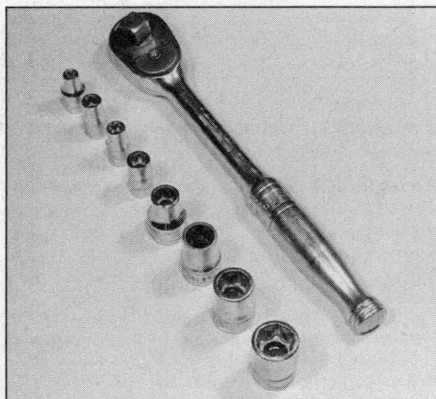

Sockets and reversible ratchet drive

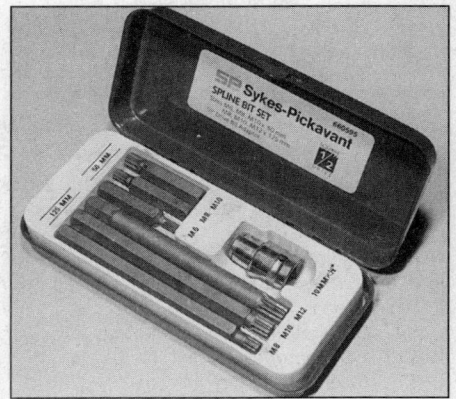

Spline bit set

Tools and Working Facilities

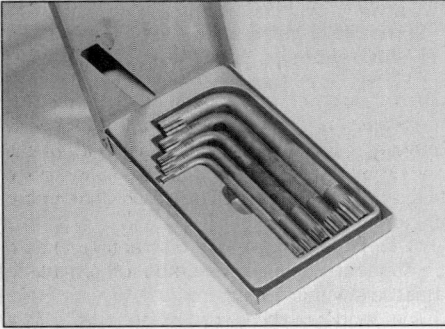

Spline key set

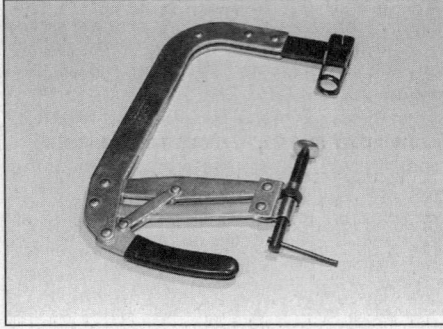

Valve spring compressor

Piston ring compressor

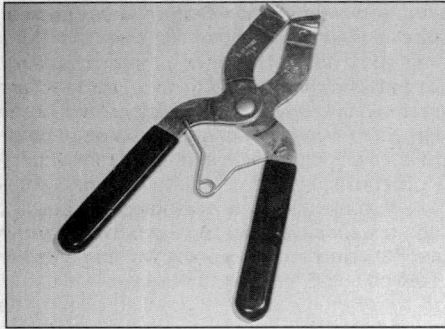

Piston ring removal/installation tool

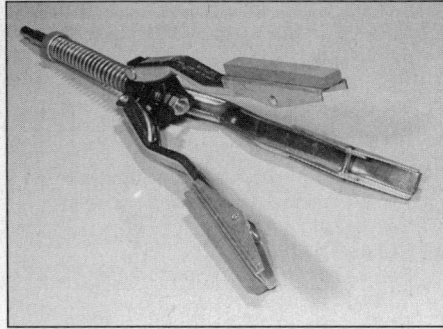

Cylinder bore hone

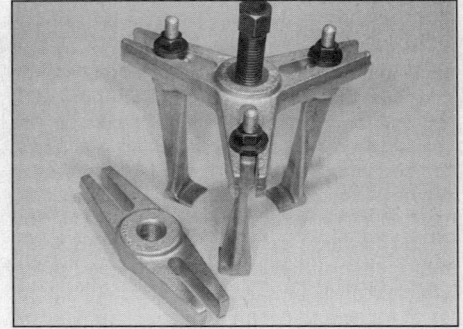

Three-legged hub and bearing puller

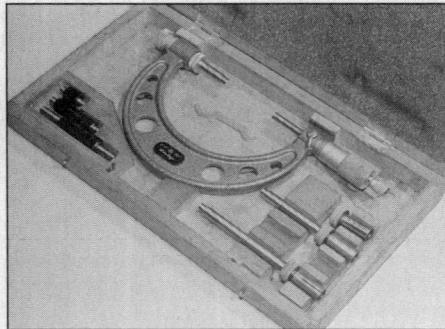

Micrometer set

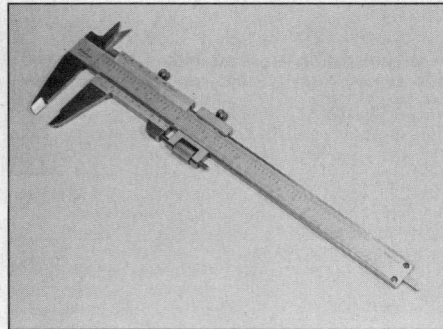

Vernier calipers

Dial test indicator and magnetic stand

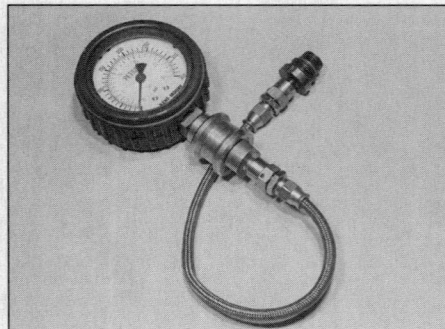

Compression testing gauge

Clutch plate alignment set

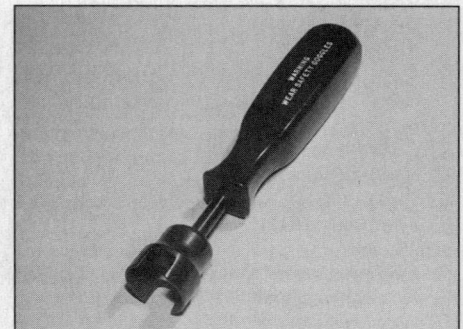

Brake shoe steady spring cup removal tool

☐ Impact screwdriver
☐ Micrometer and/or vernier calipers **(see illustrations)**
☐ Dial gauge **(see illustration)**
☐ Universal electrical multi-meter
☐ Cylinder compression gauge **(see illustration)**
☐ Clutch plate alignment set **(see illustration)**
☐ Brake shoe steady spring cup removal tool **(see illustration)**
☐ Bush and bearing removal/installation set **(see illustration)**
☐ Stud extractors **(see illustration)**
☐ Tap and die set **(see illustration)**
☐ Lifting tackle
☐ Trolley jack

Buying tools

For practically all tools, a tool factor is the best source, since he will have a very comprehensive range compared with the average garage or accessory shop. Having said that, accessory shops often offer excellent quality tools at discount prices, so it pays to shop around.

Remember, you don't have to buy the most expensive items on the shelf, but it is always advisable to steer clear of the very cheap tools. There are plenty of good tools around at reasonable prices, but always aim to purchase items which meet the relevant national safety standards. If in doubt, ask the proprietor or manager of the shop for advice before making a purchase.

Care and maintenance of tools

Having purchased a reasonable tool kit, it is necessary to keep the tools in a clean and serviceable condition. After use, always wipe off any dirt, grease and metal particles using a clean, dry cloth, before putting the tools away. Never leave them lying around after they have been used. A simple tool rack on the garage or workshop wall for items such as screwdrivers and pliers is a good idea. Store all normal spanners and sockets in a metal box. Any measuring instruments, gauges, meters, etc, must be carefully stored where they cannot be damaged or become rusty.

Take a little care when tools are used. Hammer heads inevitably become marked, and screwdrivers lose the keen edge on their blades from time to time. A little timely attention with emery cloth or a file will soon restore items like this to a good serviceable finish.

Working facilities

Not to be forgotten when discussing tools is the workshop itself. If anything more than routine maintenance is to be carried out, some form of suitable working area becomes essential.

It is appreciated that many an owner-mechanic is forced by circumstances to remove an engine or similar item without the benefit of a garage or workshop. Having done this, any repairs should always be done under the cover of a roof.

Wherever possible, any dismantling should be done on a clean, flat workbench or table at a suitable working height.

Any workbench needs a vice; one with a jaw opening of 100 mm is suitable for most jobs. As mentioned previously, some clean dry storage space is also required for tools, as well as for any lubricants, cleaning fluids, touch-up paints and so on, which become necessary.

Another item which may be required, and which has a much more general usage, is an electric drill with a chuck capacity of at least 8 mm. This, together with a good range of twist drills, is virtually essential for fitting accessories.

Last, but not least, always keep a supply of old newspapers and clean, lint-free rags available, and try to keep any working area as clean as possible.

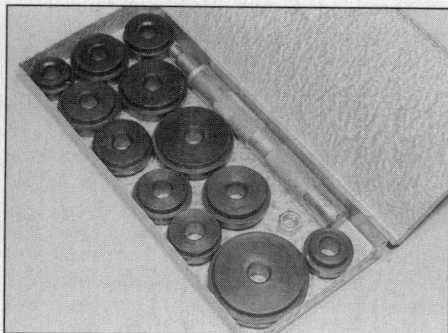

Bush and bearing removal/installation set

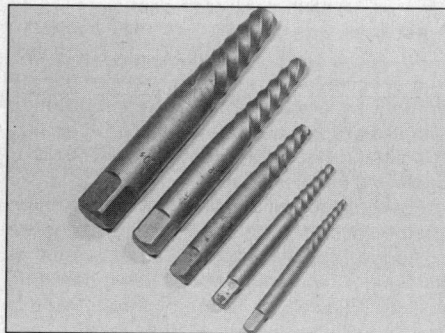

Stud extractor set

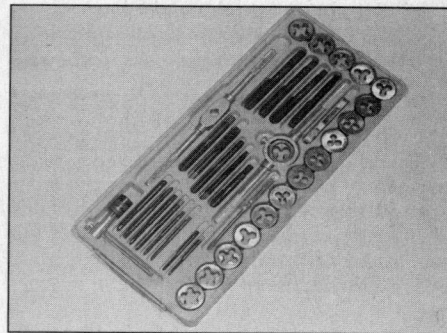

Tap and die set

General Repair Procedures

Whenever servicing, repair or overhaul work is carried out on the car or its components, it is necessary to observe the following procedures and instructions. This will assist in carrying out the operation efficiently and to a professional standard of workmanship.

Joint mating faces and gaskets

When separating components at their mating faces, never insert screwdrivers or similar implements into the joint between the faces in order to prise them apart. This can cause severe damage which results in oil leaks, coolant leaks, etc upon reassembly. Separation is usually achieved by tapping along the joint with a soft-faced hammer in order to break the seal. However, note that this method may not be suitable where dowels are used for component location.

Where a gasket is used between the mating faces of two components, ensure that it is renewed on reassembly, and fit it dry unless otherwise stated in the repair procedure. Make sure that the mating faces are clean and dry, with all traces of old gasket removed. When cleaning a joint face, use a tool which is not likely to score or damage the face, and remove any burrs or nicks with an oilstone or fine file.

Make sure that tapped holes are cleaned with a pipe cleaner, and keep them free of jointing compound, if this is being used, unless specifically instructed otherwise.

Ensure that all orifices, channels or pipes are clear, and blow through them, preferably using compressed air.

Oil seals

Oil seals can be removed by levering them out with a wide flat-bladed screwdriver or similar implement. Alternatively, a number of self-tapping screws may be screwed into the seal, and these used as a purchase for pliers or some similar device in order to pull the seal free.

Whenever an oil seal is removed from its working location, either individually or as part of an assembly, it should be renewed.

The very fine sealing lip of the seal is easily damaged, and will not seal if the surface it contacts is not completely clean and free from scratches, nicks or grooves.

Protect the lips of the seal from any surface which may damage them in the course of fitting. Use tape or a conical sleeve where possible. Lubricate the seal lips with oil before fitting and, on dual-lipped seals, fill the space between the lips with grease.

Unless otherwise stated, oil seals must be fitted with their sealing lips toward the lubricant to be sealed.

Use a tubular drift or block of wood of the appropriate size to install the seal and, if the seal housing is shouldered, drive the seal down to the shoulder. If the seal housing is unshouldered, the seal should be fitted with its face flush with the housing top face (unless otherwise instructed).

Screw threads and fastenings

Seized nuts, bolts and screws are quite a common occurrence where corrosion has set in, and the use of penetrating oil or releasing fluid will often overcome this problem if the offending item is soaked for a while before attempting to release it. The use of an impact driver may also provide a means of releasing such stubborn fastening devices, when used in conjunction with the appropriate screwdriver bit or socket. If none of these methods works, it may be necessary to resort to the careful application of heat, or the use of a hacksaw or nut splitter device.

Studs are usually removed by locking two nuts together on the threaded part, and then using a spanner on the lower nut to unscrew the stud. Studs or bolts which have broken off below the surface of the component in which they are mounted can sometimes be removed using a proprietary stud extractor. Always ensure that a blind tapped hole is completely free from oil, grease, water or other fluid before installing the bolt or stud. Failure to do this could cause the housing to crack due to the hydraulic action of the bolt or stud as it is screwed in.

When tightening a castellated nut to accept a split pin, tighten the nut to the specified torque, where applicable, and then tighten further to the next split pin hole. Never slacken the nut to align the split pin hole, unless stated in the repair procedure.

When checking or retightening a nut or bolt to a specified torque setting, slacken the nut or bolt by a quarter of a turn, and then retighten to the specified setting. However, this should not be attempted where angular tightening has been used.

For some screw fastenings, notably cylinder head bolts or nuts, torque wrench settings are no longer specified for the latter stages of tightening, "angle-tightening" being called up instead. Typically, a fairly low torque wrench setting will be applied to the bolts/nuts in the correct sequence, followed by one or more stages of tightening through specified angles.

Locknuts, locktabs and washers

Any fastening which will rotate against a component or housing in the course of tightening should always have a washer between it and the relevant component or housing.

Spring or split washers should always be renewed when they are used to lock a critical component such as a big-end bearing retaining bolt or nut. Locktabs which are folded over to retain a nut or bolt should always be renewed.

Self-locking nuts can be re-used in non-critical areas, providing resistance can be felt when the locking portion passes over the bolt or stud thread. However, it should be noted that self-locking stiffnuts tend to lose their effectiveness after long periods of use, and in such cases should be renewed as a matter of course.

Split pins must always be replaced with new ones of the correct size for the hole.

When thread-locking compound is found on the threads of a fastener which is to be re-used, it should be cleaned off with a wire brush and solvent, and fresh compound applied on reassembly.

Special tools

Some repair procedures in this manual entail the use of special tools such as a press, two or three-legged pullers, spring compressors, etc. Wherever possible, suitable readily-available alternatives to the manufacturer's special tools are described, and are shown in use. Unless you are highly-skilled and have a thorough understanding of the procedures described, never attempt to bypass the use of any special tool when the procedure described specifies its use. Not only is there a very great risk of personal injury, but expensive damage could be caused to the components involved.

Environmental considerations

When disposing of used engine oil, brake fluid, antifreeze, etc, give due consideration to any detrimental environmental effects. Do not, for instance, pour any of the above liquids down drains into the general sewage system, or onto the ground to soak away. Many local council refuse tips provide a facility for waste oil disposal, as do some garages. If none of these facilities are available, consult your local Environmental Health Department for further advice.

With the universal tightening-up of legislation regarding the emission of environmentally-harmful substances from motor vehicles, most current vehicles have tamperproof devices fitted to the main adjustment points of the fuel system. These devices are primarily designed to prevent unqualified persons from adjusting the fuel/air mixture, with the chance of a consequent increase in toxic emissions. If such devices are encountered during servicing or overhaul, they should, wherever possible, be renewed or refitted in accordance with the vehicle manufacturer's requirements or current legislation.

OIL CARE

FOLLOW THE CODE

OIL BANK LINE
0800 66 33 66

Note: It is antisocial and illegal to dump oil down the drain. To find the location of your local oil recycling bank, call this number free.

Buying spare parts

Spare parts are available from many sources, Volkswagen have many dealers throughout the country and other dealers, accessory stores and motor factors will also stock Volkswagen spare parts. Our advice regarding spare part sources is as follows:

Officially appointed vehicle main dealers – This is the best source of parts which are peculiar to your vehicle and are otherwise not generally available (eg complete cylinder heads, internal transmission components, badges, interior trim, etc). It is also the only place at which you should buy parts if your vehicle is still under warranty. To be sure of obtaining the correct parts it will always be necessary to give the storeman your car's engine and chassis number, and if possible, to take the 'old' parts along for positive identification. Remember that many parts are available on a factory exchange scheme — any parts returned should always be clean ! It obviously makes good sense to go straight to the specialists on your vehicle for this type of part for they are best equipped to supply you.

Other dealers and accessory shops – These are often very good places to buy materials and components needed for the maintenance of your vehicle (eg oil filters, spark plugs, bulbs, fan belts, oils and grease, touch-up paint, filler paste, etc). They also sell general accessories, usually have convenient opening hours, charge lower prices and can often be found not far from home.

Motor factors – Good factors will stock all of the more important components which wear out relatively quickly (eg clutch components, pistons, valves, exhaust systems, brake cylinders/pipes/ hoses/seals/ shoes and pads, etc). Motor factors will often provide new or reconditioned components on a part exchange basis – this can save a considerable amount of money.

Vehicle identification numbers

Modifications are a continuing and unpublicised process in vehicle manufacture. Spare parts manuals and lists are compiled upon a numerical basis, the individual vehicle numbers being essential to identify correctly the component required.

The vehicle identification plate is located in the engine compartment on the right-hand side.

The chassis number is stamped on the rear panel of the engine compartment.

The engine number is stamped on the left-hand side of the cylinder block.

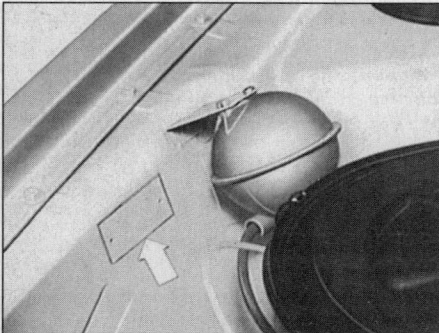

Vehicle identification plate location

Vehicle chassis number/identification plate location

Engine number location – 1.6 litre shown

Fault Finding

Introduction

The vehicle owner who does his or her own maintenance according to the recommended schedules should not have to use this section of the manual very often. Modern component reliability is such that, provided those items subject to wear or deterioration are inspected or renewed at the specified intervals, sudden failure is comparatively rare. Faults do not usually just happen as a result of sudden failure, but develop over a period of time. Major mechanical failures in particular are usually preceded by characteristic symptoms over hundreds or even thousands of miles. Those components which do occasionally fail without warning are often small and easily carried in the vehicle.

With any fault finding, the first step is to decide where to begin investigations. Sometimes this is obvious, but on other occasions a little detective work will be necessary. The owner who makes half a dozen haphazard adjustments or replacements may be successful in curing a fault (or its symptoms), but he will be none the wiser if the fault recurs and he may well have spent more time and money than was necessary. A calm and logical approach will be found to be more satisfactory in the long run. Always take into account any warning signs or abnormalities that may have been noticed in the period preceding the fault – power loss, high or low gauge readings, unusual noises or smells, etc – and remember that failure of components such as fuses or spark plugs may only be pointers to some underlying fault.

The pages which follow here are intended to help in cases of failure to start or breakdown on the road. There is also a Fault Diagnosis Section at the end of each Chapter which should be consulted if the preliminary checks prove unfruitful. Whatever the fault, certain basic principles apply. These are as follows:

Verify the fault. This is simply a matter of being sure that you know what the symptoms are before starting work. This is particularly important if you are investigating a fault for someone else who may not have described it very accurately.

Don't overlook the obvious. For example, if the vehicle won't start, is there petrol in the tank? (Don't take anyone else's word on this particular point, and don't trust the fuel gauge either!) If an electrical fault is indicated, look for loose or broken wires before digging out the test gear.

Cure the disease, not the symptom. Substituting a flat battery with a fully charged one will get you off the hard shoulder, but if the underlying cause is not attended to,the new battery will go the same way. Similarly, changing oil-fouled spark plugs for a new set will get you moving again, but remember that the reason for the fouling (if it wasn't simply an incorrect grade of plug) will have to be established and corrected.

Don't take anything for granted. Particularly, don't forget that a 'new' component may itself be defective (especially if it's been rattling round in the boot for months), and don't leave components out of a fault diagnosis sequence just because they are new or recently fitted. When you do finally diagnose a difficult fault, you'll probably realise that all the evidence was there from the start.

Electrical faults

Electrical faults can be more puzzling than straightforward mechanical failures, but they are no less susceptible to logical analysis if the basic principles of operation are understood. Vehicle electrical wiring exists in extremely unfavourable conditions – heat, vibration and chemical attack and the first things to look for are loose or corroded connections and broken or chafed wires, especially where the wires pass through holes in the bodywork or are subject to vibration.

All metal-bodied vehicles in current production have one pole of the battery 'earthed', ie connected to the vehicle bodywork, and in nearly all modern vehicles it is the negative (–) terminal. The various electrical components — motors, bulb holders, etc – are also connected to earth, either by means of a lead or directly by their mountings. Electric current flows through the component and then back to the battery via the bodywork. If the component mounting is loose or corroded, or if a good path back to the battery is not available, the circuit will be incomplete and malfunction will result. The engine and/or gearbox are also earthed by means of flexible metal straps to the body or subframe; if these straps are loose or missing, starter motor, generator and ignition trouble may result.

Assuming the earth return to be satisfactory, electrical faults will be due either to component malfunction or to defects in the current supply. Individual components are dealt with in Chapter 10. If supply wires are broken or cracked internally this results in an open-circuit, and the easiest way to check for this is to bypass the suspect wire temporarily with a length of wire having a crocodile clip or suitable connector at each end. Alternatively, a 12V test lamp can be used to verify the presence of supply voltage at various points along the wire and the break can be thus isolated.

If a bare portion of a live wire touches the bodywork or other earthed metal part, the electricity will take the low-resistance path thus formed back to the battery: this is known as a short-circuit. Hopefully a short-circuit will blow a fuse, but otherwise it may cause burning of the insulation (and possibly further short-circuits) or even a fire. This is why it is inadvisable to bypass persistently blowing fuses with silver foil or wire.

Spares and tool kit

Most vehicles are supplied only with sufficient tools for wheel changing; the *Maintenance and minor repair* tool kit detailed in *Tools and working facilities*, with the addition of a hammer, is probably sufficient for those repairs that most motorists would consider attempting at the roadside. In addition a few items which can be fitted without too much trouble in the event of a breakdown should be carried. Experience and available space will modify the list below, but the following may save having to call on professional assistance:

☐ *Spark plugs, clean and correctly gapped*
☐ *HT lead and plug cap – long enough to reach the plug furthest*
☐ *from the distributor*
☐ *Drivebelt(s) — emergency type may suffice*
☐ *Spare fuses*
☐ *Set of principal light bulbs*
☐ *Tin of radiator sealer and hose bandage*
☐ *Exhaust bandage*
☐ *Roll of insulating tape*
☐ *Length of soft iron wire*
☐ *Length of electrical flex*
☐ *Torch or inspection lamp (can double as test lamp)*
☐ *Battery jump leads*
☐ *Tow-rope*
☐ *Ignition waterproofing aerosol*
☐ *Litre of engine oil*
☐ *Sealed can of hydraulic fluid*
☐ *Emergency windscreen*
☐ *Wormdrive clips*
☐ *Tube of filler paste*

If spare fuel is carried, a can designed for the purpose should be used to minimise risks of leakage and collision damage. A first aid kit and a warning triangle, whilst not at present compulsory in the UK, are obviously sensible items to carry in addition to the above. When touring abroad it may be advisable to carry additional spares which, even if you cannot fit them yourself, could save having to wait while parts are obtained. The items below may be worth considering:

☐ *Clutch and throttle cables*
☐ *Cylinder head gasket*
☐ *Alternator brushes*
☐ *Tyre valve core*

One of the motoring organisations will be able to advise on availability of fuel, etc, in foreign countries.

Engine will not start

Engine fails to turn when starter operated

- [ ] Flat battery (recharge use jump leads or push start)
- [ ] Battery terminals loose or corroded
- [ ] Battery earth to body defective
- [ ] Engine earth strap loose or broken
- [ ] Starter motor (or solenoid) wiring loose or broken
- [ ] Automatic transmission selector in wrong position, or inhibitor switch faulty
- [ ] Ignition/starter switch faulty
- [ ] Major mechanical failure (seizure)
- [ ] Starter or solenoid internal fault (see Chapter 10)

Starter motor turns engine slowly

- [ ] Partially discharged battery (recharge, use jump leads, or push start)
- [ ] Battery terminals loose or corroded
- [ ] Battery earth to body defective
- [ ] Engine earth strap loose
- [ ] Starter motor (or solenoid) wiring loose
- [ ] Starter motor internal fault (see Chapter 10)

Starter motor spins without turning engine

- [ ] Flywheel gear teeth damaged or worn
- [ ] Starter motor mounting bolts loose

Engine turns normally but fails to start

- [ ] Damp or dirty HT leads and distributor cap (crank engine and check for spark)
- [ ] No fuel in tank (check for delivery at carburettor)
- [ ] Automatic choke faulty (carburettor engine)
- [ ] Fouled or incorrectly gapped spark plugs (remove, clean and regap)
- [ ] Other ignition system fault (see Chapter 4)
- [ ] Other fuel system fault (see Chapter 3)
- [ ] Poor compression (see Chapter 1)
- [ ] Major mechanical failure (eg camshaft drive)

Engine fires but will not run

- [ ] Automatic choke faulty (carburettor engine)
- [ ] Air leaks at carburettor or inlet manifold
- [ ] Fuel starvation (see Chapter 3)
- [ ] Ballast resistor defective, or other ignition fault (see Chapter 4)

Engine cuts out and will not restart

Engine cuts out suddenly – ignition fault

- [ ] Loose or disconnected LT wires
- [ ] Wet HT leads or distributor cap (after traversing water splash)
- [ ] Coil failure (check for spark)
- [ ] Other ignition fault (see Chapter 4)

Engine misfires before cutting out – fuel fault

- [ ] Fuel tank empty
- [ ] Fuel pump defective or filter blocked (check for delivery)
- [ ] Fuel tank filler vent blocked (suction will be evident on releasing cap)
- [ ] Carburettor needle valve sticking
- [ ] Carburettor jets blocked (fuel contaminated)
- [ ] Other fuel system fault (see Chapter 3)

Engine cuts out – other causes

- [ ] Serious overheating
- [ ] Major mechanical failure (eg camshaft drive)

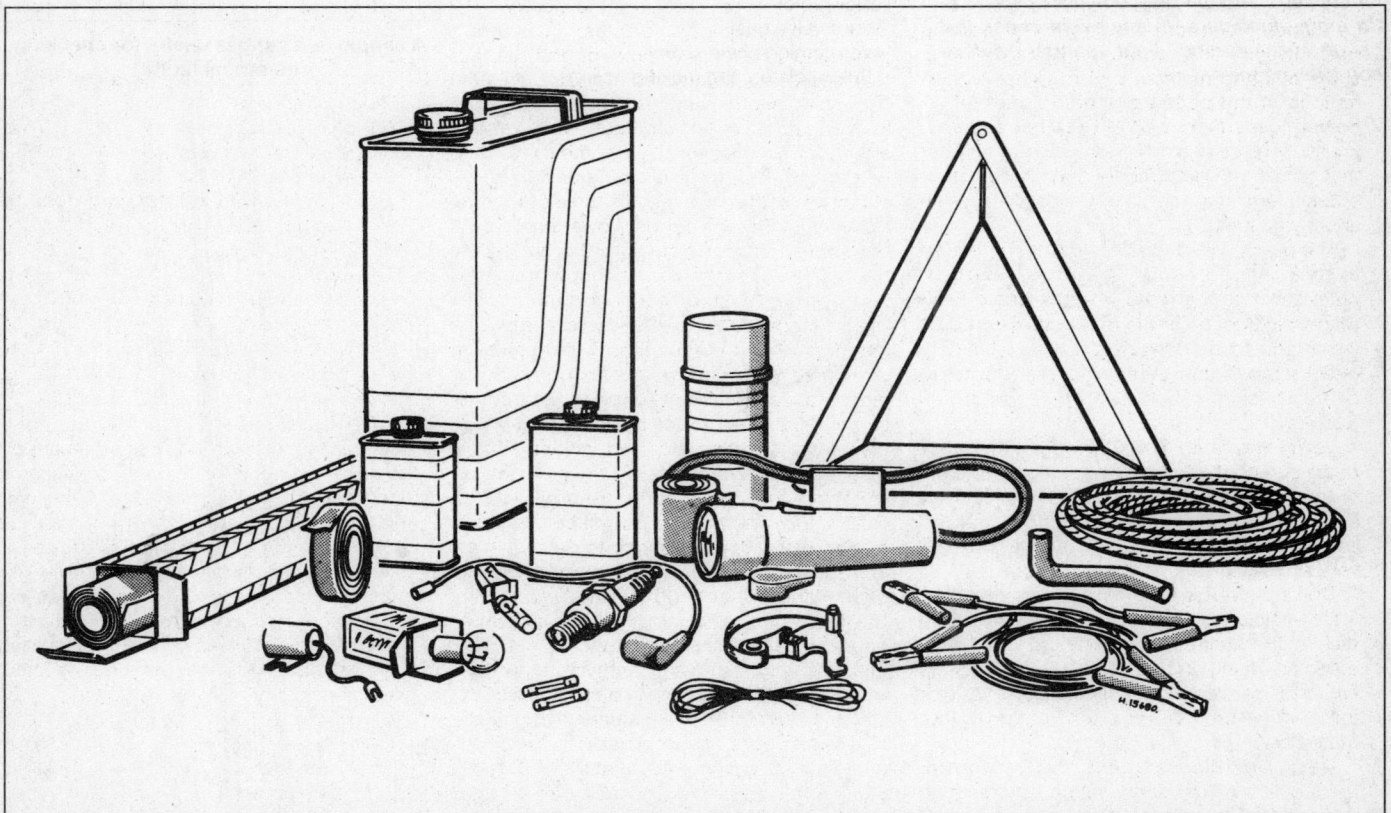

Carrying a few spares may save you a long walk!

Engine overheats

Ignition (no-charge) warning light illuminated

- [ ] Slack or broken drivebelt — retension or renew (Chapter 2)

Ignition warning light not illuminated

- [ ] Coolant loss due to internal or external leakage (see Chapter 2)
- [ ] Thermostat defective
- [ ] Low oil level
- [ ] Brakes binding
- [ ] Radiator clogged externally or internally
- [ ] Electric cooling fan not operating correctly
- [ ] Engine waterways clogged
- [ ] Ignition timing incorrect or automatic advance malfunctioning
- [ ] Mixture too weak

Note: *Do not add cold water to an overheated engine or damage may result*

Low engine oil pressure

Note: *Low oil pressure in a high-mileage engine at tickover is not necessarily a cause for concern. Sudden pressure loss at speed is far more significant. In any event check the gauge or warning light sender before condemning the engine.*

Gauge reads low or warning light illuminated with engine running

- [ ] Oil level low or incorrect grade
- [ ] Defective gauge or sender unit
- [ ] Wire to sender unit earthed
- [ ] Engine overheating
- [ ] Oil filter clogged or bypass valve defective
- [ ] Oil pressure relief valve defective
- [ ] Oil pick-up strainer clogged
- [ ] Oil pump worn or mountings loose
- [ ] Worn main or big-end bearings

Engine noises

Pre-ignition (pinking) on acceleration

- [ ] Incorrect grade of fuel
- [ ] Ignition timing incorrect
- [ ] Distributor faulty or worn
- [ ] Worn or maladjusted carburettor
- [ ] Excessive carbon build-up in engine

Whistling or wheezing noises

- [ ] Leaking vacuum hose
- [ ] Leaking carburettor or manifold gasket
- [ ] Blowing head gasket

Tapping or rattling

- [ ] Incorrect valve clearances (where applicable)
- [ ] Worn valve gear
- [ ] Worn timing chain or belt
- [ ] Broken piston ring (ticking noise)

Knocking or thumping

- [ ] Unintentional mechanical contact (eg fan blades)
- [ ] Worn drivebelt
- [ ] Peripheral component fault (generator, water pump, etc)
- [ ] Worn big-end bearings (regular heavy knocking, perhaps less under load)
- [ ] Worn main bearings (rumbling and knocking, perhaps worsening under load)
- [ ] Piston slap (most noticeable when cold)

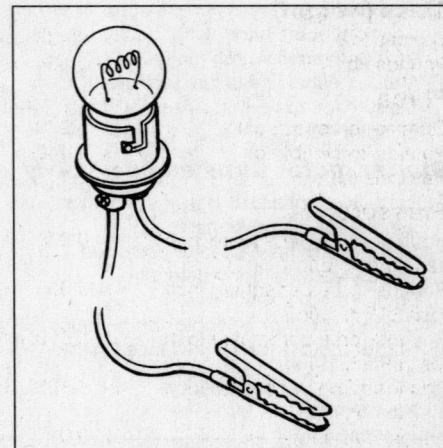

A simple test lamp is useful for checking electrical faults

Length (distance)

Inches (in)	x 25.4	= Millimetres (mm)	x 0.0394	=	Inches (in)
Feet (ft)	x 0.305	= Metres (m)	x 3.281	=	Feet (ft)
Miles	x 1.609	= Kilometres (km)	x 0.621	=	Miles

Volume (capacity)

Cubic inches (cu in; in³)	x 16.387	= Cubic centimetres (cc; cm³)	x 0.061	=	Cubic inches (cu in; in³)
Imperial pints (Imp pt)	x 0.568	= Litres (l)	x 1.76	=	Imperial pints (Imp pt)
Imperial quarts (Imp qt)	x 1.137	= Litres (l)	x 0.88	=	Imperial quarts (Imp qt)
Imperial quarts (Imp qt)	x 1.201	= US quarts (US qt)	x 0.833	=	Imperial quarts (Imp qt)
US quarts (US qt)	x 0.946	= Litres (l)	x 1.057	=	US quarts (US qt)
Imperial gallons (Imp gal)	x 4.546	= Litres (l)	x 0.22	=	Imperial gallons (Imp gal)
Imperial gallons (Imp gal)	x 1.201	= US gallons (US gal)	x 0.833	=	Imperial gallons (Imp gal)
US gallons (US gal)	x 3.785	= Litres (l)	x 0.264	=	US gallons (US gal)

Mass (weight)

Ounces (oz)	x 28.35	= Grams (g)	x 0.035	=	Ounces (oz)
Pounds (lb)	x 0.454	= Kilograms (kg)	x 2.205	=	Pounds (lb)

Force

Ounces-force (ozf; oz)	x 0.278	= Newtons (N)	x 3.6	=	Ounces-force (ozf; oz)
Pounds-force (lbf; lb)	x 4.448	= Newtons (N)	x 0.225	=	Pounds-force (lbf; lb)
Newtons (N)	x 0.1	= Kilograms-force (kgf; kg)	x 9.81	=	Newtons (N)

Pressure

Pounds-force per square inch (psi; lbf/in²; lb/in²)	x 0.070	= Kilograms-force per square centimetre (kgf/cm²; kg/cm²)	x 14.223	=	Pounds-force per square inch (psi; lbf/in²; lb/in²)
Pounds-force per square inch (psi; lbf/in²; lb/in²)	x 0.068	= Atmospheres (atm)	x 14.696	=	Pounds-force per square inch (psi; lbf/in²; lb/in²)
Pounds-force per square inch (psi; lbf/in²; lb/in²)	x 0.069	= Bars	x 14.5	=	Pounds-force per square inch (psi; lbf/in²; lb/in²)
Pounds-force per square inch (psi; lbf/in²; lb/in²)	x 6.895	= Kilopascals (kPa)	x 0.145	=	Pounds-force per square inch (psi; lbf/in²; lb/in²)
Kilopascals (kPa)	x 0.01	= Kilograms-force per square centimetre (kgf/cm²; kg/cm²)	x 98.1	=	Kilopascals (kPa)
Millibar (mbar)	x 100	= Pascals (Pa)	x 0.01	=	Millibar (mbar)
Millibar (mbar)	x 0.0145	= Pounds-force per square inch (psi; lbf/in²; lb/in²)	x 68.947	=	Millibar (mbar)
Millibar (mbar)	x 0.75	= Millimetres of mercury (mmHg)	x 1.333	=	Millibar (mbar)
Millibar (mbar)	x 0.401	= Inches of water (inH₂O)	x 2.491	=	Millibar (mbar)
Millimetres of mercury (mmHg)	x 0.535	= Inches of water (inH₂O)	x 1.868	=	Millimetres of mercury (mmHg)
Inches of water (inH₂O)	x 0.036	= Pounds-force per square inch (psi; lbf/in²; lb/in²)	x 27.68	=	Inches of water (inH₂O)

Torque (moment of force)

Pounds-force inches (lbf in; lb in)	x 1.152	= Kilograms-force centimetre (kgf cm; kg cm)	x 0.868	=	Pounds-force inches (lbf in; lb in)
Pounds-force inches (lbf in; lb in)	x 0.113	= Newton metres (Nm)	x 8.85	=	Pounds-force inches (lbf in; lb in)
Pounds-force inches (lbf in; lb in)	x 0.083	= Pounds-force feet (lbf ft; lb ft)	x 12	=	Pounds-force inches (lbf in; lb in)
Pounds-force feet (lbf ft; lb ft)	x 0.138	= Kilograms-force metres (kgf m; kg m)	x 7.233	=	Pounds-force feet (lbf ft; lb ft)
Pounds-force feet (lbf ft; lb ft)	x 1.356	= Newton metres (Nm)	x 0.738	=	Pounds-force feet (lbf ft; lb ft)
Newton metres (Nm)	x 0.102	= Kilograms-force metres (kgf m; kg m)	x 9.804	=	Newton metres (Nm)

Power

Horsepower (hp)	x 745.7	= Watts (W)	x 0.0013	=	Horsepower (hp)

Velocity (speed)

Miles per hour (miles/hr; mph)	x 1.609	= Kilometres per hour (km/hr; kph)	x 0.621	=	Miles per hour (miles/hr; mph)

Fuel consumption*

Miles per gallon (mpg)	x 0.354	= Kilometres per litre (km/l)	x 2.825	=	Miles per gallon (mpg)

Temperature

Degrees Fahrenheit = (°C x 1.8) + 32

Degrees Celsius (Degrees Centigrade; °C) = (°F - 32) x 0.56

It is common practice to convert from miles per gallon (mpg) to litres/100 kilometres (l/100km), where mpg x l/100 km = 282

Glossary of Technical Terms

A

ABS (Anti-lock brake system) A system, usually electronically controlled, that senses incipient wheel lockup during braking and relieves hydraulic pressure at wheels that are about to skid.

Air bag An inflatable bag hidden in the steering wheel (driver's side) or the dash or glovebox (passenger side). In a head-on collision, the bags inflate, preventing the driver and front passenger from being thrown forward into the steering wheel or windscreen.

Air cleaner A metal or plastic housing, containing a filter element, which removes dust and dirt from the air being drawn into the engine.

Air filter element The actual filter in an air cleaner system, usually manufactured from pleated paper and requiring renewal at regular intervals.

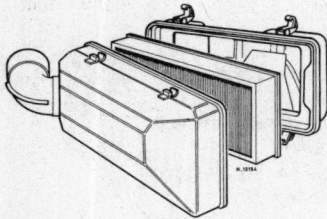

Air filter

Allen key A hexagonal wrench which fits into a recessed hexagonal hole.

Alligator clip A long-nosed spring-loaded metal clip with meshing teeth. Used to make temporary electrical connections.

Alternator A component in the electrical system which converts mechanical energy from a drivebelt into electrical energy to charge the battery and to operate the starting system, ignition system and electrical accessories.

Alternator (exploded view)

Ampere (amp) A unit of measurement for the flow of electric current. One amp is the amount of current produced by one volt acting through a resistance of one ohm.

Anaerobic sealer A substance used to prevent bolts and screws from loosening. Anaerobic means that it does not require oxygen for activation. The Loctite brand is widely used.

Antifreeze A substance (usually ethylene glycol) mixed with water, and added to a vehicle's cooling system, to prevent freezing of the coolant in winter. Antifreeze also contains chemicals to inhibit corrosion and the formation of rust and other deposits that would tend to clog the radiator and coolant passages and reduce cooling efficiency.

Anti-seize compound A coating that reduces the risk of seizing on fasteners that are subjected to high temperatures, such as exhaust manifold bolts and nuts.

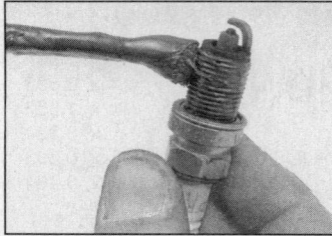

Anti-seize compound

Asbestos A natural fibrous mineral with great heat resistance, commonly used in the composition of brake friction materials. Asbestos is a health hazard and the dust created by brake systems should never be inhaled or ingested.

Axle A shaft on which a wheel revolves, or which revolves with a wheel. Also, a solid beam that connects the two wheels at one end of the vehicle. An axle which also transmits power to the wheels is known as a live axle.

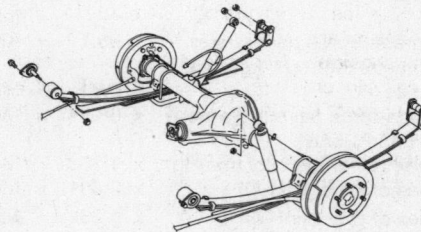

Axle assembly

Axleshaft A single rotating shaft, on either side of the differential, which delivers power from the final drive assembly to the drive wheels. Also called a driveshaft or a halfshaft.

B

Ball bearing An anti-friction bearing consisting of a hardened inner and outer race with hardened steel balls between two races.

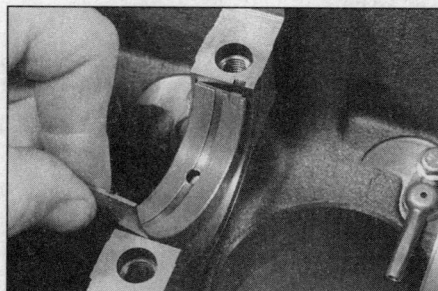

Bearing

Bearing The curved surface on a shaft or in a bore, or the part assembled into either, that permits relative motion between them with minimum wear and friction.

Big-end bearing The bearing in the end of the connecting rod that's attached to the crankshaft.

Bleed nipple A valve on a brake wheel cylinder, caliper or other hydraulic component that is opened to purge the hydraulic system of air. Also called a bleed screw.

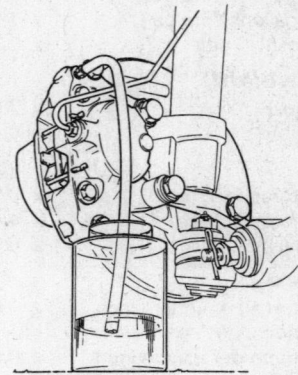

Brake bleeding

Brake bleeding Procedure for removing air from lines of a hydraulic brake system.

Brake disc The component of a disc brake that rotates with the wheels.

Brake drum The component of a drum brake that rotates with the wheels.

Brake linings The friction material which contacts the brake disc or drum to retard the vehicle's speed. The linings are bonded or riveted to the brake pads or shoes.

Brake pads The replaceable friction pads that pinch the brake disc when the brakes are applied. Brake pads consist of a friction material bonded or riveted to a rigid backing plate.

Brake shoe The crescent-shaped carrier to which the brake linings are mounted and which forces the lining against the rotating drum during braking.

Braking systems For more information on braking systems, consult the *Haynes Automotive Brake Manual*.

Breaker bar A long socket wrench handle providing greater leverage.

Bulkhead The insulated partition between the engine and the passenger compartment.

C

Caliper The non-rotating part of a disc-brake assembly that straddles the disc and carries the brake pads. The caliper also contains the hydraulic components that cause the pads to pinch the disc when the brakes are applied. A caliper is also a measuring tool that can be set to measure inside or outside dimensions of an object.

Camshaft A rotating shaft on which a series of cam lobes operate the valve mechanisms. The camshaft may be driven by gears, by sprockets and chain or by sprockets and a belt.

Canister A container in an evaporative emission control system; contains activated charcoal granules to trap vapours from the fuel system.

Canister

Carburettor A device which mixes fuel with air in the proper proportions to provide a desired power output from a spark ignition internal combustion engine.

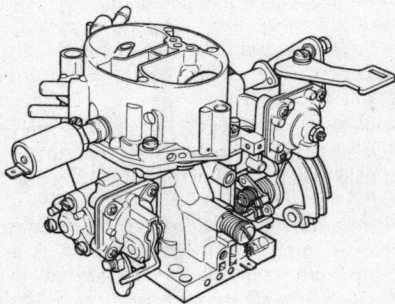

Carburettor

Castellated Resembling the parapets along the top of a castle wall. For example, a castellated balljoint stud nut.

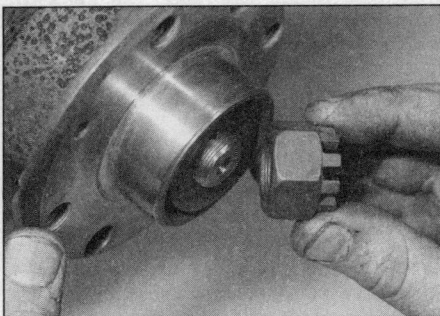

Castellated nut

Castor In wheel alignment, the backward or forward tilt of the steering axis. Castor is positive when the steering axis is inclined rearward at the top.

Catalytic converter A silencer-like device in the exhaust system which converts certain pollutants in the exhaust gases into less harmful substances.

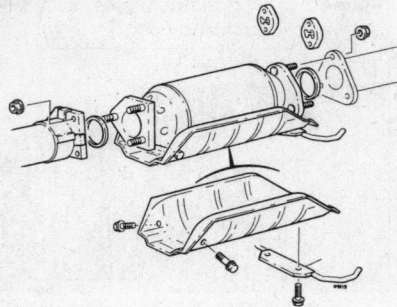

Catalytic converter

Circlip A ring-shaped clip used to prevent endwise movement of cylindrical parts and shafts. An internal circlip is installed in a groove in a housing; an external circlip fits into a groove on the outside of a cylindrical piece such as a shaft.

Clearance The amount of space between two parts. For example, between a piston and a cylinder, between a bearing and a journal, etc.

Coil spring A spiral of elastic steel found in various sizes throughout a vehicle, for example as a springing medium in the suspension and in the valve train.

Compression Reduction in volume, and increase in pressure and temperature, of a gas, caused by squeezing it into a smaller space.

Compression ratio The relationship between cylinder volume when the piston is at top dead centre and cylinder volume when the piston is at bottom dead centre.

Constant velocity (CV) joint A type of universal joint that cancels out vibrations caused by driving power being transmitted through an angle.

Core plug A disc or cup-shaped metal device inserted in a hole in a casting through which core was removed when the casting was formed. Also known as a freeze plug or expansion plug.

Crankcase The lower part of the engine block in which the crankshaft rotates.

Crankshaft The main rotating member, or shaft, running the length of the crankcase, with offset "throws" to which the connecting rods are attached.

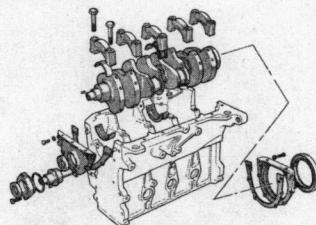

Crankshaft assembly

Crocodile clip See Alligator clip

D

Diagnostic code Code numbers obtained by accessing the diagnostic mode of an engine management computer. This code can be used to determine the area in the system where a malfunction may be located.

Disc brake A brake design incorporating a rotating disc onto which brake pads are squeezed. The resulting friction converts the energy of a moving vehicle into heat.

Double-overhead cam (DOHC) An engine that uses two overhead camshafts, usually one for the intake valves and one for the exhaust valves.

Drivebelt(s) The belt(s) used to drive accessories such as the alternator, water pump, power steering pump, air conditioning compressor, etc. off the crankshaft pulley.

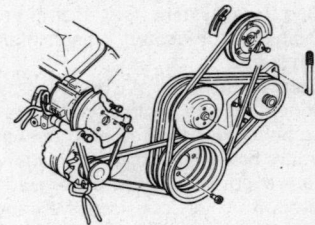

Accessory drivebelts

Driveshaft Any shaft used to transmit motion. Commonly used when referring to the axleshafts on a front wheel drive vehicle.

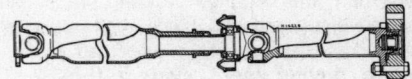

Driveshaft

Drum brake A type of brake using a drum-shaped metal cylinder attached to the inner surface of the wheel. When the brake pedal is pressed, curved brake shoes with friction linings press against the inside of the drum to slow or stop the vehicle.

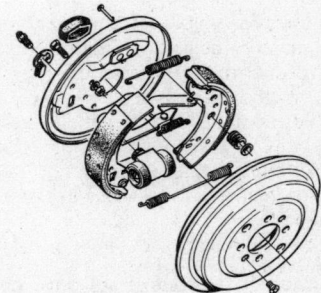

Drum brake assembly

Glossary of Technical Terms

E

EGR valve A valve used to introduce exhaust gases into the intake air stream.

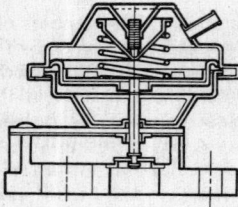

EGR valve

Electronic control unit (ECU) A computer which controls (for instance) ignition and fuel injection systems, or an anti-lock braking system. For more information refer to the *Haynes Automotive Electrical and Electronic Systems Manual.*

Electronic Fuel Injection (EFI) A computer controlled fuel system that distributes fuel through an injector located in each intake port of the engine.

Emergency brake A braking system, independent of the main hydraulic system, that can be used to slow or stop the vehicle if the primary brakes fail, or to hold the vehicle stationary even though the brake pedal isn't depressed. It usually consists of a hand lever that actuates either front or rear brakes mechanically through a series of cables and linkages. Also known as a handbrake or parking brake.

Endfloat The amount of lengthwise movement between two parts. As applied to a crankshaft, the distance that the crankshaft can move forward and back in the cylinder block.

Engine management system (EMS) A computer controlled system which manages the fuel injection and the ignition systems in an integrated fashion.

Exhaust manifold A part with several passages through which exhaust gases leave the engine combustion chambers and enter the exhaust pipe.

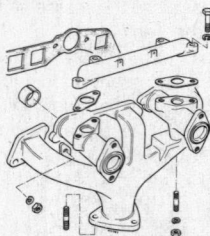

Exhaust manifold

F

Fan clutch A viscous (fluid) drive coupling device which permits variable engine fan speeds in relation to engine speeds.

Feeler blade A thin strip or blade of hardened steel, ground to an exact thickness, used to check or measure clearances between parts.

Feeler blade

Firing order The order in which the engine cylinders fire, or deliver their power strokes, beginning with the number one cylinder.

Flywheel A heavy spinning wheel in which energy is absorbed and stored by means of momentum. On cars, the flywheel is attached to the crankshaft to smooth out firing impulses.

Free play The amount of travel before any action takes place. The "looseness" in a linkage, or an assembly of parts, between the initial application of force and actual movement. For example, the distance the brake pedal moves before the pistons in the master cylinder are actuated.

Fuse An electrical device which protects a circuit against accidental overload. The typical fuse contains a soft piece of metal which is calibrated to melt at a predetermined current flow (expressed as amps) and break the circuit.

Fusible link A circuit protection device consisting of a conductor surrounded by heat-resistant insulation. The conductor is smaller than the wire it protects, so it acts as the weakest link in the circuit. Unlike a blown fuse, a failed fusible link must frequently be cut from the wire for replacement.

G

Gap The distance the spark must travel in jumping from the centre electrode to the side

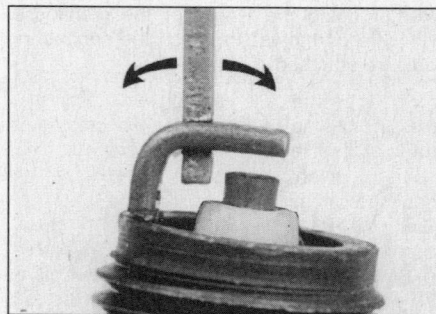

Adjusting spark plug gap

electrode in a spark plug. Also refers to the spacing between the points in a contact breaker assembly in a conventional points-type ignition, or to the distance between the reluctor or rotor and the pickup coil in an electronic ignition.

Gasket Any thin, soft material - usually cork, cardboard, asbestos or soft metal - installed between two metal surfaces to ensure a good seal. For instance, the cylinder head gasket seals the joint between the block and the cylinder head.

Gasket

Gauge An instrument panel display used to monitor engine conditions. A gauge with a movable pointer on a dial or a fixed scale is an analogue gauge. A gauge with a numerical readout is called a digital gauge.

H

Halfshaft A rotating shaft that transmits power from the final drive unit to a drive wheel, usually when referring to a live rear axle.

Harmonic balancer A device designed to reduce torsion or twisting vibration in the crankshaft. May be incorporated in the crankshaft pulley. Also known as a vibration damper.

Hone An abrasive tool for correcting small irregularities or differences in diameter in an engine cylinder, brake cylinder, etc.

Hydraulic tappet A tappet that utilises hydraulic pressure from the engine's lubrication system to maintain zero clearance (constant contact with both camshaft and valve stem). Automatically adjusts to variation in valve stem length. Hydraulic tappets also reduce valve noise.

I

Ignition timing The moment at which the spark plug fires, usually expressed in the number of crankshaft degrees before the piston reaches the top of its stroke.

Inlet manifold A tube or housing with passages through which flows the air-fuel mixture (carburettor vehicles and vehicles with throttle body injection) or air only (port fuel-injected vehicles) to the port openings in the cylinder head.

J

Jump start Starting the engine of a vehicle with a discharged or weak battery by attaching jump leads from the weak battery to a charged or helper battery.

L

Load Sensing Proportioning Valve (LSPV) A brake hydraulic system control valve that works like a proportioning valve, but also takes into consideration the amount of weight carried by the rear axle.

Locknut A nut used to lock an adjustment nut, or other threaded component, in place. For example, a locknut is employed to keep the adjusting nut on the rocker arm in position.

Lockwasher A form of washer designed to prevent an attaching nut from working loose.

M

MacPherson strut A type of front suspension system devised by Earle MacPherson at Ford of England. In its original form, a simple lateral link with the anti-roll bar creates the lower control arm. A long strut - an integral coil spring and shock absorber - is mounted between the body and the steering knuckle. Many modern so-called MacPherson strut systems use a conventional lower A-arm and don't rely on the anti-roll bar for location.

Multimeter An electrical test instrument with the capability to measure voltage, current and resistance.

N

NOx Oxides of Nitrogen. A common toxic pollutant emitted by petrol and diesel engines at higher temperatures.

O

Ohm The unit of electrical resistance. One volt applied to a resistance of one ohm will produce a current of one amp.

Ohmmeter An instrument for measuring electrical resistance.

O-ring A type of sealing ring made of a special rubber-like material; in use, the O-ring is compressed into a groove to provide the sealing action.

O-ring

Overhead cam (ohc) engine An engine with the camshaft(s) located on top of the cylinder head(s).

Overhead valve (ohv) engine An engine with the valves located in the cylinder head, but with the camshaft located in the engine block.

Oxygen sensor A device installed in the engine exhaust manifold, which senses the oxygen content in the exhaust and converts this information into an electric current. Also called a Lambda sensor.

P

Phillips screw A type of screw head having a cross instead of a slot for a corresponding type of screwdriver.

Plastigage A thin strip of plastic thread, available in different sizes, used for measuring clearances. For example, a strip of Plastigage is laid across a bearing journal. The parts are assembled and dismantled; the width of the crushed strip indicates the clearance between journal and bearing.

Plastigage

Propeller shaft The long hollow tube with universal joints at both ends that carries power from the transmission to the differential on front-engined rear wheel drive vehicles.

Proportioning valve A hydraulic control valve which limits the amount of pressure to the rear brakes during panic stops to prevent wheel lock-up.

R

Rack-and-pinion steering A steering system with a pinion gear on the end of the steering shaft that mates with a rack (think of a geared wheel opened up and laid flat). When the steering wheel is turned, the pinion turns, moving the rack to the left or right. This movement is transmitted through the track rods to the steering arms at the wheels.

Radiator A liquid-to-air heat transfer device designed to reduce the temperature of the coolant in an internal combustion engine cooling system.

Refrigerant Any substance used as a heat transfer agent in an air-conditioning system. R-12 has been the principle refrigerant for many years; recently, however, manufacturers have begun using R-134a, a non-CFC substance that is considered less harmful to

the ozone in the upper atmosphere.

Rocker arm A lever arm that rocks on a shaft or pivots on a stud. In an overhead valve engine, the rocker arm converts the upward movement of the pushrod into a downward movement to open a valve.

Rotor In a distributor, the rotating device inside the cap that connects the centre electrode and the outer terminals as it turns, distributing the high voltage from the coil secondary winding to the proper spark plug. Also, that part of an alternator which rotates inside the stator. Also, the rotating assembly of a turbocharger, including the compressor wheel, shaft and turbine wheel.

Runout The amount of wobble (in-and-out movement) of a gear or wheel as it's rotated. The amount a shaft rotates "out-of-true." The out-of-round condition of a rotating part.

S

Sealant A liquid or paste used to prevent leakage at a joint. Sometimes used in conjunction with a gasket.

Sealed beam lamp An older headlight design which integrates the reflector, lens and filaments into a hermetically-sealed one-piece unit. When a filament burns out or the lens cracks, the entire unit is simply replaced.

Serpentine drivebelt A single, long, wide accessory drivebelt that's used on some newer vehicles to drive all the accessories, instead of a series of smaller, shorter belts. Serpentine drivebelts are usually tensioned by an automatic tensioner.

Serpentine drivebelt

Shim Thin spacer, commonly used to adjust the clearance or relative positions between two parts. For example, shims inserted into or under bucket tappets control valve clearances. Clearance is adjusted by changing the thickness of the shim.

Slide hammer A special puller that screws into or hooks onto a component such as a shaft or bearing; a heavy sliding handle on the shaft bottoms against the end of the shaft to knock the component free.

Sprocket A tooth or projection on the periphery of a wheel, shaped to engage with a chain or drivebelt. Commonly used to refer to the sprocket wheel itself.

Starter inhibitor switch On vehicles with an

Glossary of Technical Terms

automatic transmission, a switch that prevents starting if the vehicle is not in Neutral or Park.

Strut See MacPherson strut.

T

Tappet A cylindrical component which transmits motion from the cam to the valve stem, either directly or via a pushrod and rocker arm. Also called a cam follower.

Thermostat A heat-controlled valve that regulates the flow of coolant between the cylinder block and the radiator, so maintaining optimum engine operating temperature. A thermostat is also used in some air cleaners in which the temperature is regulated.

Thrust bearing The bearing in the clutch assembly that is moved in to the release levers by clutch pedal action to disengage the clutch. Also referred to as a release bearing.

Timing belt A toothed belt which drives the camshaft. Serious engine damage may result if it breaks in service.

Timing chain A chain which drives the camshaft.

Toe-in The amount the front wheels are closer together at the front than at the rear. On rear wheel drive vehicles, a slight amount of toe-in is usually specified to keep the front wheels running parallel on the road by offsetting other forces that tend to spread the wheels apart.

Toe-out The amount the front wheels are closer together at the rear than at the front. On front wheel drive vehicles, a slight amount of toe-out is usually specified.

Tools For full information on choosing and using tools, refer to the *Haynes Automotive Tools Manual*.

Tracer A stripe of a second colour applied to a wire insulator to distinguish that wire from another one with the same colour insulator.

Tune-up A process of accurate and careful adjustments and parts replacement to obtain the best possible engine performance.

Turbocharger A centrifugal device, driven by exhaust gases, that pressurises the intake air. Normally used to increase the power output from a given engine displacement, but can also be used primarily to reduce exhaust emissions (as on VW's "Umwelt" Diesel engine).

U

Universal joint or U-joint A double-pivoted connection for transmitting power from a driving to a driven shaft through an angle. A U-joint consists of two Y-shaped yokes and a cross-shaped member called the spider.

V

Valve A device through which the flow of liquid, gas, vacuum, or loose material in bulk may be started, stopped, or regulated by a movable part that opens, shuts, or partially obstructs one or more ports or passageways. A valve is also the movable part of such a device.

Valve clearance The clearance between the valve tip (the end of the valve stem) and the rocker arm or tappet. The valve clearance is measured when the valve is closed.

Vernier caliper A precision measuring instrument that measures inside and outside dimensions. Not quite as accurate as a micrometer, but more convenient.

Viscosity The thickness of a liquid or its resistance to flow.

Volt A unit for expressing electrical "pressure" in a circuit. One volt that will produce a current of one ampere through a resistance of one ohm.

W

Welding Various processes used to join metal items by heating the areas to be joined to a molten state and fusing them together. For more information refer to the *Haynes Automotive Welding Manual*.

Wiring diagram A drawing portraying the components and wires in a vehicle's electrical system, using standardised symbols. For more information refer to the *Haynes Automotive Electrical and Electronic Systems Manual*.

Note: *References throughout this index relate to Chapter•page number*

Haynes Manuals – The Complete List

Title	Book No.
ALFA ROMEO	
Alfa Romeo Alfasud/Sprint (74 - 88)	0292
Alfa Romeo Alfetta (73 - 87)	0531
AUDI	
Audi 80 (72 - Feb 79)	0207
Audi 80, 90 (79 - Oct 86) & Coupe (81 - Nov 88)	0605
Audi 80, 90 (Oct 86 - 90) & Coupe (Nov 88 - 90)	1491
Audi 100 (Oct 76 - Oct 82)	0428
Audi 100 (Oct 82 - 90) & 200 (Feb 84 - Oct 89)	0907
AUSTIN	
Austin Ambassador (82 - 84)	0871
Austin/MG Maestro 1.3 & 1.6 (83 - 95)	0922
Austin Maxi (69 - 81)	0052
Austin/MG Metro (80 - May 90)	0718
Austin Montego 1.3 & 1.6 (84 - 94)	1066
Austin/MG Montego 2.0 (84 - 95)	1067
Mini (59 - 69)	0527
Mini (69 - 96)	0646
Austin/Rover 2.0 litre Diesel Engine (86 - 93)	1857
BEDFORD	
Bedford CF (69 - 87)	0163
Bedford Rascal (86 - 93)	3015
BL	
BL Princess & BLMC 18-22 (75 - 82)	0286
BMW	
BMW 316, 320 & 320i (4-cyl) (75 - Feb 83)	0276
BMW 320, 320i, 323i & 325i (6-cyl) (Oct 77 - Sept 87)	0815
BMW 3-Series (Apr 91 - 96)	3210
BMW 3-Series (sohc) (83 - 91)	1948
BMW 520i & 525e (Oct 81 - June 88)	1560
BMW 525, 528 & 528i (73 - Sept 81)	0632
BMW 5-Series (sohc) (81 - 93)	1948
BMW 1500, 1502, 1600, 1602, 2000 & 2002 (59 - 77)	0240
CITROEN	
Citroen 2CV, Ami & Dyane (67 - 90)	0196
Citroen AX Petrol & Diesel (87 - 94)	3014
Citroen BX (83 - 94)	0908
Citroen CX (75 - 88)	0528
Citroen Visa (79 - 88)	0620
Citroen Xantia Petrol & Diesel (93 - Oct 95)	3082
Citroen ZX Diesel (91 - 93)	1922
Citroen ZX Petrol (91 - 94)	1881
Citroen 1.7 & 1.9 litre Diesel Engine (84 - 96)	1379
COLT	
Colt 1200, 1250 & 1400 (79 - May 84)	0600
Colt Galant (74 - 78) & Celeste (76 - 81)	0236
DAIMLER	
Daimler Sovereign (68 - Oct 86)	0242
Daimler Double Six (72 - 88)	0478
DATSUN (see also **Nissan**)	
Datsun 120Y (73 - Aug 78)	0228
Datsun 1300, 1400 & 1600 (69 - Aug 72)	0123
Datsun Cherry (79 - Sept 82)	0679
Datsun Pick-up (75 - 78)	0277
Datsun Sunny (Aug 78 - May 82)	0525
Datsun Violet (78 - 82)	0430

Title	Book No.
FIAT	
Fiat 126 (73 - 87)	0305
Fiat 127 (71 - 83)	0193
Fiat 500 (57 - 73)	0090
Fiat 850 (64 - 81)	0038
Fiat Panda (81 - 95)	0793
Fiat Punto (94 - 96)	3251
Fiat Regata (84 - 88)	1167
Fiat Strada (79 - 88)	0479
Fiat Tipo (88 - 91)	1625
Fiat Uno (83 - 95)	0923
Fiat X1/9 (74 - 89)	0273
FORD	
Ford Capri II (& III) 1.6 & 2.0 (74 - 87)	0283
Ford Capri II (& III) 2.8 & 3.0 (74 - 87)	1309
Ford Cortina Mk III 1600 & 2000 (70 - 76)	0295
Ford Cortina Mk IV (& V) 1.6 & 2.0 (76 - 83)	0343
Ford Cortina Mk IV (& V) 2.3 V6 (77 - 83)	0426
Ford Escort (75 - Aug 80)	0280
Ford Escort (Sept 80 - Sept 90)	0686
Ford Escort (Sept 90 - 96)	1737
Ford Escort Mk II Mexico, RS 1600 & RS 2000 (75 - 80)	0735
Ford Fiesta (inc. XR2) (76 - Aug 83)	0334
Ford Fiesta (inc. XR2) (Aug 83 - Feb 89)	1030
Ford Fiesta (Feb 89 - 93)	1595
Ford Granada (Sept 77 - Feb 85)	0481
Ford Granada (Mar 85 - 94)	1245
Ford Mondeo 4-cyl (93 - 96)	1923
Ford Orion (83 - Sept 90)	1009
Ford Orion (Sept 90 - 93)	1737
Ford Sierra 1.3, 1.6, 1.8 & 2.0 (82 - 93)	0903
Ford Sierra 2.3, 2.8 & 2.9 (82 - 91)	0904
Ford Scorpio (Mar 85 - 94)	1245
Ford Transit Petrol (Mk 1) (65 - Feb 78)	0377
Ford Transit Petrol (Mk 2) (78 - Jan 86)	0719
Ford Transit Petrol (Mk 3) (Feb 86 - 89)	1468
Ford Transit Diesel (Feb 86 - 95)	3019
Ford 1.6 & 1.8 litre Diesel Engine (84 - 96)	1172
Ford 2.1, 2.3 & 2.5 litre Diesel Engine (77 - 90)	1606
Ford Vehicle Carburettors	1783
FREIGHT ROVER	
Freight Rover Sherpa (74 - 87)	0463
HILLMAN	
Hillman Avenger (70 - 82)	0037
Hillman Minx & Husky (56 - 66)	0009
HONDA	
Honda Accord (76 - Feb 84)	0351
Honda Accord (Feb 84 - Oct 85)	1177
Honda Civic 1300 (80 - 81)	0633
Honda Civic (Feb 84 - Oct 87)	1226
Honda Civic (Nov 91 - 96)	3199
JAGUAR	
Jaguar E Type (61 - 72)	0140
Jaguar MkI & II, 240 & 340 (55 - 69)	0098
Jaguar XJ6, XJ & Sovereign (68 - Oct 86)	0242
Jaguar XJ12, XJS & Sovereign (72 - 88)	0478
JEEP	
Jeep Cherokee Petrol (93 - 96)	1943

Title	Book No.
LADA	
Lada 1200, 1300, 1500 & 1600 (74 - 91)	0413
Lada Samara (87 - 91)	1610
LAND ROVER	
Land Rover 90, 110 & Defender Diesel (83 - 95)	3017
Land Rover Discovery Diesel (89 - 95)	3016
Land Rover Series IIA & III Diesel (58 - 85)	0529
Land Rover Series II, IIA & III Petrol (58 - 85)	0314
MAZDA	
Mazda 323 fwd (Mar 81 - Oct 89)	1608
Mazda 323 rwd (77 - Apr 86)	0370
Mazda 626 fwd (May 83 - Sept 87)	0929
Mazda B-1600, B-1800 & B-2000 Pick-up (72 - 88)	0267
Mazda RX-7 (79 - 85)	0460
MERCEDES-BENZ	
Mercedes-Benz 190 & 190E (83 - 87)	0928
Mercedes-Benz 200, 240, 300 Diesel (Oct 76 - 85)	1114
Mercedes-Benz 250 & 280 (68 - 72)	0346
Mercedes-Benz 250 & 280 (123 Series) (Oct 76 - 84)	0677
Mercedes-Benz 124 Series (85 - Aug 93)	3253
MG	
MGB (62 - 80)	0111
MG Maestro 1.3 & 1.6 (83 - 95)	0922
MG Metro (80 - May 90)	0718
MG Midget & AH Sprite (58 - 80)	0265
MG Montego 2.0 (84 - 95)	1067
MITSUBISHI	
Mitsubishi 1200, 1250 & 1400 (79 - May 84)	0600
Mitsubishi Shogun & L200 Pick-Ups (83 - 94)	1944
MORRIS	
Morris Ital 1.3 (80 - 84)	0705
Morris Marina 1700 (78 - 80)	0526
Morris Marina 1.8 (71 - 78)	0074
Morris Minor 1000 (56 - 71)	0024
NISSAN (See also Datsun)	
Nissan Bluebird 160B & 180B rwd (May 80 - May 84)	0957
Nissan Bluebird fwd (May 84 - Mar 86)	1223
Nissan Bluebird (T12 & T72) (Mar 86 - 90)	1473
Nissan Cherry (N12) (Sept 82 - 86)	1031
Nissan Micra (K10) (83 - Jan 93)	0931
Nissan Micra (93 - 96)	3254
Nissan Primera (90 - Oct 96)	1851
Nissan Stanza (82 - 86)	0824
Nissan Sunny (B11) (May 82 - Oct 86)	0895
Nissan Sunny (Oct 86 - Mar 91)	1378
Nissan Sunny (Apr 91 - 95)	3219
OPEL	
Opel Ascona & Manta (B Series) (Sept 75 - 88)	0316
Opel Ascona (81 - 88)	3215
Opel Astra (Oct 91 - 96)	3156
Opel Corsa (83 - Mar 93)	3160
Opel Corsa (Mar 93 - 94)	3159
Opel Kadett (Nov 79 - Oct 84)	0634
Opel Kadett (Oct 84 - Oct 91)	3196
Opel Omega & Senator (86 - 94)	3157

Title	Book No.
Opel Rekord (Feb 78 - Oct 86)	0543
Opel Vectra (88 - Oct 95)	3158
PEUGEOT	
Peugeot 106 Petrol & Diesel (91 - June 96)	1882
Peugeot 205 (83 - 95)	0932
Peugeot 305 (78 - 89)	0538
Peugeot 306 Petrol & Diesel (93 - 95)	3073
Peugeot 309 (86 - 93)	1266
Peugeot 405 Petrol (88 - 96)	1559
Peugeot 405 Diesel (88 - 96)	3198
Peugeot 505 (79 - 89)	0762
Peugeot 1.7 & 1.9 litre Diesel Engines (82 - 96)	0950
Peugeot 2.0, 2.1, 2.3 & 2.5 litre Diesel Engines (74 - 90)	1607
PORSCHE	
Porsche 911 (65 - 85)	0264
Porsche 924 & 924 Turbo (76 - 85)	0397
RANGE ROVER	
Range Rover V8 (70 - Oct 92)	0606
RELIANT	
Reliant Robin & Kitten (73 - 83)	0436
RENAULT	
Renault 5 (72 - Feb 85)	0141
Renault 5 (Feb 85 - 96)	1219
Renault 6 (68 - 79)	0092
Renault 9 & 11 (82 - 89)	0822
Renault 12 (70 - 80)	0097
Renault 15 & 17 (72 - 79)	0763
Renault 16 (65 - 79)	0081
Renault 18 (79 - 86)	0598
Renault 19 Petrol (89 - 94)	1646
Renault 19 Diesel (89 - 95)	1946
Renault 21 (86 - 94)	1397
Renault 25 (84 - 86)	1228
Renault Clio Petrol (91 - 93)	1853
Renault Clio Diesel (91 - June 96)	3031
Renault Espace (85 - 96)	3197
Renault Fuego (80 - 86)	0764
Renault Laguna (94 - 96)	3252
ROVER	
Rover 111 & 114 (95 - 96)	1711
Rover 213 & 216 (84 - 89)	1116
Rover 214 & 414 (Oct 89 - 92)	1689
Rover 216 & 416 (Oct 89 - 92)	1830
Rover 820, 825 & 827 (86 - 95)	1380
Rover 2000, 2300 & 2600 (77 - 87)	0468
Rover 3500 (76 - 87)	0365
Rover Metro (May 90 - 94)	1711
Rover 2.0 litre Diesel Engine (86 - 93)	1857
SAAB	
Saab 95 & 96 (66 - 76)	0198
Saab 99 (69 - 79)	0247
Saab 90, 99 & 900 (79 - Oct 93)	0765
Saab 9000 (4-cyl) (85 - 95)	1686
SEAT	
Seat Ibiza & Malaga (85 - 92)	1609

Title	Book No.
SIMCA	
Simca 1100 & 1204 (67 - 79)	0088
Simca 1301 & 1501 (63 - 76)	0199
SKODA	
Skoda 1000 & 1100 (64 - 78)	0303
Skoda Estelle 105, 120, 130 & 136 (77 - 89)	0604
Skoda Favorit (89 - 92)	1801
SUBARU	
Subaru 1600 (77 - Oct 79)	0237
Subaru 1600 & 1800 (Nov 79 - 90)	0995
SUZUKI	
Suzuki SJ Series, Samurai & Vitara (82 - 94)	1942
Suzuki Supercarry (86 - Oct 94)	3015
TALBOT	
Talbot Alpine, Solara, Minx & Rapier (75 - 86)	0337
Talbot Horizon (78 - 86)	0473
Talbot Samba (82 - 86)	0823
TOYOTA	
Toyota 2000 (75 - 77)	0360
Toyota Celica (78 - Jan 82)	0437
Toyota Celica (Feb 82 - Sept 85)	1135
Toyota Corolla (fwd) (Sept 83 - Sept 87)	1024
Toyota Corolla (rwd) (80 - 85)	0683
Toyota Corolla (Sept 87 - 92)	1683
Toyota Hi-Ace & Hi-Lux (69 - Oct 83)	0304
Toyota Starlet (78 - Jan 85)	0462
TRIUMPH	
Triumph Acclaim (81 - 84)	0792
Triumph GT6 (62 - 74)	0112
Triumph Herald (59 - 71)	0010
Triumph Spitfire (62 - 81)	0113
Triumph Stag (70 - 78)	0441
Triumph TR2, TR3, TR3A, TR4 & TR4A (52 - 67)	0028
Triumph TR7 (75 - 82)	0322
Triumph Vitesse (62 - 74)	0112
VAUXHALL	
Vauxhall Astra (80 - Oct 84)	0635
Vauxhall Astra & Belmont (Oct 84 - Oct 91)	1136
Vauxhall Astra (Oct 91 - 96)	1832
Vauxhall Carlton (Oct 78 - Oct 86)	0480
Vauxhall Carlton (Nov 86 - 94)	1469
Vauxhall Cavalier 1300 (77 - July 81)	0461
Vauxhall Cavalier 1600, 1900 & 2000 (75 - July 81)	0315
Vauxhall Cavalier (81 - Oct 88)	0812
Vauxhall Cavalier (Oct 88 - Oct 95)	1570
Vauxhall Chevette (75 - 84)	0285
Vauxhall Corsa (Mar 93 - 94)	1985
Vauxhall Nova (83 - 93)	0909
Vauxhall Rascal (86 - 93)	3015
Vauxhall Senator (Sept 87 - 94)	1469
Vauxhall Victor & VX4/90 (FD Series) (67 - 72)	0053
Vauxhall Viva HC (70 - 79)	0047
Vauxhall/Opel 1.5, 1.6 & 1.7 litre Diesel Engines (82 - 96)	1222
VOLKSWAGEN	
VW Beetle 1200 (54 - 77)	0036
VW Beetle 1300 & 1500 (65 - 75)	0039

Title	Book No.
VW Beetle 1302 & 1302S (70 - 72)	0110
VW Beetle 1303, 1303S & GT (72 - 75)	0159
VW Golf Mk 1 1.1 & 1.3 (74 - Feb 84)	0716
VW Golf Mk 1 1.5, 1.6 & 1.8 (74 - 85)	0726
VW Golf Mk 1 Diesel (78 - Feb 84)	0451
VW Golf Mk 2 (Mar 84 - Feb 92)	1081
VW Golf Mk 3 Petrol & Diesel (Feb 92 - 96)	3097
VW Jetta Mk 1 1.1 & 1.3 (80 - June 84)	0716
VW Jetta Mk 1 1.5, 1.6 & 1.8 (80 - June 84)	0726
VW Jetta Mk 1 Diesel (81 - June 84)	0451
VW Jetta Mk 2 (July 84 - 92)	1081
VW LT vans & light trucks (76 - 87)	0637
VW Passat (Sept 81 - May 88)	0814
VW Passat (May 88 - 91)	1647
VW Polo & Derby (76 - Jan 82)	0335
VW Polo (82 - Oct 90)	0813
VW Polo (Nov 90 - Aug 94)	3245
VW Santana (Sept 82 - 85)	0814
VW Scirocco Mk 1 1.5, 1.6 & 1.8 (74 - 82)	0726
VW Scirocco (82 - 90)	1224
VW Transporter 1600 (68 - 79)	0082
VW Transporter 1700, 1800 & 2000 (72 - 79)	0226
VW Transporter with air-cooled engine (79 - 82)	0638
VW Type 3 (63 - 73)	0084
VW Vento Petrol & Diesel (Feb 92 - 96)	3097
VOLVO	
Volvo 66 & 343, Daf 55 & 66 (68 - 79)	0293
Volvo 142, 144 & 145 (66 - 74)	0129
Volvo 240 Series (74 - 93)	0270
Volvo 262, 264 & 260/265 (75 - 85)	0400
Volvo 340, 343, 345 & 360 (76 - 91)	0715
Volvo 440, 460 & 480 (87 - 92)	1691
Volvo 740 & 760 (82 - 91)	1258
Volvo 850 (92 - 96)	3260
Volvo 940 (90 - 96)	3249
YUGO/ZASTAVA	
Yugo/Zastava (81 - 90)	1453

TECH BOOKS	
Automotive Brake Manual	3050
Automotive Electrical & Electronic Systems	3049
Automotive Tools Manual	3052
Automotive Welding Manual	3053

CAR BOOKS	
Automotive Fuel Injection Systems	9755
Car Bodywork Repair Manual	9864
Caravan Manual (2nd Edition)	9894
Ford Vehicle Carburettors	1783
Haynes Technical Data Book (87 - 96)	1996
In-Car Entertainment Manual (2nd Edition)	9862
Japanese Vehicle Carburettors	1786
Pass the MOT!	9861
Small Engine Repair Manual	1755
Solex & Pierburg Carburettors	1785
SU Carburettors	0299
Weber Carburettors (to 79)	0393
Weber Carburettors (79 - 91)	1784

01/10/96

Preserving Our Motoring Heritage

The Model J Duesenberg Derham Tourster. Only eight of these magnificent cars were ever built – this is the only example to be found outside the United States of America

Almost every car you've ever loved, loathed or desired is gathered under one roof at the Haynes Motor Museum. Over 300 immaculately presented cars and motorbikes represent every aspect of our motoring heritage, from elegant reminders of bygone days, such as the superb Model J Duesenberg to curiosities like the bug-eyed BMW Isetta. There are also many old friends and flames. Perhaps you remember the 1959 Ford Popular that you did your courting in? The magnificent 'Red Collection' is a spectacle of classic sports cars including AC, Alfa Romeo, Austin Healey, Ferrari, Lamborghini, Maserati, MG, Riley, Porsche and Triumph.

A Perfect Day Out

Each and every vehicle at the Haynes Motor Museum has played its part in the history and culture of Motoring. Today, they make a wonderful spectacle and a great day out for all the family. Bring the kids, bring Mum and Dad, but above all bring your camera to capture those golden memories for ever. You will also find an impressive array of motoring memorabilia, a comfortable 70 seat video cinema and one of the most extensive transport book shops in Britain. The Pit Stop Cafe serves everything from a cup of tea to wholesome, home-made meals or, if you prefer, you can enjoy the large picnic area nestled in the beautiful rural surroundings of Somerset.

John Haynes O.B.E., Founder and Chairman of the museum at the wheel of a Haynes Light 12.

Graham Hill's Lola Cosworth Formula 1 car next to a 1934 Riley Sports.

The Museum is situated on the A359 Yeovil to Frome road at Sparkford, just off the A303 in Somerset. It is about 40 miles south of Bristol, and 25 minutes drive from the M5 intersection at Taunton.
Open 9.30am - 5.30pm (10.00am - 4.00pm Winter) 7 days a week, *except Christmas Day, Boxing Day and New Years Day*
Special rates available for schools, coach parties and outings Charitable Trust No. 292048